AMERICAN MATHEMATICAL SOCIETY
TRANSLATIONS

Series 2

Volume 118

Sixteen Papers on Differential Equations

by

D. M. Galin

O. A. Oleĭnik

E. V. Radkevič

V. M. Petkov

P. R. Popivanov

A. N. Šošitaĭšvili

D. A. Silaev

A. I. Suslov

M. I. Višik

A. I. Komeč

Ju. V. Egorov

C. V. Rangelov

Ju. S. Il'jašenko

T. F. Kalugina

V. Ju. Kiselev

A. A. Lokšin

N. O. Maksimova

A. V. Fursikov

M. A. Šubin

AMERICAN MATHEMATICAL SOCIETY
PROVIDENCE, RHODE ISLAND

Edited by LEV J. LEIFMAN

1980 *Mathematics Subject Classification.* Primary 34C05, 35A07, 35A30, 35B45, 35B65, 35F25, 35H05, 35J40, 35J55, 35K10, 35L15, 35L45, 35L99, 35P20, 35R60, 35S05, 42A75, 47G05, 57R25, 58F05, 58F14, 65M20, 65N40, 76D10; Secondary 35A10, 35A30, 35B40, 35F15, 35H05, 35J30, 35L25, 35P99, 35Q99, 35R99, 35S05, 35S15, 41A60, 45D05, 47A10, 47C15, 47F05, 58F19, 58F21, 70K10, 76D15.

Library of Congress Cataloging in Publication Data

Sixteen papers on differential equations.
    (American Mathematical Society translations; ser. 2, v. 118)
    1. Differential equations — Addresses, essays, lectures. 2. Differential equations, Partial — Addresses, essays, lectures. I. Galin, D. M. II. Title: 16 papers on differential equations.
III. Series.
QA3.A572   ser. 2, vol. 118 [QA371]   510s   82-20595
ISBN 0-8218-3073-2   [515.3′5]

Table of Contents

Russian Table of Contents*

---

*The American Mathematical Society scheme for transliteration of Cyrillic may be found at the end of index issues of *Mathematical Reviews*.

iv

Amer. Math. Soc. Transl.
(2) Vol. **118**, 1982

Versal Deformations
of Linear Hamiltonian Systems*

D. M. GALIN

The subject of this paper is families of linear Hamiltonian systems of differential equations that smoothly depend on parameters. The main result (Theorems 1 and 2) describes normal forms to which such a family can be reduced in a neighborhood of a point in the parameter space by means of a linear symplectic transformation that depends smoothly on the parameters.

The main technical device for finding normal forms is the study of the centralizer of a Hamiltonian matrix; that is, the set of all Hamiltonian matrices which commute with it. In particular, the dimension of the centralizer is calculated in relation to the Jordan form of the given Hamiltonian matrix.

Further, bifurcations of eigenvalues of Hamiltonian matrices are studied for one and two parameter families in general position.

§1. Linear Hamiltonian systems

Consider a mechanical system with n degrees of freedom. Hamilton's canonical variables $p_1,\ldots,p_n$ and $q_1,\ldots,q_n$, will be considered as coordinates in $2n$-dimensional Euclidean space R^{2n}, by setting

$$p_i = x_i, \quad q_i = x_{n+i} \qquad (i = 1, 2, \ldots, n). \tag{1}$$

We assume that the given mechanical system has the Hamiltonian

$$H = \tfrac{1}{2}(Ax, x), \tag{2}$$

where A is a real symmetric matrix of order $2n$. If one introduces the square matrix

$$I = \begin{pmatrix} 0 & -E_n \\ E_n & 0 \end{pmatrix}, \tag{3}$$

1980 *Mathematics Subject Classification*. Primary 58F05; Secondary 58F19.
* Translation of Trudy Sem. Petrovsk. **1** (1975), 63–74. MR **56** #6724.

where E_n is the unit matrix of order n, then Hamilton's canonical equations can be written in the form

$$\frac{dx}{dt} = IAx. \tag{4}$$

A system of the form (4) will be called a *linear Hamiltonian system*, and its matrix IA a *Hamiltonian matrix*.

§2. Symplectic transformations

DEFINITION. A *skew-symmetric inner product* in R^{2n} is a nondegenerate bilinear skew-symmetric function of two vectors which will be denoted by

$$[x, y] = -[y, x]. \tag{5}$$

DEFINITION. A *symplectic basis* in R^{2n} is a system of $2n$ vectors such that their symmetric inner products have the form

$$\begin{cases} [e_i, e_j] = 0 & (i, j = 1, 2, \ldots, n), \\ [e_{n+i}, e_{n+j}] = 0 & (i, j = 1, 2, \ldots, n), \\ [e_i, e_{n+j}] = 0 & (i, j = 1, 2, \ldots, n;\ i \neq j), \\ [e_i, e_{n+i}] = 1 & (i = 1, 2, \ldots, n). \end{cases} \tag{t}$$

DEFINITION. A linear transformation $S: R^{2n} \to R^{2n}$ is called *symplectic* if

$$[Sx, Sy] = [x, y], \forall x, y \in R^{2n}. \tag{7}$$

DEFINITION. The matrix of a symplectic transformation relative to a symplectic basis is called a *symplectic matrix*.

§3. Williamson normal form of a symmetric matrix

Let us consider an arbitrary Hamiltonian matrix IA. It can be shown that, if $z \neq 0$ is an eigenvalue of IA, then $\bar{z}$, $-z$ and $-\bar{z}$ are also eigenvalues, and the Jordan blocks, in the Jordan form of IA, corresponding to these eigenvalues have the same dimension. For each odd integer k, there exist in the Jordan form of IA an even number of Jordan blocks of dimension k which correspond to the eigenvalue $z = 0$.

WILLIAMSON'S THEOREM [1]. *If A is a real symmetric matrix of order $2n$, then there exists a symplectic matrix S such that*

$$S^*AS = A_0, \tag{8}$$

where A_0 is the Williamson normal form of A which depends on the Jordan form of the matrix IA and is determined by the formulas (12)–(18) given below.

COROLLARY. *There exists a symplectic transformation*

$$x = Sy, \qquad x \in R^{2n}, \quad y \in R^{2n}, \tag{9}$$

which transforms the system (4) to the form

$$\frac{dy}{dt} = IA_0 y. \tag{10}$$

The system (10) will be called the *normal form* of the Hamiltonian system (4).

In order to describe the matrix A_0 it is, of course, sufficient to exhibit the general form of the Hamiltonian

$$H_0 = \tfrac{1}{2}(A_0 x, x). \tag{11}$$

According to [1], we can write

$$H_0 = \sum_{j=1}^{s} H_0^{(j)}, \tag{12}$$

where each of the Hamiltonians $H_0^{(j)}$ corresponds either to a pair of Jordan blocks with real eigenvalues $\pm a$, or a pair of Jordan blocks with purely imaginary eigenvalues $\pm ai$, or a quadruple of Jordan blocks with eigenvalues $\pm a \pm bi$, or a pair of Jordan blocks of odd dimension with eigenvalue 0, or a Jordan block of even dimension with eigenvalue 0; all these Jordan blocks are taken from the Jordan form of IA.

To a pair of Jordan blocks of dimension k with eigenvalues $\pm a$ there corresponds a Hamiltonian

$$H_0 = -a \sum_{i=1}^{k} p_i q_i + \sum_{i=1}^{k-1} p_i q_{i+1}. \tag{13}$$

To a quadruple of Jordan blocks of dimension k with eigenvalues $\pm a \pm bi$ there corresponds a Hamiltonian

$$H_0 = -a \sum_{i=1}^{2k} p_i q_i + b \sum_{i=1}^{k} (p_{2i-1} q_{2i} - p_{2i} q_{2i-1}) + \sum_{i=1}^{2k-2} p_i q_{i+2}. \tag{14}$$

To a pair of Jordan blocks of odd dimension k with eigenvalue 0 there corresponds a Hamiltonian

$$H_0 = \sum_{i=1}^{k-1} p_i q_{i+1}. \tag{15}$$

To a Jordan block of even dimension k with eigenvalue 0 there corresponds a Hamiltonian

$$H_0 = \pm \frac{1}{2} \left(\sum_{i=1}^{l-1} p_i p_{l-i} - \sum_{i=1}^{l} q_i q_{l+1-i} \right) - \sum_{i=1}^{l-1} p_i q_{i+1}, \tag{16}$$

where $l = k/2$.

To a pair of Jordan blocks of odd dimension k with eigenvalues $\pm ai$ there corresponds a Hamiltonian

$$H_0 = \mp \frac{1}{2} \left[\sum_{i=1}^{l} \left(a^2 p_{2i-1} p_{2l-1-2i} + q_{2i-1} q_{2l+1-2i} \right) \right.$$
$$\left. - \sum_{i=1}^{l-1} \left(a^2 p_{2i} p_{2l-2i} + q_{2i} q_{2l-2i} \right) \right] - \sum_{i=1}^{2l-2} p_i q_{i+1}, \tag{17}$$

where $l = (k + 1)/2$.

To a pair of Jordan blocks of even dimension k with eigenvalues $\pm ia$ there corresponds a Hamiltonian

$$
H_0 = \mp \frac{1}{2}\left[\sum_{i=1}^{l-1} \left(a^2 p_{2i+1}p_{2l+1-2i} + p_{2i+2}p_{2l+2-2i} \right) \right.
$$
$$
\left. - \sum_{i=1}^{l} \left(\frac{1}{a^2} q_{2i-1}q_{2l+1-2i} + q_{2i}q_{2l+2-2i} \right) \right] - a^2 \sum_{i=1}^{l} p_{2i-1}q_{2i} + \sum_{i=1}^{l} p_{2i}q_{2i-1},
$$

$$(18)$$

where $l = k/2$.

REMARK 1. In each of the Hamiltonians $H_0^{(j)}$ appearing in (12) one has to change the labelling of the variables in such a way that each pair p_i, q_i appears in only one of the $H_0^{(j)}$; in this way, to each $H_0^{(j)}$ there corresponds a subspace of R^{2n}. If, for a summation symbol Σ, the upper index is smaller than the lower one, the corresponding sum is set equal zero. If in the expression for $H_0^{(j)}$ there appear two signs, then the choice of sign is uniquely determined by the original matrix A according to [1].

REMARK 2. Williamson's Theorem is valid also for a complex matrix A. In this case the Hamiltonians $H_0^{(j)}$ can have only the form (13), (15), or (16), and the numbers $\pm a$ appearing in (13) are in general complex.

§4. Versal deformations of Hamiltonian matrices

DEFINITION. A *deformation* of a Hamiltonian matrix IA_0 is a Hamiltonian matrix $IA(\lambda)$ whose elements are power series in an arbitrary number of real parameters $\{\lambda_i\}$, which are convergent in some neighborhood of the point $\lambda = 0$, and such that $IA(0) = IA_0$.

DEFINITION. A deformation $IA(\lambda)$ of a Hamiltonian matrix IA_0 is called *versal* if, for any other deformation $IB(\mu)$ of the matrix IA_0, there exist a map φ: $\{\mu_i\} \to \{\lambda_i\}$, smooth in a neighborhood of the point $\mu = 0$, and a symplectic matrix $C(\mu)$, depending smoothly on $\{\mu_i\}$, such that

$$
\varphi(0) = 0, \quad C(0) = E, \tag{19}
$$

$$
IB(\mu) = C(\mu) \cdot IA[\varphi(\mu)] \cdot C^{-1}(\mu). \tag{20}
$$

Let us consider the manifold M of the Hamiltonian matrices of order $2n$, the dimension of which is $n(2n + 1)$, and the Lie group G of symplectic matrices of the same order. We define an action of G on M by the formula

$$
\mathrm{Ad}_s(IA) = SIAS^{-1}, \qquad IA \in M, S \in G. \tag{21}
$$

DEFINITION. The *centralizer* of a Hamiltonian matrix IA is the set Z of all Hamiltonian matrices IC such that

$$
IA \cdot IC = IC \cdot IA. \tag{22}
$$

Computation of a versal deformation of an arbitrary Hamiltonian matrix can be performed in a way analogous to the method of computing a versal deformation of an arbitrary complex matrix indicated in [2]. In order to apply this method, one has to compute the centralizers of the Hamiltonian matrices IA_0, where A_0 has Williamson normal form. These centralizers have been found explicitly; however, we do not write them down because they are cumbersome. As a result of computing the centralizers, we get

LEMMA 1. *The dimension of the centralizer of a Hamiltonian matrix IA_0 depends on the Jordan form of the matrix, and is given by the formula*

$$\dim Z = \frac{1}{2} \sum_{z \neq 0} \sum_{j=1}^{s(z)} (2j - 1)n_j(z) + \frac{1}{2} \sum_{j=1}^{u} (2j - 1)m_j$$

$$+ \sum_{j=1}^{v} \left[2(2j - 1)\tilde{m}_j + 1 \right] + 2 \sum_{j=1}^{u} \sum_{k=1}^{v} \min\{m_j, \tilde{m}_k\}, \qquad (23)$$

where $n_1(z) \geqslant n_2(z) \geqslant \cdots \geqslant n_s(z)$ are the dimensions of the Jordan blocks with eigenvalue $z \neq 0$, $m_1 \geqslant m_2 \geqslant \cdots \geqslant m_u$, $\tilde{m}_1 \geqslant \tilde{m}_2 \geqslant \cdots \geqslant \tilde{m}_v$ are the dimensions of the Jordan blocks with eigenvalue $z = 0$, the numbers m_j being even, while $\tilde{m}_j$ are odd (only one block out of each pair of blocks of odd dimension is taken into account).

As an example we write down the centralizer in the case when the matrix IA_0 has only a pair of real eigenvalues $\pm a$, to each of which there correspond Jordan blocks of dimensions $n_1 \geqslant n_2 \geqslant \cdots \geqslant n_s$ in the Jordan form of IA_0.

According to (12) and (13), we have

$$A_0 = \begin{pmatrix} 0 & B \\ B^* & 0 \end{pmatrix}, \qquad (24)$$

where B is the Jordan form of the matrix with eigenvalue $-a$ and blocks of dimensions $n_1 \geqslant n_2 \geqslant \cdots n_s$. The calculations show that the centralizer of IA_0 consists of all matrices of the form IC, where

$$C = \begin{pmatrix} 0 & D \\ D^* & 0 \end{pmatrix}, \qquad (25)$$

and D commutes with B. The general form of a matrix D is shown in [2].

Consider a deformation $IA(\lambda)$ of a Hamiltonian matrix IA_0 as a mapping from the parameter space to the space of Hamiltonian matrices. Also, consider the orbit N of the matrix IA_0 under the action of the group G defined by (21). The following two lemmas are formulated and proved the same way as the analogous lemmas in [2].

LEMMA 2. *A deformation $IA(\lambda)$ of an arbitrary Hamiltonian matrix IA_0 is versal if and only if the mapping $IA(\lambda)$ is transversal to N at the point $\lambda = 0$.*

LEMMA 3. *The dimension of the centralizer of an arbitrary Hamiltonian matrix IA_0 is equal to the codimension of its orbit in the space M of the Hamiltonian matrices.*

The above information makes it possible to construct a versal deformation of an arbitrary Hamiltonian matrix.

6   D. M. GALIN

§5. Formulation of the main result

THEOREM 1. *Each Hamiltonian matrix has a versal deformation, the minimal number of parameters of which coincides with the codimension d of the orbit N of the matrix, and is determined from (23) by replacing* dim *Z by d on the left-hand side.*

By virtue of the Corollary to Williamson's Theorem it suffices to describe versal deformations of matrices IA_0, where A_0 has Williamson normal form. In addition, one can restrict one's attention to the case when the number of terms s in (12) does not exceed 2 (otherwise it suffices to consider all possible pairs of subspaces in R^{2n} corresponding to the Hamiltonians $H_0^{(j)}$ and to obtain a description of the whole versal deformation by using the results obtained for $s \leqslant 2$). In order to describe a versal deformation $IA(\lambda)$ of a matrix IA_0 it clearly suffices to exhibit the general form of the Hamiltonian

$$H(\lambda) = \tfrac{1}{2}(A(\lambda)x, x). \tag{26}$$

In the following we shall denote each of the possible forms of IA_0 by products of the determinants of Jordan blocks, as was done in [2]. We introduce abbreviated notation

$$(+a)^k(-a)^k = (\pm a)^k, \tag{27}$$

$$(+ai)^k(-ai)^k = (\pm ai)^k, \tag{28}$$

$$(+a+bi)^k(+a-bi)^k(-a+bi)^k(-a-bi)^k = (\pm a \pm bi)^k. \tag{29}$$

THEOREM 2. *There exists a versal deformation of a form such that $H(\lambda)$ is a sum of the Hamiltonian* (11) *and a quadratic form in the Hamiltonian canonical variables. The nonvanishing coefficients of the quadratic form are the parameters λ_j $(j = 1,\ldots,d)$.*

As indicated above, for the cases it is sufficient to consider, one can choose the generators of this pencil of quadratic forms that are shown in the table below. If, in (12), $s = 2$, *and the Hamiltonians*

$$H_0^{(1)} = H_0^{(1)}(p_1, p_2,\ldots,p_k, q_1, q_2,\ldots,q_k),$$

$$H_0^{(2)} = H_0^{(2)}(p_{k+1}, p_{k+2},\ldots,p_n, q_{k+1}, q_{k+2},\ldots,q_n)$$

correspond to Jordan blocks with different eigenvalues, then all the coefficients of $p_i p_j$, $q_i q_j$, $p_i q_j$ and $p_j q_i$ $(i = 1,\ldots,k; j = k+1,\ldots,n)$ in $H(\lambda)$ can be set equal to zero, and the remaining coefficients clearly can be determined from the cases considered before.

REMARK 3. If one considers complex deformations of complex Hamiltonian matrices, Theorems 1 and 2 remain valid. Theorem 2 is even simplified if one takes into account Remark 2 about the Hamiltonians $H_0^{(j)}$, the number of different forms of which is smaller in the complex case than in the real one.

§6. Application of the main result to linear Hamiltonian systems

Let $IA(\lambda)$ and $IB(\mu)$ be deformations of the same Hamiltonian matrix IA_0, and let $IA(\lambda)$ be a versal deformation. The fundamental result implies

THEOREM 3. *A linear Hamiltonian system*

$$\frac{dx}{dt} = IB(\mu)x, \tag{30}$$

smoothly depending on the parameters $\{\mu_i\}$, *can be brought, by a change of parameters*

$$\lambda = \varphi(\mu), \quad \varphi(0) = 0, \tag{31}$$

and variables

$$x = C(\mu)y, \quad C(0) = E, \tag{32}$$

to the normal form

$$\frac{dy}{dt} = IA(\lambda)y, \tag{33}$$

with both changes smoothly depending on the parameters $\{\mu_i\}$ *in a neighborhood of* $\mu = 0$.

The system (33) will be called a *versal deformation* of the Hamiltonian system (30).

COROLLARY. *If the symmetric matrix* $B(\mu)$ *in the system* (30) *is such that* $B(0)$ *has Williamson normal form, then this system can be reduced to the normal form* (33) *with the matrix* $IA(\lambda)$ *described in* §5.

REMARK 4. Theorems 1, 2, and 3 will also hold if one generalizes the definition of a deformation of a Hamiltonian matrix given in §4 by allowing as elements of a deformation any functions of the parameters $\{\lambda_i\}$ which are infinitely differentiable in a neighborhood of the point $\lambda = 0$.

§7. Hamiltonians in general position

We partition the space of all Hamiltonian matrices of a given order $2n$ into subsets corresponding to the dimensions of the Jordan blocks and the presence of eigenvalues of different types; these subsets will be called *bundles of Hamiltonian matrices*. We shall denote a bundle as in [2], by the product of the determinants of Jordan blocks with multiple eigenvalues, using the notation (27)–(29).

All the bundles which have no multiple eigenvalues, and only these, have codimension $k = 0$. Bundles of codimension 1 are as follows: $(\pm a)^2$—a pair of Jordan blocks of second order with real eigenvalues; $(\pm ia)^2$—a pair of Jordan blocks of second order with purely imaginary eigenvalues; and 0^2—a Jordan block of second order with eigenvalue 0.

Bundles of codimension $k = 2$ are as follows:

$$(\pm a)^3, \quad (\pm ai)^2, \quad (\pm a \pm bi)^2, \quad 0^4, \quad (\pm a)^2(\pm b)^2,$$

$$(\pm ai)^2(\pm bi)^2, \quad (\pm a)^2(\pm bi)^2, \quad (\pm a)^2 0^2, \quad (\pm ai)^2 0^2.$$

8   **D. M. GALIN**

Table

Form of IA_0	Generators of the sheaf of quadratic forms
$(\pm a)^n$	$p_1q_1,\ p_2q_1,\ \ldots,\ p_nq_1$
$(\pm a)^k (\pm a)^{n-k},$ $k \geqslant \dfrac{n}{2}$	$p_1q_1,\ p_2q_1,\ldots,\ p_kq_1;\ p_{k+1}q_{k+1},\ p_{k+2}q_{k+1},\ \ldots\, p_nq_{k+1};$ $p_kq_{k+1},\ p_kq_{k+2},\ \ldots,\ p_kq_n;\ p_{k+1}q_1,\ p_{k+2}q_1,\ \ldots,\ p_nq_1$
$(\pm a \pm bi)^{\frac{n}{2}}$	$p_1q_1,\ p_2q_1,\ \ldots,\ p_nq_1$
$(\pm a \pm bi)^{\frac{k}{2}} (\pm a \pm bi)^{\frac{n-k}{2}},$ $k \geqslant \dfrac{n}{2}$	$p_1q_1,\ p_2q_1,\ \ldots,p_kq_1;\ p_{k+1}q_{k+1},\ p_{k+2}q_{k+1},\ \ldots,p_nq_{k+1};$ $p_kq_{k+1},\ p_kq_{k+2},\ \ldots,\ p_kq_n;\ p_{k+1}q_1,\ p_{k+2}q_1,\ \ldots,\ p_nq_1$
$0^n0^n,$ $n - \text{odd}$	$p_1q_1,\ p_2q_1,\ \ldots,\ p_nq_1;\ p_1p_n,\ p_3p_n,\ \ldots,\ p_n^2;$ $q_1^2,\ q_1q_3,\ \ldots,\ q_1q_n$
$0^k0^k0^{n-k}0^{n-k},$ $k \geqslant \dfrac{n}{2};$ k and $(n-k)$ odd	$p_1q_1,\ p_2q_1,\ \ldots,\ p_kq_1;\ p_{k+1}q_{k+1},\ p_{k+2}q_{k+1},\ \ldots,\ p_nq_{k+1};$ $p_1p_k,\ p_3p_k,\ \ldots,\ p_k^2;\ q_1^2,\ q_1q_3,\ \ldots,\ q_1q_k;$ $p_{k+1}p_n,\ p_{k+3}p_n,\ \ldots,\ p_n^2;\ q_{k+1}^2\cdot q_{k+1}q_{k+3},\ \ldots,\ q_{k+1}q_n;$ $p_kq_{k+1},\ p_kq_{k+2},\ \ldots,\ p_kq_n;\ p_{k+1}q_1,\ p_{k+2}q_1,\ \ldots,\ p_nq_1;$ $p_kp_{k+1},\ p_kp_{k+2},\ \ldots,\ p_kp_n;\ q_1q_{k+1},\ q_1q_{k+2},\ \ldots,\ q_1q_n$
$(\pm ai)^n,\ n$ odd	$p_2q_1,\ p_4q_1,\ \ldots,\ p_{n-1}q_1;\ q_1^2,\ q_1q_3,\ \ldots,\ q_1q_n$
$(\pm ai)^k (\pm ai)^{n-k},\ k \geqslant \dfrac{n}{2};$ k and $(n-k)$ odd	$p_2q_1,\ p_4q_1,\ldots,\ p_{k-1}q_1;\ p_{k+2}q_{k+1},\ p_{k+4}q_{k+1},\ldots,\ p_{n-1}q_{k+1};$ $q_1^2,\ q_1q_3,\ \ldots,\ q_1q_k;\ q_{k+1}^2,\ q_{k+1}q_{k+3},\ \ldots,\ q_{k+1}q_n;$ $p_{k+1}q_1,\ p_{k+2}q_1,\ \ldots,\ p_nq_1;\ q_1q_{k+1},\ q_1q_{k+2},\ \ldots,\ q_1q_n$
0^{2n}	$p_1p_n,\ p_2p_n,\ \ldots,\ p_n^2$
$0^{2k}0^{2(n-k)},\ k \geqslant \dfrac{n}{2}$	$p_1p_k,\ p_2p_k,\ \ldots,\ p_k^2;\ p_{k+1}p_n,\ p_{k+2}p_n,\ \ldots,\ p_n^2;$ $p_{k+1}q_1,\ p_{k+2}q_1,\ \ldots,\ p_nq_1;\ p_kp_{k+1},\ p_kp_{k+2},\ \ldots,\ p_kp_n$
$(\pm ai)^n,\ n$ even	$p_1q_2,\ p_1q_4,\ \ldots,\ p_1q_n;\ p_1^2,\ p_1p_3,\ \ldots,\ p_1p_{n-1}$
$(\pm ai)^k (\pm ai)^{n-k},\ k \geqslant \dfrac{n}{2};$ k and $(n-k)$ even	$p_1q_2,\ p_1q_4,\ \ldots,\ p_1q_k;\ p_{k+1}q_{k+2},\ p_{k+1}q_{k+4},\ \ldots,\ p_{k+1}q_n;$ $p_1^2,\ p_1p_3,\ \ldots,\ p_1p_{k-1};\ p_{k+1}^2,\ p_{k+1}p_{k+3},\ \ldots,\ p_{k+1}p_{n-1};$ $p_1q_{k+1},\ p_1q_{k+2},\ \ldots,\ p_1q_n;\ p_1p_{k+1},\ p_1p_{k+2},\ \ldots,\ p_1p_n$
$0^{2k}0^{n-k}0^{n-k},\ k \geqslant \dfrac{n}{3};$ $(n-k)$ odd	$p_1p_k,\ p_2p_k,\ \ldots,\ p_k^2;\ p_{k+1}q_{k+1},\ p_{k+2}q_{k+1},\ \ldots,\ p_nq_{k+1};$ $p_{k+1}p_n,\ p_{k+3}p_n,\ \ldots,\ p_n^2;\ q_{k+1}^2,\ q_{k+1}q_{k+3},\ \ldots,\ q_{k+1}q_n;$ $p_kq_{k+1},\ p_kq_{k+2},\ \ldots,\ p_kq_n;\ p_kp_{k+1},\ p_kp_{k+2},\ \ldots,\ p_kp_n$

Table continued

Form of IA_0	Generators of the sheaf of quadratic forms
$0^{2k}0^{n-k}0^{n-k},\ k < \dfrac{n}{3};$ $(n-k)\,\text{odd}$	$p_1p_k,\ p_2p_k,\ \ldots,\ p_k^2;\ p_{k+1}q_{k+1},\ p_{k+2}q_{k+1},\ \ldots,\ p_nq_{k+1};$ $p_{k+1}p_n,\ p_{k+3}p_n,\ \ldots,\ p_n^2;\ q_{k+1}^2,\ q_{k+1}q_{k+3},\ \ldots,\ q_{k+1}q_n;$ $p_1q_{k+1},\ p_2q_{k+1},\ \ldots,\ p_kq_{k+1};\ p_1q_{k+2},\ p_2q_{k+2},\ \ldots,\ p_kq_{k+2};$ $p_1p_{n-1},\ p_2p_{n-1},\ \ldots,\ p_kp_{n-1};\ p_1p_n,\ p_2p_n,\ \ldots,\ p_kp_n$
$(\pm ai)^k(\pm ai)^{n-k},\ k \geqslant \dfrac{n}{2};$ $k\ \text{even}$ $(n-k)\ \text{odd}$	$p_1q_2,\ p_1q_4,\ \ldots,\ p_1q_k;\ p_1^2,\ p_1p_3,\ \ldots,\ p_1p_{k-1};$ $p_{k+2}q_{k+1},\ p_{k+4}q_{k+1},\ \ldots,\ p_{n-1}q_{k+1};$ $q_{k+1}^2,\ q_{k+1}q_{k+3},\ \ldots,\ q_{k+1}q_n;$ $p_1q_{k+1},\ p_1q_{k+2},\ \ldots,\ p_1q_n;\ p_2q_{k+1},\ p_2q_{k+2},\ \ldots,\ p_2q_n$
$(\pm ai)^k(\pm ai)^{n-k},\ k < \dfrac{n}{2};$ $k\ \text{even}$ $(n-k)\ \text{odd}$	$p_1q_2,\ p_1q_4,\ \ldots,\ p_1q_k;\ p_1^2,\ p_1p_3,\ \ldots,\ p_1p_{k-1};$ $p_{k+2}q_{k-1},\ p_{k+4}q_{k+1},\ \ldots,\ p_{n-1}q_{k+1};$ $q_{k+1}^2,\ q_{k+1}q_{k+3},\ \ldots,\ q_{k+1}q_n;$ $p_1q_{k+1},\ p_2q_{k+1},\ \ldots,\ p_kq_{k+1};\ p_1q_{k+2},\ p_2q_{k+2},\ \ldots,\ p_kq_{k+2}$

Bundles of codimension $k = 3$ are as follows:

$$(\pm a)(\pm a),\quad (\pm a)^4,\quad (\pm ai)(\pm ai),\quad (\pm ai)^4,\quad 00,\quad 0^6,$$
$$(\pm a)^2(\pm b)^2(\pm c)^2,\quad (\pm a)^2(\pm bi)^2(\pm ci)^2,\quad (\pm a)^2(\pm b)^2(\pm ci)^2,$$
$$(\pm ai)^2(\pm bi)^2(\pm ci)^2,\quad (\pm a)^2(\pm b)^20^2,\quad (\pm a)^2(\pm bi)^20^2,$$
$$(\pm ai)^2(\pm bi)^20^2,\quad (\pm a)^3(\pm b)^2,\quad (\pm a)^3(\pm bi)^2,\quad (\pm a)^30^2,$$
$$(\pm ai)^3(\pm b)^2,\quad (\pm ai)^3(=bi)^2,\quad (\pm ai)^30^2,\quad (\pm a)^20^4,\quad (\pm ai)^20^4,$$
$$(\pm a \pm bi)^2(\pm c)^2,\quad (\pm a \pm bi)^2(\pm ci)^2,\quad (\pm a \pm bi)^20^2.$$

DEFINITION. A Hamiltonian of the form (26) is called a Hamiltonian *in general position* for the bundle of the matrix $IA(0)$ if the corresponding mapping $IA(\lambda)$ of the parameter space into the space of Hamiltonian matrices is transversal to the bundle at the point $\lambda = 0$.

THEOREM 4. *If a symmetric matrix $A(0)$ has Williamson normal form, and the bundle of the Hamiltonian matrix $IA(0)$ has codimension $k \leqslant 2$, then, depending on the bundle, a Hamiltonian in general position can be taken in one of the following forms:*

$$H(\lambda) = -a(p_1q_1 + p_2q_2) + p_1q_2 + \lambda_1 p_2q_1, \tag{34}$$

$$H(\lambda) = p_2q_1 - a^2 p_1q_2 \pm \frac{1}{2}\left(\frac{1}{a^2}q_1^2 + q_2^2\right) + \frac{\lambda_1}{2}p_1^2, \tag{35}$$

$$H(\lambda) = \mp\frac{1}{2}q_1^2 + \frac{\lambda_1}{2}p_1^2, \tag{36}$$

$$H(\lambda) = -a(p_1q_1 + p_2q_2 + p_3q_3) + (p_1q_2 + p_2q_3)$$
$$+\lambda_1 p_2 q_1 + \lambda_2 p_3 q_1, \tag{37}$$

$$H(\lambda) = -(p_1q_2 + p_2q_3) \mp \frac{1}{2}(2a^2 p_1 p_3 - a^2 p_2^2 + 2q_1 q_3 - q_2^2)$$
$$+\lambda_1 p_2 q_1 + \frac{\lambda_2}{2} q_1^2, \tag{38}$$

$$H(\lambda) = -a(p_1q_1 + p_2q_2 + p_3q_3 + p_4q_4) + b(p_1q_2 - p_2q_1$$
$$+p_3q_4 - p_4q_3) + (p_1q_3 + p_2q_4) + \lambda_1 p_3 q_1 + \lambda_2 p_4 q_1, \tag{39}$$

$$H(\lambda) = -p_1q_2 \pm \frac{1}{2}(p_1^2 - 2q_1q_2) + \lambda_1 p_1 p_2 + \frac{\lambda_2}{2} p_2^2, \tag{40}$$

$$H(\lambda) = -a(p_1q_1 + p_2q_2) + p_1q_2 - b(p_3q_3 + p_4q_4)$$
$$+p_3q_4 + \lambda_1 p_2 q_1 + \lambda_2 p_4 q_3, \tag{41}$$

$$H(\lambda) = p_2q_1 - a^2 p_1 q_2 \pm \frac{1}{2}\left(\frac{1}{a^2} q_1^2 + q_2^2\right) + p_4 q_3$$
$$-b^2 p_3 q_4 \pm \frac{1}{2}\left(\frac{1}{b^2} q_3^2 + q_4^2\right) + \frac{\lambda_1}{2} p_1^2 + \frac{\lambda_2}{2} p_3^2, \tag{42}$$

$$H(\lambda) = -a(p_1q_1 + p_2q_2) + p_1q_2 + p_4q_3$$
$$-b^2 p_3 q_4 \pm \frac{1}{2}\left(\frac{1}{b^2} q_3^2 + q_4^2\right) + \lambda_1 p_2 q_1 + \frac{\lambda_2}{2} p_3^2, \tag{43}$$

$$H(\lambda) = -a(p_1q_1 + p_2q_2) + p_1q_2 \mp \frac{1}{2} q_3^2 + \lambda_1 p_2 q_1 + \frac{\lambda_2}{2} p_3^2, \tag{44}$$

$$H(\lambda) = p_2q_1 - a^2 p_1 q_2 \pm \frac{1}{2}\left(\frac{1}{a^2} q_1^2 + q_2^2\right) \mp \frac{1}{2} q_3^2 + \frac{\lambda_1}{2} p_1^2 + \frac{\lambda_2}{2} p_3^2. \tag{45}$$

In addition, if there are simple eigenvalues, then one should add to $H(\lambda)$ a parameter-independent Hamiltonian (12) in which the summation is taken only over the indices j corresponding to the Jordan blocks of $IA(0)$ with simple eigenvalues.

The order of formulas (34)–(45) corresponds to the order in which the bundles with codimension $k = 1$ and $k = 2$ were listed above.

§8. Bifurcation diagrams

Let $IA(\lambda)$ be a family of Hamiltonian matrices depending smoothly on the parameters $\{\lambda_i\}$, with $IA(0) = IA_0$.

DEFINITION. The *bifurcation diagram* of the family $IA(\lambda)$ is the partition of a neighborhood of the point $\lambda = 0$ in the parameter space into inverse images of the bundles under the mapping IA.

Let $IA(\lambda)$ be a family such that the bundle of the matrix $IA(0)$ has codimension $k = 1$ or $k = 2$, the matrix $A(0)$ has Williamson normal form, and the Hamiltonian of the form (26) corresponding to the family is in general position with respect to the bundle of the matrix $IA(0)$ and is given by one of the formulas (34)–(40), depending on the bundle. The bifurcation diagrams of such families are presented in Figure 1 under the numbers 1)–7), respectively.

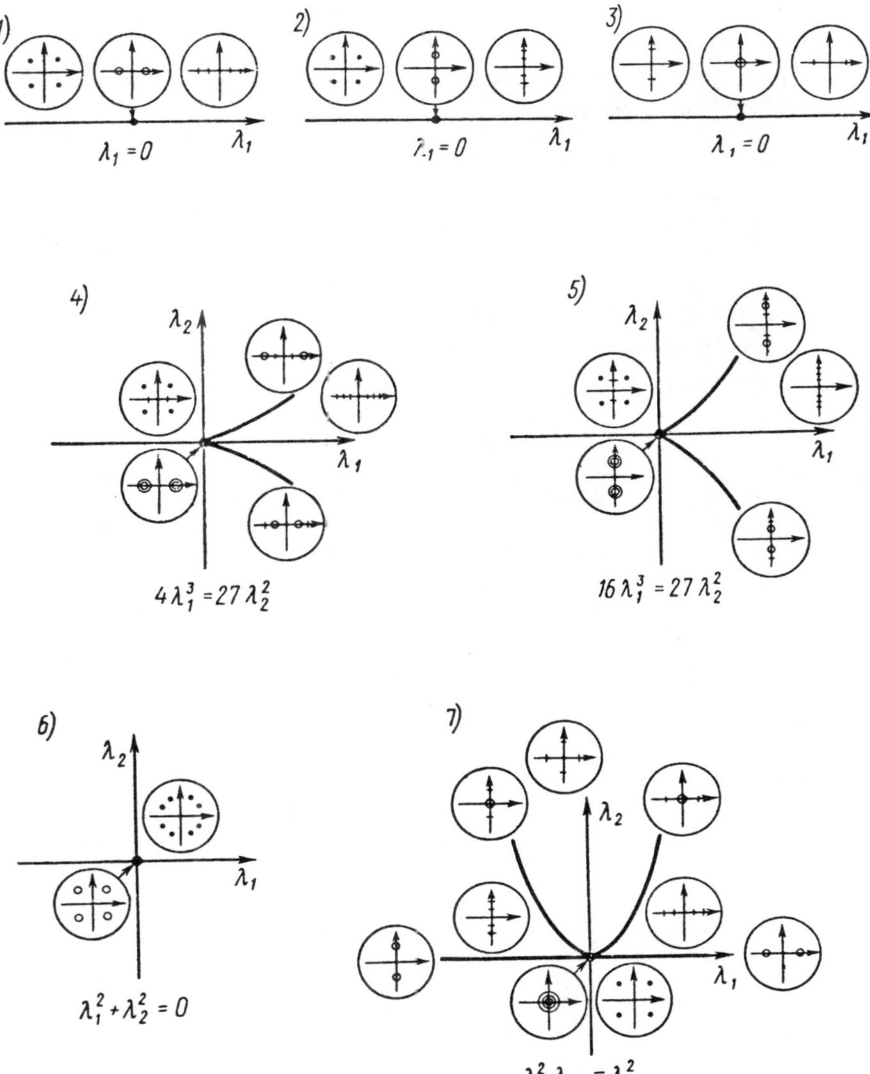

FIGURE 1

EXAMPLE 1. The second diagram in Figure 1 shows that, if the Jordan form of the matrix $IA(0)$ has a pair of blocks of second order with purely imaginary eigenvalues, then, for all matrices of the family, when $\lambda_1 > 0$ this pair turns into two different pairs of blocks of first order with purely imaginary eigenvalues, and, when $\lambda_1 < 0$, into a quadruple of blocks of first order with pairwise complex conjugate eigenvalues.

EXAMPLE 2. The fourth diagram in Figure 1 shows that, if the Jordan form of the matrix $IA(0)$ has a pair of blocks of third order with real eigenvalues, then, for all matrices of the family, when $4\lambda_1^3 = 27\lambda_2^2$ this pair turns into two pairs of blocks with different real eigenvalues, where one pair is of second order while the other is of first order; when $4\lambda_1^3 > 27\lambda_2^2$ three different pairs of first order blocks with real eigenvalues appear, and, when $4\lambda_1^3 < 27\lambda_2^2$, a pair of blocks of first order with real eigenvalues and a quadruple of blocks of first order with pairwise complex conjugate eigenvalues.

THEOREM 5. *If the bundle of the matrix IA_0 has codimension $k = 1$, then the bifurcation diagram of any one-parameter family $IA(\lambda)$, $IA(0) = IA_0$, transversal to the bundle is diffeomorphic to one of the diagrams 1)–3) in Figure 1.*

THEOREM 6. *If the bundle of the matrix IA_0 has codimension $k = 2$, then the bifurcation diagram of any two-parameter family $IA(\lambda)$, $IA(0) = IA_0$, transversal to the bundle is diffeomorphic to the direct product of any two of the diagrams 1)–3), except the product of the diagram 3) with itself.*

BIBLIOGRAPHY

1. John Williamson, *On the algebraic problem concerning the normal forms of linear dynamical systems*, Amer. J. Math. **58** (1936), 141–163.

2. V. I. Arnol'd, *On matrices depending on parameters*, Uspehi Mat. Nauk **26** (1971), no. 2 (158), 101–114; English transl. in Russian Math. Surveys **26** (1971).

Translated by T. M. ŚNIATYCKI

Amer. Math. Soc. Transl.
(2) Vol. **118**, 1982

On the Analyticity of Solutions
of Linear Second Order
Partial Differential Equations*

O. A. OLEĬNIK AND E. V. RADKEVIČ

For second order linear equations of the form

$$L(u) \equiv \sum_{k,j=1}^{n} a^{kj}(x)u_{x_k x_j} + \sum_{k=1}^{n} b^{k}(x)u_{x_k} + c(x)u = f(x) \tag{1}$$

with real analytic coefficients in a domain $\Omega \subset R^n = \{x = (x_1,\ldots,x_n)\}$, a necessary and sufficient condition for hypoellipticity was given in [1], provided that

$$\sum_{k=1}^{n} \left(a^{kk}(x) + |b^{k}(x)| \right) \neq 0$$

everywhere in Ω. Here we study the question of which hypoelliptic equations of the form (1) have only analytic solutions.

1. The case of many independent variables. We shall say that the operator L is analytic at the point $x_0 \in \Omega$ if in any sufficiently small neighborhood $Q \subset \Omega$ of x_0 any solution of the equation $L(u) = g$ in $D'(Q)$ is analytic in Q when g is analytic in Q.

LEMMA 1. *Let analytic real functions $a(x)$ and $H(x)$ be given in a neighborhood Ω of the point $x = 0$ in R^n, such that $H(0) = 0$, grad $H(0) \neq 0$, and $a(x) = 0$ at any point x such that $H(x) = 0$. Then in some neighborhood Ω_1 of $x = 0$ there exists an analytic real function $H_1(x)$ such that*

$$a(x) = H(x)H_1(x) \tag{2}$$

in Ω_1, and $\Omega_1 \subset \Omega$.

1980 *Mathematics Subject Classification.* Primary 35H05, 35B65; Secondary 35A10.
*Translation of Trudy Sem. Petrovsk. **1** (1975), 163–173. MR **55** #3505.

If $D^\alpha a(x) = 0$ at points x where $H(x) = 0$, for any multi-index α with $|\alpha| < k$, and if $D^{\alpha^p} a(0) \neq 0$ for any multi-index α^0 with $|\alpha^0| = k$, then

$$a(x) = H^k(x)H_2(x),\tag{3}$$

where $H_2(x)$ is real analytic in Ω_1 and $H_2(0) \neq 0$.

PROOF. Since grad $H(0) \neq 0$, we may assume that $H_{x_1}(0) \neq 0$ and $a(x) \not\equiv 0$ for $x_2 = \cdots = x_n = 0$. By virtue of the Weierstrass preparation theorem (see [2], p. 289, Theorem 5) there exist analytic functions $a_j(x')$, $a_j(0) = 0$, $j = 0,\ldots,l-1$, $x' = (x_2,\ldots,x_n)$ and $h(x)$, $h(0) \neq 0$, such that in some neighborhood of Ω_1

$$a(x) = \left(\sum_{j=0}^{l-1} a_j(x')x_1^j + x_1^l \right)h(x).\tag{4}$$

Exactly the same way, we obtain that in Ω_1

$$H(x) = (x_1 - h_1(x'))h_2(x),\tag{5}$$

where h_1 and h_2 are analytic and $h_1(0) = 0$, $h_2(0) \neq 0$. Replacing x_1 by $(x_1 - h_1(x')) + h_1(x')$ in (4) we obtain

$$a(x) = \left(\sum_{j=0}^{l-1} \tilde{a}_j(x')(x_1 - h_1(x'))^j + (x_1 - h_1(x'))^l \right)h(x).\tag{6}$$

Since $a(x) = 0$ for $x_1 = h_1(x')$ and $(h_1(x'), x') \in \Omega_1$, it follows from (6) that representation (2) holds. If $D^\alpha a(x) = 0$ for $x_1 = h_1(x')$ and $(h_1(x'), x') \in \Omega_1$ for any multi-index α such that $|\alpha| < k$, we obtain by successively differentiating (6) with respect to x_1 that $\tilde{a}_j(x') \equiv 0$ in $\Omega_1, j = 0, 1,\ldots,k-1$. Since $D^{\alpha^0}a(0) \neq 0$ for some α^0 with $|\alpha^0| = k$, it follows from (6) that $l = k$ and therefore that (3) holds. The lemma is proved.

THEOREM 1. *Let a second order differential operator L with real analytic coefficients of the form* (1) *be given in a neighborhood Ω of the point $x = 0$ in R^n. Suppose that for any $x \in \Omega$ and any $\xi \in R^n$*

$$\sum_{k,j=1}^{n} a^{kj}(x)\xi_k\xi_j \geqslant 0.\tag{7}$$

Suppose that the coefficients $a^{kj}(x)$ do not vanish simultaneously at $x = 0$, and that there exists a real analytic function $H(x)$ such that grad $H(0) \neq 0$, $H(0) = 0$, *and*

$$\sum_{k,j=1}^{n} a^{kj}(x)H_{x_k}(x)H_{x_j}(x) = 0\tag{8}$$

for each point $x \in \Omega$ such that $H(x) = 0$. Then the operator (1) *is not analytic at $x = 0$.*

PROOF. Suppose $H_{x_1}(0) \neq 0$. In a neighborhood Ω_1 of $x = 0$ where $H_{x_1} \neq 0$, we consider an analytic coordinate transformation $y_1 = H(x)$, $y_2 = x_2,\ldots,y_n = x_n$.

In the new coordinates, the operator L has the form

$$L(u) \equiv \sum_{k,j=1}^{n} A^{kj}(y)u_{y_k y_j} + \sum_{k=1}^{n} B^k(y)u_{y_k} + c(y)u,$$

where

$$A^{11} = \sum_{k,j=1}^{n} a^{kj}H_{x_k}H_{x_j}, \qquad A^{1j} = \sum_{k=1}^{n} a^{kj}H_{x_k}, \qquad j = 2,\ldots,n,$$

and by virtue of Lemma 1 and conditions (7) and (8) we have

$$A^{11} = y_1^2 a(y), \qquad A^{1j} = y_1 a_j(y), \quad j = 2,\ldots,n.$$

If

$$B^1 \equiv \sum_{k,j=1}^{n} a^{kj}H_{x_k x_j} + \sum_{k=1}^{n} b^k H_{x_k} \equiv 0$$

for $y_1 = 0$, then the Lie algebra spanned by the system of operators

$$\left\{ \sum_{j=1}^{n} A^{kj}D_{y_j}, \, k = 1,\ldots,n; \; \sum_{k=1}^{n} B^k D_{y_k} - \sum_{k,j=1}^{n} A_{y_j}^{kj}D_{y_k} \right\},$$

has rank less than n at $x = 0$, and therefore, because of the results of [1], the operator L is not hypoelliptic in a neighborhood of $x = 0$, and hence not analytic at that point.

Let $B^1 \not\equiv 0$ in $\dot{\Omega}_1$ for $y_1 = 0$. We shall assume that $A^{22} \neq 0$ in Ω_1. We may also suppose that $B^1(0, y_2, 0,\ldots,0) \not\equiv 0$, since this condition may always be attained by a small rotation of the coordinate axes. Performing the change of variable $u = (4 - e^{\mu y_2})v$ and assuming that $|\mu y_2| < 1$ in Ω_1, we obtain

$$L_\mu(v) \equiv (4 - e^{\mu y_2})^{-1}L(u)$$

$$\equiv \sum_{k,j=1}^{n} A^{kj}v_{y_k y_j} + \sum_{k=1,k\neq 2}^{n} B^k v_{y_k} + \left[B^2 - 2\mu A^{22}e^{\mu y_2}(4 - e^{\mu y_2})^{-1} \right]v_{y_2}$$

$$- \left[(A^{22}\mu^2 + B^2\mu)e^{\mu y_2}(4 - e^{\mu y_2})^{-1} - c \right]v = 0. \tag{9}$$

It is clear that if μ is sufficiently large and the domain Ω_1 is so small that $|\mu y_2| < 1$ in Ω_1, then the coefficient of v in (9) is negative.

Since the function B^1 is analytic in Ω_1, in any ball $Q \subset \Omega_1$ with center at $x = 0$ there exists a ball $Q^0 \subset Q$ with center at $y^0 = (0, y_2^0, 0,\ldots,0)$ such that $B^1(y) \neq 0$ in Q^0. Let $B^1 > 0$ in Q^0. We consider the domain $Q^* = Q^0 \cap \{y_1 < \delta\}$, where δ is a sufficiently small constant. Moreover, by G we denote any domain containing Q^* and the point $y = 0$, whose boundary is infinitely differentiable and contains the entire boundary of Q^* lying in the halfspace $y_1 < \delta$. By S we denote the boundary of G. Let γ be a sufficiently small neighborhood of one of the points lying on the y_2 axis and belonging to the boundary of Q^0, so that for points of S belonging to γ the condition

$$\sum_{k,j=1}^{n} A^{kj} \nu_k \nu_j > 0 \tag{10}$$

holds, with $\nu = (\nu_1, \ldots, \nu_n)$ a vector normal to S. The existence of such a neighborhood follows from the fact that $A^{22} \neq 0$ in Q^0. Let the function $a_1^\varepsilon(y) \in C^\infty(\Omega_1)$ and let a_1^ε be positive in the ε-neighborhood $S \cap \{|y_1| < \varepsilon\} \setminus \gamma$ and equal to zero outside this neighborhood. Let the function $a_2^\varepsilon(y) \in C^\infty(\Omega_1)$ be positive in the $\varepsilon/4$-neighborhood of the boundary S not belonging to the set $\overline{Q}^0 \cap \{|y_1| < \varepsilon\}$, and $a_2^\varepsilon(y) = 0$ outside this neighborhood. If ε is sufficiently small, it is easy to see that on the entire boundary S the condition

$$a_1^\varepsilon(y) \sum_{j=2}^{n} \nu_j^2 + a_2^\varepsilon(y) \sum_{j=1}^{n} \nu_j^2 + \sum_{kj=1}^{n} A^{kj}(y)\nu_k\nu_j > 0$$

holds, where $\nu = (\nu_1, \ldots, \nu_n)$ is a vector normal to S. In G we consider the equation

$$L_\mu(v) + a_1^\varepsilon \sum_{j=2}^{n} v_{y_j y_j} + a_2^\varepsilon \sum_{j=1}^{n} v_{y_j y_j} = 0 \tag{11}$$

with boundary condition

$$v|_S = \varphi, \tag{12}$$

where φ is an infinitely differentiable function with $\varphi = 0$ on S for $y_1 \leq 0$ and $\varphi > 0$ on S for $y_1 > 0$. According to Theorem 1.9.2 in [3], a solution of (11), (12) exists and is a function of class $C^2(\overline{G})$, if μ is sufficiently large. Since the boundary of the domain $G \cap Q^0 \cap \{y_1 < 0\}$ lying in the plane $y_1 = 0$ belongs to Σ_1, according to Theorem 1.1.2 of [3] on the maximum principle, $v = 0$ in this domain. Since $v \not\equiv 0$ on γ and for sufficiently small ε the domain where (11) coincides with (9) contains a subdomain where $v \neq 0$, the point $y = 0$, and some subdomain of $G \cap Q^0 \cap \{y_1 < 0\}$ where $v \equiv 0$, it follows that $u = v(4 - e^{\mu y_2})$ is a nonanalytic solution of (1) in some domain containing the origin. The theorem is proved.

2. The case of two independent variables. We examine (1) in the plane (x_1, x_2). Let

$$L(u) \equiv a^{11} u_{x_1 x_1} + 2a^{12} u_{x_1 x_2} + a^{22} u_{x_2 x_2} + b^1 u_{x_1} + b^2 u_{x_2} + cu = 0, \tag{13}$$

where a^{11}, a^{12}, a^{22}, b^1, b^2 and c are real analytic functions defined in some neighborhood Ω of the origin. Above, we proved that if there exists a characteristic $H_1(x_1, x_2) = 0$ of the equation $L(u) = 0$ passing through the origin, then in an arbitrarily small neighborhood of this point there exists a nonanalytic solution of this equation, i.e. the operator L is not analytic at the origin.

We now consider the case when there is a curve h passing through the origin, given by the equation $H(x_1, x_2) = 0$, at points of which $A = a^{11}a^{22} - (a^{12})^2 = 0$, this curve not, however, being a characteristic for the equation $L(u) = 0$, and the inequality

$$a^{11}H_{x_1}^2 + 2a^{12}H_{x_1}H_{x_2} + a^{22}H_{x_2}^2 > 0 \tag{14}$$

holding at points of h. Moreover, we assume that $a^{11}a^{22} - (a^{12})^2 > 0$ at points of Ω not belonging to h. In this case, the following assertion was proved in [4].

THEOREM 2. *Let the function* $A(x_1, x_2)$ *be defined in the neighborhood* Ω *of the origin, and satisfy* $A(x_1, x_2) \geqslant 0$ *in* Ω *and* $D^\alpha A = 0$ *wherever* $H(x_1, x_2) = 0$ *and* $|\alpha| < 2k$. *Moreover, let* $D^\alpha A(0,0) \neq 0$ *for some multi-index* α *with* $|\alpha| = 2k$. *Assume that* $H(0,0) = 0$, grad $H(0,0) \neq 0$, *and* $H(x_1, x_2)$ *is an analytic function of* x_1 *and* x_2 *in* Ω. *Then in some neighborhood* Ω_1 *of the origin there exists a transformation of independent variables*

$$x = \psi_1(x_1, x_2), \qquad y = \psi_2(x_1, x_2),$$

where ψ_1 *and* ψ_2 *are analytic functions in* Ω_1 *such that, in the new coordinates* (x, y), *(13) takes the form*

$$L(u) = u_{xx} + x^{2k}u_{yy} + a_1 u_x + a_2 u_y + cu = 0, \tag{15}$$

where a_1 *and* a_2 *are analytic functions in* Ω_1.

We note that we may assume that $a_1 \equiv 0$ in Ω_1. In fact, if $a_1 \not\equiv 0$ in Ω_1, then we may pass to a new function $v(x, y)$ such that

$$u = v \exp\left[-\frac{1}{2}\int_C^x a_1(s, y)\, ds\right].$$

Therefore in our case we may limit attention to (15) with $a_1 \equiv 0$.

THEOREM 3. *In a neighborhood* Ω *of the origin in the* (x, y)-*plane, let*

$$L(u) \equiv u_{xx} + x^{2k}u_{yy} + d(x, y)u_y + c(x, y)u \tag{16}$$

be a second order differential operator with real analytic coefficients. In the domain $Q_\delta(\Omega) = \{(x, y + it); (x, y) \in \Omega, |t| < \delta\}$ *for some* $\delta > 0$, *let the coefficients* $c(x, y)$ *and* $d(x, y)$ *satisfy one of the following conditions:*

1) $$|\operatorname{Im} c(x, y + it)|^2 \leqslant M_1 |\operatorname{Im} d(x, y + it)|, \tag{17}$$

$$\operatorname{Im} d(x, y + it) \cdot t \leqslant 0, \tag{18}$$

2) $$|\operatorname{Im} c(x, y + it)| \leqslant M_2 |x|^k, \quad \operatorname{Im} d(x, y + it) \cdot t \leqslant 0, \tag{19}$$

3) $$|\operatorname{Im} c(x, y + it)| \leqslant M_3 |x|^k, \quad |\operatorname{Im} d(x, y + it)| \leqslant M_4 |x|^{2k}, \tag{20}$$

where M_j, $j = 1, \dots, 4$, *are constants. Then any solution of* $L(u) = f$ *in* $D'(\Omega)$ *is analytic in* Ω *if* $f(x, y)$ *is analytic in* Ω.

PROOF. It suffices to prove the theorem for the case $f \equiv 0$, since, by Kowalewski's theorem, in a neighborhood of any point of Ω there exists an analytic solution $v(x, y)$ of $L(v) = f$ and therefore in place of $u(x, y)$ we may consider the function $u - v$, which satisfies the equation $L(u - v) = 0$.

Since the equation $L(u) = 0$ is elliptic in Ω for $x \neq 0$, it suffices to prove that a solution $u(x, y)$ of $L(u) = 0$ in $D'(\Omega)$ is analytic in a neighborhood of any point P of the line $x = 0$. We take such a point to be the origin of our coordinate system. In Chapter II of [3] it was shown that (16) is hypoelliptic, i.e. any solution $u(x, y)$ in $D'(\Omega)$ of the equation

$$L(u) = 0 \tag{21}$$

is infinitely differentiable in Ω.

In (21), we pass to a new function $w(x, y)$ such that

$$u = w \cdot (4 - e^{\mu x}), \qquad \mu = \text{const} > 0.$$

For this function w we obtain an equation of the form

$$L_\mu(w) \equiv L(w) - 2\mu(4e^{-\mu x} - 1)^{-1}w_x - \mu^2(4e^{-\mu x} - 1)^{-1}w = 0. \tag{22}$$

Consider the domain Ω_μ, bounded by an ellipse S_μ:

$$y^2 + \mu x^2 = h^2, \qquad h = \text{const} < 1.$$

Let $\varphi_\mu(x, y) = 1$ in a neighborhood of S_μ, $\varphi_\mu = 0$ in some neighborhood $\tilde{\Omega}_\mu \subset \Omega_\mu$ of the origin, and $\varphi_\mu \in C^\infty(\Omega_\mu)$. Clearly, the function $\tilde{w} = w - \varphi_\mu w$ satisfies the equation

$$L_\mu(\tilde{w}) = -L_\mu(\varphi_\mu w) \equiv F_\mu(x, y), \tag{23}$$

in Ω_μ, and the boundary condition

$$\tilde{w}\big|_{S_\mu} = 0, \tag{24}$$

where $F_\mu(x, y) \in C^\infty(\Omega_\mu)$ and $F_\mu \equiv 0$ in $\tilde{\Omega}_\mu$. We now consider, in Ω_μ, an equation for a function w_ε of the form

$$L_\mu(w) + \varepsilon(w_{xx} + w_{yy}) = F_\mu(x, y), \qquad \varepsilon = \text{const} > 0, \tag{25}$$

with boundary condition

$$w_\varepsilon\big|_{S_\mu} = 0. \tag{26}$$

For sufficiently large μ, solutions $w_\varepsilon(x, y)$ of the Dirichlet problem for (25) under condition (26) in the domain Ω_μ exist and are bounded uniformly in ε together with their derivatives up to order N inclusive; here $w_\varepsilon \in C^\infty(\Omega_\mu)$; and N is any number prescribed in advance. This follows from known theorems on the solvability of the Dirichlet problem for elliptic second order equations (see, for example, [5]) and Lemmas 1.8.2 and 1.8.4 in [3].

We verify that the hypotheses of Lemmas 1.8.2 and 1.8.4 in [3] are satisfied. The functions B_l, defined in [3] by (1.8.12), are negative in Ω_μ for $l \leq N$ and for sufficiently large μ depending on N, since the coefficient of the derivative w_y in (22) does not depend on μ and the coefficient of w_x, equal to

$$b^\mu \equiv -2\mu(4e^{-\mu x} - 1)^{-1},$$

satisfies the condition $b_x < 0$ for $|\mu x| < 1$. The coefficient of w in (22) tends to $-\infty$ as $\mu \to \infty$ when $|\mu x| \leq 1$. It is clear that for sufficiently large $\mu > 0$ the boundary S_μ of Ω_μ belongs to Σ_3.

Thus, the functions w_ε and their derivatives through order N are bounded in Ω_μ, uniformly with respect to ε, for some sufficiently large $\mu = \mu_0$, and therefore as $\varepsilon \to 0$ the functions w_ε converge uniformly, together with their derivatives through order $N - 1$, in the domain $\tilde{\Omega}_{\mu_0}$ to the function $w = u(4 - e^{\mu x})^{-1}$.

In (25) we now pass from the functions $w_\varepsilon(x, y)$ to new function $u_\varepsilon(x, y)$ such that $w_\varepsilon = u_\varepsilon \Phi_\varepsilon$, where $\Phi_\varepsilon(x) = (4 - e^{\mu x})^{-1/(1+\varepsilon)}$. For u_ε, in the domain $\tilde{\Omega}_{\mu_0}$, we obtain

$$(1 + \varepsilon)u_{\varepsilon xx} + (\varepsilon + x^{2k})u_{\varepsilon yy} + d(x, y)u_{\varepsilon y} + c_\varepsilon(x, y)u_\varepsilon = 0, \qquad (27)$$

where

$$c_\varepsilon = c(x, y) + \psi_\varepsilon(x, y) \quad \text{and} \quad \psi_\varepsilon = \frac{\varepsilon}{1+\varepsilon}\mu^2 e^{2\mu x}(4 - e^{\mu x})^{-1/(1+\varepsilon)-2}.$$

It is clear that $u_\varepsilon = w_\varepsilon \Phi_\varepsilon^{-1}(x)$ converge as $\varepsilon \to 0$, uniformly in $\tilde{\Omega}_{\mu_0}$, together with their derivatives up to order $N - 1$, to the function $u(x, y)$.

We show that the functions u_ε are analytic functions of the variable $y + it$ in the domain

$$Q_{\delta_0} = \left\{ |x| \leq \delta_0, |y| \leq \delta_0, |t| \leq \delta_0 \right\}$$

for some $\delta_0 > 0$ independent of ε. For this we continue the solutions $u_\varepsilon(x, y)$ of (27) into Q_{δ_0} for complex values of $y + it$, so that the u_ε satisfy the Cauchy-Riemann conditions in Q_{δ_0} and are bounded uniformly in ε.

We consider the Cauchy problem for the equation

$$P_\varepsilon U_\varepsilon \equiv (1 + \varepsilon)U_{\varepsilon xx} - (\varepsilon + x^{2k})U_{\varepsilon tt} - id(x, y + it)U_{\varepsilon t}$$
$$+ (c(x, y + it) + \psi_\varepsilon(x, y + it))U_\varepsilon = 0 \quad (28)$$

with initial conditions

$$U_\varepsilon|_{t=0} = u_\varepsilon(x, y), \qquad U_{\varepsilon t}|_{t=0} = iu_{\varepsilon y}(x, y) \qquad (29)$$

for $|x| \leq X_0$ and for any value of the parameter y such that $|y| \leq Y_0$, where X_0 and Y_0 are sufficiently small numbers, and the domain $\{|x| \leq X_0, |y| \leq Y_0\}$ lies in $\tilde{\Omega}_{\mu_0}$. It is easy to see that for suffciently large γ_1 and γ_2 the domain $G = \{|x| \leq X_0, |\gamma_1 t \pm x| \leq X_0, |t| \leq 1/\gamma_2\}$ is a space type domain for the operator P_ε for any ε such that $0 < \varepsilon \leq 1$, and for $|y| \leq Y_0$, since at the points of the boundary of this domain we have for sufficiently large γ_1 and γ_2

$$(1 + \varepsilon)\nu_1^2 - (\varepsilon + x^{2k})\nu_2^2 < 0,$$

where (ν_1, ν_2) is a vector normal to the boundary G.

Equation (28) is hyperbolic in G for $|y| \leq Y_0$ and $0 < \varepsilon \leq 1$. Thus for $0 < \varepsilon \leq 1$ there exists a unique solution U_ε of (28), (29) (see, for example, [6]). We show that in the domain $Q' = \{(x, t) \in G, |y| \leq Y_0\}$ the functions U_ε are analytic in $y + it$. Let

$$W_\varepsilon = \bar{\delta}U_\varepsilon \equiv U_{\varepsilon y} + iU_{\varepsilon t}.$$

We show that $W_\varepsilon \equiv 0$ in Q'. It is easy to see that, in Q', W_ε satisfies the equation

$$P_\varepsilon(W_\varepsilon) \equiv P_\varepsilon(\delta U_\varepsilon) = \delta P_\varepsilon U_\varepsilon = 0 \tag{30}$$

and the initial conditions

$$W_\varepsilon\big|_{t=0} = 0, \qquad W_{\varepsilon t}\big|_{t=0} = 0. \tag{31}$$

By virtue of the uniqueness theorem for the solution of the Cauchy problem for a hyperbolic equation, it follows from (30) and (31) that $W_\varepsilon \equiv 0$ in Q'. This means that $U_\varepsilon(x, y, t)$ is an analytic function of the variable $y + it$.

We now consider the solution V_ε of the mixed problem

$$P_\varepsilon V_\varepsilon = 0 \quad \text{in } G_1 = \left\{|x| \leqslant X_0, |t| \leqslant \delta_1\right\} \tag{32}$$

with initial conditions

$$V_\varepsilon\big|_{t=0} = u_\varepsilon(x, y), \qquad V_{\varepsilon t}\big|_{t=0} = iu_{\varepsilon y}(x, y)$$

and boundary conditions

$$V_\varepsilon\big|_{x=X_0} = 0, \qquad V_\varepsilon\big|_{x=-X_0} = 0,$$

where y is considered a parameter, $|y| \leqslant Y_0$, with δ_1 some number such that $\delta_1 \geqslant 1/\gamma_2$. It is clear that in the domain G for $|y| \leqslant Y_0$ we have $V_\varepsilon = U_\varepsilon$.

We now estimate V_ε uniformly in ε in G_1 for $|y| \leqslant Y_0$.

Let $Q_\tau^+ = \{|x| \leqslant X_0, 0 < t < \tau\}$ and $\sigma_\tau = \{|x| \leqslant X_0, t = \tau\}$. We multiply (32) by $\overline{V}_{\varepsilon t}$ and integrate it over the domain Q_τ^+, $0 < \tau \leqslant \delta_1$. Setting the real part of the left side of the resulting equation equal to zero and transforming its terms by integration by parts, we obtain

$$\frac{1}{2}\int_{\sigma_\tau}\left[(1 + \varepsilon)|V_{\varepsilon x}|^2 + (\varepsilon + x^{2k})|V_{\varepsilon t}|^2 - \operatorname{Re} c_\varepsilon|V_\varepsilon|^2\right] dx$$

$$- \int_{Q_\tau^+} \operatorname{Im} d|V_{\varepsilon t}|^2 \, dx dt + \int_{Q_\tau^+} \operatorname{Im} c \cdot \operatorname{Im}\left(V_\varepsilon \overline{V}_{\varepsilon t}\right) dx dt + \int_{Q_\tau^+} (\operatorname{Re} c)_t|V_\varepsilon|^2 \, dx dt$$

$$= \frac{1}{2}\int_{|x| \leqslant X_0}\left[(1 + \varepsilon)|u_{\varepsilon x}|^2 + (\varepsilon + x^{2k})|u_{\varepsilon y}|^2 - \operatorname{Re} c_\varepsilon|u_\varepsilon|^2\right] dx.$$

$$\tag{33}$$

By the Schwarz inequality,

$$\left|\int_{Q_\tau^+} \operatorname{Im} c \cdot \operatorname{Im}\left(V_\varepsilon \overline{V}_{\varepsilon t}\right) dx dt\right| \leqslant \lambda\int_{Q_\tau^+} |\operatorname{Im} c|^2|V_{\varepsilon t}| \, dx dt + \frac{1}{\lambda}\int_{Q_\tau^+} |V_\varepsilon|^2 \, dx dt$$

for any $\lambda > 0$. If condition 1) of Theorem 3 is fulfilled, then, setting $\lambda = 1/M_1$, we deduce from (33) that for any $0 < \tau \leqslant \delta_1$ and $|y| \leqslant Y_0$

$$\frac{1}{2}\int_{\sigma_\tau}\left[(1 + \varepsilon)|V_{\varepsilon x}|^2 + (\varepsilon + x^{2k})|V_{\varepsilon t}|^2 - \operatorname{Re} c_\varepsilon|V_\varepsilon|^2\right] dx$$

$$\leqslant C_1\left\{\int_{Q_\tau^+} |V_\varepsilon|^2 \, dx dt + \int_{|\tau| \leqslant X_0}\left[(1 + \varepsilon)|u_{\varepsilon x}|^2 + \left(\varepsilon + |x|^{2k}\right)|u_{\varepsilon y}|^2 + |u_\varepsilon|^2\right] dx\right\}. \tag{34}$$

Since $V_\varepsilon = 0$ for $x = -X_0$, we have

$$V_\varepsilon(x, y + it) = \int_{-X_0}^x V_{\varepsilon x}(s, y + it)\, ds, \tag{35}$$

$$\int_{-X_0}^{X_0} |V_\varepsilon|^2\, dx \leq 2 X_0 \int_{-X_0}^{X_0} |V_{\varepsilon x}|^2\, dx, \tag{36}$$

and from (34) and (36) it follows that for sufficiently small X_0

$$\int_{\sigma_\tau} \left[(1 + \varepsilon)|V_{\varepsilon x}|^2 + (\varepsilon + x^{2k})|V_{\varepsilon t}|^2 + |V_\varepsilon|^2 \right] dx$$

$$\leq M_3 \int_{|x| \leq X_0} \left[|u_{\varepsilon x}|^2 + |u_{\varepsilon y}|^2 + |u_\varepsilon|^2 \right] dx, \tag{37}$$

where the constant M_3 does not depend on ε, τ, or y.

Effecting the transformation $t = -t'$, we obtain an estimate of the form (37) for $-\delta_1 < \tau < 0$. In the same manner, we obtain (37) if condition 2) or 3) of Theorem 3 is fulfilled.

Since $V_\varepsilon = U_\varepsilon$ in Q' and U_ε satisfies an elliptic equation $\bar\delta U_\varepsilon = 0$ in Q', known a priori estimates (see [7]) imply that

$$\max_{Q} \sum_{|\alpha| \leq m} |D_y^{\alpha_1} D_t^{\alpha_2} U_\varepsilon| \leq M_4 \sup_{Q'} |U_\varepsilon|, \qquad M_4 = \mathrm{const} > 0, \tag{38}$$

for any subdomain Q of Q' and any m. From this and from (35) it follows that

$$\sum_{|\alpha| \leq m} |D_y^{\alpha_1} D_t^{\alpha_2} U_\varepsilon| \leq M_5 \quad \text{in } Q, \tag{39}$$

where M_5 does not depend on ε. The function U_ε satisfies in Q a Hölder condition in x, since

$$|V_\varepsilon(x + \Delta x, y + it) - V_\varepsilon(x, y + it)| \leq \sqrt{|\Delta x|} \left(\int_{-X_0}^{X_0} |V_{\varepsilon x}|^2\, dx \right)^{1/2} \leq M_6 \sqrt{|\Delta x|}.$$

$$\tag{40}$$

Moreover, the derivatives $U_{\varepsilon t}$ and $U_{\varepsilon y}$ are bounded in Q uniformly in ε by virtue of (39), and therefore the functions U_t form a compact family in Q in the sense of uniform convergence. The limit function $U(x, y + it)$ is continuous and analytic in $y + it$ in Q, since it is the limit of a uniformly converging sequence of functions analytic in $y + it$. Since $U_\varepsilon = u_\varepsilon$ for $t = 0$ and u_ε converges $u(x, y)$ as $\varepsilon \to 0$, the function $U(x, y + it)$ is the analytic continuation of the solution $u(x, y)$ of the equation $L(u) = 0$.

We now show that $u_{\varepsilon x}$ is also an analytic function of $y + it$ in Q. From (28) it follows that $U_{\varepsilon x x}$ is bounded in Q uniformly in ε. Since

$$\int_{\sigma_\tau} |V_{\varepsilon x}(x, y + i\tau)|^2\, dx \leq C_2,$$

where C_2 does not depend on ε, it follows that for any t and y there is a point $\tilde{x}(y, t)$ on the segment $\sigma_t \cap Q$, where $|V_{\varepsilon x}|^2 \leqslant C_2/2 X_0$. Therefore from the representation

$$V_{\varepsilon x}(x, y + it) = V_{\varepsilon x}(\tilde{x}, y + it) + \int_{\tilde{x}}^{x} V_{\varepsilon xx} \, ds$$

it follows that $U_{\varepsilon x}$ is bounded in Q uniformly in ε. From the fact that $U_{\varepsilon x}$ satisfies the Cauchy-Riemann system, we obtain the estimate

$$\sup_{\tilde{Q}} |D_y^{\alpha_1} D_t^{\alpha_2} U_{\varepsilon x}| \leqslant C \sup_{Q} |U_{\varepsilon x}|, \qquad \alpha_1 + \alpha_2 \leqslant m,$$

where $\tilde{Q}$ is any interior subdomain of Q. This and the estimate for $U_{\varepsilon xx}$ imply that $U_{\varepsilon x}$ converges uniformly in $\tilde{Q}$ over some subsequence $\varepsilon \to 0$. The limit function U_x is continuous in $\tilde{Q}$ and analytic as a function of $y + it$. For $t = 0$ we have $U_x = u_x$, since $U_{\varepsilon x}|_{t=0} = u_{\varepsilon x}$ and $u_{\varepsilon x} \to u_x$ as $\varepsilon \to 0$. By virtue of the Kowalewski theorem, there exists a solution U' of the equation $L(u) = 0$, analytic in x and y, such that

$$U'|_{x=0} = u(0, y), \qquad U'_x|_{x=0} = u_x(0, y).$$

By Holmgren's theorem such a solution is unique, and therefore U' must coincide with the solution $u(x, y)$ being considered. The theorem is proved.

REMARK. The results of [8] imply the analyticity of all generalized solutions of the equation

$$L(u) = u_{xx} + 2x^l a^{12} u_x z_y + |x|^{2l} a^{22}(x, y) u_{yy}$$
$$+ b^1(x, y) u_x + x^l b^2(x, y) u_y + c(x, y) u = 0,$$

where $l \geqslant 0$ is an integer, b^1, b^2 and c are analytic complex valued functions, a^{12} and a^{22} are real analytic functions of x and y, and, moreover, $a^{22} - (a^{12})^2 > 0$ in Ω. From [9], Theorem 6, it follows that an equation of the form

$$u_{xx} + x^{2k} u_{yy} + i\lambda x^{k-1} u_y = 0, \qquad \lambda = \text{const}, \tag{41}$$

for some real λ has a solution which is not analytic in y, in a neighborhood of the origin. In fact, it follows from Lemma 1 of [9] that for some $\lambda > 0$ there exists a solution $v(t)$ in $L^2(R^1)$ of the equation

$$v''(t) - t^{2k} v(t) - \lambda t^{k-1} v(t) = 0.$$

The example of (41) shows that condition (18) on the coefficient d in (16) is essential for the validity of the theorem.

Received 2/NOV/73

BIBLIOGRAPHY

1. O. A. Oleĭnik and E. V. Radkevič, *The local smoothness of generalized solutions, and the hypoellipticity of second order differential equations*, Uspehi Mat. Nauk **26** (1971), no. 2(158), 265–281; English transl. in Russian Math. Surveys **26** (1971).

2. B. V. Šabat, *An introduction to complex analysis*, "Nauka", Moscow, 1969. (Russian)

3. O. A. Oleĭnik and E. V. Radkevič, *Equations of second order with nonnegative characteristic form*, Itogi Nauki: Mat. Anal. **1969**, VINITI, Moscow, 1971; English transl., Plenum Press, New York, 1973.

4. Maria Cibrario, *Sulla riduzione a forma canonica delle equazioni lineari alle derivate parziali di secondo ordine di tipo misto*, Ist. Lombardo Sci. Lett. Rend. (2) **65** (1932), 889–906.

5. Carlo Miranda, *Equazioni alle derivate parziali di tipo ellittico*, Springer-Verlag, 1955; English transl., 1970.

6. S. L. Sobolev, *Some applications of functional analysis in mathematical physics*, Izdat. Leningrad. Gos. Univ., Leningrad, 1950; English transl., Amer. Math. Soc., Providence, R. I., 1963.

7. Avron Douglis and Louis Nirenberg, *Interior estimates for elliptic systems of partial differential equations*, Comm. Pure Appl. Math. **8** (1955), 503–538

8. V. V. Grušin, *On a class of elliptic pseudodifferential operators degenerate on a submanifold*, Mat. Sb. **84(126)** (1971), 163–195; English transl. in Math. USSR Sb. **13** (1971).

9. O. A. Oleĭnik and E. V. Radkevič, *On the analyticity of solutions of linear partial differential equations*, Mat. Sb. **90(132)** (1973), no. 4, 592–606; English transl. in Math. USSR Sb. **19** (1973).

Translated by P. C. FIFE

Amer. Math. Soc. Transl.
(2) Vol. **118**, 1982

Necessary Conditions for the Cauchy Problem
for Nonsymmetrizable Hyperbolic Systems
to be Well-Posed*

V. M. PETKOV

Introduction

The Cauchy problem for first order hyperbolic systems with variable coefficients has been studied in many papers. The majority of them are to a greater or lesser extent connected with symmetrizable systems. In this paper we consider nonsymmetrizable systems; that is, systems for which the principal symbol is not similar to a diagonal matrix at every point. For such systems under certain assumptions we prove necessary conditions for the Cauchy problem to be well-posed; in a number of cases these conditions are also sufficient.

We consider the system

$$Pu \equiv D_{x_0}u + \sum_{\nu=1}^{l} A_\nu(x)D_{x_\nu}u + B(x)u = f, \tag{1}$$

where $x = (x_0,\ldots,x_l)$, $D_{x_j} = -i\partial/\partial x_j$, and $A_\nu(x)$ and $B(x)$ are $d \times d$ matrices with infinitely smooth elements. We denote by $\lambda_j(x, \xi')$ the eigenvalues of the matrix

$$A(x, \xi') = \sum_{\nu=1}^{l} A_\nu(x)\xi_\nu, \qquad \xi' = (\xi_1,\ldots,\xi_l) \in \mathbf{R}^l.$$

Let G be a bounded open domain in $\mathbf{R}^{l+1}$ and $W = \{(x, \xi'): x \in G, |\xi'| = 1\}$. It is well known [6], [8], [1] that if at some point $(\hat{x}, \hat{\xi}') \in W$ the matrix $A(\hat{x}, \hat{\xi}')$ has a nonreal eigenvalue $\lambda(\hat{x}, \hat{\xi}')$, then the Cauchy problem for (1) is not

---

*Translation of Trudy Sem.Petrovsk. **1** (1975), 211–236. MR **55** #10861.
1980 *Mathematics Subject Classification*. Primary 35L45.

well-posed. Therefore, in what follows we assume that all eigenvalues $\lambda_j(x, \xi')$ are real for $x \in G$ and $\xi' \in \mathbf{R}^l$. The systems for which this property is satisfied are called *hyperbolic in G*.

In his classical paper [14] Petrovskiĭ considered hyperbolic systems whose coefficients are periodic in $x' = (x_1, \ldots, x_l)$. Under the assumption that all the eigenvalues $\lambda_j(x, \xi')$ are of multiplicity 1 he proved the fundamental energy L_2-estimate

$$\|u(x_0, x')\|_{L_2(\mathbf{R}')} \leqslant C(K)\left\{\|u(T, x')\|_{L_2(\mathbf{R}')} + \int_T^{x_0}\|Pu(\tau, x')\|_{L_2(\mathbf{R}')}\,d\tau\right\},$$

$$T < x_0, \qquad u \in C_0^\infty(K), \tag{2}$$

where $K \subset G$ is a compactum.

Based on the theory of pseudodifferential operators and using Petrovskiĭ's ideas, Calderón [2], Mizohata [7] and Yamaguti [21] have proved the estimate (2) for all systems for which there is a smooth matrix $s(x, \xi')$, homogeneous in ξ' of degree zero and satisfying the following conditions:

1) $s(x, \xi') = s^*(x, \xi') > 0, (x, \xi') \in W$,

2) $s(x, \xi')A(x, \xi') = A^*(x, \xi')s(x, \xi')$,

where a^* denotes the conjugate matrix of a.

The systems for which there is a matrix $s(x, \xi')$ satisfying conditions 1) and 2) are called *symmetrizable*. For these systems the Cauchy problem is well-posed for any smooth matrices $B(x)$ (see [2] and [5]).

Kreiss [5] and Friedrichs [19] have proved that the systems for which $A(x, \xi')$ is similar to a diagonal matrix for $(x, \xi') \in W$ and for which the multiplicity of the eigenvalues $\lambda_j(x, \xi')$ does not depend on x and ξ' are symmetrizable.

It has been proved in [1] and [3] that if (2) holds, then there is a matrix $Q(x, \xi')$ such that the matrix

$$Q^{-1}(x, \xi')A(x, \xi')Q(x, \xi') \tag{3}$$

is a diagonal for $(x, \xi') \in W$. Furthermore, in [1] it has also been proved that for any compactum $K \subset G$ the matrices $Q(x, \xi')$ and $Q^{-1}(x, \xi')$ are uniformly bounded for $x \in K$ and $|\xi'| = 1$. In the case where the coefficients of the system are constant, these conditions are necessary and sufficient for estimate (2) to hold (see [4], [16] and [17]). As shown by Strang's example [15], in the general case (2) does not follow from these conditions. Thus, the essential role in the question of whether the Cauchy problem is well-posed is played not by the multiplicities of the eigenvalues $\lambda_j(x, \xi')$, but by the elementary divisors of $A(x, \xi')$.

The systems for which the matrix $A(x, \xi')$ is not similar to a diagonal matrix at some point $(x, \xi') \in W$ are called *nonsymmetrizable*. No necessary or sufficient conditions have so far been known for the Cauchy problem for nonsymmetrizable systems with variable coefficients to be well-posed.

We introduce the following notation. If $\mathcal{D} \subset \mathbf{R}^{l+1}$, then $\overline{\mathcal{D}}$ is the closure of $\mathcal{D}$, $\overset{\circ}{\mathcal{D}}$ the interior of $\mathcal{D}$, $\mathcal{D}_t^+ = \{x \in \mathcal{D}: x_0 \geqslant t\}$, $\mathcal{D}_t^- = \{x \in \mathcal{D}: x_0 \leqslant t\}$, and $\mathcal{D}_t = \{x \in \mathcal{D}: x_0 = t\}$. We assume that all the elements of the matrices $A_\nu(x)$ and $B(x)$

belong to $C^\infty(G)$. By $K \subset\subset G$ we denote the fact that K is precompact in G and $\overline{K} \subset G$. Also, $\{\ ,\ \}$ denotes the inner product in $\mathbf{R}^l$ and $\langle\ ,\ \rangle$ the inner product in $\mathbf{C}^d$.

DEFINITION 1. The Cauchy problem

$$Pu = f \quad \text{in } G, \tag{4}$$

$$u \in \mathcal{D}'(G), \qquad \operatorname{supp} u \subset \overline{G}_T^+ \tag{5}$$

is said to be *well-posed in* $G_T^+ \neq G$ if the following conditions are satisfied:

(E). For any $f \in C_0^\infty(G)$, $\operatorname{supp} f \subset \overline{G}_T^+$, there is a generalized function $u \in \mathcal{D}'(G)$ satisfying (4) and (5).

(U). If $u \in \mathcal{D}'(G)$ satisfies (4) and (5) and $Pu = 0$ in G_t^-, then $u = 0$ in G_t^- for any $t > T$.

If the Cauchy problem is well-posed in G_T^+, then P satisfies some a priori estimate (see Lemma 2.1 in [1]). Therefore, the proof of necessary conditions for the Cauchy problem to be well-posed can be reduced to the construction of an asymptotic solution of the operator P.

Before we formulate our results we consider an example. Let

$$P = D_{x_0} + A D_{x_1} + B(x), \tag{6}$$

where

$$A = \begin{pmatrix} 0 & 1 & 0 \\ 0 & 0 & 1 \\ 0 & 0 & 0 \end{pmatrix}, \qquad B(x) = \{b_{\mu,\nu}(x)\}_{\mu,\nu=1}^{3}, \quad x = (x_0, x_1).$$

We assume that $b_{33}(x) \neq 0$ in some neighborhood V of the point $\hat{x} \in G$. For the operator (6) we construct in V an asymptotic solution of the form

$$u_\rho(x) = \sum_{n=0}^{N} v_n(x)\rho^{-n/3} \exp i\big(x\rho + l^1(x)\rho^{2/3} + l^2(x)\rho^{1/3}\big), \tag{7}$$

where l^1 and l^2 are scalar-valued functions and the v_n are vector-valued functions.

Let

$$R = (1,0,0), \qquad h_1 = (0,1,0), \qquad h_2 = L = (0,0,1),$$

$$M(\sigma) = \sigma_{x_0} I + \sigma_{x_1} A,$$

where I is the unit matrix and σ a scalar-valued function. In (7) we put

$$v_0 = \sigma^0(x)R,$$

where by $\sigma^0(x), \sigma^1(x), \dots$ we denote smooth scalar-valued functions. For determining v_1 we obtain the system

$$-Av_1 = M(l^1)\sigma^0 R = l_{x_0}^1 \sigma^0 R.$$

This system is solvable for any σ^0 and l^1, and

$$v_1 = -l_{x_0}^1 \sigma^0 h_1 + \sigma^1 R.$$

The vector-valued function v_2 is determined from the system

$$-Av_2 = M(l^1)v_1 + M(l^2)v_0 = -\left(l^1_{x_0}\right)^2 \sigma^0 h_1 + \left(l^2_{x_0}\sigma^0 + l^1_{x_0}\sigma^1 - l^1_{x_0}l^1_{x_1}\sigma^0\right)R.$$

This system is also solvable for any σ^0, σ^1, l^1 and l^2, and we obtain

$$v_2 = \left(l^1_{x_0}\right)^2 \sigma^0 h_2 - \left(l^2_{x_0}\sigma^0 + l^1_{x_0}\sigma^1 - l^1_{x_0}l^1_{x_1}\sigma^0\right)h_1 + \sigma^2 R.$$

Finally, v_3 is determined from the system

$$-Av_3 = M(l^1)v_2 + M(l^2)v_1 - iM(\sigma^0)R + \sigma^0 BR. \tag{8}$$

The solvability condition for (8) has the form

$$\left(l^1_{x_0}\right)^3 + b_{33} = 0.$$

Since $b_{33}(x) \neq 0$ in V, we can find a simple root $\eta_0 = F(x)$ of the equation $\eta_0^3 + b_{33} = 0$ for which $\operatorname{Im} F(x) < 0$. Then we define l^1 as the solution of the Cauchy problem

$$\begin{cases} l^1_{x_0} = F(x), \\ l^1\big|_{x_0=\hat{x}_0} = ix_1^2, \end{cases}$$

and v_3 as the solution of (8).

The construction of the solution u_ρ continues in the standard way, and it is shown that the Cauchy problem (4), (5) for the operator (6) is not well-posed in G_T^+, $T < \hat{x}_0$.

We show that, besides the condition $b_{33} = 0$, for the Cauchy problem for the operator (6) to be well-posed it is also necessary that

$$b_{21} + b_{32} = 0, \qquad b_{11}b_{32} + b_{21}b_{33} = 0.$$

These conditions are also sufficient for the Cauchy problem to be well-posed [11], [12].

In this paper we assume that all the eigenvalues $\lambda_j(x, \xi')$ are of constant multiplicity r_j for $(x, \xi') \in G \times (\mathbf{R}^l \setminus \{0\})$, and that the rank of every matrix

$$\mathfrak{A}_j(x, \xi') = -\lambda_j(x, \xi')I + A(x, \xi')$$

is constant in $G \times (\mathbf{R}^l \setminus \{0\})$ and equal to $d - r_j + \kappa_j$.

We now define certain vector-valued functions involved in the formulation of the main theorem which establishes necessary conditions for the Cauchy problem to be well-posed.

Let $(\hat{x}, \hat{\xi}') \in G \times (\mathbf{R}^l \setminus \{0\})$ and suppose that in some neighborhood U of $\hat{x}$ the function $\varphi_j(x)$ satisfies

$$\frac{\partial \varphi_j}{\partial x_0} = -\lambda_j\left(x, \Phi_j(x)\right),$$

$$\Phi_j(x) = \operatorname{grad}_{x'} \varphi_j(x), \qquad \Phi_j(\hat{x}) = \hat{\xi}'. \tag{9}$$

We denote by $\operatorname{Ker} a$ the kernel of the matrix a. If $\operatorname{rank} \mathfrak{A}_j(\hat{x}, \hat{\xi}') = d - 1$ and $\kappa_j > 0$, then in the neighborhood of $\hat{x}$ there are smooth vector-valued functions

$R_j(x, \Phi_j) \neq 0$ and $L_j(x, \Phi_j) \neq 0$ belonging to Ker $\mathfrak{A}_j(x, \Phi_j)$ and Ker $\mathfrak{A}_j^*(x, \Phi_j)$, respectively. We can show that

$$\langle L_j(x, \Phi_j), R_j(x, \Phi_j)\rangle = 0.$$

Consequently, in the neighborhood of $\hat{x}$ there is a smooth vector-valued function $h_j(x, \Phi_j)$ for which

$$\mathfrak{A}_j(x, \Phi_j)h_j(x, \Phi_j) = R_j(x, \Phi_j). \tag{10}$$

Furthermore, if $\langle L_j(x, \Phi_j), P(R_j(x, \Phi_j))\rangle = 0$, then we denote by $H_j(x, \Phi_j)$ the solution of the system

$$\mathfrak{A}_j(x, \Phi_j)H_j(x, \Phi_j) = P(R_j(x, \Phi_j)). \tag{11}$$

The smooth functions $h_j(x, \Phi_j)$ and $H_j(x, \Phi_j)$ are defined in the neighborhood of $\hat{x}$ to within a term of the form $\sigma(x)R_j(x, \Phi_j)$, where $\sigma(x)$ is an arbitrary smooth scalar-valued function.

Let rank $\mathfrak{A}_j(\hat{x}, \hat{\xi}') = d - 2$ and $\kappa_j > 0$. Then in the neighborhood of $\hat{x}$ there are smooth vector-valued functions $R_{k,j}(x, \Phi_j)$ and $L_{k,j}(x, \Phi_j)$, $k = 1, 2$, forming bases in Ker $\mathfrak{A}_j(x, \Phi_j)$ and Ker $\mathfrak{A}_j^*(x, \Phi_j)$, respectively, and satisfying the equality

$$\begin{pmatrix} \langle L_{1,j}, R_{1,j}\rangle & \langle L_{1,j}, R_{2,j}\rangle \\ \langle L_{2,j}, R_{1,j}\rangle & \langle L_{2,j}, R_{2,j}\rangle \end{pmatrix} (x, \Phi_j) = \begin{pmatrix} 0 & 0 \\ 0 & 1 \end{pmatrix}. \tag{12}$$

The following statement is the main result of this paper.

THEOREM 1. *If the Cauchy problem* (4), (5) *is well-posed in* G_T^+, *then for any point* $(\hat{x}, \hat{\xi}') \in G_T^+ \times (\mathbf{R}^l \setminus \{0\})$ *and any function* $\varphi_j(x)$ *which satisfies* (9) *the following conditions are satisfied in the neighborhood of* $\hat{x}$:
(i) *if* $r_j = 2$ *and* $\kappa_j = 1$, *then*

$$\langle L_j(x, \Phi_j), P(R_j(x, \Phi_j))\rangle = 0; \tag{13}$$

(ii) *if* $r_j = 3$ *and* $\kappa_j = 2$, *then* (13) *holds and*

$$\langle L_j(x, \Phi_j), P(h_j(x, \Phi_j)) + H_j(x, \Phi_j)\rangle = 0, \tag{14}$$

$$\langle L_j(x, \Phi_j), P(H_j(x, \Phi_j))\rangle = 0; \tag{15}$$

(iii) *if* $r_j = 3$ *and* $\kappa_j = 1$, *then*

$$\langle L_{1,j}(x, \Phi_j), P(R_{1,j}(x, \Phi_j))\rangle = 0, \tag{16}$$

$$\langle L_{1,j}(x, \Phi_j), P(R_{2,j}(x, \Phi_j))\rangle\langle L_{2,j}(x, \Phi_j), P(R_{1,j}(x, \Phi_j))\rangle$$
$$= \mathcal{L}_{1,2,j}\mathcal{L}_{2,1,j} = 0. \tag{17}$$

We show that the conditions in Theorem 1 do not depend on the choice of the functions $R_j, \ldots, L_{k,j}$.

Conditions (13)–(17) in Theorem 1 can also be rewritten in a different form which is the analogue of Levi's condition for equations with eigenvalues of constant multiplicity (see [1], [18] and [20]).

DEFINITION 2. *Let $(\hat{x}, \hat{\xi}') \in G \times (\mathbf{R}^{l} \setminus \{0\})$. We say that the operator P satisfies condition $(H_{\hat{x},\hat{\xi}'})$ if for any function $\varphi_{j}(x)$ satisfying condition (9) in a neighborhood U of $\hat{x}$ and any $f \in C_{0}^{\infty}(U)$, in some neighborhood U_{1} of $\hat{x}$ there are a smooth scalar-valued function $g_{j}(x, \varphi_{j}, f)$ and smooth vector-valued functions $v_{k,j}(x, \varphi_{j}, f)$, $k = 1, \ldots, \kappa_{j}$, such that*

$$P\left[\left(fR_{1,j} + g_{j}R_{2,j} + \sum_{k=1}^{\kappa_{j}} v_{k,j}\rho^{-k}\right)e^{i\rho\varphi_{j}}\right] = O(\rho^{-\kappa_{j}}), \qquad \rho \to +\infty, \quad (18)$$

with $R_{2,j} \equiv 0$ in the cases when rank $\mathfrak{A}_{j} = d - 1$, $\kappa_{j} > 0$. The operator P satisfies condition (H) if it satisfies condition $(H_{x,\xi'})$ for all $(x, \xi') \in G \times (\mathbf{R}^{l} \setminus \{0\})$.

From Theorem 1 we obtain the following corollaries.

COROLLARY 1. *Condition (H) is necessary for the Cauchy problem to be well-posed in cases* (i) *and* (ii). *Condition $(H_{\hat{x},\hat{\xi}'})$ is necessary for the Cauchy problem to be well-posed in case* (iii) *if $|\mathcal{L}_{1,2,j}|^{2} + |\mathcal{L}_{2,1,j}|^{2} \neq 0$ for any $\varphi_{j}(x)$ satisfying* (9). *If the coefficients of the operator P are analytic in G and the functions $\varphi_{j}(x)$ are analytic, then condition (H) is necessary for the Cauchy problem to be well-posed in case* (iii).

COROLLARY 2. *Suppose that all the eigenvalues $\lambda_{j}(x, \xi')$ are of constant multiplicity not exceeding three. If the Cauchy problem for the operator $P + B_{1}$ is well-posed in G_{T}^{+} for any smooth matrix $B_{1}(x)$, then the matrix $A(x, \xi')$ is similar to a diagonal matrix at every point $(x, \xi') \in G_{T}^{+} \times (\mathbf{R}^{l} \setminus \{0\})$.*

In Corollary 2 we do not assume that the rank of the matrices $\mathfrak{A}_{j}(x, \xi')$ is constant.

Sufficient conditions for the Cauchy problem for nonsymmetrizable systems to be well-posed will be considered in another paper. Condition (H) permits us to construct the parametrix of the Cauchy problem and to prove that the Cauchy problem is well-posed (see [11]–[13]).

Theorem 1 is proved by the method of constructing an asymptotic solution for the operator P, of the form

$$u_{\rho}(x) = \sum_{n=0}^{N} v_{n}(x)\rho^{-n/\tau} \exp i\left(\varphi(x)\rho + \sum_{j=0}^{\nu-1} l^{j+1}(x)\rho^{(\nu-j)/\tau}\right), \quad (19)$$

where ν, $\tau = 2, 3$, the v_{n} are vector-valued functions and φ and the l^{j} are scalar-valued functions. As in the above example, the first few systems for determining the functions v_{n} are always solvable. However, one must take into account the dependence of the solutions of these systems on unknown scalar-valued functions. After a few such "neutral" steps we obtain a system which is not always solvable. From the solvability conditions for this system we determine the function l^{1} with the required properties; then we use mathematical induction. The number of "neutral" steps depends on r_{j}, κ_{j} and the form of the asymptotic solution $u_{\rho}(x)$.

Using this method of constructing an asymptotic solution and the method of the asymptotic change of variables [1], we can obtain necessary conditions for the

Cauchy problem for nonsymmetrizable systems with eigenvalues of variable multiplicity to be well-posed.

The presentation in this paper is as follows. In §1 it is shown that there are smooth functions $R_j, \ldots, L_{k,j}$ defined in a neighborhood of every point $x \in G$. It is then proved that the conditions in Theorem 1 are invariant with respect to a change of variables and the choice of $R_j, \ldots, L_{k,j}$. After a suitable change of variables we rectify the bicharacteristics and construct an asymptotic solution of the form (19) in the new variables. In §§2 and 3 we prove that the conditions (13), (16) and (14), (15), (17), respectively, are necessary for the Cauchy problem to be well-posed.

§1. Prerequisites for the proof of Theorem 1

First we prove an algebraic lemma from which it follows that smooth vector-valued functions $R_j, \ldots, L_{k,j}$ are defined in a neighborhood of every point $x \in G$.

LEMMA 1.1. *Suppose that in the domain* $\omega \subset \mathbf{R}^{l+1}$ *all the elements of the* $d \times d$ *matrix* $a(y)$ *belong to* $C^\infty(\omega)$. *Suppose also that zero is an eigenvalue of* $a(y)$ *in* ω *of constant multiplicity* $r \geq 2$, *and that the rank of* $a(y)$ *is constant. Then the following assertions are true:*

1) *If* rank $a(y) = d - 1$, *then there are smooth vector-valued functions* $h_j(y) \neq 0$, $j = 0, \ldots, r - 1$, *defined in a neighborhood of every point* $\hat{y} \in \omega$ *and such that*

$$a(y)h_0(y) = 0,$$

$$a(y)h_j(j) = h_{j-1}(y), \qquad j = 1, \ldots, r - 1. \tag{1.1}$$

2) *If* $r = 3$ *and* rank $a(y) = d - 2$, *then there are smooth linearly independent vector-valued functions* $h_0^{(1)}(y)$, $h_0^{(2)}(y)$ *and* $h_1^{(1)}(y)$ *defined in a neighborhood of every point* $\hat{y} \in \omega$ *and such that*

$$a(y)h_0^{(k)}(y) = 0, \qquad k = 1, 2,$$

$$a(y)h_1^{(1)}(y) = h_0^{(1)}(y). \tag{1.2}$$

PROOF. Let $\hat{y} \in \omega$ and let V be a neighborhood of $\hat{y}$, $V \subset\subset \omega$. In the complex z-plane there is an annulus $K = \{z : 0 < \delta_1 \leq |z| \leq \delta_2\}$ which contains all nonzero eigenvalues of the matrix $a(y)$, for $y \in \overline{V}$.

We consider the projections

$$\pi_1(y) = \frac{1}{2\pi i} \int_\gamma (zI - a(y))^{-1} \, dz, \qquad \pi_2(y) = I - \pi_1(y),$$

where γ is the boundary of the domain $\{z : \delta_1 - \varepsilon < |z| < \delta_2 + \varepsilon\}$, $0 < \varepsilon < \delta_1$. We choose rank minors in the matrices $\pi_1(\hat{y})$ and $\pi_2(\hat{y})$. If $V_1 \subset\subset V$ is a sufficiently small neighborhood of $\hat{y}$, then these minors are also rank minors for $y \in \overline{V}_1$. Let $F(y)$ be the matrix consisting of the columns which contain these rank minors. In $\overline{V}_1$ we then obtain

$$F^{-1}(y)a(y)F(y) = C(y) = \begin{pmatrix} C_1(y) & 0 \\ 0 & C_2(y) \end{pmatrix}, \tag{1.3}$$

where $C_1(y)$ is a smooth $r \times r$ matrix whose eigenvalues are all zero and $C_2(y)$ a smooth $(d - r) \times (d - r)$ matrix with nonzero eigenvalues. Besides, if rank $a(y) = d - \nu$, then

$$\text{rank } C_1(y) = r - \nu, \qquad \text{rank } C_2(y) = d - r.$$

Let us consider case 1). Then it is easy to show that

$$C_1^r(y) \equiv 0, \qquad \text{rank } C_1^{r-1}(y) \equiv 1.$$

In some smaller neighborhood of $\hat{y}$ we define a smooth vector-valued function $g_{r-1}(y)$ such that

$$C_1^{r-1}(y)g_{r-1}(y) \neq 0.$$

We put

$$g_j(y) = C_1^{r-j-1}(y)g_{r-1}(y), \qquad j = 0, 1, \ldots, r - 1,$$

and transform all these r-vectors into d-vectors by supplementing them with components $g_{j,\mu}(y) \equiv 0$ for $j = 0, 1, \ldots, r - 1$ and $\mu = r + 1, \ldots, l$. Then it is obvious that the functions

$$h_j(y) = F(y)g_j(y), \qquad j = 0, 1, \ldots, r - 1,$$

satisfy (1.1).

We now consider case 2). Then rank $C_1(y) \equiv 1$ and $C_1^2(y) \equiv 0$. In some neighborhood of $\hat{y}$ we define a smooth vector-valued function $g_1^{(1)}(y)$ such that

$$g_0^{(1)}(y) = C_1(y)g_1^{(1)}(y) \neq 0.$$

Let $C_1(y) = \{c_{\mu,\nu}(y)\}_{\mu,\nu=1}^3$. Without loss of generality we assume that $c_{11}(y) \neq 0$ in some smaller neighborhood of $\hat{y}$. Let

$$g_0^{(2)}(y) = (\alpha_1(y), \alpha_2, \alpha_3),$$

where α_2 and α_3 are constants. We can obviously choose α_2 and α_3 so that $g_0^{(1)}(y)$ and $g_0^{(2)}(y)$ are linearly independent. We then determine $\alpha_1(y)$ from the equation

$$c_{11}(y)\alpha_1(y) + \alpha_2 c_{12}(y) + \alpha_3 c_{13}(y) = 0.$$

Then the equalities (1.2) are satisfied if

$$h_0^{(k)}(y) = F(y)g_0^{(k)}(y), \quad k = 1, 2, \qquad h_1^{(1)}(y) = F(y)g_1^{(1)}(y).$$

Lemma 1.1 is proved.

Suppose that the function $\varphi_j(x)$ satisfies (9). We consider the cases $r_j = 2$, $\kappa_j = 1$; $r_j = 3$, $\kappa_j = 2$; $r_j = 3$, $\kappa_j = 1$ separately and define the vector-valued functions $R_j, \ldots, L_{k,j}$ in a neighborhood of the point $\hat{x} \in G$.

Let $r_j = 2$ and $\kappa_j = 1$. We apply Lemma 1.1 to the matrix $\mathfrak{A}_j(x, \xi')$ and obtain that in a neighborhood of $(\hat{x}, \hat{\xi}')$ smooth functions $R_j(x, \xi')$ and $h_{1,j}(x, \xi')$ are defined for which

$$\mathfrak{A}_j(x, \xi')R_j(x, \xi') = 0, \tag{1.4}$$

$$\mathfrak{A}_j(x, \xi')h_{1,j}(x, \xi') = R_j(x, \xi'). \tag{1.5}$$

Similarly, we obtain that there is a smooth function $L_j(x, \xi')$ such that

$$\mathfrak{A}_j^*(x, \xi')L_j(x, \xi') = 0. \tag{1.6}$$

It is easy to show that in a neighborhood of $(\hat{x}, \hat{\xi}')$

$$\langle L_j(x, \xi'), h_{1,j}(x, \xi')\rangle \neq 0.$$

For if $\langle L_j(\hat{z}), h_{1,j}(\hat{z})\rangle = 0$ at some point $\hat{z}$, then we consider the solution w of the system

$$\mathfrak{A}_j(\hat{z})w = h_{1,j}(\hat{z}).$$

Then

$$g_0(\tilde{z}) = F^{-1}(\hat{z})R_j(\hat{z}) = F^{-1}(\hat{z})\mathfrak{A}_j^2(\hat{z})w = C^2(\hat{z})F^{-1}(\hat{z})w,$$

where $g_0(\hat{z})$, $F(\hat{z})$ and $C(\hat{z})$ are the same as in the proof of Lemma 1.1. Since in our case $C_1^2(z) \equiv 0$, we obtain from the last equality that $g_0(\hat{z}) = 0$, which is impossible. Thus, we can without loss of generality assume that

$$\langle L_j(x, \xi'), h_{1,j}(x, \xi')\rangle \equiv 1. \tag{1.7}$$

Let $r_j = 3$ and $\kappa_j = 2$. From Lemma 1.1 we then obtain that in a neighborhood of $(\hat{x}, \hat{\xi}')$ the functions $R_j(x, \xi')$ and $h_{1,j}(x, \xi')$ are defined, satisfying (1.4) and (1.5), respectively, and a smooth function $h_{2,j}(x, \xi')$ for which

$$\mathfrak{A}_j(x, \xi')h_{2,j}(x, \xi') = h_{1,j}(x, \xi'). \tag{1.8}$$

In place of (1.7) we have

$$\langle L_j(x, \xi'), h_{1,j}(x, \xi')\rangle = 0. \tag{1.9}$$

Moreover, we can without loss of generality assume that

$$\langle L_j(x, \xi'), h_{2,j}(x, \xi')\rangle \equiv 1. \tag{1.10}$$

Finally, we consider the case when $r_j = 3$ and $\kappa_j = 1$. Applying Lemma 1.1 to the matrix $\mathfrak{A}_j(x\,\xi')$, we obtain that in a neighborhood of (x, ξ') smooth linearly independent functions $R_{k,j}(x, \xi')$, $k = 1\,2$, and $h_{1,j}(x, \xi')$ are defined for which

$$\mathfrak{A}_j(x, \xi')R_{k,j}(x, \xi') = 0, \qquad k = 1, 2, \tag{1.11}$$

$$\mathfrak{A}_j(x, \xi')h_{1,j}(x, \xi') = R_{1,j}(x, \xi'). \tag{1.12}$$

Similarly, we find smooth linearly independent functions $L_{k,j}(x, \xi')$, $k = 1, 2$, and $q_{1,j}(x, \xi')$ such that

$$\mathfrak{A}_j^*(x, \xi')L_{k,j}(x, \xi') = 0, \qquad k = 1, 2, \tag{1.13}$$

$$\mathfrak{A}_j^*(x, \xi')q_{1,j}(x, \xi') = L_{1,j}(x, \xi'). \tag{1.14}$$

It is easy to show that $\langle L_{2,j}(x, \xi'), R_{2,j}(x, \xi')\rangle \neq 0$. Thus, without loss of generality we can assume that in a neighborhood of $(\hat{x}, \hat{\xi}')$ smooth functions $R_{k,j}(x, \xi')$ and $L_{k,j}(x, \xi')$, $k = 1, 2$, are defined, that form bases in $\operatorname{Ker} \mathfrak{A}_j(x, \xi')$ and $\operatorname{Ker} \mathfrak{A}_j^*(x, \xi')$, respectively, and satisfy

$$\begin{pmatrix} \langle L_{1,j}, R_{1,j}\rangle & \langle L_{1,j}, R_{2,j}\rangle \\ \langle L_{2,j}, R_{1,j}\rangle & \langle L_{2,j}, R_{2,j}\rangle \end{pmatrix}(x, \xi') = \begin{pmatrix} 0 & 0 \\ 0 & 1 \end{pmatrix}. \tag{1.15}$$

After all these considerations it is obvious that in a neighborhood of $\hat{x}$ smooth functions $R_j(x, \Phi_j)$, $L_j(x, \Phi_j)$, $h_{k,j}(x, \Phi_j)$, $R_{k,j}(x, \Phi_j)$, and $L_{k,j}(x, \Phi_j)$, $k = 1, 2$ are defined. Moreover, R_j and L_j are defined to within a factor $\sigma(x)$, and $h_{1,j}$ and $h_{2,j}$ to within a term $\sigma(x)R_j(x, \Phi_j)$, where $\sigma(x)$ is a smooth scalar-valued function. The functions $R_{k,j}$ and $L_{k,j}$ are not uniquely defined either. If $R'_{k,j}(x, \Phi_j)$ and $L'_{k,j}(x, \Phi_j)$, $k = 1, 2$, are another collection of vector-valued functions forming smooth bases in Ker $\mathfrak{A}_j(x, \Phi_j)$ and Ker $\mathfrak{A}_j^*(x, \Phi_j)$, respectively, and satisfying (12), then

$$R_{1,j} = \mu_1 R_{1,j}, \qquad R'_{2,j} = \mu_{21} R_{1,j} + \mu_{22} R_{2,j},$$
$$L'_{1,j} = \nu_1 L_{1,j}, \qquad L'_{2,j} = \nu_{21} L_{1,j} + \nu_{22} R_{2,j}, \tag{1.16}$$

where μ_1, μ_{21}, μ_{22}, ν_1, ν_{21} and ν_{22} are smooth scalar-valued functions such that $\mu_1 \nu_1 \neq 0$ and $\mu_{22} \nu_{22} \equiv 1$. Thus, we have shown that condition (H) does not depend on the choice of the functions R_j and $R_{k,j}$.

We introduce the following notation:

$$V_{x_\mu}(x, \xi') = \frac{\partial}{\partial x_\mu} V(x, \xi'), \qquad V^{(\mu)}(x, \xi') = \frac{\partial}{\partial \xi_\mu} V(x, \xi'),$$

$$\frac{\partial}{\partial \Phi_\mu} V(x, \Phi) = \frac{\partial}{\partial(\partial\varphi/\partial x_\mu)} V(x, \Phi), \qquad \Phi = \mathrm{grad}_{x'}\varphi,$$

$$M(\sigma) = \sigma_{x_0} I + A_\nu(x)\sigma_{x_\nu};$$

here and until the end of the proof of Theorem 1 we adopt the convention of summation over repeated Greek indices from 1 to l. By σ, σ^0, σ^1, σ_1^0, $\sigma_1^1, \ldots$ we denote smooth scalar-valued functions. The derivative in the direction of the bicharacteristic corresponding to the eigenvalue $\lambda_j(x, \xi')$ and the function $\varphi_j(x)$ is denoted by

$$\frac{d\sigma}{ds_j} = \sigma_{s_j} = \sigma_{x_0} + \lambda_j^{(\mu)}(x, \Phi_j)\sigma_{x_\mu}, \qquad \Phi_j = \mathrm{grad}_{x'}\varphi_j.$$

The following identity holds:

$$M(\sigma)V = \sigma_{s_j} V - \sigma_{x_\mu} \mathfrak{A}_j \frac{\partial V}{\partial \Phi_\mu} + \sigma_{x_\mu} \frac{\partial}{\partial \Phi_\mu}(\mathfrak{A}_j V). \tag{1.17}$$

In particular,

$$M(\sigma)R_j = \sigma_{s_j} R_j - \sigma_{x_\mu} \mathfrak{A}_j R_j^{(\mu)}, \tag{1.18}$$

$$M(\sigma)h_{1,j} = \sigma_{s_j} h_{1,j} - \sigma_{x_\mu} \mathfrak{A}_j h_{1,\mu}^{(\mu)} + \sigma_{x_\mu} R_j^{(\mu)}. \tag{1.19}$$

To show that the conditions in Theorem 1 do not depend on the choice of $R_j, \ldots, L_{k,j}$ we prove the following lemma.

LEMMA 1.2. *Condition (H) is equivalent to condition (13) in case* (i) *and to conditions (13)–(15) in case* (ii). *If the function $\varphi_j(x)$ satisfying (9) also satisfies* $|\mathfrak{L}_{1,2,j}|^2 + |\mathfrak{L}_{2,1,j}|^2 \neq 0$, *then condition $(H_{\hat{x},\hat{\xi}'})$ is equivalent to conditions (16) and (17) in case* (iii).

PROOF. In cases (i) and (ii) it follows from (1.18) that for any $f \in C^\infty$

$$\langle L_j, P(fR_j)\rangle = f\langle L_j, P(R_j)\rangle. \tag{1.20}$$

From this we obtain immediately that (H) and (13) are equivalent in case (i).

Let us consider case (ii). Then (18) takes the form

$$P\big[(fR + v_1\rho^{-1} + v_2\rho^{-2})e^{i\rho\varphi}\big] = O(\rho^{-2}), \qquad \rho \to +\infty,$$

where for brevity we omit the index j. Condition (13) is necessary and sufficient for the existence of v_1. Here

$$v_1 = i\big(f_s h_1 - f_{x_\mu}R^{(\mu)}\big) - fH + \sigma R,$$

where

$$\mathfrak{A} H = P(R).$$

Applying (1.17) with $V = H$ and using (1.18) and (1.19), we rewrite $P(v_1)$ in the form

$$P(v_1) = \mathfrak{A}\left\{ f_{x_\mu,x_\nu}R^{(\mu,\nu)} - f_{s,x_\mu}h_1^{(\mu)} - if_{x_\mu}\frac{\partial H}{\partial \Phi_\mu} + i\sigma_{x_\mu}R^{(\mu)}\right\}$$

$$+f_{ss}h_1 + if_s(P(h_1) + H) - i\sigma_s R - fP(H) + f_{x_{\mu_1},x_\nu}\big(\lambda^{(\nu)}R^{(\mu)} - A_\nu R^{(\mu)}\big)$$

$$+f_{x_\mu}\left\{\lambda_{x_\nu}^{(\mu)}R^{(\nu)} + i\left[\frac{\partial}{\partial \Phi_\mu}, P\right](R)\right\}, \tag{1.21}$$

where as usual $[P, Q]$ denotes the commutator of the operators P and Q.

We differentiate (1.4) first with respect to ξ_μ, then with respect to ξ_ν, μ, $\nu = 1,\ldots,l$, and put $\xi' = \Phi$. This yields

$$-\lambda^{(\mu,\nu)}R - \lambda^{(\mu)}R^{(\nu)} - \lambda^{(\nu)}R^{(\mu)} + A_\mu R^{(\nu)} + A_\nu R^{(\mu)} + \mathfrak{A}R^{(\mu,\nu)} = 0. \tag{1.22}$$

Multiplying by f_{x_μ,x_ν} and summing, we obtain

$$\langle L, f_{x_\mu,x_\nu}\big(\lambda^{(\nu)}R^{(\mu)} - A_\nu R^{(\mu)}\big)\rangle = 0. \tag{1.23}$$

Taking (9) into account, we transform the last term in (1.21) as follows:

$$i\left[\frac{\partial}{\partial \Phi_\mu}, P\right](R) = \left[\frac{\partial}{\partial \Phi_\mu}, \frac{\partial}{\partial x_0}\right](R)$$

$$= \frac{\partial}{\partial \Phi_\mu}(\varphi_{x_0,x_\nu})R^{(\nu)} = -\lambda_{x_\nu}^{(\mu)}R^{(\nu)}. \tag{1.24}$$

From (1.21), (1.23) and (1.24) it follows that

$$\langle L, P(v_1)\rangle = if_s\langle L, P(h_1) + H\rangle + f\langle L, P(H)\rangle. \tag{1.25}$$

This implies that (H) is equivalent to conditions (13)–(15).

Finally, we consider case (iii). If $|\mathcal{L}_{1,2,j}|^2 + |\mathcal{L}_{2,1,j}|^2 \neq 0$, then either $\mathcal{L}_{1,2,j} \neq 0$, or $\mathcal{L}_{2,1,j} \neq 0$ in the neighborhood of the point $\hat{x}$ where $R_{k,j}$ and $L_{k,j}$ are defined. The equalities

$$\langle L_k, P(fR_1 - gR_2)\rangle = 0, \qquad k = 1,2 \tag{1.26}$$

(in what follows we omit the index j) are necessary and sufficient for the existence of the function v_1 in (18).

If $\mathcal{L}_{1,2} \neq 0$, then it follows from (16) and (17) that

$$\langle L_k, P(R_1)\rangle = 0, \qquad k = 1, 2.$$

Using (1.18) with R_j replaced by R_1, we then obtain that (1.26) holds for $g \equiv 0$.

If $\mathcal{L}_{2,1} \neq 0$, it follows from (16) and (17) that

$$\langle L_1, P(R_k)\rangle = 0, \qquad k = 1, 2.$$

Consequently, for any f and g we have

$$\langle L_1, P(fR_1 + gR_w)\rangle = 0.$$

We consider the equation

$$\langle L_2, P(fR_1 + gR_2)\rangle = g_s + g_s\langle L_2, P(R_2)\rangle + f\langle L_2, P(R_1)\rangle = 0. \quad (1.27)$$

It is obvious that for any $f \in C^\infty$ we can determine g from (1.27) after integrating along the bicharacteristics.

Thus, we have shown that condition $(H_{\hat{x},\hat{\xi}'})$ follows from (16) and (17).

Suppose now that $(H_{\hat{x},\hat{\xi}'})$ is satisfied. We assume that $\langle L_1, P(R_1)\rangle \neq 0$ in a neighborhood of $\hat{x}$. From (1.26) we then obtain

$$\mathcal{L}_{1,2} \neq 0, \qquad g = -(\mathcal{L}_{1,2})^{-1} f\langle L_1, P(R_1)\rangle.$$

Substituting g in (1.27), we have

$$f_s\frac{\langle L_1, P(R_1)\rangle}{\mathcal{L}_{1,2}} + f\left\{ \frac{\langle L_2, P(R_2)\rangle\langle L_1, P(R_1)\rangle}{\mathcal{L}_{1,2}} + \left(\frac{\langle L_1, P(R_1)\rangle}{\mathcal{L}_{1,2}}\right)_s - \mathcal{L}_{2,1}\right\} = 0.$$

This is impossible, because f is an arbitrary scalar-valued function. Consequently, (16) holds.

We now assume that (17) does not hold. From the first equality in (1.26) we then obtain that $g \equiv 0$. This implies that the second equality in (1.26) is not satisfied. Lemma 1.2 is proved.

We observe that if the coefficients of the operator P and the $\varphi_j(x)$ are analytic functions, then condition (H) follows from (16) and (17) in case (iii).

It is easy to show that in case (iii) conditions (16) and (17) do not depend on the choice of the functions $R_{k,j}$ and $L_{k,j}$ even if $|\mathcal{L}_{1,2,j}|^2 + |\mathcal{L}_{2,1,j}|^2 = 0$. To obtain this, it is necessary to use (1.18) with R_j replaced by $R_{k,j}$, $k = 1, 2$. It is also easy to show that conditions (16) and (17) are invariant with respect to a change of variables. The invariance of (13)–(15) follows from Lemma 1.2.

In the following sections we construct an asymptotic solution of the form (19) for P under the assumption that for some function $\varphi_k(x)$ satisfying (9) one of the conditions (13)–(17) does not hold in a neighborhood of the point $\hat{x}$. Without loss of generality we assume that $\hat{x} = 0$ and $\hat{\xi}' = \zeta = (0,\ldots,0,1)$.

We may assume that $\varphi_k(x)$ is the solution of the Cauchy problem

$$\begin{cases} \dfrac{\partial \varphi_k}{\partial x_0} = -\lambda_k(x, \operatorname{grad}_{x'} \varphi_k), \\[2mm] \varphi_k|_{x_0=0} = x_l. \end{cases} \qquad (1.28)$$

Otherwise, we can achieve this by the change of variables

$$\tilde{x}_\nu = x_\nu, \qquad \nu = 0, 1, \dots, l-1,$$

$$\tilde{x}_l = \varphi_k(0, x').$$

In order to rectify the bicharacteristics corresponding to the eigenvalue $\lambda_k(x, \xi')$ and the function $\varphi_k(x)$ we perform the following change of variables in a neighborhood of the origin:

$$y_0 = x_0, \qquad y_\nu = y_\nu(x), \quad \nu = 1, \dots, l, \qquad (1.29)$$

where $y(x) = (y_1(x), \dots, y_l(x))$ is the solution of the Cauchy problem

$$\begin{cases} \dfrac{\partial y}{\partial x_0} = -I\left(\dfrac{y}{x'}\right) \operatorname{grad}_{\xi'} \lambda_k\left(x, I*\left(\dfrac{y}{x'}\right)\zeta\right), \\[2mm] y|_{x_0=0} = x', \end{cases}$$

in which $I(y/x')$ is the Jacobian $\{\partial y_\mu(x)/\partial x_\nu\}^l_{\mu,\nu=1}$.

In the new variables the symbol $\xi_0 + \lambda_k(x, \xi')$ becomes $\eta_0 + \tilde{\lambda}_k(y, \eta')$, where

$$\tilde{\lambda}_k(y, \eta') = \lambda_k\left(x(y), I*\left(\dfrac{y}{x'}\right)\eta'\right)$$

$$- \left\{\operatorname{grad}_{\xi'} \lambda_k\left(x(y), I*\left(\dfrac{y}{x'}\right)\zeta\right), I*\left(\dfrac{y}{x'}\right)\eta'\right\},$$

with $x = x(y)$ denoting the inverse transformation of variables. From this we obtain

$$\operatorname{grad}_{\eta'} \tilde{\lambda}_k(y, \eta')\big|_{\eta'=\zeta} = 0.$$

Using Euler's identity for homogeneous functions, we obtain from the last equality that in a neighborhood of zero

$$\tilde{\lambda}_k(y, \zeta) = 0. \qquad (1.30)$$

Therefore, after the change of variables (1.29) the function $\varphi_k(x)$ satisfying (1.28) becomes y_l and the derivatives along the bicharacteristics corresponding to $\lambda_k(x, \xi')$ and $\varphi_k(x)$ become derivatives along the y_0-axis.

Consequently, without loss of generality we may assume that $\lambda_k(x, \xi')$ satisfies (1.30) and that $\varphi_k(x)$ has the form $\varphi_k(x) \equiv x_l$.

The following result is proved in [1].

LEMMA 1.3. *Suppose that the Cauchy problem* (4), (5) *is well-posed in* G_T^+. *Then for any compactum* $K \subset\subset G$ *there are a constant* $C > 0$ *and integers* p *and* q *such that for all* $t > T$

$$\|u\|_{H_p(K_t^-)} \leqslant C\|P_u\|_{H_q(K_t^-)}, \qquad u \in C_0^\infty(K_T^+). \qquad (1.31)$$

Let

$$E(x, \rho) = \exp i\left(\varepsilon x_l \rho + \sum_{j=0}^{\nu-1} l^{j+1}(x)\rho^{(\nu-j)/\tau} \right), \qquad \varepsilon = \pm 1.$$

Using (1.31), we can show (see §2 in [1]) that proving that the Cauchy problem is well-posed reduces to the construction of an asymptotic solution $u_\rho(x)$ of the form (19) for which $\varphi = \varepsilon x_l$, $\varepsilon = \pm 1$, $v_0(0) = 1$ and $v_n \in C^\infty(W)$ (where W is a neighborhood of zero), the functions $l^j \in C^\infty(W)$ satisfy

$$\mathrm{Im}\left(\sum_{j=0}^{\nu-1} l^{j+1}\rho^{(\nu-j)/\tau} \right) \geqslant \rho^{\nu/\tau}(\delta - o(1))\left(-x_0 + |x'|^2\right), \qquad \delta > 0, \quad (1.32)$$

for $x \in W_0^-$, and for any fixed $N_1 > 0$

$$E^{-1}(x, \rho)Pu_\rho = O(\rho^{-N_1}) \tag{1.33}$$

holds in W. If in some neighborhood of zero the equation

$$\Phi_1(x, \eta_0) = 0 \tag{1.34}$$

has a simple root $\eta_0 = F(x)$ such that

$$\mathrm{Im}\, F(0) < 0, \tag{1.35}$$

then we define $l^1(x)$ as the solution of the Cauchy problem

$$\begin{cases} l_{x_0}^1 = F(x), \\ l^1(0, x') = i\,|x'|^2. \end{cases}$$

Then it is easy to show that if the functions l^j, $j = 2,\ldots,\nu$, satisfy $l^j(0, x') = 0$, then (1.32) holds in some neighborhood of zero.

Consequently, the proof of Theorem 1 reduces to establishing that (1.34) has a simple root satisfying (1.35) and to constructing an asymptotic solution u_ρ for which (1.33) holds.

§2. Conditions (13) and (16) are necessary for the
Cauchy problem to be well-posed

In this and the next sections we keep all the notation introduced in §1. We construct for P an asymptotic solution of the form (19) in a neighborhood of zero under the assumption that one of the conditions (13)–(17) is not satisfied there for the function $\varphi_k(x) \equiv x_l$.

We consider the vector-valued functions $R_k,\ldots,L_{2,k}$ and their derivatives at points (x, ζ_0), and omit the index k. The same also applies to the function $\lambda_k(x, \xi')$. By $\sigma^0, \sigma^1, \sigma_1^0,\ldots$ we denote smooth scalar-valued functions of x, by $\Phi_j(l^1, l^2, \sigma^0,\ldots)$ and $\Psi_j(l^1, l^2, \sigma^0,\ldots)$ smooth scalar-valued functions of $x, l^1, l^2, \sigma^0,\ldots$, and by $V(l^1, l^2, \sigma^0,\ldots)$ smooth vector-valued functions of $x, l^1, l^2, \sigma^0,\ldots$.

We introduce the notation

$$\sigma_x \frac{d}{dl_x^1}Q(l^1) = \sum_{\nu=0}^{l} \sigma_{x_\nu} \frac{\partial}{\partial l_{x_\nu}^1}Q\left(l_{x_0}^1,\ldots,l_{x_l}^1\right),$$

where $\sigma(x)$ is a scalar-valued function and $Q(l^1)$ a vector- or scalar-valued function. Furthermore, if the system

$$A_l(x)w = V(x, \zeta_0)$$

is solvable, then we write $V \equiv 0 \pmod{A_l}$.

As in the case of (1.17), at points (x, ζ_0) we obtain

$$M(\sigma)V = \sigma_{x_0}V - \sigma_{x_\mu}A_l V^{(\mu)} + \sigma_{x_\mu}(A_l V)^{(\mu)}. \tag{2.1}$$

In particular,

$$M(\sigma)R = \sigma_{x_0}R - \sigma_{x_\mu}A_l R^{(\mu)}, \tag{2.2}$$

$$M(\sigma)h_1 = \sigma_{x_0}h_1 - \sigma_{x_\mu}A_l h_1^{(\mu)} + \sigma_{x_\mu}R^{(\mu)}. \tag{2.3}$$

We denote by K_n the coefficient of $\rho^{n/\tau}$ in the expression of $P(u_\rho)$. We determine successively the functions l^j and v_n, $n = 0, \ldots, N$, so that $K_n \equiv 0$ in a neighborhood of zero for $n > -N_1$, where $N_1 > 0$ is any fixed number. We consider the various cases when each one of the conditions (13)–(17) is in turn not satisfied.

(1°). $r = 3$, $\kappa = 2$, $\langle L, P(R) \rangle \neq 0$.

In this case we put $\nu = 2$, $\tau = 3$ and $\varphi = x_l$ in (19). Then

$$P(u_\rho) = \left\{ \left[A_l \rho + M(l^1)\rho^{2/3} + M(l^2)\rho^{1/3} \right] \sum_{n=0}^{N} v_n \rho^{-n/3} + \sum_{n=0}^{N} P(v_n)\rho^{-n/3} \right\}$$

$$\times \exp\left(x_l \rho + l^1 \rho^{2/3} + l^2 \rho^{1/3} \right)i.$$

Let $v_0 = \sigma^0 R$. From $K_2 = 0$ we obtain a system for v_1:

$$-A_l v_1 = \sigma^0 M(l^1)R = \sigma^0 Q_1.$$

By virtue of (2.2), from this system we can determine the function

$$v_1 = \sigma^0 V_1(l^1) + \sigma^1 R,$$

where

$$V_1(\sigma) = \sigma_{x_\mu}R^{(\mu)} - \sigma_{x_0}h_1.$$

The equality $K_1 = 0$ leads to a system for v_2:

$$-A_l v_2 = M(l^1)v_1 + \sigma^0 M(l^2)R$$

$$= \sigma^0 \left[M(l^1)V_1(l^1) + M(l^2)R \right] + \sigma^1 Q_1 = \sigma^0 Q_2 + \sigma^1 Q_1. \tag{2.4}$$

Remarking that $\lambda^{(\mu)}(x, \zeta_0) = 0$, we obtain from (1.22)

$$A_\nu R^{(\mu)} = \tfrac{1}{2}\left(\lambda^{(\mu,\nu)}R - A_l R^{(\mu,\nu)} \right) \equiv 0 \qquad \pmod{A_l}. \tag{2.5}$$

This means that the system

$$A_l V(\sigma) = \sigma_{x_\mu, x_\nu}A_\nu R^{(\mu)}$$

is solvable and

$$V(\sigma) = \tfrac{1}{2}\sigma_{x_\mu, x_\nu}\left(\lambda^{(\mu,\nu)}h_1 - R^{(\mu,\nu)} \right).$$

By (2.3) and (2.5) we find that

$$M(l^1)V_1(l^1) = l^1_{x_0}\left(l^1_{x_\mu}R^{(\mu)} - l^1_{x_0}h_1\right) + l^1_{x_\mu}A_\mu\left(l^1_{x_\nu}R^{(\nu)} - l^1_{x_0}h_1\right) \equiv 0 \qquad (\mathrm{mod}\ A_l).$$

$$(2.6)$$

Consequently, the solvability condition for system (2.4) is satisfied, and we can find the function

$$v_2 = \sigma^0\left[V_2(l^1) + V_1(l^2)\right] + \sigma^1 V_1(l^1) + \sigma^2 R,$$

where

$$V_2(\sigma) = (\sigma_{x_0})^2 h_2 - \sigma_{x_0}\sigma_{x_\mu}h_1^{(\mu)} - V(\sigma).$$

For determining v_3 we then obtain the system

$$-A_l v_3 = M(l^1)v_2 + M(l^2)v_1 - iM(\sigma^0)R + \sigma^0 P(R)$$

$$= \sigma^0\left\{M(l^1)\left[V_2(l^1) + l_x^2\frac{d}{dl_x^1}V_1(l^1)\right] + M(l^2)V_1(l^1) + P(R)\right\}$$

$$+\sigma^1 Q_2 + \sigma^2 Q_1 - iM(\sigma)R$$

$$\equiv \sigma^0\left\{M(l^1)V_2(l^1) + l_x^2\frac{d}{dl_x^1}Q_2 + P(R)\right\} \equiv \sigma^0 Q_3 \qquad (\mathrm{mod}\ A_l).\ (2.7)$$

It follows from (2.6) that

$$\left\langle L, l_x^2\frac{d}{dl_x^1}Q_2\right\rangle = l_x^2\frac{d}{dl_x^1}(\langle L, Q_2\rangle) = 0.$$

Therefore, (2.7) is solvable if

$$M(l^1)V_2(l^1) + P(R) \equiv 0 \qquad (\mathrm{mod}\ A_l). \tag{2.8}$$

We show that

$$M(l^1)V_2(l^1) \equiv \left(l^1_{x_0}\right)^3 h_2 \qquad (\mathrm{mod}\ A_l). \tag{2.9}$$

Then (2.8) holds if l^1 satisfies the equation

$$\left(l^1_{x_0}\right)^3 + \langle L, P(R)\rangle = 0.$$

Since in our case $\langle L, P(R)\rangle \neq 0$, in some neighborhood of zero there is a simple root $\eta_0 = F(x)$ of the equation

$$\eta_0^3 + \langle L, P(R)\rangle = 0,$$

such that $\mathrm{Im}\ F(0) < 0$. Consequently, we can find l^1 with the required properties.

Applying (2.1) to h_2, we obtain

$$\left(l^1_{x_0}\right)^2 M(l^1)h_2 \equiv \left(l^1_{x_0}\right)^3 h_2 + \left(l^1_{x_0}\right)^2 l^1_{x_\mu}h_1^{(\mu)} \qquad (\mathrm{mod}\ A_l). \tag{2.10}$$

We differentiate (1.5) twice with respect to ξ_μ and ξ_ν, $\mu, \nu = 1,\ldots,l$, and put $\xi' = \xi_0$; this yields

$$-\lambda^{(\mu,\nu)}h_1 + A_\mu h_1^{(\nu)} + A_\nu h_1^{(\mu)} + A_l h_1^{(\mu,\nu)} = R^{(\mu,\nu)}. \tag{2.11}$$

Then

$$l^1_{x_0} l^1_{x_\nu} M(l^1) h_1^{(\nu)} \equiv \left(l^1_{x_0}\right)^2 l^1_{x_\nu} h_1^{(\nu)} - l^1_{x_0} V(l^1) \qquad (\text{mod } A_l). \qquad (2.12)$$

In the equality

$$M(l^1) V(l^1) \equiv l^1_{x_0} V(l^1) + \tfrac{1}{2} l^1_{x_\mu} l^1_{x_\nu} l^1_{x_j} \left(\lambda^{(\nu,j)} R^{(\mu)} - A_\mu R^{(\nu,j)}\right) \qquad (\text{mod } A_l)$$

$$(2.13)$$

we transform the last two terms. With this aim we differentiate (1.4) with respect to ξ_μ, ξ_ν and ξ_j, and put $\xi' = \zeta_0$; we obtain

$$-\lambda^{(\mu,\nu,j)} R - \left[\lambda^{(\mu,\nu)} R^{(j)} + \lambda^{(\mu,j)} R^{(\nu)} + \lambda^{(\nu,j)} R^{(\mu)}\right]$$

$$+ A_\mu R^{(\nu,j)} + A_\nu R^{(\mu,j)} + A_j R^{(\mu,\nu)} + A_l R^{(\mu,\nu,j)} = 0. \qquad (2.14)$$

We multiply (2.14) by $l^1_{x_\mu} l^1_{x_\nu} l^1_{x_j}$; summing, we obtain

$$l^1_{x_\mu} l^1_{x_\nu} l^1_{x_j} \left[\lambda^{(\nu,j)} R^{(\mu)} - A_\mu R^{(\nu,j)}\right] \equiv 0 \qquad (\text{mod } A_l). \qquad (2.15)$$

From (2.10), (2.12), (2.13) and (2.15) we obtain (2.9).

From (2.7) we now determine the function

$$v_3 = \sigma^0 \left[V_3(l^1) + l^2_x \frac{d}{dl^1_x} V_2(l^1)\right] - iV_1(\sigma^0)$$

$$+ \sigma^1 \left[V_2(l_1) + V_1(l^2)\right] + \sigma^2 V_1(l^1) + \sigma^3 R,$$

where the explicit expression of $V_3(l^1)$ is not needed in the sequel.

For v_4 we obtain the system

$$-A_l v_4 = M(l^1) v_3 + M(l^2) v_2 + P(v_1)$$

$$\equiv \sigma^0 \left\{ M(l^1) V_3(l^1) + P\left(V_1(l^1)\right) + l^2_x \frac{d}{dl^1_x} \left(M(l^1) V_2(l^1)\right) + M(l^2) V_1(l^2) \right\}$$

$$- i\sigma^0_x \frac{d}{dl^1_x} \left(M(l^1) V_1(l^1)\right)$$

$$\equiv \sigma^0 \left[3\left(l^1_{x_0}\right)^2 l^2_{x_0} h_2 + V_4(l^1)\right] \qquad (\text{mod } A_l); \qquad (2.16)$$

here we have used the fact that, after replacing l^1 by l^2, (2.6) implies

$$M(l^2) V_1(l^2) \equiv 0 \qquad (\text{mod } A_l).$$

From the solvability condition for (2.16)

$$3\left(l^1_{x_0}\right)^2 l^2_{x_0} + \langle L, V_4(l^1)\rangle = 0$$

we determine the function l^2, by imposing the additional requirement $l^2(0, x') = 0$. Then

$$v_4 = \sigma^0 V_5(l^1, l^2) - i\left[\sigma^0_x \frac{d}{dl^1_x} V_2(l^1) + V_1(\sigma^1)\right]$$

$$+ \sigma^1(\ldots) + \sigma^2(\ldots) + \sigma^3(\cdots) + \sigma^4 R.$$

We do not need to write the terms containing σ^1, σ^2 and σ^3. We only mention that v_4 depends on σ^1, σ^2, σ^3 and σ^4 in exactly the same way as v_3 depends on σ^0, σ^1, σ^2 and σ^3, respectively. Such dependence is also encountered in the subsequent cases, and we shall make no special mention of it. Thus, it suffices to find the equation from which we determine σ^0. The other functions σ^1, $\sigma^2, \ldots$ are determined from the same equation, only with a different right-hand side.

From the solvability condition for the system

$$-A_l v_5 = M(l^1)v_4 + M(l^2)v_3 + P(v_2)$$
$$\equiv -i\sigma_x^0 \frac{d}{dl_x^1}\left(M(l^1)V_2(l^1)\right) + \sigma^0 M(l^1)V_5(l^1, l^2) \qquad (\mathrm{mod}\ A_l)$$

we obtain the equation

$$3\left(l_{x_0}^1\right)^2 \sigma_{x_0}^0 + \Phi_1(l^1, l^2)\sigma^0 = 0, \tag{2.17}$$

which we solve with the initial condition

$$\sigma^0(0, x') = 1. \tag{2.18}$$

After this we reason by induction. At every step, from the solvability condition for the system $K_{-(n+2)} = 0$ we obtain

$$3\left(l_{x_0}^1\right)^2 \sigma_{x_0}^n + \Phi_1(l^1, l^2)\sigma^n + \Psi_n\left(l^1, l^2, \sigma_0, \ldots, \sigma^{n-1}\right) = 0,$$

from which we determine σ^n with the initial condition

$$\sigma^n(0, x') = 0. \tag{2.19}$$

From the system $K_{-(n+2)} = 0$ we then find v_{n+5} to within a term $\sigma^{n+5}R$.

(2°). $r = 3$, $\kappa = 1$, $\langle L_1, P(R_1)\rangle \neq 0$.

We put $\nu = 1$, $\tau = 2$ and $\varphi = \varepsilon x_l\ (\varepsilon = \pm 1)$ in (19). We have

$$P(u_\rho) = \left\{\left(\varepsilon A_l \rho + M(l^1)\rho^{1/2}\right) \sum_{n=0}^{N} v_n \rho^{-n/2} + \sum_{n=0}^{N} P(v_n)\rho^{-n/2}\right\}$$
$$\times \exp i\left(\varepsilon x_l \rho + l^1 \rho^{1/2}\right).$$

We now find a solution of the system

$$\mathfrak{A}(x, \xi')h_1(x, \xi') = R_1(x, \xi'), \tag{2.20}$$

such that

$$\langle L_1(x, \xi'), h_1(x, \xi')\rangle \equiv 1, \tag{2.21}$$
$$\langle L_2(x, \xi'), h_1(x, \xi')\rangle \equiv 0 \tag{2.22}$$

in a neighborhood of $(0, \zeta_0)$.

Let $h_1' = h_1 - \langle L_2, h_1\rangle R_2$, where h_1 is a solution of (2.20). Then L_2 and h_1' satisfy (2.22). If $\langle L_1(\hat{z}), h_1'(\hat{z})\rangle = 0$ at some point $\hat{z}$, then the system $\mathfrak{A}(\hat{z})w = h_1'(\hat{z})$ is solvable. As in the proof of (1.7), we establish that this is impossible. Putting $L_1' = (\langle L_1, h_1'\rangle)^{-1}L_1$, we obtain (2.21) for L_1' and h_1'. Until the end of case (2°) we write again L_1 and h_1 instead of L_1' and h_1', respectively.

We remark that, similarly to (2.2) and (2.5), we have

$$M(\sigma)R_1 \equiv 0 \qquad (\mathrm{mod}\ A_l), \tag{2.23}$$

$$\sigma_{x_\mu,x_\nu}A_\nu R_1^{(\mu)} \equiv 0 \qquad (\mathrm{mod}\ A_l). \tag{2.24}$$

We proceed to determine the functions l^1 and v_n in (19). We put $v_0 = \sigma^0 R_1$. From the equation

$$-\varepsilon A_l v_1 = \sigma_1^0 M(l^1)R_1 = \sigma_1^0 Q_1$$

we find that

$$v_1 = \varepsilon\sigma_1^0 V_1(l^1) + \sigma_1^1 R_1 + \sigma_2^1 R_2,$$

where

$$V_1(\sigma) = \sigma_{x_\mu}R_1^{(\mu)} - \sigma_{x_0}h_1.$$

For v_2 we obtain the system

$$-\varepsilon A_l v_2 = M(l^1)v_1 - iM(\sigma_1^0)R_1 + \sigma_1^0 P(R_1)$$
$$\equiv \sigma_1^0\big[\varepsilon M(l^1)V_1(l^1) + P(R_1)\big] + \sigma_2^1 l_{x_0}^1 R_2 \qquad (\mathrm{mod}\ A_l). \tag{2.25}$$

From the first solvability condition for (2.25)

$$\varepsilon\langle L_1, M(l^1)V_1(l^1)\rangle + \langle L_1, P(R_1)\rangle = 0$$

we determine l^1. As in the case of (2.6), we can show that

$$\langle L_1, M(l^1)V_1(l^1)\rangle = -\big(l_{x_0}^1\big)^2.$$

Taking the inequality $\langle L_1, P(R_1)\rangle \neq 0$ into account and selecting the sign of ε, we obtain that in a neighborhood of zero the equation

$$\eta_0^2 - \varepsilon\langle L_1, P(R_1)\rangle = 0$$

has a simple root $\eta_0 = F(x)$ such that $\mathrm{Im}\, F(0) < 0$. We now determine l^1 in the standard way.

The function σ_2^1 is found from the second solvability condition for (2.25):

$$\sigma_0^1\big[\varepsilon\langle L_2, M(l^1)V_1(l^1)\rangle + \mathcal{L}_{2,1}\big] + \sigma_2^1 l_{x_0}^1 = \sigma_1^0 \mathcal{L}_{2,1} + \sigma_2^1 l_{x_0}^1 = 0,$$

where in the next to last equality we have used (2.22). We put

$$\sigma_2^1 = -\sigma_1^0\big(l_{x_0}^1\big)^{-1}\mathcal{L}_{2,1} \tag{2.26}$$

and from (2.25) we determine

$$v_2 = \sigma_1^0 V_6(l^1) + \varepsilon\sigma_1^1 V_1(l^1) - i\varepsilon V_1(\sigma_1^0) + \sigma_1^2 R_1 + \sigma_2^2 R_2,$$

where $V_6(l^1)$ is the solution of the system

$$A_l V_6(l^1) = \big(l_{x_0}^1\big)^2 h_1 - \varepsilon P(R_1)$$

$$+ \tfrac{1}{2}l_{x_\mu}^1 l_{x_\nu}^1\big(A_l R_1^{(\mu,\nu)} - \lambda^{(\mu,\nu)}\dot{R}_1\big) + \varepsilon\mathcal{L}_{2,1}\left\{R_2 - \sum_{k=1,2}\frac{A_l\big(l_{x_\nu}^1 R_k^{(\nu)}\big)}{l_{x_0}^1}\right\}.$$

From the first solvability condition for the system

$$-\varepsilon A_l v_3 = M(l^1)v_2 + P(v_1)$$

$$= \sigma_1^0 M(l^1)V_6(l^1) - i\varepsilon\sigma_{1,x}^0 \frac{d}{dl_x^1}\left(M(l^1)V_1(l^1)\right)$$

$$+ \sum_{k=1,2} \left\{\sigma_k^1 P(R_k) - iM(\sigma_k^1)R_k + M(l^1)\sigma_k^2 R_k\right\} \qquad (2.27)$$

we determine σ_1^0. Using (2.26), we obtain that this condition has the form

$$2l_{x_0}^1\sigma_{1,x_0}^0 + \varepsilon\Phi_2(l^1)\sigma_1^0 = 0.$$

As usual, we determine σ_1^0 with the initial condition (2.18).

From the second solvability condition for (2.27) we find that

$$\sigma_2^2 = -\sigma_1^1\left(l_{x_0}^1\right)^{-1}\mathcal{L}_{2,1} + \Phi_3\left(l_1, \sigma_1^0\right).$$

After this we reason by induction. At every step we determine the function σ_1^n from the equation

$$2l_{x_0}^1\sigma_{1,x_0}^n + \varepsilon\Phi_2(l^1)\sigma_1^n + \Psi_n\left(l^1, \sigma_1^0,\ldots,\sigma_1^{n-1}, \sigma_2^{n-1}\right) = 0$$

with the initial condition (2.19). The function σ_2^{n+2} has the form

$$\sigma_2^{n+2} = -\sigma_1^{n+1}\left(l_{x_0}^1\right)^{-1}\mathcal{L}_{2,1} + \Phi_{n+2}\left(l^1, \sigma_1^0,\ldots,\sigma_1^n, \sigma_2^n\right),$$

and v_{n+3} is found from the system $K_{-(n+1)} = 0$ to within a term $\sigma_1^{n+3}R_1 + \sigma_2^{n+3}R_2$.

We observe that u_ρ can be constructed in the same way in the case $r = 2$, $\kappa = 1$, $\langle L, PR\rangle \neq 0$. There we have only one solvability condition for every system $K_{-n} = 0$, and the functions σ_2^n are identically equal to zero for all n.

This proves that conditions (13) and (16) are necessary for the Cauchy problem to be well-posed.

§3. Conditions (14), (15) and (17) are necessary for the Cauchy problem to be well-posed

First we consider the following case.

($3°$). $r = 3$, $\kappa = 2$, $\langle L, P(h_1) + H\rangle \neq 0$.

We keep all the notation used in case ($1°$) in §2. Furthermore, we assume that (13) holds for all functions $\varphi_k(x)$ satisfying (9). If this is not so, then from what we proved in case ($1°$) it will follow that the Cauchy problem is not well-posed.

We consider again the asymptotic solution (19), where $\nu = 1$, $\tau = 2$ and $\varphi = \varepsilon x_l$ ($\varepsilon = \pm 1$). We define the functions v_0 and v_1 as in case ($2°$). We only need to replace σ_1^0 by σ^0, R_1 by R, and put $\sigma_2^1 \equiv 0$. Then it follows from (2.6) that the system (2.25) with $R_2 \equiv 0$ is solvable and

$$v_2 = \sigma^0\left(V_2(l^1) - \varepsilon H\right) + \varepsilon\sigma^1(l^1) - i\varepsilon\sigma_x^0\frac{d}{dl_x^1}V_1(l^1) + \sigma^2 R,$$

where $A_l H = P(R)$.

We consider the system

$$-\varepsilon A_l v_3 = M(l^1)v_2 + P(v_1)$$

$$= -i\varepsilon\sigma_x^0 \frac{d}{dl_x^1}\big(M(l^1)V_1(l^1)\big) + \sigma^0\big[M(l^1)(V_2(l^1) - \varepsilon H) + \varepsilon P(V_1(l^1))\big]$$

$$\equiv \sigma^0\Big[\big(l_{x_0}^1\big)^3 h_2 - \varepsilon M(l^1)H + \varepsilon P(V_1(l^1))\Big] \qquad (\mathrm{mod}\ A_l). \qquad (3.1)$$

We show that

$$P\big(V_1(l^1)\big) - M(l^1)H \equiv -l_{x_0}^1\big(P(h_1) + H\big) \qquad (\mathrm{mod}\ A_l). \qquad (3.2)$$

Then the solvability condition for (3.1) takes the form

$$\big(l_{x_0}^1\big)^3 - \varepsilon l_{x_0}^1\langle L,\, P(h_1) + H\rangle = 0.$$

Using the fact that $\langle L,\, P(h_1) + H\rangle \neq 0$, from this equation we can determine l^1, as in the previous cases.

We transform $P(V_1(l^1))$ by means of (2.3) and (2.6). We have

$$P\big(V_1(l^1)\big) = P\big(l_{x_\mu}^1 R^{(\mu)} - l_{x_0}^1 h_1\big)$$

$$= -l_{x_0}^1 P(h_1) + iM\big(l_{x_0}^1\big)h_1 - iM\big(l_{x_\mu}^1\big)R^{(\mu)} + l_{x_\mu}^1 P(R^{(\mu)})$$

$$\equiv -l_{x_0}^1 P(h_1) + l_{x_\mu}^1 P(R^{(\mu)}) \qquad (\mathrm{mod}\ A_l).$$

Consequently, (3.2) will be proved if we establish that

$$P(R^{(\mu)}) - A_\mu H \equiv 0 \qquad (\mathrm{mod}\ A_l),\ \mu = 1,\dots,l. \qquad (3.3)$$

We consider the system

$$\mathfrak{A}_k(x, \Phi_k)H_k(x, \xi') = P(R_k(x, \Phi_k)), \qquad (3.4)$$

where $\varphi_k(x, \xi')$ is the solution of the Cauchy problem

$$\begin{cases} \partial\varphi_k/\partial x_0 = -\lambda_k(x, \Phi_k), \\ \varphi_k(0, x') = \{x', \xi'\}, \end{cases}$$

and $\Phi_k = \mathrm{grad}_{x'}\,\varphi_k(x, \xi')$. It is easy to show that $\varphi_k^{(\nu)}(x, \xi')$ satisfies the relations

$$\begin{cases} \dfrac{\partial}{\partial x_0}\varphi_k^{(\nu)} = -\lambda_k^{(\mu)}(x, \Phi_k)\varphi_{k,x_\mu}^{(\nu)}, \\ \varphi_k^{(\nu)}(0, x') = x_\nu, \qquad \nu = 1,\dots,l. \end{cases}$$

Therefore, for $\xi' = \zeta_0$ we obtain

$$\varphi_k^{(\nu)}(x, \zeta_0) = \lambda_\nu. \qquad (3.5)$$

This permits us to derive (3.3) from (3.4), after differentiating with respect to ξ_μ and substituting $\xi' = \zeta_0$. Thus, (3.3) is proved and from (3.1) we can determine

$$v_3 = -i\sigma_x^0 \frac{d}{dl_x^1} V_2(l^1) + \sigma^0 V_7(l^1) + \sigma^1\big(V_2(l^1) - \varepsilon H\big)$$

$$-i\varepsilon V_1(\sigma^1) + \varepsilon\sigma^2 V_1(l^1) + \sigma^3 R.$$

For v_4 we obtain the system

$$-\varepsilon A_l v_4 = M(l^1)v_3 + P(v_2)$$

$$\equiv -i\sigma_x^0 \frac{d}{dl_x^1}\big(M(l^1)V_2(l^1)\big) - i\varepsilon P\big(V_1(\sigma^0)\big) + i\varepsilon M(\sigma^0)H \qquad (3.6)$$

$$+\sigma^0 V_8(l^1) \qquad (\mathrm{mod}\ A_l).$$

If we replace l^1 by σ^0 in (3.2), we have

$$P\big(V_1(\sigma^0)\big) - M(\sigma^0)H \equiv -\sigma_{x_0}^0\big(P(h_1) + H\big) \qquad (\mathrm{mod}\ A_l). \qquad (3.7)$$

Using (3.7), it is easy to show that the solvability condition for (3.6) has the form

$$2\varepsilon\langle L, P(h_1) + H\rangle\sigma_{x_0}^0 + \Phi_3(l^1)\sigma^0 = 0. \qquad (3.8)$$

From (3.8) we find σ^0 with the initial condition (2.18). After this we use a standard inductive argument.

(4°). $r = 3$, $\kappa = 2$, $\langle L, P(H)\rangle \neq 0$.

We assume that conditions (13) and (14) holds in a neighborhood of zero for all $\varphi_k(x)$ satisfying (9). Othersise, from cases (1°) and (3°) we shall obtain that the Cauchy problem is not well-posed.

We put $\nu = 1$, $\tau = 3$ and $\varphi = x_l$ in (19). Then

$$P(u_\rho) = \left\{\big(A_l\rho + M(l^1)\rho^{1/3}\big)\sum_{n=0}^{N} v_n\rho^{-n/3} + \sum_{n=0}^{N} P(v_n)\rho^{-n/3}\right\} \exp i\big(x_l\rho + l^1\rho^{1/3}\big).$$

Let $v_0 = \sigma^0 R$ and $v_1 = \sigma^1 R$. For v_2 and v_3 we obtain the systems

$$-A_l v_2 = \sigma^0 M(l^1)R,$$

$$-A_l v_3 = M(l^1)v_1 + P(v_0) = \sigma^1 M(l^1)R + \sigma^0 P(R) - iM(\sigma^0)R.$$

The solvability conditions for these systems are obviously satisfied, and

$$v_2 = \sigma^0 V_1(l^1) + \sigma^2 R,$$

$$v_3 = \sigma^1 V_1(l^1) - iV_1(\sigma^0) - \sigma^0 H + \sigma^3 R.$$

We determine v_4 and v_5 from the systems

$$-A_l v_4 = M(l^1)v_2 + P(v_1) = \sigma^0 M(l^1)V_1(l^1) + \sigma^1 P(R) - iM(\sigma^1)R + \sigma^2 M(l^1)$$

$$\equiv 0 \qquad (\mathrm{mod}\ A_l),$$

$$-A_l v_5 = M(l^1)v_3 + P(v_2)$$

$$= -i\sigma_x^0 \frac{d}{dl_x^1}\big(M(l^1)V_1(l^1)\big) + \sigma^0\big[P\big(V_1(l^1)\big) - M(l^1)H\big] + \sigma^1 M(l^1)V_1(l^1)$$

$$+\sigma^2 P(R) - iM(\sigma^2)R + \sigma^3 M(l^1)R$$

$$\equiv l_{x_0}^1 \sigma^0\big(P(h_1) + H\big) \equiv 0 \qquad (\mathrm{mod}\ A_l).$$

We have

$$v_4 = \sigma^0 V_2(l^1) - iV_1(\sigma^1) - \sigma^2 H + \sigma^2 V_1(l^1) + \sigma^4 R,$$

$$v_5 = -i\sigma_x^0 \frac{d}{dl_x^1}(l^1) + \sigma^0 V_9(l^1) + \sigma^1 V_2(l^1) - \sigma^2 H - iV_1(\sigma^2) + \sigma^3 V_1(l^1) + \sigma^5 R.$$

We consider the system

$$-A_l v_6 = M(l^1)v_4 + P(v_3) \equiv \sigma^0\left[\left(l_{x_0}^1\right)^3 h_2 - P(H)\right] \qquad (\mathrm{mod}\ A_l). \qquad (3.9)$$

Since $\langle L, P(H)\rangle \neq 0$, the equation

$$\eta_0^3 - \langle L, P(H)\rangle = 0$$

has a simple root $\eta_0 = F(x)$ such that $\mathrm{Im}\ F(0) < 0$. Therefore, we can find l^1 in a standard way and then determine v_6 from (3.9).

Finally, we consider the solvability conditions for the system

$$-Av_7 = M(l^1)v_5 + P(v_4)$$

$$\equiv -i\sigma_x^0 \frac{d}{dl_x^1}\left(M(l^1)V_2(l^1)\right) + \sigma^1\left[M(l^1)V_2(l^1) - P(H)\right] + \sigma^0 V_{10}(l^1) \qquad (\mathrm{mod}\ A_l).$$

We obtain

$$3\left(l_{x_0}^1\right)^2 \sigma_{x_0}^C + \Phi_4(l^1)\sigma^0 = 0.$$

From this equation we find σ^0 with the initial condition (2.18), and then use a standard inductive argument.

(5°). $r = 3$, $\kappa = 1$, $\mathcal{L}_{1,2}\mathcal{L}_{2,1} \neq 0$.

We keep all the notation we used in the case (2°) in §2. In particular, (2.21) and (2.22) hold. Furthermore, we assume that (16) is satisfied for $\varphi_k(x)$. If this is not so, then case (2°) in §2 implies that the Cauchy problem is not well-posed.

We consider the same asymptotic solution as in case (4°). We define the functions

$$v_0 = \sigma_1^0 R_1, \qquad v_1 = \sigma_1^1 R_1 + \sigma_2^1 R_2, \qquad v_2 = \sigma_1^0 V_1(l^1) + \sigma_1^2 R_1 + \sigma_2^2 R_2.$$

For v_3 we obtain the system

$$-A_l v_3 = \sum_{k=1,2} \sigma_k^1 M(l^1) R_k + \sigma_0^1 P(R_1) - M(\sigma_1^0) R_1. \qquad (3.10)$$

By (16), the first solvability condition for (3.10) is satisfied. The second is satisfied if we put

$$\sigma_2^1 = -\sigma^0\left(l_{x_0}^1\right)^{-1}\mathcal{L}_{2,1}. \qquad (3.11)$$

Then

$$v_3 = \sigma_1^0\left[H_{1,2} - \left(l_{x_0}^1\right)^{-1}l_{x_\mu}^1 \mathcal{L}_{2,1} R_2^{(\mu)}\right] + \sigma_1^1 V_1(l^1) + \sigma_1^3 R_1 + \sigma_2^3 R_2,$$

where

$$A_l H_{1,2} = \mathcal{L}_{2,1} R_2 - P(R_1).$$

We consider the system

$$-A_l v_4 = M(l^1) v_2 + P(v_1)$$
$$= \sigma_1^0 M(l^1) V_1(l^1) + \sum_{k=1,2} \left\{ \sigma_k^2 M(l^1) R_k + \sigma_k^1 P(R_k) - iM(\sigma_k^1) R_k \right\}.$$

$$(3.12)$$

By (12), (2.21) and (3.11), the first solvability condition for (3.12) has the form

$$\left(l_{x_0}^1 \right)^2 + \left(l_{x_0}^1 \right)^{-1} \mathcal{L}_{1,2} \mathcal{L}_{2,1} = 0$$

or

$$\left(l_{x_0}^1 \right)^3 + \mathcal{L}_{1,2} \mathcal{L}_{2,1} = 0.$$

From this equation we determine l^1 in the standard way, and $l_{x_0}^1 \neq 0$ in a neighborhood of zero. From the second solvability condition for (3.12)

$$\sigma_2^2 l_{x_0}^1 - i\sigma_{2,x_0}^1 + \sum_{k=1,2} \sigma_k^1 \langle L_2, P(R_k) \rangle = 0$$

we then determine

$$\sigma_2^2 = -\left(l_{x_0}^1 \right)^{-1} \left\{ \sum_{k=1,2} \sigma_k^1 \langle L_2, P(R_k) \rangle - i\sigma_{2,x_0}^1 \right\}, \qquad (3.13)$$

and from (3.12) we find v_4. Since v_4 does not occur in the subsequent equations, we do not write it out.

Finally, we consider the system

$$-A_l v_5 = M(l^1) v_3 + P(v_2)$$
$$= \sum_{k=1,2} \left[\sigma_k^2 P(R_k) - iM(\sigma_k^2) R_k \right] - iM(\sigma_1^0) V_1(l^1)$$
$$+ \sigma_1^0 V_{11}(l^1) + M(l^1) \left[\sum_{k=1,2} \sigma_k^3 R_k + \sigma_1^1 V_1(l^1) \right]. \qquad (3.14)$$

If in (3.14) we substitute the function σ_2^2 determined in (3.13) and use (3.11), then the solvability condition for (3.14) takes the form

$$2l_{x_0}^1 \sigma_{1,x_0}^0 - \left(l_{x_0}^1 \right)^{-2} \mathcal{L}_{1,2} \mathcal{L}_{2,1} \sigma_{1,x_0}^0 + \Phi_5(l^1) \sigma_1^0$$
$$= -3 \left(l_{x_0}^1 \right)^{-2} \mathcal{L}_{1,2} \mathcal{L}_{2,1} \sigma_{1,x_0}^0 + \Phi_5(l^1) \sigma_1^0 = 0.$$

From this equation we find σ_1^0, and from the second solvability condition for (3.14) we obtain

$$\sigma_2^3 = -\left(l_{x_0}^1 \right)^{-1} \left\{ \sum_{k=1,2} \sigma_k^2 \langle L_2, P(R_k) \rangle - i\sigma_{2,x_0}^2 + \Phi_6(l^1, \sigma_1^0) \right\}.$$

After this we use a standard inductive argument.

Thus, we have proved that conditions (14), (15) and (17) are necessary for the Cauchy problem to be well-posed. This concludes the proof of Theorem 1.

Corollary 1 follows immediately from Theorem 1 and Lemma 1.2.

PROOF OF COROLLARY 2. Suppose that at some point $\hat{z}$ the matrix $A(\hat{z})$ is not similar to a diagonal matrix. Then $\kappa_{j_0} > 0$ for some j_0. If rank $\mathfrak{A}_{j_0}(\hat{z}) = d - 1$, then the rank of $\mathfrak{A}_{j_0}(z)$ is also equal to $d - 1$ in some neighborhood of $\hat{z}$. If rank $\mathfrak{A}_{j_0}(\hat{z}) = d - 2$, then there are two possibilities: either rank $\mathfrak{A}_{j_0}(\hat{z}) = d - 2$ in some neighborhood of $\hat{z}$, or rank $\mathfrak{A}_{j_0}(z_1) = d - 1$ at some point z_1. In the latter case we choose a neighborhood of z_1 in which rank $\mathfrak{A}_{j_0}(z) = d - 1$.

Consequently, we can always find a domain W' in which the rank of $\mathfrak{A}_{j_0}$ is constant. As we say in the proof of Theorem 1, if any of the conditions (13)–(17) is violated, then in every one of the cases (1°)–(5°) we can construct an asymptotic solution provided that the rank of the corresponding matrix $\mathfrak{A}_j$ is constant in a neighborhood of some point (x, ξ'). Consequently, there are smooth matrices $B_1(x)$ for which the Cauchy problem (4), (5) for the operator $(P + B_1(x))$ is not well-posed in G_T^+. This contradicts our assumption. Corollary 2 is proved.

Received 14/SEPT/73

BIBLIOGRAPHY

1. V. Ja. Ivriĭ and V. M. Petkov, *Necessary conditions for the Cauchy problem for nonstrictly hyperbolic equations to be well-posed*, Uspeh: Mat. Nauk **29** (1974), no. 5(179), 3–70; English transl. in Russian Math. Surveys **29** (1974).

2. Alberto P. Calderón, *Integrales singulares y sus aplicaciones a equaciones differenciales hiperbolicas*, Cursos y Sem. Mat., Fasc. 3, Univ. Buenos Aires, Buenos Aires, 1960.

3. Tadayoshi Kano, *Les systèmes strictement hyperboliques. Une condition nécessaire*, C. R. Acad. Sci. Paris Sér. A-B **274** (1972), A1116–A1119.

4. Koji Kasahara and Masaya Yamaguti, *Strongly hyperbolic systems of linear partial differential equations with constant coefficients*, Mem. Coll. Sci. Univ. Kyoto Ser. A Math. **33** (1960), 1–23.

5. Heinz-Otto Kreiss, *Über sachgemässe Cauchyprobleme*, Math. Scand. **13** (1963), 109–128.

6. Peter D. Lax, *Asymptotic solutions of oscillatory initial value problem*, Duke Math. J. **24** (1957), 627–646.

7. Sigeru Mizohata, *Note sur le traitement par les opérateurs d'intégrale singulière du problème de Cauchy*, J. Math. Soc. Japan **11** (1959), 234–240.

8. ______, *Some remarks on the Cauchy problem*, J. Math. Kyoto Univ. **1** (1961/62), 109–127.

9. V. M. Petkov, *The Cauchy problem for symmetrizable systems and for nonstrictly hyperbolic equations*, Uspehi Mat. Nauk **26** (1971), no. 6(162), 251–252. (Russian)

10. ______, *Necessary conditions for the Cauchy problem for hyperbolic systems with multiple characteristics to be well-posed*, Uspehi Mat. Nauk **27** (1972), no. 4 (166), 221–222.

11. ______, *The Cauchy problem for hyperbolic operators with multiple characteristics*, Dissertation, Moscow State Univ., Moscow, 1972. (Russian)

12. ______, *On the Cauchy problem for first-order hyperbolic systems with multiple characteristics*, Dokl. Akad. Nauk SSSR **209** (1973), 795–797; English transl. in Soviet Math. Dokl. **14** (1973).

13. ______, *The Cauchy problem for nonsymmetrizable hyerbolic systems*. Math. and Math. Education (Proc. Second Spring Conf. Bulgarian Math. Soc., Vidin, 1973), Izdat. B″lgar. Akad. Nauk.. Sofia, 1974, pp. 167–173. (Bulgarian)

14. I. Petrovsky [I. G. Petrovskiĭ], *Über das Cauchysche problem für Systeme von partiellen Differentialgleichungen*, Mat. Sb. **2(44)** (1937), 815–870.

15. Gilbert Strang, *Necessary and insufficient conditions for well-posed Cauchy problems*, J. Differential Equations **2** (1966), 107–114.

16. ______, *On strong hyperbolicity*, J. Math. Kyoto Univ. **6** (1967), 397–417.

17. S. Leif Svensson, *Necessary and sufficient conditions for the hyperbolicity of polynomials with hyperbolic principal part*, Ark. Mat. **8** (1969), 145–162.

18. Hermann Flaschka and Gilbert Strang, *The correctness of the Cauchy problem*, Advances in Math. **6** (1971), 347–379.

19. K. O. Friedrichs, *Pseudo-differential operators. An introduction*, Courant Inst. Math. Sci., New York Univ., New York, 1968.

20. Jacques Chazarain, *Une classe d'opérateurs à caractéristiques de multiplicité constante*, Équations aux Dérivées Partielles Linéaires (Colloq. Internat. C.N.R.S., Univ. Paris-Sud, Orsay, 1972), Astérisque no. 2/3, Soc. Math. France, Paris, 1973, pp. 135–142.

21. Masaya Tamaguti, *Sur l'inégalité d'énergie pour le système hyperbolique*, Proc. Japan Acad. **35** (1959), 37–41.

Translated by D. B. C. CONSTANDA

Amer. Math. Soc. Transl.
(2) Vol. **118**, 1982

On the Local Solvability
of a Class of Pseudodifferential Equations
with Double Characteristics*

P. R. POPIVANOV

§1. Introduction

The problem of the solvability of linear differential operators is one of the main problems of the theory of differential equations. Essential progress has been achieved in this direction in the last 25 years. In this connection, the creation and systematic use of the distribution theory and pseudodifferential operators have played a principal role. Thanks to the efforts of such mathematicians as Lewy, Hörmander, Nirenberg, Trèves, Ju. Egorov, and others, a definitive theory of the local solvability of pseudodifferential equations with simple characteristics was created. The solvability of pseudodifferential, even differential, operators with multiple characteristics has been studied so far only slightly. Our aim in this paper is to study a fairly special class of operators with characteristics of multiplicity identically two.

We consider an equation of the form $P(x, D)u = f(x)$, where $P(x, D)$ is a differential or pseudodifferential operator with a smooth symbol (defined in [2] and [3]). We assume that the principal part of the symbol of the operator $p^0(x, \xi)$ has the form

$$p^0_{2m}(x, \xi) = p^{0^2}_m(x, \xi),$$

where p^0_m is a symbol of principal type, i.e.

$$p^0_m(x, \xi) = 0, \qquad \xi \neq 0 \Rightarrow \mathrm{grad}_\xi \, p^0_m(x, \xi) \neq 0.$$

1980 *Mathematics Subject Classification*. Primary 35S05, 47G05; Secondary 35H05, 35S15.
*Translation of Trudy Sem. Petrovsk. 1 (1975), 237–278. MR **55** #841.

We recall that an operator $P(x, D)$ is said to be *locally solvable* in a neighborhood ω of a point x^0 if for an arbitrary function $f \in C_0^\infty(\omega)$ there exists a distribution $u \in \mathcal{E}'(\Omega)$ with compact support such that $P(x, D)u = f$ in ω.

We describe all the conditions of local solvability in the invariant language of the zero bicharacteristics of the symbol p_m^0 and in the language of the subprincipal symbol.

Let $\operatorname{Re} p_m^0 = a$ and $\operatorname{Im} p_m^0 = b$. Let (x^0, ξ^0) be a point from $T^*(\Omega) \setminus 0$ at which $p_m^0(x^0, \xi^0) = 0$. We speak of such points as *characteristic points*. By virtue of the condition of principal type, we have $\operatorname{grad}_\xi p_m^0(x^0, \xi^0) \neq 0$. There is no loss of generality if we assume that $\operatorname{grad}_\xi a(x^0, \xi^0) \neq 0$. Then a piece of the surface $a(x, \xi) = 0$ is fibered into the zero bicharacteristics of the function $a(x, \xi)$, i.e., the integral curves $x = x(t)$, $\xi = \xi(t)$ of the Hamiltonian system

$$\dot{x} = \operatorname{grad}_\xi a, \qquad \dot{\xi} = -\operatorname{grad}_x a.$$

Throughout the sequel we shall denote the zero bicharacteristic of the function a passing through the point (x^0, ξ^0) by $\mathcal{L}$.

Following Hörmander and Duistermaat, we define a subprincipal symbol p'_{2m-1} of an operator $P(x, D)$ as follows:

$$p'_{2m-1}(x, \xi) = p_{2m-1}(x, \xi) + \frac{i}{2} \sum_{j=1}^n \frac{\partial^2}{\partial \xi_j \partial x_j} p_{2m}^0(x, \xi),$$

where p_{2m}^0 is the leading symbol and $p_{2m-1}(x, \xi)$ is a term of order of homogeneity $2m - 1$ in the asymptotic expansion of the symbol $p(x, \xi)$. It is readily verified that $p'_{2m-1}(x, \xi)$ is invariant with respect to a change of variables at those points for which $\operatorname{grad}_\xi p_m^0(x, \xi) = 0$.

§2. Main results

We now state the main results of our paper.

I. *Necessary conditions.*

THEOREM 1. *Let p_m^0 be a real symbol of a pseudodifferential operator of principal type. Consider a pseudodifferential operator $P(x, D)$ with leading symbol $p_{2m}^0 = p_m^{0^2}$. Let $(x^0, \xi^0) \in T^*(\Omega) \setminus 0$ be a characteristic point for $p_m^0(x, \xi)$. The operator $P(x, D)$ is then locally unsolvable in a neighborhood of the point x^0 if the following conditions are satisfied:*

1) $\operatorname{Re} p'_{2m-1}(x^0, \xi^0) < 0$.

2) $\operatorname{Im} p'_{2m-1}(x, \xi)$ has a zero of finite odd order along the zero bicharacteristic $\mathcal{L}$ of the function p_m^0, passing through the point (x^0, ξ^0).

We remark that in the case when $P(x, D)$ is a differential operator the condition 1) can be replaced by $\operatorname{Re} p'_{2m-1}(x^0, \xi^0) \neq 0$.

The following theorem holds for a pseudodifferential operator with a complex-valued principal symbol.

THEOREM 2. *Let $P(x, D)$ be a pseudodifferential operator the principal part of whose symbol has the representation $p^0_{2m+r} = p^{0^2}_m q_r$. Here $p^0_m(x, \xi) = a(x, \xi) + ib(x, \xi)$ is a symbol of principal type and $q_r(x, \xi)$ is a symbol elliptic in a neighborhood of the point (x^0, ξ^0). Assume that the derivative along $\mathcal{L}$ of $b(x, \xi)$ at (x^0, ξ^0) is greater than zero and that $p^0_{2m+r}(x, \xi)$ and $p^0_{2m+r-1}(x, \xi)$ are analytically continuable with respect to ξ in some neighborhood of ξ^0. The operator $P(x, D)$ is then unsolvable in a neighborhood of x^0 if $p'_{2m+r-1}(x^0, \xi^0) \neq 0$.*

A proof of Theorem 2 in the case of a differential operator was given by V. Ja. Ivriĭ in [18].

II. *Sufficient conditions.*

THEOREM 3. *Let $p_m(x, \xi)$ be a symbol of principal type such that $\operatorname{Im} p_m(x, \xi)$ does not change sign from minus to plus in moving in a positive direction along any zero bicharacteristic of $\operatorname{Re} p_m(x, \xi)$. Then the operator $P(x, D)$ with leading symbol $p_m^2(x, \xi)$ is locally solvable if $p_{2m-1} \equiv 0 \pmod{p_m}$. In addition, for any real s*

$$\|u\|_s \leqslant C(\omega)\|Pu\|_{s-2m+2} + C_1\|u\|_{s-1}, \qquad \forall u \in C_0^\infty(\omega).$$

If $s \geqslant \max(-n/2, m - 1 - n/2)$, we can assume that $C_1 = 0$. Here the constant $C(\omega) \to 0$ as $\operatorname{diam} \omega \to 0$ $(\omega \ni x^0)$.

For a class of differential equations with characteristics of multiplicity identically two, we prove a theorem on improving the smoothness of a solution.

THEOREM 4. *Let p_m be the symbol of a differential operator of principal type for which $\operatorname{Im} p_m$ preserves its sign along the zero bicharacteristics of $\operatorname{Re} p_m$. We consider the differential operator $P(x, D)$ with the principal symbol p_m^2 and subprincipal symbol $p_{2m-1} \equiv 0 \pmod{p_m}$. Then for any function $f \in H^k$ there exists a neighborhood $\omega^k \ni x^0$, which depends on k but not on f, in which the equation $P(x, D)u = f$ has a solution $u \in H^{k+2m-2}$.*

For an operator $P(x, D)$ with real coefficients in its leading part we can obtain a more precise sufficient condition of local solvability.

THEOREM 5. *Let $p_m(x, \xi)$ be a real symbol of principal type. Consider a pseudodifferential operator $P(x, D)$ with leading symbol $p_m^2(x, \xi)$. The operator $P(x, D)$ is then locally solvable if at least one of the following conditions is satisfied.*

a) $\operatorname{Re} p_{2m-1}(x, \xi) \geqslant 0$ *at the characteristic points of $p_m(x, \xi)$. Then*

$$\|u\|_{m-1} \leqslant C(\omega)\|Pu\|_{1-m}, \qquad \forall u \in C_0^\infty(\omega);$$

and, in addition, $C(\omega) \to 0$ as $\operatorname{diam} \omega \to 0$.

b) $\operatorname{Re} p_{2m-1}(x, \xi) > 0$ *at the characteristic points of $p_m(x, \xi)$. Then*

$$\|u\|_{m-1/2} \leqslant C\|Pu\|_{1/2-m}, \qquad \forall u \in C_0^\infty(\omega).$$

c) $\operatorname{Im} p_{2m-1}(x, \xi) > 0$ *everywhere, and $m \geqslant 1$. The estimate is the same as in* b).

We note that in case a) we obtain an estimate with a 2-unit loss of smoothness in comparison with elliptic operators, while in cases b) and c) there is a 1-unit loss. Using Theorems 1 and 5, we can readily find precise necessary and sufficient conditions for the local solvability of some differential and pseudodifferential operators. Actually, it is easy to show that, by virtue of our results, operators with the symbols

$$p_2 = \xi_1^2 - \left(\alpha - i\beta x_1^{2k+1}\right)|\xi|, \qquad q_2 = \xi_1^2 - \left(\alpha - i\beta x_1^{2k+1}\right)\xi_2,$$

where α and β are real constants, $\beta \neq 0$, are solvable in a neighborhood of the origin if and only if $\alpha \geq 0$ for p_2, and $\alpha = 0$ for q_2.

The main results of this paper were published without proofs in the notes [22] and [23].

The author expresses his thanks to Ju. V. Egorov for his constant interest and encouragement.

§3. Some results from the general theory of pseudodifferential operators

Theorems 1 and 2 are proved following the classical scheme of Hörmander in [19]. Indeed, the assumption of local solvability of the operator $P(x, D)$ in the neighborhood of a point x^0 implies the following statement.

There exist a constant $C > 0$, an integer $R \geq 0$, and a function $\psi \in C_0^\infty(\omega)$, such that for all pairs of functions $v, f \in C_0^\infty(\omega)$ the estimate

$$\left|\int f(x)v(x)\,dx\right| \leq C \sup\left(\sum_{|\alpha|\leq R} |D^\alpha f(x)|\right) \sup\left(\sum_{|\beta|\leq R} |D_x^\beta(\psi(x){}^t Pv)|\right) \quad (1)$$

$$\left(D_{x_j} = (1/i)(\partial/\partial x_j)\right)$$

holds. Here ${}^t P$ is an operator formally adjoint to P and ω is some neighborhood of x^0.

Under the assumptions of Theorems 1 and 2, we prove that the inequality (1) is not satisfied for some pair $f(x)$, $v(x) \in C_0^\infty(\omega)$. Speaking more precisely, we construct the functions

$$f(x) = \sum_{j=0}^{N} \varphi_j \lambda^{-j} e^{-\lambda^2 g - \lambda h}, \qquad v(x) = \overline{f(x)},$$

where $\varphi_j \in C_0^\infty(\omega)$, $g(x), h(x) \in C^\infty(\omega)$, and $\lambda \geq 1$ is a parameter, and we show that for N sufficiently large and $\lambda \to \infty$ the estimate (1) is certainly violated. Consequently, we need to study the action of an arbitrary pseudodifferential operator on a function of the form $\varphi(x)e^{-\lambda^2 g - \lambda h}$, $\varphi \in C_0^\infty(\omega)$. Finally, we note, that in contrast to the proof of the necessary conditions for the local solvability of pseudodifferential operators with simple characteristics, in the case of Theorems 1 and 2 the two phase functions g and h are employed.

We consider now an operator $P(x, D)$ with the symbol $p_m(x, \xi) \in S_{1,0}^m$, where $S_{1,0}^m$ is the class of symbols introduced by Hörmander in [5]. In formulating the following lemma we assume that ω is a neighborhood of 0.

LEMMA 1. *Let* $\varphi \in C_0^\infty(\omega)$ *and* $g(x) = i(\xi^0, x) + ig_1'(x)$, *where* $g_1'(x) \in C^\infty(\omega)$, $g_1'(x)$ *is a real function, and* $g_1'(x) = O(|x|^2)$, $x \to 0$. *Let* $h \in C^\infty(\omega)$ *and* $\operatorname{Re} h(x_1, y) \geqslant c_0(x_1^{2k} + |y|^2)$ *in* ω, *where* $c_0 = \text{const} > 0$ *and* $k \geqslant 1$. *Then for any pair of numbers* M, M' *there is a number* $N' > 0$ *such that all derivatives of order not exceeding* M *of the function*

$$\lambda^{M'}\left\{ P(x, D)\left(\varphi e^{-\lambda^2 g - \lambda h}\right) - e^{-i\lambda^2(x,\xi^0)} \sum_{|\alpha| \leqslant N'} \frac{1}{\alpha!} p_m^{(\alpha)}\left(x, -\lambda^2\xi^0\right) D_x^\alpha\left(\varphi e^{-\lambda^2 g_1 - \lambda h}\right)\right\}$$

(2)

are bounded in the space of the variables x *for* $\lambda \geqslant 1$. *Here,* $g_1(x) = ig_1'(x)$, $x = (x_1, y)$, $y \in \mathbf{R}^{n-1}$.

We show, first of all, that in $\operatorname{supp} \varphi$ we have

$$\left| D^\gamma\left(e^{-\lambda^2 g_1 - \lambda h}\right)\right| \leqslant c_\gamma \lambda^{2|\gamma| - |\gamma|/2k} e^{-\lambda c_{\gamma_0}(x_1^{2k} + |y|^2)},$$

(3)

where c_γ and c_{γ_0} are positive constants and γ is an arbitrary multi-index.

This inequality is readily proved by induction.

Indeed, let $|\gamma| = 1$. Then

$$\left| D_{x_j}\left(e^{-\lambda^2 g_1 - \lambda h}\right)\right| = \left|\left(-\lambda^2 D_{x_j} g_1 - \lambda D_{x_j} h\right) e^{-\lambda^2 g_1 - \lambda h}\right|$$

$$\leqslant c\left[\lambda^2(|x_1| + |y|) + \lambda\right] e^{-\lambda c_0(x_1^{2k} + |y|^2)},$$

(4)

Let $A = (|x_1| + |y|)\exp(-\lambda c_0(x_1^{2k} + |y|^2)/2)$. Then

$$A^{2k} \leqslant \text{const}\left(x_1^{2k} + |y|^2\right) e^{-\lambda k c_0(x_1^{2k} + |y|^2)} \leqslant \text{const } \lambda^{-1} \Rightarrow A \leqslant \text{const } \lambda^{-1/2k}. \quad (5)$$

Combining (4) and (5), we find that

$$\left| D_{x_j}\left(e^{-\lambda^2 g_1 - \lambda h}\right)\right| \leqslant \text{const } \lambda^{2 - 1/2k} e^{-\lambda c_0(x_1^{2k} + |y|^2)/2}.$$

Let us assume now that (3) has been proved for all multi-indices of order γ, $|\gamma| \leqslant l$, $l > 1$, and let $|\gamma| = l$. Applying Leibniz's formula and using the inductive assumption, we have

$$\left| D_{x_j} D^\gamma\left(e^{-\lambda^2 g_1 - \lambda h}\right)\right| = \left| D^\gamma\left(-\lambda^2 D_{x_j} g_1 - \lambda D_{x_j} h\right) e^{-\lambda^2 g_1 - \lambda h}\right|$$

$$\leqslant \text{const}\left(\lambda^2 |x| + \lambda\right)\lambda^{2|\gamma| - |\gamma|/2k} e^{-\lambda c_\gamma(x_1^{2k} + |y|^2)}$$

$$+ \text{const } \lambda^2 \sum_{|\beta| \leqslant |\gamma| - 1} \lambda^{2|\beta| - |\beta|/2k} e^{-\lambda c_\gamma'(x_1^{2k} + |y|^2)}$$

$$\leqslant c_{\gamma+1} \lambda^{2(|\gamma| + 1) - (|\gamma| + 1)/2k} e^{-\lambda c_{\gamma-1}'(x_1^{2k} + |y|^2)},$$

Q.E.D.

Following this small digression, we return to the proof of Lemma 1.

We have

$$P(x, D)\left(\varphi e^{-i\lambda^2(x,\xi^0)-\lambda^2 g_1(x)-\lambda h(x)}\right)$$

$$= \int e^{i(x,\xi)} p_m(x, \xi)\left(\widetilde{\varphi e^{-i\lambda^2(x,\xi^0)-\lambda^2 g_1-\lambda h}}\right)(\xi)\, d\xi$$

$$= \int e^{i(x,\xi)} p_m(x, \xi)\left(\widetilde{\varphi e^{-\lambda^2 g_1-\lambda h}}\right)(\xi + \lambda^2\xi^0)\, d\xi$$

$$= e^{-i\lambda^2(x,\xi^0)} \int e^{i(x,\xi)} p_m(x, \xi - \lambda^2\xi^0)\left(\widetilde{\varphi e^{-\lambda^2 g_1-\lambda h}}\right)(\xi)\, d\xi$$

$$= e^{-i\lambda^2(x,\xi^0)} \sum_{|\alpha|\leqslant N'} \frac{1}{\alpha!} p_m^{(\alpha)}(x, -\lambda^2\xi^0) D_x^\alpha\left(\varphi e^{-\lambda^2 g_1-\lambda h}\right)$$

$$- e^{-i\lambda^2(x,\xi^0)} \int e^{i(x,\xi)} R_{N'}(x, \xi, \lambda)\left(\widetilde{\varphi e^{-\lambda^2 g_1-\lambda h}}\right)(\xi)\, d\xi$$

$$\left(\tilde{u}(\xi) = \int e^{-i(x,\xi)} u(x)\, dx\right).$$

Here we denote by $R_{N'}(x, \xi, \lambda)$ the remainder in the Taylor expansion of $p_m(x, \xi - \lambda^2\xi^0)$ in a neighborhood of the point $-\lambda^2\xi^0$:

$$R_{N'}(x, \xi, \lambda) = p_m(x, \xi - \lambda^2\xi^0) - \sum_{|\alpha|\leqslant N'} \frac{1}{\alpha!} p_m^{(\alpha)}(x, -\lambda^2\xi^0)\xi^\alpha$$

$$= \sum_{|\beta|=N'+1} c_\beta \xi^\beta \int_0^1 p_m^{(\beta)}(x, -\lambda^2\xi^0 + (1 - s)\xi) s^{N'}\, ds.$$

Thus, the proof of the lemma reduces to estimating $R_{N'}(x, \xi, \lambda)$.

Since

$$\left| \int e^{i(x,\xi)} R_{N'}(x, \xi, \lambda)\left(\widetilde{\varphi e^{-\lambda^2 g_1-\lambda h}}\right)(\xi)\, d\xi \right|$$

$$\leqslant C(N') \iint_0^1 (1 + |\lambda^2\xi^0 + (s - 1)\xi|)^{m-N'-1} |\xi|^{N'+1} |\left(\widetilde{\varphi e^{-\lambda^2 g_1-\lambda h}}\right)(\xi)|\, ds\, d\xi,$$

$$(6)$$

it is sufficient for us to estimate the integral on the right side of (6). To do this, we break it up into an integral over $|\xi| < \lambda^2|\xi^0|/2$ and an integral over $|\xi| \geqslant \lambda^2|\xi^0|/2$.

In the first case, using (3), we obtain

$$\int_0^1 \int_{|\xi|<\lambda^2|\xi^0|/2} (1 + |\lambda^2\xi^0 + (s - 1)\xi|)^{m-N'-1} |\xi|^{N'+1} |\left(\widetilde{\varphi e^{-\lambda^2 g_1-\lambda h}}\right)(\xi)|\, ds\, d\xi$$

$$\leqslant \mathrm{const}(1 + \lambda^2)^{m-N'-1+2} \sup | D^\alpha\left(\varphi e^{-\lambda^2 g_1-\lambda h}\right) |$$

$$\leqslant \mathrm{const}\, \lambda^{2m+2n-(N'+1)/2k} \qquad (|\alpha| = N' + 1).$$

From (3) it also follows that for any multi-index γ

$$|\xi|^{|\gamma|} \xi|^{(8k-1)|\gamma|} |\left(\widetilde{\varphi e^{-\lambda^2 g_1-\lambda h}}\right)(\xi)| \leqslant \mathrm{const}\, \lambda^{16k|\gamma|-4|\gamma|},$$

whence we obtain, for an arbitrary ξ, $|\xi| \geq \lambda^2 |\xi^0|/2$, the estimate

$$|\xi|^{|\gamma|} \left| \widehat{\left(\varphi e^{-\lambda^2 g_1 - \lambda h} \right)}(\xi) \right| \leq \text{const } \lambda^{-2|\gamma|}.$$

Now choosing $|\gamma| = N' + 2 + n$, we find that the integral over the set $|\xi| \geq \lambda^2 |\xi^0|/2$ does not exceed

$$\text{const } \lambda^{-2(N'+1+n)} \int_{|\xi| \geq |\xi^0|/2} |\xi|^{-n-1} \, d\xi.$$

From this we conclude that the right side of (6) may be estimated in terms of $\lambda^{2m+2n-(N'+1)/2k}$. Consequently,

$$\left| \int e^{i(x,\xi)} R_{N'}(x, \xi, \lambda) \widehat{\left(\varphi e^{-\lambda^2 \varepsilon_1 - \lambda h} \right)}(\xi) \, d\xi \right| \leq \text{const } \lambda^{2m+2n-(N'+1)/2k}.$$

Similar considerations with D_x^p, $|p| \leq M$, give us, for the derivatives of the function (2) of order not exceeding M, the majorant $\lambda^{2M+M'+2m+2n-(N'+1)/2k}$. These estimates are independent of x and λ if $N' - 1 \geq 2k(2M + 2n + 2m + M')$. Thus we have proved Lemma 1.

For the proof of Theorem 2 we need the following lemma.

LEMMA 1′. *Let*

$$\varphi(x) \in C_0^\infty(\omega), \qquad g(x) = g_1(x) + i(\xi^0, x),$$

where

$$g_1(x) \in C^\infty(\omega), \quad g = O(|x|^2), \quad x \to 0 \qquad \text{and}$$
$$c_0 |x|^2 \leq \text{Re } g_1 \leq c_1 |x|^2, \quad c_0, c_1 > 0.$$

Let

$$h(x) \in C^\infty(\omega) \quad \text{and} \quad h(x) = O(|x|), \qquad x \to 0.$$

Then for any pair of positive integers M, M' there exists a positive integer N such that all the derivatives of order not exceeding M of the function

$$\lambda M' \left\{ P(x, D) \left(\varphi e^{-\lambda^2 g_1 - \lambda h} \right) - e^{-i\lambda^2(x,\xi^0)} \sum_{|\alpha| \leq N'} \frac{1}{\alpha!} p_m^{(\alpha)}(x, -\lambda^2 \xi^0) D_x^\alpha \left(\varphi e^{-\lambda^2 g_1 - \lambda h} \right) \right\}$$

are bounded in the space of x for all $x \in \omega$ and $\lambda \geq 1$.

The proof of this lemma follows word for word the proof of Lemma 1. The only difference is that the function g_1 is complex-valued and the estimate for $|D_x^\alpha(e^{-\lambda^2 g_1 - \lambda h})|$ on supp φ is obtained from the following reasoning. It is not hard to see that for any $\varepsilon > 0$

$$|h|/\lambda \leq \varepsilon |x|^2 + 1/c\lambda^2\varepsilon \qquad (c \text{ does not depend on } \lambda \text{ or } \varepsilon).$$

Then, obviously,

$$\text{Re } g_1 + \text{Re } h/\lambda \leq \text{const}(\varepsilon)\left(|x|^2 + 1/\lambda^2\right),$$
$$\text{Re } g_1 + \text{Re } h/\lambda \geq c_0 |x|^2 - \varepsilon |x|^2 - 1/c\lambda^2\varepsilon.$$

Choosing $0 < \varepsilon < c_0/2$, we find that

$$\operatorname{Re} g_1 + \operatorname{Re} h/\lambda \geqslant (c_0/2)\big(|x|^2 - c'(\varepsilon)/\lambda^2\big)$$

and, consequently,

$$c_3 e^{-c_4 \lambda^2 |x|^2} \leqslant |e^{-\lambda^2 g_1 - \lambda h}| \leqslant c_5 e^{-c_6 \lambda^2 |x|^2},$$

where $c_3, \ldots, c_6$ are positive constants and $x \in \omega$. Thus,

$$\big|D_x^\alpha\big(e^{-\lambda^2 g_1 - \lambda h}\big)\big| \leqslant c \sum_{|\gamma| + |\beta| = |\alpha|} |D^\gamma(e^{-\lambda^2 g_1})| \, |D^\beta(e^{-\lambda h})|$$

$$\leqslant c' \sum_{|\gamma| + |\beta| = |\alpha|} \lambda^{|\gamma| + |\beta|} \leqslant c' \lambda^{|\alpha|}$$

(since $D_{x_j} g_1 = O(|x|) \Rightarrow |D^\gamma e^{-\lambda^2 g_1}| \leqslant c_\gamma \lambda^{|\gamma|} e^{-\lambda^2 c_\gamma' |x|^2}$, c_γ, $c_\gamma' > 0$). We complete the proof as we did in the previous lemma.

§4. Proof of Theorem 1 for the case in which
Im $p'_{2m-1}(x, \xi)$ has a zero of the first order
along the zero bicharacteristic of the function $p_m^0(x, \xi)$

As we have already said, we are going to prove that inequality (1) is not satisfied for some function φ of the form

$$\varphi = \sum_{j=0}^{L} \varphi_j \lambda^{-j} e^{-\lambda^2 g - \lambda h},$$

where $\varphi_j \in C_0^\infty(\Omega)$ and $\lambda \geqslant 1$ is a parameter. Thus, we need to determine in succession the functions g, h, and φ_j. We first construct g. Since $p_m(x, \xi)$ is a real symbol of principal type, it follows from [19] that there exists a unique real function $g_1'(x) \in C^\infty(\omega)$ possessing the following properties:

$$p_m\big(x, \xi^0 + \operatorname{grad} g_1'\big) = 0,$$

$$\operatorname{grad} g_1'(0) = 0, \qquad g_1'\big|_{x_1 = 0} = |y^2|, \qquad x = (x_1, y)$$

(i.e. $g_1'(x) = O(|x|^2)$, $x \to 0$).

Introducing the function $g''(x) = g_1'(x) + (x, \xi^0)$, we find that $p_m(x, \operatorname{grad} g'') = 0$. Finally, for technical convenience, we let

$$g(x) = ig''(x) \Rightarrow p_m(x, Dg) = 0, \qquad Dg(0) = \xi^0; \qquad D_{x_j} = i^{-1} \partial/\partial x_j.$$

We now examine the action of the operator ${}^tP(x, D)$ on the function $\varphi(x) e^{-\lambda^2 g - \lambda h}$. We recall (see [3]) that the symbol of the adjoint operator ${}^tP$ has the form

$$\,^t p(x, \xi) \sim p_m^2(x, -\xi) + i \sum_{j=1}^{n} \frac{\partial^2}{\partial x_j \partial \xi_j}(p_m^2)(x, -\xi)$$

$$+ p_{2m-1}(x, -\xi) + p_{2m-2}(x, -\xi),$$

wherein p_{2m-2} denotes the sum of the terms of order of homogeneity not exceeding $2m-2$.

We assume that $h(x)$ satisfies all the requirements of Lemma 1. We then immediately obtain the following asymptotic expansions:

$$p_m^2(x, -D)\left(\varphi(x)e^{-\lambda^2 g(x) - \lambda h(x)}\right)$$

$$\sim e^{-i\lambda^2(x,\xi^0)} \sum_\alpha \frac{(-1)^{|\alpha|}}{\alpha!}(p_m^2)^{(\alpha)}(x,\lambda^2\xi^0)D_x^\alpha\left(\varphi(x)e^{-\lambda^2 g_1(x) - \lambda h(x)}\right),$$

$$\sum_{j=1}^n \frac{\partial^2(p_m^2)}{\partial x_j \partial \xi_j}(x,-D)\left(\varphi(x)e^{-\lambda^2 g(x) - \lambda h(x)}\right)$$

$$\sim e^{-i\lambda^2(x,\xi^0)} \sum_{j=1}^n \sum_\alpha \frac{(-1)^{|\alpha|}}{\alpha!}\left(\frac{\partial^2(p_m^2)}{\partial x_j \partial \xi_j}\right)^{(\alpha)}(x,\lambda^2\xi^0)D_x^\alpha\left(\varphi(x)e^{-\lambda^2 g_1(x) - \lambda h(x)}\right),$$

$$p_{2m-1}(x,-D)\left(\varphi(x)e^{-\lambda^2 g(x) - \lambda h(x)}\right)$$

$$\sim e^{-i\lambda^2(x,\xi^0)} \sum_\alpha \frac{(-1)^{|\alpha|}}{\alpha!}p_{2m-1}^{(\alpha)}(x,\lambda^2\xi^0)D_x^\alpha\left(\varphi(x)e^{-\lambda^2 g_1(x) - \lambda h(x)}\right).$$

A similar expansion holds for p_{2m-2}, since $p_{2m-2} \in S_{1,0}^{2m-2}$.

The precise meaning of these expansions is the following. For any positive integers M and M' we can select a positive integer N such that in a small neighborhood of the origin all the derivatives of order not exceeding M of the function

$$\lambda^{M'}\left\{{}^t P\left(\varphi e^{-\lambda^2 g - \lambda h}\right) - e^{-\lambda^2 q - \lambda h}\left(S_N^1 + S_N^2 + S_N^3\right)\right\}$$

are bounded independently of $\lambda \geq 1$. Here

$$S_N^1 = \sum_{|\alpha| \leq N} \frac{(-1)^{|\alpha|}}{\alpha!}(p_m^2)^{(\alpha)}(x,\lambda^2\xi^0)D_x^\alpha\left(\varphi(x)e^{-\lambda^2 g(x) - \lambda h(x)}\right)e^{\lambda^2 g_1(x) + \lambda h(x)},$$

$$S_N^2 = i\sum_{j=1}^n \sum_{|\alpha| \leq N} \frac{(-1)^{|\alpha|}}{\alpha!}\left(\frac{\partial^2(p_m^2)}{\partial x_j \partial \xi_j}\right)^{(\alpha)}(x,\lambda^2\xi^0)D_x^\alpha\left(\varphi e^{-\lambda^2 g_1 \lambda h}\right)e^{\lambda^2 g(x) + \lambda h(x)},$$

$$S_N^3 = \left\{\sum_{|\alpha| \leq N} \frac{(-1)^{|\alpha|}}{\alpha!}p_{2m-1}^{(\alpha)}(x,\lambda^2\xi^0)D_x^\alpha\left(\varphi e^{-\lambda^2 g_1 - \lambda h}\right)\right.$$

$$\left. + \sum_{|\alpha| \leq N} \frac{(-1)^{|\alpha|}}{\alpha!}p_{2m-2}^{(\alpha)}(x,\lambda^2\xi^0)D_x^\alpha\left(\varphi e^{-\lambda^2 g_1 - \lambda h}\right)\right\}e^{\lambda^2 g + \lambda h}.$$

The terms in S_N^1, S_N^2, S_N^3 are polynomials in powers of $1/\lambda$. It is easy to see that the term containing λ to the power $4m$ arises only from S_N^1 and that it has the form

$$\varphi \sum_{|\alpha| \leqslant N} \frac{(-1)^{|\alpha|}}{\alpha!} \left(p_m^2 \right)^{(\alpha)}(x, \lambda^2 \xi^0)(-\lambda^2)^{|\alpha|} (D_x g_1)^\alpha$$

$$= \lambda^{4m} \varphi \sum_{|\alpha| \leqslant N} \frac{1}{\alpha!} \left(p_m^2 \right)^{(\alpha)}(x, \xi^0)(\operatorname{grad} g_1')^\alpha$$

$$= \lambda^{4m} \varphi \left[\left(p_m^2 \right)(x, \operatorname{grad} g'') + O(|x|^N) \right] = \lambda^{4m} \varphi O(|x|^N), \qquad x \to 0.$$

We come now to the determination of the coefficient of λ^{4m-1}. It arises from the following term:

$$\sum_{|\alpha| \leqslant N-1} \sum_{j=1}^n \frac{(-1)^{|\alpha|}}{\alpha!} \frac{\partial}{\partial \xi_j} \left(p_m^2 \right)^{(\alpha)}(x, \lambda^2 \xi^0) D_x^\alpha \left[\left(-\lambda^2 D_{x_j} g_1 - \lambda D_{x_j} h \right) \varphi e^{-\lambda^2 g_1 - \lambda h} \right].$$

Consequently, the coefficient in question is equal to

$$\frac{1}{\lambda^{4m-1}} \varphi(x) \sum_{|\alpha| \leqslant N-1} \sum_{j=1}^n \frac{(-1)^{|\alpha|}}{\alpha!} \frac{\partial}{\partial \xi_j} \left(p_m^2 \right)^{(\alpha)}(x, \lambda^2 \xi^0) \lambda D_{x_j} h (-\lambda^2)^{|\alpha|} (D_x g_1)^\alpha$$

$$= \varphi(x) \left[\sum_{j=1}^n \left(\frac{\partial}{\partial \xi_j} p_m^2 \right)(x, \operatorname{grad} g'') D_{x_j} h(x) + O(|x|^N) \right] = \varphi(x) O(|x|^N),$$

$$x \to 0. \quad (7)$$

The coefficient of λ^{4m-2} is obtained from the leading symbol and from the lowest symbols of order of homogeneity $2m - 1$. In the first place, the following term arises from the principal symbol:

$$\varphi(x) \sum_{|\alpha| \leqslant N-2} \sum_{j,k=1}^n \frac{1}{2} \frac{(-1)^{|\alpha|}}{\alpha!}$$

$$\times \left(\frac{\partial^2 \left(p_m^2 \right)}{\partial \xi_k \partial \xi_j} \right)^{(\alpha)}(x, \xi^0) \lambda^{4m-2|\alpha|-4} D_x^\alpha \left(-\lambda^2 D_{x_j} D_{x_k} g_1(x) \cdot e^{-\lambda^2 g_1(x) - \lambda h(x)} \right).$$

Gathering together all the terms in this summand containing λ^{4m-2}, we have

$$-\frac{\varphi(x)}{2} \lambda^{4m-2} \sum_{j,k=1}^n \sum_{|\alpha| \leqslant N-2} \frac{1}{\alpha!} \left(\frac{\partial^2 \left(p_m^2 \right)}{\partial \xi_k \partial \xi_j} \right)^{(\alpha)}(x, \xi^0)(\operatorname{grad} g_1')^\alpha D_{x_j} D_{x_k} g_1$$

$$= \frac{i}{2} \varphi(x) \lambda^{4m-2} \left[\sum_{j,k=1}^n \frac{\partial^2 \left(p_m^2 \right)}{\partial \xi_k \partial \xi_j}(x, \operatorname{grad} g'') \frac{\partial^2 g''}{\partial x_j \partial x_k} + O(|x|^{N-1}) \right]. \quad (8)$$

Indeed,

$$D_{x_j} D_{x_k} g_1 = i D_{x_j} D_{x_k} g_1' = i D_{x_j} D_{x_k} g'' = -i \partial^2 g'' / \partial x_j \partial x_k.$$

The expression (8) is simplified substantially with the application of an elementary identity:

$$\sum_{j,k=1}^{n} \frac{\partial^2 p}{\partial \xi_j \partial \xi_k}(x, \operatorname{grad} \varphi) \frac{\partial^2 \varphi}{\partial x_j \partial x_k}$$

$$= \sum_{j=1}^{n} \frac{\partial}{\partial x_j}\left(\frac{\partial p}{\partial \xi_j}(x, \operatorname{grad} \varphi)\right) - \sum_{j=1}^{n} \frac{\partial^2 p}{\partial x_j \partial \xi_j}(x, \operatorname{grad} \varphi). \tag{9}$$

Consequently, (8) is transformed into

$$\frac{i}{2}\varphi(x)\lambda^{4m-2}\left[-\sum_{j=1}^{n} \frac{\partial^2(p_m^2)}{\partial x_j \partial \xi_j}(x, \operatorname{grad} g'') + O(|x|^{N-1})\right]. \tag{8'}$$

We note that in the coefficient of λ^{4m-2} there arise two additional terms from the principal symbol $p_m^2(x, -\xi)$. We write them out:

$$-\sum_{j=1}^{n} \sum_{|\alpha| \leqslant N-1} \frac{(-1)^{|\alpha|}}{\alpha!}\left(\frac{\partial(p_m^2)}{\partial \xi_j}\right)^{(\alpha)}(x, \xi^0)\lambda^{4m-2|\alpha|-2}D_{x_j}\varphi(-\lambda^2)^{|\alpha|}(\operatorname{grad} g_1)^{\alpha}$$

$$= -\lambda^{4m-2}\left[\sum_{j=1}^{n} \frac{\partial(p_m^2)}{\partial \xi_j}(x, \operatorname{grad} g'')D_{x_j}\varphi(x) + O(|x|^{N-1})\right]$$

$$= -\lambda^{4m-2}O(|x|^{N}), \qquad x \to 0; \tag{10}$$

$$\varphi(x)\sum_{j,k=1}^{n} \sum_{|\alpha| \leqslant N-2} \frac{1}{2}\frac{(-1)^{|\alpha|}}{\alpha!}\left(\frac{\partial^2(p_m^2)}{\partial \xi_k \partial \xi_j}\right)^{(\alpha)}(x, \xi^0)\lambda^{4m-2|\alpha|-4}(-\lambda^2)^{|\alpha|}$$

$$\times (\operatorname{grad} g_1')^{\alpha}\lambda^2 D_{x_j}h(x)D_{x_k}h(x)$$

$$= \varphi(x)\lambda^{4m-2}\left[\frac{1}{2}\sum_{j,k=1}^{n} \frac{\partial^2(p_m^2)}{\partial \xi_k \partial \xi_j}(x, \operatorname{grad} g'')D_{x_j}hD_{x_k}h + O(|x|^{N-1})\right]$$

$$= -\varphi(x)\lambda^{4m-2}\left[\left(\sum_{j=1}^{n} \frac{\partial p_m}{\partial \xi_j}(x, \operatorname{grad} g'')\frac{\partial h}{\partial x_j}\right)^2 + O(|x|^{N-1})\right]. \tag{11}$$

It is easy to see that the contribution of the lower terms of the order of homogeneity $2m - 1$ will have the form

$$\lambda^{4m-2}\varphi(x)\left[p_{2m-1}(x, \operatorname{grad} g'') - i\sum_{j=1}^{n} \frac{\partial^2(p_m^2)}{\partial x_j \partial \xi_j}(x, \operatorname{grad} g'') + O(|x|^{N-1})\right].$$

$$\tag{12}$$

Thus we find, from (8′), (10), (11), and (12), that the coefficient of λ^{4m-2} is

$$
\varphi(x)\left[p_{2m-1}(x, \operatorname{grad} g'') + \frac{i}{2} \sum_{j=1}^{n} \frac{\partial^2 (p_m^2)}{\partial x_j \partial \xi_j}(x, \operatorname{grad} g'') \right.
$$

$$
\left. - \left(\sum_{j=1}^{n} \frac{\partial p_m}{\partial \xi_j}(x, \operatorname{grad} g'') \frac{\partial h}{\partial x_j} \right)^2 + O(|x|^{N-1}) \right] + O(|x|^{N-1}), \qquad x \to 0.
$$

We shall show below that in some neighborhood of the origin there exists an approximate solution $h \in C^\infty$ of the Cauchy problem for the nonlinear equation with infinitely smooth coefficients:

$$
\left(\sum_{j=1}^{n} \frac{\partial p_m}{\partial \xi_j}(x, \operatorname{grad} g'') \frac{\partial h}{\partial x_j} \right)^2 = p'_{2m-1}(x, \operatorname{grad} g''),
$$

$$
h\big|_{x_1=0} = \sum_{j,k=1}^{n} \alpha_{jk} x_j x_k \big|_{x_1=0}, \tag{12′}
$$

$$
\left(p'_{2m-1}(x, \xi) = p_{2m-1}(x, \xi) + \frac{i}{2} \sum_{j=1}^{n} \frac{\partial^2 (p_m^2)}{\partial x_j \partial \xi_j} \right),
$$

for which $\operatorname{Re} h \geqslant c_0 |x|^2$, $c_0 > 0$. Finally, we turn to λ^{4m-3}. It is easy to see that the term containing $\partial \varphi / \partial x_j$ in the coefficient of λ^{4m-3}, $j = 1,\ldots,n$, is obtained only from $p_m^2(x, -\xi)$. This follows from the fact that when φ is differentiated the power of λ is lowered by 2 units, while the maximum possible power of λ arising from the lower terms is equal to λ^{4m-2}. Consequently, to determine the coefficient of λ^{4m-3} containing $\partial \varphi / \partial x_j$, $j = 1,\ldots,n$, it is sufficient to find the quadratic form in $\partial \varphi / \partial x_j$ and $\partial h / \varphi x_k$.

Thus, we obtain

$$
\frac{1}{2} \sum_{j,k=1}^{n} \sum_{|\alpha| \leqslant N-2} \frac{(-1)^{|\alpha|}}{\alpha!} \left(\frac{\partial^2 (p_m^2)}{\partial \xi_k \partial \xi_j} \right)^{(\alpha)} (x, \xi^0) \lambda^{4m-2|\alpha|-4} D_{x_j} \varphi(x)(-\lambda)
$$

$$
\times D_{x_k} h(x)(\operatorname{grad} g_1')^\alpha (-\lambda^2)^{|\alpha|}
$$

$$
= \frac{1}{2} \lambda^{4m-3} \left[\sum_{j,k=1}^{n} \frac{\partial^2 (p_m^2)}{\partial \xi_j \partial \xi_k}(x, \operatorname{grad} g'') \frac{\partial \varphi}{\partial x_j} \frac{\partial h}{\partial x_k} + O(|x|^{N-1}) \right], \qquad x \to 0.
$$

$$
\tag{13}
$$

All the remaining terms containing λ^{4m-3} have the form $\lambda^{4m-3} B\varphi$, where B is an infinitely smooth function depending on $h(x)$. Since an explicit expression for B is of no interest here, we shall not write one down.

We note that the first term in (13) is transformed as follows:

$$\frac{1}{2}\sum_{j,k=1}^{n}\frac{\partial^2\left(p_m^2\right)}{\partial\xi_j\partial\xi_k}(x,\operatorname{grad} g'')\frac{\partial\varphi}{\partial x_j}\frac{\partial h}{\partial x_k}$$

$$=\sum_{j,k=1}^{n}\frac{\partial p_m}{\partial\xi_j}(x,\operatorname{grad} g'')\frac{\partial p_m}{\partial\xi_k}(x,\operatorname{grad} g'')\frac{\partial\varphi}{\partial x_j}\frac{\partial h}{\partial x_k}$$

$$=\left(\sum_{j=1}^{n}\frac{\partial p_m}{\partial\xi_j}(x,\operatorname{grad} g'')\frac{\partial\varphi}{\partial x_j}\right)\left(\sum_{k=1}^{n}\frac{\partial p_m}{\partial\xi_k}(x,\operatorname{grad} g'')\frac{\partial h}{\partial x_k}\right). \tag{14}$$

Since $p_m(x,\xi)$ is a symbol of principal type, we lose no generality in assuming that $\partial p_m(0,\xi^0)/\partial\xi_1\neq 0$.

If we define $h(x)$ as an approximate solution of (12$'$), then, keeping in mind that $p'_{2m-1}(0,\xi^0)\neq 0$ and that $\operatorname{grad} g''(0)=\xi^0$, we find that in some neighborhood of the origin in x-space

$$\sum_{k=1}^{n}\frac{\partial p_m}{\partial\xi_k}(x,\operatorname{grad} g'')\frac{\partial h}{\partial x_k}\neq 0.$$

Consequently, the coefficient of $\partial\varphi/\partial x_1$ in (14) will be different from zero if $|x|$ is sufficiently small.

We come now to an approximate solution of (12$'$). For this we use Lemma 6.1.4 of [19]. First of all, we expand the coefficients on the right and left sides of (12$'$) in a Taylor series up to some sufficiently large order M. Thus we can assume that they are analytic since segments of a Taylor series satisfy the same assumptions as do the coefficients. Allowing a certain liberty, we retain for the Taylor series segments the very same notation which we employed for the functions. This should not lead to any misunderstanding. After this, we consider the equation

$$\sum_{j=1}^{n}\frac{\partial p_m}{\partial\xi_j}(x,\operatorname{grad} g'')\frac{\partial h}{\partial x_j}=\sqrt{p'_{2m-1}(x,\operatorname{grad} g'')}\,; \tag{15}$$

since $p'_{2m-1}(0,\xi^0)\neq 0$, there is a branch of the square root. We select this branch in the complex plane with a cut for $\operatorname{Re} z>0$ in such a way that

$$\sqrt{-1}=i,\quad\text{if } A=\sum_{k=1}^{n}\frac{\partial\lim p'_{2m-1}(0,\xi^0)}{\partial x_k}\frac{\partial p_m(0,\xi^0)}{\partial\xi_k}$$

$$-\frac{\partial p_m(0,\xi^0)}{\partial x_k}\frac{\partial\operatorname{Im} p'_{2m-1}(0,\xi^0)}{\partial\xi_k}>0.$$

$$\sqrt{-1}=-i,\quad\text{if }\sum_{k=1}^{n}\left[\frac{\partial\operatorname{Im} p'_{2m-1}(0,\xi^0)}{\partial x_k}\cdot\frac{\partial p_m(0,\xi^0)}{\partial\xi_k}\right.$$

$$\left.-\frac{\partial p_m(0,\xi^0)}{\partial x_k}\cdot\frac{\partial\operatorname{Im} p'_{2m-1}(0,\xi^0)}{\partial\xi_k}\right]<0$$

$$\left(\text{since } A=\frac{\partial\operatorname{Im} p'_{2m-1}}{\partial\ell}(0,\xi^0)\neq 0\right). \tag{16}$$

We note that if $m \geqslant 1$ the conditions (16) are satisfied simultaneously for the coefficients of (15) and for the segments of their Taylor series.

From the Cauchy-Kovalevskaja theorem it follows that the Cauchy problem for equation (15) has a unique analytic solution with the initial condition

$$h\big|_{x_1=0} = \sum_{j,k=1}^{n} \alpha_{jk} x_j x_k \big|_{x_1=0}, \tag{17}$$

where the symmetric matrix (α_{jk}) has a positive real part. It is obvious that

$$h(0) = \partial h(0)/\partial x_j = 0, \quad \text{if } j = 2,\ldots,n.$$

Our problem consists in constructing the matrix (α_{jk}) so that $\operatorname{Re} h$ will obey the estimate

$$c_0 |x|^2 \leqslant \operatorname{Re} h \leqslant c_1 |x|^2. \tag{17'}$$

To do this, we determine the linear and quadratic terms in the Taylor series expansion of h in a neighborhood of the origin. We equate the coefficients of x to the zero power and of x_j in the expansions of the functions appearing on the left and right sides of (15).

Thus, equating the free terms in (15), we obtain

$$\frac{\partial p_m(0, \xi^0)}{\partial \xi_1} \frac{\partial h(0)}{\partial x_1} = \sqrt{p'_{2m-1}(0, \xi^0)} \Rightarrow \frac{\partial h(0)}{\partial x_1} = \frac{\sqrt{p'_{2m-1}(0, \xi^0)}}{(\partial p_m/\partial \xi_1)(0, \xi^0)}, \tag{18}$$

i.e. the number $\partial h(0)/\partial x_1$ is pure imaginary $(\operatorname{Re} p'_{2m-1}(0, \xi^0) < 0,$ $\operatorname{Im} p'_{2m-1}(0, \xi^0) = 0)$. Next, we equate the coefficients of x_k, $k = 1,\ldots,n$. To do this, we differentiate both sides of (15) with respect to x_k and put $x = 0$. Then from (17) and (18) we obtain

$$\frac{\partial^2 p_m(0, \xi^0)}{\partial \xi_1 \partial x_k} \cdot \frac{\sqrt{p'_{2m-1}(0, \xi^0)}}{\partial p_m(0, \xi^0)/\partial \xi_1} + \sum_{l=1}^{n} \frac{\partial^2 p_m(0, \xi^0)}{\partial \xi_1 \partial \xi_l}$$

$$\times \frac{\partial^2 g''(0)}{\partial x_k \partial x_l} \cdot \frac{\sqrt{p'_{2m-1}(0, \xi^0)}}{\partial p_m(0, \xi^0)/\partial \xi_1} + \sum_{j=1}^{n} \frac{\partial p_m(0, \xi^0)}{\partial \xi_j} \frac{\partial^2 h(0)}{\partial x_j \partial x_k}$$

$$= \frac{1}{2} \frac{\partial p'_{2m-1}(0, \xi^0)/\partial x_k + \sum_{l=1}^{n}(\partial p'_{2m-1}(0, \xi^0)/\partial \xi_l)(\partial^2 g''(0)/\partial x_k \partial x_l)}{\sqrt{p'_{2m-1}(0, \xi^0)}}$$

$$(k = 1, 2,\ldots,n). \tag{19}$$

We now construct the symmetric matrix (α_{jk}) with a positive real part so that h will satisfy (17').

Thus, we already know that (19) is satisfied for the symmetric matrix $\partial^2 h(0)/\partial x_j \partial x_k$. We now consider the following algebraic system:

$$\frac{\partial^2 p_m(0, \xi^0)}{\partial \xi_1 \partial x_k} \cdot \frac{\sqrt{p'_{2m-1}(0, \xi^0)}}{\partial p_m(0, \xi^0)/\partial \xi_1} + \sum_{l=1}^{n} \frac{\partial^2 p_m(0, \xi^0)}{\partial \xi_1 \partial \xi_l} \cdot \frac{\partial^2 g''(0)}{\partial x_k \partial x_l}$$

$$\times \frac{\sqrt{p'_{2m-1}(0, \xi^0)}}{\partial p_m(0, \xi^0)/\partial \xi_1} + \sum_{j=1}^{n} \frac{\partial p_m(0, \xi^0)}{\partial \xi_j} \alpha_{jk}$$

$$= \frac{1}{2} \frac{\partial p'_{2m-1}(0, \xi^0)/\partial x_k + \Sigma_{l=1}^{n}(\partial p'_{2m-1}(0, \xi^0)/\partial \xi_l)(\partial^2 g''(0)/\partial x_k \partial x_l)}{\sqrt{p'_{2m-1}(0, \xi^0)}}$$

$$(k = 1, 2, \ldots, n). \tag{20}$$

It follows from Lemma 6.1.4 of [19] that a symmetric matrix with a positive real part for which (20) holds exists if and only if

$$\operatorname{Re} \sum_{k=1}^{n} \left(\frac{(\partial p'_{2m-1}/\partial x_k)(0, \xi^0) + \Sigma_{l=1}^{n}(\partial p'_{2m-1}(0, \xi^0)/\partial \xi_l)(\partial^2 g''(0)/\partial x_k \partial x_l)}{\sqrt{p'_{2m-1}(0, \xi^0)}} \right)$$

$$\times \frac{\partial p_m(0, \xi^0)}{\partial \xi_k} > 0. \tag{21}$$

We transform (21) using the fact that $p_m(x, \operatorname{grad} g'') = 0$. Differentiating this identity with respect to x_k and putting $x = 0$, we find that

$$\frac{\partial p_m(0, \xi^0)}{\partial x_k} + \sum_{l=1}^{n} \frac{\partial p_m(0, \xi^0)}{\partial \xi_l} \cdot \frac{\partial^2 g''(0)}{\partial x_k \partial x_l} = 0.$$

Consequently, (21) becomes

$$\operatorname{Re} \sum_{k=1}^{n} \left(\frac{\partial p'_{2m-1}(0, \xi^0)}{\partial x_k} \cdot \frac{\partial p_m(0, \xi^0)}{\partial \xi_k} - \frac{\partial p_m(0, \xi^0)}{\partial x_k} \cdot \frac{\partial p'_{2m-1}(0, \xi^0)}{\partial \xi_k} \right)$$

$$\times \frac{1}{\sqrt{p'_{2m-1}(0, \xi^0)}} > 0,$$

so that

$$\operatorname{Re} \sum_{k=1}^{n} \frac{i}{\sqrt{p'_{2m-1}(0, \xi^0)}} \left(\frac{\partial \operatorname{Im} p'_{2m-1}(0, \xi^0)}{\partial x_k} \cdot \frac{\partial p_m(0, \xi^0)}{\partial \xi_k} \right.$$

$$\left. - \frac{\partial \operatorname{Im} p'_{2m-1}(0, \xi^0)}{\partial \xi_k} \cdot \frac{\partial p_m(0, \xi^0)}{\partial x_k} \right) > 0.$$

By virtue of choosing the branch (16) of the square root this inequality is satisfied, and so the matrix (α_{jk}) with the required properties exists.

From the Cauchy data we find that $\partial^2 h(0)/\partial x_j \partial x_k = \alpha_{jk}$; $j, k = 2, \ldots, n$. If in (19) we put, successively, $k = 2, \ldots, n$, and use the fact that $\partial p_m(0, \xi^0)/\partial \xi_1 \neq 0$, we find that $\partial^2 h(0)/\partial x_1 \partial x_k = \alpha_{1k}$; $k = 2, \ldots, n$. Finally, from (19) we find that $\partial^2 h(0)/\partial x_1^2 = \alpha_{11}$. Thus we have completed the construction of $h(x)$. Obviously,

$$\operatorname{Re} h = \operatorname{Re} \sum_{j,k=1}^{n} \alpha_{jk} x_j x_k + O(|x|^3),$$

whence it follows that, with a constant $C > 0$,

$$C^{-1}|x|^2 \leqslant \operatorname{Re} h(x) \leqslant C|x|^2; \qquad |x| < \varepsilon, \varepsilon > 0.$$

The proof of the theorem is completed in the standard way. We find the function

$$\varphi(x) = \sum_{j=0}^{L} \varphi_j(x) \lambda^{-j} e^{-\lambda^2 g(x) - \lambda h(x)},$$

$\varphi_i \in C_0^\infty(\omega)$, for which (1) is not satisfied. Moreover, we can choose φ_0 so that $\varphi_0(0) = 1$. For $f(x)$ we choose $\bar{\varphi}$. A simple calculation then shows that a constant $C_1 > 0$ exists such that the left side of (1) is not less than $C_1 \lambda^{-n/2}$. It is obvious, moreover, that the first factor on the right side of (1) does not exceed $C\lambda^{2R}$, $C = \text{const} > 0$.

Let the positive integer M be such that $M > 2R + n/2$. Using Lemma 1, we conclude that all the derivatives up to order R of the expression

$$p_m^2(x, -D)(\varphi(x)) - e^{-i\lambda^2(x, \xi^0)} \sum_{j=0}^{L} \sum_{|\alpha| \leqslant N} \frac{(-1)^{|\alpha|}}{\alpha!} (p_m^2)^{(\alpha)}(x, \lambda^2 \xi^0)$$

$$\times D_x^\alpha \left(\varphi_j(x) e^{-\lambda^2 g_1(x) - \lambda h(x)} \lambda^{-j} \right)$$

may be estimated uniformly in $x \in \omega$ with constant λ^{-M}, if $N \geqslant N_0$ (where N_0 is a sufficiently large positive integer). Consequently, we must study the action only of finite segments of the Taylor series expansion of the symbol of the operator $p_m^2(x, -D)$ in a neighborhood of the point $-\lambda^2 \xi^0$. A similar remark also applies to the lower terms. Thus, we have "approximated" the operator $P(x, D)$ sufficiently well by means of differential operators with variable coefficients.

We see at once that

$$e^{<i\lambda^2(x,\xi^0)} \sum_{j=0}^{L} \sum_{|\alpha| \leqslant N} \frac{(-1)^{|\alpha|}}{\alpha!} p(x, \lambda^2 \xi^0) D_x^\alpha \left(\varphi_j(x) \lambda^{-j} e^{-\lambda^2 g_1 - \lambda h} \right)$$

$$= e^{-\lambda^2 g(x) - \lambda h(x)} \lambda^{4m} \sum_{j=0}^{L+2N} a_j(x) \lambda^{-j}. \tag{22}$$

From (7), (8′), (10), (11), (12′), and (13) it follows that

$$a_0(x) = O(|x|^N), \quad x \to 0, \qquad a_1(x) = O(|x|^{N-1}), \quad x \to 0,$$

$$a_2(x) = O(|x|^{N-1})$$

$$+ \varphi_0(x)\left[p'_{2m-1}(x, \operatorname{grad} g'') - \left(\sum_{j=1}^{n} \frac{\partial p_m}{\partial \xi_j}(x, \operatorname{grad} g'') \frac{\partial h}{\partial x_j} \right)^2 \right]$$

$$= O(|x|^{N-1}), \quad x \to 0,$$

$$a_3(x) = O(|x|^{N-1}) + \left(\sum_{j=1}^{n} \frac{\partial h}{\partial x_j} \cdot \frac{\partial p_m}{\partial \xi_j}(x, \operatorname{grad} g'') \right)$$

$$\times \left(\sum_{k=1}^{n} \frac{\partial p_m}{\partial \xi_k}(x, \operatorname{grad} g'') \frac{\partial \varphi_0}{\partial x_k} \right) + B(x)\varphi_0(x),$$

$$a_\mu(x) = O(|x|^{N-1}) + \left(\sum_{j=1}^{n} \frac{\partial p_m(x, \operatorname{grad} g'')}{\partial \xi_j} \cdot \frac{\partial h}{\partial x_j} \right)$$

$$\times \left(\sum_{k=1}^{n} \frac{\partial p_m}{\partial \xi_k}(x, \operatorname{grad} g'') \frac{\partial \varphi_{\mu-3}}{\partial x_k} \right) + B(x)\varphi_{\mu-3} + C_\mu(x),$$

where $C_\mu(x)$ is a linear combination of the functions φ_ν and their derivatives $\nu \leqslant \mu - 4$.

In a standard way (see [19]) we construct φ_0 as an approximate solution of the following Cauchy problem:

$$S\varphi_0 \equiv \sum_{j=1}^{n} \frac{\partial p_m}{\partial \xi_j}(x, \operatorname{grad} g'') \frac{\partial \varphi_0}{\partial x_j} + \frac{B\varphi_0}{\Sigma_{j=1}^{n}(\partial p_m/\partial \xi_j)(x, \operatorname{grad} g'')\partial h/\partial x_j}$$

$$= O(|x|^{N-4}) \qquad \varphi_0(0, y) = 1, x = (x_1, y).$$

In order to give the function $\varphi_0 \in C^\infty_{(\omega)}$ compact support, we must multiply it by a cut-off function $\eta(x) \in C^\infty_0(\omega)$ equal to one in the domain $\omega' \subset \omega$. It is easy to see that $S(\eta\varphi_0) = O(|x|^{N-4})$. In a similar way we construct the remaining functions $\varphi_{\mu-3}$, $\mu \geqslant 4$, such that $a_\mu(x) = O(|x|^{N-\mu})$. For this it is sufficient to find an approximate solution of the Cauchy problem for the equation

$$\sum_{j=1}^{n} \frac{\partial p_m}{\partial \xi_j}(x, \operatorname{grad} g'') \frac{\partial \varphi_{\mu-3}}{\partial x_j} + \frac{B\varphi_{\mu-3} + C_\mu}{\Sigma_{j=1}^{n}(\partial p_m/\partial \xi_j)(x, \operatorname{grad} g'')\partial h/\partial x_j}$$

$$= O(|x|^{N-\mu-1}),$$

$$\varphi_{\mu-3}(0, y) = 0.$$

We now prove a lemma from which the violation of (1) is an immediate consequence.

LEMMA 2. *Let $\varphi(x) = O(|x|^N)$, $x \to 0$, $x \in \omega$, where ω is a neighborhood of the origin. Then for any multi-index γ, $|\gamma| \leqslant N$,*

$$\left| D^\gamma\left(\varphi(x)e^{-\lambda^2 g(x) - \lambda h(x)}\right)\right| \leqslant C_\gamma \lambda^{2|\gamma| - N/2}; \qquad C_\gamma > 0, x \in \omega,$$

Actually, from Leibniz's formula it is easy to obtain the following chain of inequalities:

$$\left| D^\gamma\left(\varphi(x)e^{-\lambda^2 g - \lambda h}\right)\right| \leqslant C \sum_{|\alpha+\beta|=|\gamma|} |x|^{N-|\alpha|} \lambda^{2|\beta|} e^{-\lambda c_0|x|^2}$$

$$\leqslant C' \sum_{|\alpha+\beta|=|\gamma|} \lambda^{2|\beta|} \lambda^{(|\alpha|-N)/2} \leqslant C'' \lambda^{2|\gamma| - N/2}.$$

From Lemma 2 it follows that

$$\lambda^{-\mu}\left| D_x^\alpha a_\mu'(x)\right| \leqslant C\lambda^{-\mu} \cdot \lambda^{2R-(N-\mu)/2} \leqslant C\lambda^{2R-N/2},$$

where $a_\mu'(x) = a_\mu(x)e^{-\lambda^2 g - \lambda h}$ and $|\alpha| \leqslant R$. Thus, choosing $N > \max(N_0, 8R + n + 8m)$, we find that (1) is certainly not satisfied as $\lambda \to \infty$.

§5. Proof of Theorem 1 (general case)

To prove Theorem 1 for the case in which the order of contact $k > 1$ we need a result from [10]. The idea is that a subprincipal symbol of a pseudodifferential operator $P(x, D)$ of order m is invariant with respect to a change of variables at those points for which

$$p_m^{0(j)}(x, \xi) = 0 \qquad \left(p_m^{0(j)}(x, \xi) = \partial p_m^0/\partial \xi_j\right), j = 1, 2, \ldots, n.$$

A geometrical proof of this fact was given in [10]; however, it appears to us that this invariance is more easily proved by means of a direct calculation using the formula (2.17) from [5].

In the case under consideration the principal symbol has the form p_m^2, from which it follows that $\operatorname{grad}_\xi p_m^2 \equiv 0 \pmod{p_m}$.

Since the zero bicharacteristics of the function p_m are invariant with respect to diffeomorphisms, we conclude that the assumptions of Theorem 1 are preserved after any change of variables. Below we shall make essential use of this invariance.

Let the characteristic point (x^0, ξ^0), in the neighborhood of which we consider $p_m(x, \xi)$, be such that $x^0 = 0$, $\xi_1^0 = 0$ and $|\xi'^0| > 0$. From the implicit function theorem we find that in some neighborhood of $(0, \xi^0)$ the following factorization of $p_m(x, \xi)$ holds:

$$p_m(x, \xi) = (\xi_1 - a(x, \xi'))Q(x, \xi), \tag{23}$$

where $Q(x, \xi)$ is an elliptic symbol, homogeneous of degree $m - 1$. It is obvious, moreover, that $a(0, \xi'^0) = 0$. (Assuming that $\partial p_m(0, \xi^0)/\partial \xi_1 \neq 0$, we first prove the representation (23) on the unit sphere $|\xi| = 1$, and then extend the functions a and Q by homogeneity.)

We now make a change of variables under which the bicharacteristics straighten out. For this purpose we solve the Cauchy problem for a system with smooth coefficients

$$\partial y/\partial x_1 = J(y/x')\,\mathrm{grad}_{\xi'}\,a\big(x_1, x', J^*(y/x)\xi'^0\big), \qquad y\big|_{x_1=0} = x' \qquad (24)$$

(here we denote the arguments of the function $a(x, \xi')$ by $x = (x_1, x')$ and ξ'; $J(y/x')$ is the Jacobi matrix and $J^*(y/x')$ is its adjoint). By virtue of (24) the Jacobi matrix satisfies the condition $J(y/x')\big|_{x_1=0} = E$. Consequently, the substitution $y = y(x_1, x')$, $x_1 = t$, determines a transformation of coordinates in a neighborhood of the origin. Then

$$\xi' = J^*(y/x')\eta', \qquad \xi_1 = \tau + \langle \partial y/\partial x_1, \eta' \rangle,$$

where $\eta = (\tau, \eta')$ are dual variables to (t, y). From [6] it is known that in the new coordinates the symbol $\xi_1 - a(x, \xi')$ goes over into $\tau - \tilde{a}(t, y, \eta')$; moreover, $\tilde{a}(t, y, \xi'^0) = \mathrm{grad}_{t,y,\eta'}\tilde{a}(t, y, \xi'^0) \equiv 0$ in some neighborhood of the origin.

We now consider the Hamilton equations for the bicharacteristics of p_m passing through the point $(t^0, y^0, 0, \xi'^0)$ when (t^0, y^0) is sufficiently close to the origin. It is obvious that in the new coordinates p_m has the form $\tilde{Q}(t, y, \eta)(\tau - \tilde{a}(t, y, \eta'))$. Consequently,

$$\mathrm{grad}_{t,y,\eta}\,p_m(t, y, \eta) = (\tau - \tilde{a})\,\mathrm{grad}_{t,y,\eta}\,\tilde{Q} + \tilde{Q}\,\mathrm{grad}_{t,y,\eta}(\tau - \tilde{a}).$$

Since $(\tau - \tilde{a}(t, y, \eta'))\big|_{\tau=0,\eta'=\xi'^0} = 0$, we find that the equations of these bicharacteristics have the form

$$dx_1/ds = dt/ds = \tilde{Q}\big(t, y^0, 0, \xi'^0\big), \qquad t(0) = t^0, \qquad (25)$$

$y = y^0$, $\tau = 0$, $\eta'^0 = \xi'^0$ (s is a parameter of the curve). Equation (25) for the zero bicharacteristic of p_m, passing through the point $(t^0 = 0, y^0 = 0, \xi^0)$, obviously has the solution

$$s = \int_0^t \frac{dl}{\tilde{Q}(l, 0, \xi^0)}.$$

From this it is evident that $t = s\tilde{Q}(0, \xi^0) + O(s^2)$, $s \to 0$; $\tilde{Q}(0, \xi^0) \neq 0$. Thus, in the new coordinates

$$\mathrm{Im}\, p'_{2m-1}\big(t, 0, \xi^0\big) = Ct^k + O(t^{k+1}), \qquad t \to 0,$$

where k is an odd number and $C = \mathrm{const} \neq 0$. From this point on, we return to the old notation (i.e. we work in the new coordinates with the previous symbolism).

We can immediately verify that the function $\varphi = (x, \xi^0)$ is a solution of the characteristic equation $p_m(x, \mathrm{grad}\,\varphi) = 0$. Since we are following the outline of the proof from §4, we need to find a solution of the equation

$$\sum_{j=1}^n \frac{\partial p_m}{\partial \xi_j}(x, \mathrm{grad}\,\varphi)\frac{\partial h}{\partial x_j} = \sqrt{p'_{2m-1}(x, \mathrm{grad}\,\varphi)}, \qquad (27)$$

such that $\operatorname{Re} h(x_1, y) \geqslant c_0(x_1^{2k} + |y|^2)$, where $x = (x_1, y) \in \omega$. In these coordinates the problem is readily solved. In fact,

$$(\partial p_m/\partial \xi_j)(x, \operatorname{grad} \varphi) = (\partial p_m/\partial \xi_j)(x, \xi^0) = 0 \quad \text{for } j \geqslant 2,$$

$$(\partial p_m/\partial \xi_1)(x, \xi^0) = Q(x, \xi^0) \quad \text{for } j = 1.$$

Consequently (27) is transformed into

$$Q(x, \xi^0)\partial h/\partial x_1 = \sqrt{p'_{2m-1}(x, \xi^0)}, \tag{27'}$$

where

$$\operatorname{Re} p'_{2m-1}(0, \xi^0) < 0, \qquad \operatorname{Im} p'_{2m-1}(x_1, 0, \xi^0) = Cx_1^k + O(|x_1|^{k+1}), \qquad x_1 \to 0;$$

$$C \neq 0 \qquad k > 1, \qquad k \text{ odd.}$$

In this situation the following two cases are possible:

a)

$$\left|\operatorname{grad}_x \operatorname{Im} p'_{2m-1}(x_1, 0, \xi^0)\right|^2 \leqslant C\left|\operatorname{Im} p'_{2m-1}(x_1, 0, \xi^0)\right|, \tag{28}$$

when $|x_1| < \varepsilon, \varepsilon > 0$ and $C = \text{const} > 0$.

b) The inequality a) is not satisfied on any interval $|x_1| < \varepsilon, \varepsilon > 0$.

In case b) we can assume without loss of generality that on any interval $|x_1| < \varepsilon, \varepsilon > 0$, the inequality

$$\left|\frac{\partial \operatorname{Im} p'_{2m-1}(x_1, 0, \xi^0)}{\partial x_2}\right|^2 \leqslant C|\operatorname{Im} p'_{2m-1}(x_1, 0, \xi^0)| \tag{28'}$$

is not satisfied.

Expanding the function $\operatorname{Im} p'_{2m-1}(x_1, x_2, 0, \xi^0)$ in a Taylor series and taking (28') into account, we find that

$$\operatorname{Im} p'_{2m-1}(x_1, x_2, 0, \xi^0)$$

$$= \operatorname{Im} p'_{2m-1}(x_1, 0, 0, \xi^0) + x_2 \frac{\partial \operatorname{Im} p'_{2m-1}(x_1, 0, 0, \xi^0)}{\partial x_2} + O(x_2^2)$$

$$= Cx_1^k + Dx_2 x_1^l + O(|x_1|^{k+1} + |x_2||x_1|^{l+1} + x_2^2),$$

$$(x_1, x_2) \to (0, 0), C \neq 0, \qquad D \neq 0, \qquad l \leqslant [k/2]. \tag{29}$$

Thus, applying the Lemma 6 from [15] to $\operatorname{Im} p'_{2m-1}(x_1, x_2, 0, \xi^0)$, we conclude that there exists a sequence of points $(x_1^r, x_2^r) \underset{r \to \infty}{\to} (0, 0)$ such that $\operatorname{Im} p'_{2m-1}(x_1^r, x_2^r, 0, \xi^0) = 0$ and

$$\partial \operatorname{Im} p'_{2m-1}(x_1^r, x_2^r, 0, \xi^0)/\partial x_1 \neq 0.$$

We now consider the zero bicharacteristic of p_m, passing through the point $(x_1^r, x_2^r, 0, \xi^0)$. Since

$$(d/ds)\operatorname{Im} p'_{2m-1}(x_1(s), x_2^r, 0, \xi^0) = (\partial/\partial x_1)\operatorname{Im} p'_{2m-1} dx_1/ds \neq 0 \quad \text{and}$$

$$x_2^r \equiv x_2(s),$$

it follows that $\operatorname{Im} p'_{2m-1}$, along this bicharacteristic, has a first order zero at the point $(x_1^r, x_2^r, 0, \xi^0)$. Then from the assertion proved in §4 it follows immediately that $P(x, D)$ is unsolvable at the point $(x_1^r, x_2^r, 0)$ and a fortiori unsolvable at the origin. Consequently, we can limit ourselves to a study of case a).

In what follows we construct a solution of $(27')$ for which $\operatorname{Re} h(x_1, y) \geqslant c_0(x_1^{2k} + |y|^2)$. Here, for convenience, we denote $x' = (x_2, \dots, x_n)$ by y. With no loss of generality, we can assume that Q and $p'_{2m-1}(x_1, y, \xi^0)$ are analytic functions. It follows from the theory of differential equations that there exists a unique analytic solution of the following Cauchy problem:

$$Q(x_1, y, \xi^0)\partial h/\partial x_1 = \sqrt{\operatorname{Re} p'_{2m-1}(x_1, y, \xi^0) + i \operatorname{Im} p'_{2m-1}(x_1, y, \xi^0)},$$

$$h\big|_{x_1=0} = |y|^2, \qquad (x_1, y) \in \omega, \qquad 0 \in \omega.$$

Here we select the branch of the square root as follows: $\sqrt{-1} = i$ if $\operatorname{sgn} C \operatorname{sgn} Q(0, \xi^0) > 0$, and $\sqrt{-1} = -i$ otherwise; C is the constant from (29).

We now find a two-sided estimate for $\operatorname{Re} h$. From (27) we obtain

$$\frac{\partial h}{\partial x_1} = \frac{\sqrt{\operatorname{Re} p'_{2m-1}(x_1, y, \xi^0)}}{Q(x_1, y, \xi^0)} \sqrt{1 + i \frac{\operatorname{Im} p'_{2m-1}(x_1, y, \xi^0)}{\operatorname{Re} p'_{2m-1}(x_1, y, \xi^0)}}.$$

Expanding the second root in a series, we find that

$$\frac{\partial h}{\partial x_1} = \frac{\sqrt{\operatorname{Re} p'_{2m-1}}}{Q} \left(1 + \frac{i}{2} \cdot \frac{\operatorname{Im} p'_{2m-1}}{\operatorname{Re} p'_{2m-1}} + O\left(\left(\frac{\operatorname{Im} p'_{2m-1}}{\operatorname{Re} p'_{2m-1}}\right)^2\right)\right), \qquad (30)$$

$$(x_1, y) \to 0.$$

From the assumption a) it follows that

$$\operatorname{Im} p'_{2m-1}(x_1, y, \xi^0) = Cx_1^{2k-1} + \sum_{j=1}^{n} D_j y_j x_1^l + O(|y|^2 + x_1^{2k}),$$

where $C \neq 0$, $k \geqslant 3$ and $l > [(2k - 1)/2]$ (we have made the substitution $k \to 2k - 1$, since k is odd).

The last equality may be written as follows:

$$\operatorname{Im} p'_{2m-1}(x_1, y, \xi^0) = Cx_1^{2k-1} + O(|y|^2 + x_1^{2k}), \qquad (x_1, y) \to 0. \quad (31)$$

By virtue of (31), for (30) we have

$$\frac{\partial h}{\partial x_1} = \frac{\sqrt{\operatorname{Re} p'_{2m-1}}}{Q} \left(1 + \frac{i}{2} \frac{Cx_1^{2k-1}}{\operatorname{Re} p'_{2m-1}} + O(|y|^2 + x_1^{2k})\right), \qquad (x_1, y) \to 0.$$

Integrating from 0 to x_1, we obtain

$$h(x_1, y) = |y|^2 + \int_0^{x_1} \frac{\sqrt{\operatorname{Re} p'_{2m-1}(s, y, \xi^0)}}{Q(s, y, \xi^0)}$$

$$\times \left(1 + \frac{i}{2} C \frac{s^{2k-1}}{\operatorname{Re} p'_{2m-1}(s, y, \xi^0)} + O(|y|^2 + s^{2k})\right) ds,$$

since $h(0, y) = |y|^2$. Since $\sqrt{\operatorname{Re} p'_{2m-1}(s, y, \xi^0)}$ takes on pure imaginary values, it follows that

$$\operatorname{Re} h(x_1, y) = |y|^2 + \int_0^{x_1} \frac{iC}{2} \frac{s^{2k-1}\, ds}{Q(s, y, \xi^0)\sqrt{\operatorname{Re} p'_{2m-1}(s, y, \xi^0)}}$$

$$+ \operatorname{Re} \int_0^{x_1} \frac{O(|y|^2 + s^{2k})\, ds}{Q(s, y, \xi^0)\sqrt{\operatorname{Re} p'_{2m-1}(s, y, \xi^0)}}. \tag{32}$$

We now estimate the second integral on the right side of (32):

$$\left| \int_0^{x_1} \frac{O(|y|^2 + s^{2k})\, ds}{Q(s, y, \xi^0)\sqrt{\operatorname{Re} p'_{2m-1}(s, y, \xi^0)}} \right| \leqslant \text{const} \left| \int_0^{x_1} (|y|^2 + s^{2k})\, ds \right|$$

$$\leqslant \text{const}(|x_1||y|^2 + |x_1|^{2k+1}).$$

We now find a two-sided estimate for the quantity

$$\frac{iC}{2} \int_0^{x_1} \frac{s^{2k-1}\, ds}{Q\sqrt{\operatorname{Re} p'_{2m-1}}}.$$

By the choice of the branch of the square root the function $iC/2Q\sqrt{\operatorname{Re} p'_{2m-1}}$ is positive in a small neighborhood of the origin. Consequently, there exist constants $C_1, C_2 > 0$ such that

$$C_1 x_1^{2k} \leqslant \frac{iC}{2} \int_0^{x_1} \frac{s^{2k-1}\, ds}{Q\sqrt{\operatorname{Re} p_{2m-1}}} \leqslant C_2 x_1^{2k}.$$

From this we deduce that if $|x_1|$ and $|y|$ are sufficiently small, then $\operatorname{Re} h(x_1, y) \geqslant c_0(x_1^{2k} + |y|^2)$. It can be shown even more simply that $\operatorname{Re} h(x_1, y) \leqslant c_1(x_1^{2k} + |y|^2)$.

The proof of the theorem is completed as in §4. The only difference is that, instead of applying Lemma 2, we apply

LEMMA 2'. *Let* $\varphi(x) = O(|x|^N)$, $x \to 0$; $x \in \omega$, *where* ω *is a neighborhood of the origin. Then for any multi-index* γ, $|\gamma| \leqslant N$,

$$\left| D^\gamma\left(\varphi(x) e^{-\lambda^2 g(x) - \lambda h(x)}\right) \right| \leqslant C_\gamma \lambda^{2|\gamma| - N/2k}, \qquad C_\gamma > 0.$$

We do not give the details.

REMARK 1. In the case of a differential operator the condition $\operatorname{Re} p'_{2m-1}(x^0, \xi^0) < 0$ can be replaced by $\operatorname{Re} p'_{2m-1}(x^0, \xi^0) \neq 0$ (since $\operatorname{Re} p'_{2m-1}(x, \xi)$ is an odd function of ξ).

REMARK 2. Let $r_t(x, \xi)$ be a real elliptic symbol in a neighborhood of (x^0, ξ^0), and let the leading symbol of the pseudodifferential operator $P(x, D)$ have the form $(p_m^2 r_t)(x, \xi)$. Then the assertion of Theorem 1 holds when the inequality

$$\operatorname{Re} p'_{2m+t-1}(x^0, \xi^0)/r_t(x^0, \xi^0) < 0$$

is satisfied and condition b) is retained. This is immediately evident from the fact that h is determined from the equation

$$\sum_{j=1}^{n} \frac{\partial p_m}{\partial \xi_j}(x, \operatorname{grad} g'')\frac{\partial h}{\partial x_j} = \sqrt{\frac{p'_{2m+t-1}(x, \operatorname{grad} g'')}{r_t(x, \operatorname{grad} g'')}},$$

$$h\big|_{x_1=0} = |y|^2.$$

§6. Proof of Theorem 2

The proof of this theorem is carried out with the aid of Lemma 1′ similarly to the proof of Theorem 1. Therefore we outline only the main steps in the proof.

First of all, we solve the characteristic equation for the principal symbol $p_m(x, Dg) = 0$. We note that, as Hörmander showed in [4], there exists a solution of the equation

$$\sum_{|\alpha| \leqslant N} \frac{1}{\alpha!} p_m^{(\alpha)}(x, \xi^0)(D_x g_1)^\alpha = O(|x|^N),$$

for which $c_0|x|^2 \leqslant \operatorname{Re} g_1 \leqslant c_1|x|^2$, $g_1 = O(|x|^2)$; $c_0, c_1 > 0$. From the fact that $p_m(x, \xi)$ is analytically continuable with respect to ξ in a neighborhood of ξ^0 it follows that

$$p_m(x, Dg) = O(|x|^N) \Rightarrow p_{2m+r}(x, Dg) = O(|x|^N),$$

where

$$p_{2m+r}(x, \xi) = (p_m^2 q_r)(x, \xi), \quad \text{and} \quad g = g_1 + i(\xi^0, x).$$

Second, we need to find at least an approximate solution of the equation for h:

$$q_r(x, Dg)\left(\sum_{k=1}^{n} \frac{\partial p_m}{\partial \xi_k}(x, Dg)\frac{\partial h}{\partial x_k}\right)^2 = p'_{2m-r-1}(x, Dg).$$

To do this, we replace the infinitely differentiable functions $p'_{2m+r-1}(x, Dg)$, $\partial p_m(x, Dg)/\partial \xi_k$ and $q_r(x, Dg)$ by their Taylor series expansions up to some sufficiently high order. Keeping the same notation for Taylor series segments as for the functions themselves, we solve the following Cauchy problem for a differential operator with analytic coefficients:

$$\sum_{k=1}^{n} \frac{\partial p_m}{\partial \xi_k}(x, Dg)\frac{\partial h}{\partial x_k} = \sqrt{\frac{p'_{2m+r-1}(x, Dg)}{q_r(x, Dg)}}, \qquad h\big|_{x_1=0} = 0 \qquad (32)$$

(a branch of the square root exists, since $p'_{2m+r-1}(x^0, \xi^0) \neq 0$, but its selection is of no significance in the course of the proof). We assume that $\partial p_m(x^0, \xi^0)/\partial \xi_1 \neq 0$. Since $\partial h(0)/\partial x_k = 0$ for $k \geqslant 2$,

$$\partial h(0)/\partial x_1 \neq 0 \Rightarrow h(x) = O(|x|), \qquad h \neq O(|x|^2).$$

We use the following obvious lemma in completing the proof.

LEMMA 3. *Let $\varphi(x) = O(|x|^N)$, $x \to 0$, $x \in \omega$, where ω is a neighborhood of the origin. Then for any multi-index γ, $|\gamma| \leqslant N$,*

$$\left| D^\gamma\left(\varphi(x) e^{-\lambda^2 g(x) - \lambda h(x)} \right) \right| \leqslant c_\gamma \lambda^{2|\gamma| - N}, \qquad c_\gamma > 0.$$

We now make some remarks.

REMARK 1. We employ the analyticity of p_{2m+r} and p_{2m+r-1} in determining the coefficients of λ^{4m+2r}, $\lambda^{4m+2r-1}$, $\lambda^{4m+2r-2}$ and $\lambda^{4m+2r-3}$ in the representation (22). For example, the coefficient of λ^{4m+r-1} has the form

$$\varphi \sum_{|\alpha| \leqslant N-1} \sum_{j=1}^{n} \frac{1}{\alpha!} \frac{\partial}{\partial \xi_j} p^{(\alpha)}_{2m+r-1}(x, \xi^0)(D_x g_1)^\alpha D_{x_j} h$$

$$= \varphi \left(\sum_{j=1}^{n} \frac{\partial p_{2m+r}}{\partial \xi_j}(x, Dg) D_{x_j} h + O(|x|^N) \right).$$

Since

$$(\partial p_{2m+r}/\partial \xi_j)(x, Dg) = 2 p_m(x, Dg)(\partial p_m/\partial \xi_j)(x, Dg) q_r(x, Dg) + O(|x|^{2N}),$$

we find that this coefficient behaves as $O(|x|^N)$ when $x \to 0$.

REMARK 2. In fact, the proof of the theorem makes it possible for us to establish unsolvability in other cases also besides the one considered. Thus, we have the following assertion:

ASSERTION 1. *Let $P(x, D)$ be a pseudodifferential operator, the principal part of whose symbol satisfies the condition $p_{2m+r}(x, \xi) = (p_m^2 q_r)(x, \xi)$. Here q_r is a symbol which is elliptic in a neighborhood of the point (x^0, ξ^0) and which is of order of homogeneity r. Let*

$$p_m(x^0, \xi^0) = 0 \Rightarrow \operatorname{grad}_\xi \operatorname{Re} p_m(x^0, \xi^0) \neq 0$$

and let p_{2m+r} and p_{2m+r-1} be analytically continuable with respect to ξ in some neighborhood of ξ^0. Assume that $\operatorname{Im} p_m$, along the zero bicharacteristic $\mathcal{L}$ of the function $\operatorname{Re} p_m$ passing through (x^0, ξ^0), has a third order zero and changes sign from $-$ to $+$ in moving along the oriented curve $\mathcal{L}$. Then the operator $P(x, D)$ is unsolvable in a neighborhood of x^0 if $p'_{2m+r-1}(x^0, \xi^0) \neq 0$ and the quantity

$$\sqrt{\frac{p'_{2m+r-1}(x^0, \xi^0)}{q_r(x^0, \xi^0)}} \cdot \frac{1}{(\partial p_m / \partial \xi_1)(x^0, \xi^0)}$$

is pure imaginary (the branch of the quadratic root here is of no signficance).

Indeed, in this situation two possibilities arise (see [6] and [15]).

a) There exists a sequence of points

$$\left(x^k, \xi^k \right) \underset{k \to \infty}{\to} \left(x^0, \xi^0 \right),$$

characteristic for p_m, such that $\operatorname{Im} p_m$ has a first order zero along the zero bicharacteristic of $\operatorname{Re} p_m$ passing through (x^k, ξ^k), and $\operatorname{Im} p_m$ changes sign from $-$ to $+$ in moving along this oriented curve.

b) For any smooth vector-valued function $\alpha(x, \xi)$ such that $(\alpha(x, \xi), \operatorname{grad} \operatorname{Re} p_m(x, \xi)) = 0$, the inequality

$$\varlimsup_{\substack{(x,\xi)\to(x^0,\xi^0) \\ (x,\xi)\in\mathcal{C}}} |(\alpha(x, \xi), \operatorname{grad} \operatorname{Im} p_m(x, \xi))| \, |\operatorname{Im} p_m(x, \xi)|^{-1/2} < \infty$$

is satisfied.

In the first case, it follows from Theorem 2, which we have already proved, that the operator P is unsolvable at the points x^k and, all the more, at the point x^0. In case b) we first derive an elliptic factorization of (22) and then apply a change of variables of the form (23), straightening the bicharacteristic passing through (x^0, ξ^0). It then follows from [15] and [21] that in the new coordinates there exists a function $g_1 \in C^\infty$,

$$c_0\big(x_1^4 + |y|^2\big) \leqslant \operatorname{Re} g_1(x_1, y) \leqslant c_1\big(x_1^4 + |y|^2\big),$$

which is an approximate solution of the characteristic equation

$$p_m(x, Dg) = O\big(|x|^N\big), \qquad g = g_1 + i(\xi^0, x).$$

We note that $p'_{2m+r-1}(z^0, \eta^0) \neq 0$, due to the invariance of the subprincipal symbol modulo p_m at the image (z^0, η^0) of the point (x^0, ξ^0) under the diffeomorphism in question.

In determining the function h from (32) we find that $\partial h(0)/\partial x_1$ is pure imaginary, and therefore

$$h(x) = ikx_1 + O\big(|x|^2\big), \qquad x \to 0,$$

where $k \neq 0$ is a real constant. An elementary calculation shows that in a small neighborhood of the origin the following inequalities hold:

$$c_0 e^{-\lambda^2(x_1^4+|y|^2)} \leqslant |e^{-\lambda^2 g_1 - \lambda h}| \leqslant c_1 e^{-\lambda^2(x_1^4+|y|^2)}, \qquad c_0, c_1 > 0.$$

Modifying correspondingly the proofs of Lemmas 1 and 3, we complete the proof. Details are left to the reader.

§7. A sufficient condition for the local solvability of pseudodifferential equations with two-fold characteristics. Proof of Theorems 3 and 4

Before going on to a proof of Theorem 3, we formulate and prove the following auxiliary proposition.

PROPOSITION 1. *Let $p_{m_1}(x, \xi)$ and $p_{m_2}(x, \xi)$ be symbols of pseudodifferential operators of principal type with orders of homogeneity m_1 and m_2, respectively. Assume that*

$$p_{m_1}(x^0, \xi^0) = p_{m_2}(x^0, \xi^0) = 0, \qquad \xi^0 \neq 0$$

and that $\operatorname{Im} p_{m_1}$ does not change sign from $+$ to $-$ in moving in the positive direction along the zero bicharacteristics of $\operatorname{Re} p_{m_1}$. Then the following estimate holds for the operator $P(x, D)$ with the leading symbol $p_{m_1}(x, \xi)p_{m_2}(x, \xi)$:

$$\|u\|_s \leqslant C(\omega)\|Pu\|_{s-m_1-m_2-2} + C_1\|u\|_{s-1}, \qquad \forall u \in C_0^\infty(\omega), \qquad (33)$$

if

$$\sigma\left(P_{m_1} P_{m_2} - P_{m_1} \circ P_{m_2} + P_{m_1+m_2-1}\right) \equiv 0 \qquad (\mathrm{mod}\ p_{m_2}) \qquad\qquad \text{(a)}$$

or

$$\sigma\left(P_{m_1} P_{m_2} - P_{m_2} \circ P_{m_1} + P_{m_1+m_2-1}\right) \equiv 0 \qquad (\mathrm{mod}\ p_{m_1}). \qquad\qquad \text{(b)}$$

Here σ denotes the leading symbol of the operator in parentheses, $p_{m_1+m_2-1}$ is the term of order $m_1 + m_2 - 1$ in the expansion of the symbol $p(x, \xi)$ and s is an arbitrary real number.

Inequality (33) has the following meaning: for any real s there exists a neighborhood ω of the point x^0 in which the operator $P(x, D)$ satisfies (33). The constant $C(\omega) \to 0$ as diam $\omega \to 0$, and, in addition, we can assume that $C_1 = 0$ if $(-n/2, m_2 - 1 - n/2)$ in case (a) and if $s \geqslant \max(-n/2, m_1 - 1 - n/2)$ in case (b).

The proof is a simple consequence of Lewy's method from the theory of hyperbolic equations with multiple characteristics (see [11]).

In case (a) we find immediately that

$$\left[P_{m_1} P_{m_2}(x, D) + P_{m_1+m_2-1}(x, D) + q_{m_1+m_2-2}(x, D)\right]u = f(x),$$

$$\Leftrightarrow \left\{ P_{m_1} \circ P_{m_2} + \left[\left(P_{m_1} P_{m_2} - P_{m_1} \circ P_{m_2}\right) + P_{m_1+m_2-1}\right] + q_{m_1+m_2-2}\right\}u = f(x),$$

$$\tag{34}$$

where $q_{m_1+m_2-1} \in S_{1,0}^{m_1+m_2-2}$.

By virtue of the requirement (a)

$$P_{m_1} P_{m_2} - P_{m_1} P_{m_2} + P_{m_1+m_2-1} = v_{m_1-1} P_{m_2} + r_{m_1+m_2-2}$$

$$= v_{m_1+1} \circ P_{m_2} + \tilde{r}_{m_1+m_2-2}; \qquad r_{m_1+m_2-2},\, \tilde{r}_{m_1+m_2-2} \in S_{1,0}^{m_1+m_2-2};$$

the symbol v_{m_1-1} is homogeneous of order $m_1 - 1$.

Thus, (34) is equivalent to

$$P_{m_1} \circ P_{m_2}u + v_{m_1-1} \circ P_{m_2}u + \tilde{q}_{m_1+m_2-2}u = f, \qquad \tilde{q}_{m_1+m_2-2} \in S_{1,0}^{m_1+m_2-2}.$$

Consequently, we have obtained a pair of decoupled equations

$$P_{m_2}u = v, \tag{35}$$

$$P_{m_1}v = v_{m_1-1}v + \tilde{q}_{m_1+m_2-2}u = f. \tag{35'}$$

For (35) we apply the estimate (5) from [16]:

$$\|u\|_s \leqslant C_1(\omega)\|v\|_{s-m_2-1} + C'\|u\|_{s-1}, \qquad \forall u \in C_0^\infty(\omega), \tag{36}$$

$C_1(\omega) \to 0$ as diam $\omega \to 0$; $C' = 0$ for $s \geqslant -n/2$. Since we are studying the local solvability of the operator $P(x, D)$, we can assume from the outset that $p_{m_1} \equiv 0$ and $p_{m_2} \equiv 0$ outside some neighborhood of the point x^0. The very same estimate (5) from [16] yields for (35') the following inequality:

$$\|v\|_{s-m_2+1} \leqslant C_2(\omega)\|f - v_{m_1-1}v - \tilde{q}_{m_1+m_2-2}u\|_{s+2-m_1-m_2} + C''\|v\|_{s-m_1-1},$$

where $C_2(\omega) \to 0$ as diam $\omega \to 0$, and $C'' = 0$ if $s \geqslant m_2 - 1 - n/2$.

Thus,

$$\|v\|_{s-m_2+1} \leqslant C_2(\omega)\big(\|f\|_{s-m_1-m_2+2} + C\|v\|_{s-m_2+1} + C_1\|u\|_s\big)$$
$$+ C''\|v\|_{s-m_2-1}.$$

The constants C and C_1 are the norms of the operators v_{m_1-1} and $\tilde{q}_{m_1+m_2-2}$, and do not increase as ω decreases. Upon choosing the neighborhood ω so small that in it the inequalities $CC_2(\omega) < \frac{1}{2}$ and $C_1C_2(\omega) < \frac{1}{2}$ hold, we obtain

$$\|v\|_{s-m_2+1} \leqslant 2C_2(\omega)\big(\|f\|_{s-m_1-m_2+2} + C_1\|u\|_s\big) + 2C''\|v\|_{s-m_2-1}. \quad (37)$$

Combining (36) and (37), we find that

$$\|u\|_s \leqslant 2C_1(\omega)C_2(\omega)\|f\|_{s-m_1-m_2+2} + C_1C_1(\omega)\|u\|_s$$
$$+ C'\|u\|_{s-1} + 2C_1(\omega)C''\|v\|_{s-m_2-1}. \quad (38)$$

If ω is such that $C_1C_1(\omega) < \frac{1}{2}$, then (38) becomes

$$\|u\|_s \leqslant 4C_1(\omega)C_2(\omega)\|f\|_{s-m_1-m_2+2} + 2C'\|u\|_{s-1} + C_3C''\|u\|_{s-1},$$

since it is obvious that $\|v\|_{s-m_2-1} \leqslant \text{const}\|u\|_{s-1}$. Moreover, $C' = C'' = 0$ for $s \geqslant \max(-n/2, m_2 - 1 - n/2)$, and consequently

$$\|u\|_s \leqslant C(\omega)\|f\|_{s-m_1-m_2+2}, \quad \text{Q.E.D.}$$

We note that in the proof of solvability it is necessary to pass to the adjoint operator, and that conditions (a) and (b) are symmetric neither with respect to p_{m_1} and p_{m_2} nor with respect to P and $'P$. Therefore we consider the special case when $p_{m_1} \equiv p_{m_2}$. Then condition (a) for the lower terms (or, what amounts to the same thing, condition (b)) assumes the form

$$p_{2m-1} + i \sum_{j=1}^{n} \frac{\partial p_m}{\partial \xi_j} \frac{\partial p_{r1}}{\partial x_j} \equiv 0 \qquad (\text{mod } p_m),$$

i.e.

$$p'_{2m-1} \equiv 0 \qquad (\text{mod } p_m).$$

On the other hand (see [3], the formula for the adjoint operator),

$$'P \sim p_m^2(x,-\xi) + p_{2m-1}(x,-\xi) + i \sum_{j=1}^{n} \frac{\partial^2(p_m^2)}{\partial x_j \partial \xi_j}(x,-\xi)$$
$$+ p_{2m-2}(x,-\xi), \qquad p_{2m-2} \in S_{1,0}^{2m-2}.$$

The subprincipal symbol of the operator $'P$ is calculated on the basis of the previous formula and has the form

$$'p'_{2m-1}(x,-\xi) = p_{2m-1}(x,-\xi) + i \sum_{j=1}^{n} \frac{\partial^2(p_m^2)}{\partial x_j \partial \xi_j}(x,-\xi) - \frac{i}{2} \sum_{j=1}^{n} \frac{\partial^2(p_m^2)}{\partial x_j \partial \xi_j}(x,-\xi)$$
$$= p_{2m-1}(x,-\xi) + \frac{i}{2} \sum_{j=1}^{n} \frac{\partial^2(p_m^2)}{\partial \xi_j \partial x_j}(x,-\xi).$$

Thus, the relation $'p'_{2m-1} \equiv 0 \pmod{p_m(x, -\xi)}$ is satisfied simultaneously with the relation $p'_{2m-1} \equiv 0 \pmod{p_m}$; consequently, for the operator $'P$ (33) holds.

Finally, we prove Theorem 3.

Let $s \geqslant \max(-\frac{1}{2}n, m - 1 - \frac{1}{2}n)$ and let the neighborhood $\omega \ni x^0$ be such that in it the basic inequality (33) is satisfied. Then for any function $f \in H^{-s} \cap \mathscr{E}'(\omega)$ there exists a distribution $v \in H^{-s+2m-2} \cap \mathscr{E}'(\omega)$, for which $'Pv = f$ in ω.

Indeed, if $f \in C_0^\infty(\omega)$, then

$$\|u\|_s \leqslant \text{const} \|Pu\|_{s-2m+2}, \qquad \forall u \in C_0^\infty(\omega).$$

Thus,

$$|\langle f, u \rangle| \leqslant \text{const} \|f\|_{-s} \|u\|_s \leqslant \text{const} \|Pu\|_{s-2m+2}.$$

The Hahn-Banach theorem shows that the bilinear form $\langle f, u \rangle$ is continuously extendable onto $\mathscr{E}'(\omega) \cap H^{s-2m+2}$. Consequently, there exists a generalized function $v \in \mathscr{E}'(\omega) \cap H^{-s+2m-2}$ such that

$$\langle f, u \rangle = \langle v, Pu \rangle \Rightarrow \langle f, u \rangle = \langle 'Pv, u \rangle \Rightarrow 'Pv = f$$

in a small neighborhood of x^0. The solvability of the operator P is a consequence of the equality $'p'_{2m-1} \equiv 0 \pmod{p_m(x, -\xi)}$.

In the special case in which P is a differential operator we have already stated Theorem 4, which gives a better result than does Theorem 3 for a pseudodifferential operator.

PROOF OF THEOREM 4. The following refinement of (33) for differential operators plays the main role in the proof of this theorem.

For any real s there exist a neighborhood ω of x^0 and a constant $C > 0$ such that

$$\|u\|_s \leqslant C(s) \|Pu\|_{s-2m+2}, \qquad \forall u \in C_0^\infty(\omega). \tag{39}$$

It is evident from the method of the proof that the same estimate is valid for $'P$. Then Theorem 4, and Theorem 3 likewise, is proved by going over to the adjoint operator.

We prove (39) by contradiction. Let $x^0 = 0$. Then there exists a sequence $\{u_\nu(x)\}$, $u_\nu(x) \in C_0^\infty(\omega_\nu)$, $\|u_\nu\|_s = 1$, $\|Pu_\nu\|_{s-2m+2} \to 0$, where the ω_ν are neighborhoods of the origin for which diam $\omega_\nu \to 0$. Consequently, supp $u_\nu \to \{0\}$. Since the imbedding operator $H^s \to H^{s-1}$ is compact, there exist a sequence u_{ν_k} and a function u such that $u_{\nu_k} \to u$ as $k \to \infty$ in the H^{s-1} topology. It is clear that supp $u = \{0\}$. From the fact that $u_{\nu_k} \to u$ in the space $\mathscr{D}'$, it follows that $Pu_{\nu_k} \to Pu$ in the topology of $\mathscr{D}'$. Since $Pu_{\nu_k} \to 0$ in $\mathscr{D}'$, we immediately have $Pu = 0$.

Finally, for all sufficiently large k and, consequently, for sufficiently small ω_{ν_k}, we apply (33):

$$\|u_{\nu_k}\|_s \leqslant C \|Pu_{\nu_k}\|_{s-2m+2} + C_1 \|u_{\nu_k}\|_{s-1}.$$

If we assume that $u = 0$, we find that $\|u_{\nu_k}\|_s \to 0$ as $k \to \infty$. Thus $u \not\equiv 0$, $u \in H^{s-1}$, supp $u = \{0\}$. Applying Lemma 1.1 from [7], we obtain a contradiction. (Actually, $\operatorname{grad}_\xi p_m(0, \xi^0) \neq 0$, so that at least one of the coefficients in the leading part of p_m^2 is nonzero at the point x^0.)

REMARK 1. Solvability of the operator ${}^t P$ is an easy consequence of Proposition 1. It is also easy to verify that an operator whose leading symbol has the form $p_{m_1} p_{m_2}(x, -\xi)$, while its lower symbols have the form

$$p_{m_1+m_2-1} + i \sum_{j=1}^{n} \frac{\partial^2(p_{m_1} p_{m_2})}{\partial x_j \partial \xi_j} \, ; \quad p_{m_1+m_2-1} + i \sum_{j=1}^{n} \frac{\partial p_{m_1}}{\partial \xi_j} \frac{\partial p_{m_2}}{\partial x_j} \equiv 0 \quad \left(\operatorname{mod} p_{m_2}\right),$$

is solvable.

§5. A refinement of Gårding's inequality. Proof of Theorem 5

In proving Theorem 5 we need the following refined Gårding's inequality.

LEMMA 4. *Let $P(x, D)$ be a pseudodifferential operator with leading symbol $p_{2m-1}(x, \xi)$, considered in a neighborhood of some compactum $K \subset \mathbf{R}^n$. Let $p_m(x, \xi)$ be a real function, positive and homogeneous of degree m in ξ and such that*

$$p_m(x^0, \xi^0) = 0, \qquad \xi^0 \neq 0 \Rightarrow \operatorname{grad}_\xi p_m(x^0, \xi^0) \neq 0.$$

In addition, assume that $\operatorname{Re} p_{2m-1}(x, \xi) \geqslant 0$ at those points (x, ξ) where $p_m(x, \xi) = 0$. Then there exist a pseudodifferential operator $P_{m-1}(x, D)$ of order $m - 1$ and a constant C such that for any function $u \in C_0^\infty(K)$,

$$\operatorname{Re}\big(P_{2m-1}(x, D)u, u\big) + C\|u\|_{m-1}^2 + \operatorname{Re}\big(P_m(x, D)u, P_{m-1}(x, D)u\big) \geqslant 0.$$

We note that with no loss of generality we can assume that $p_{2m-1}(x, \xi)$ is a real symbol. Indeed,

$$\operatorname{Re}\big(p_{2m-1}(x, D)u, u\big) = \operatorname{Re}(\operatorname{Re} p_{2m-1}u, u) + \operatorname{Re} i(\operatorname{Im} p_{2m-1}u, u)$$

$$= \operatorname{Re}(\operatorname{Re} p_{2m-1}u, u) - \operatorname{Im}(\operatorname{Im} p_{2m-1}u, u).$$

(Here $\operatorname{Re} p_{2m-1}$ and $\operatorname{Im} p_{2m-1}$ are operators whose symbols are equal, respectively, to the real and imaginary parts, of the symbol p_{2m-1}.) Using the formula for the symbol of the adjoint operator (see [5]), we obtain

$$(\operatorname{Im} p_{2m-1}u, u) = \big(u, (\operatorname{Im} p_{2m-1})^* u\big) = \big(u, \operatorname{Im} p_{2m-1}u\big) + \big(u, P_{2m-2}u\big),$$

where P_{2m-2} is a pseudodifferential operator of order $2m - 2$ with real coefficients. Consequently,

$$\operatorname{Re} 2i(\operatorname{Im} p_{2m-1}u, u) = (u, P_{2m-2}u).$$

From this it follows that $|(\operatorname{Im} p_{2m-1}u, u)| \leqslant \operatorname{const}\|u\|_{m-1}^2$.

Thus, from this point on we shall asume that $p_{2m-1}(x, \xi)$ is a real symbol and that $p_{2m-1}(x, \xi) \geqslant 0$ on the characteristic manifold H of the operator P_m. We note that H is a compact subset of $K \times S^{n-1}$, where $S^{n-1} = \{\xi \mid \xi \in \mathbf{R}^n, |\xi| = 1\}$.

Let $(x^0, \xi^0) \in K \times S^{n-1}$, and suppose, for example, that $\partial p_m(x^0, \xi^0)/\partial \xi_1 \neq 0$ (by virtue of the condition of principal type). From the implicit function theorem it follows that there exist a neighborhood* $U_{x_0} \times S_{x_0}^{n-1} \ni (x^0, \xi^0)$ and a symbol $q_{m-1}(x, \xi)$, elliptic on $\overline{U}_{x_0} \times \overline{S}_{x_0}^{n-1}$, such that

$$p_m(x, \xi) = (\xi_1 - a(x, \xi'))q_{m-1}(x, \xi) \tag{40}$$

for any $(x, \xi) \in U_{x_0} \times S_{x_0}^{n-1}$. Here $a(x, \xi')$ is a real symbol, homogeneous of the first order with respect to ξ', while q_{m-1} is of order $m - 1$. Using the Borel-Lebesgue theorem, we find a finite number of points $(x_j, \xi_j), j = 1, \ldots, k, |\xi_j| = 1$, for which

$$H \subset \bigcup_{j=1}^{k} U_{x_j} \times S_{x_j}^{n-1}.$$

Since the image of a compactum under a continuous mapping is compact, it follows that $\mathrm{Pr}_\xi H$ is a compact subset of S^{n-1} while $\mathrm{Pr}_x H$ is a compact subset of K. Since $\{S_{x_j}^{n-1}\}_{j=1}^{k}$ form an open covering of $\mathrm{Pr}_\xi H$ and $\{U_{x_j}\}_1^k$ an open covering of $\mathrm{Pr}_x H$, we readily see that there exist open sets $S_{x_0}^{n-1}$ on S^{n-1} and U_{x_0} in $\mathbf{R}^n$ such that $\{S_{x_j}^{n-1}\}_{j=1}^{k}$ and $S_{x_0}^{n-1}$ cover S^{n-1} while $\{U_{x_j}\}_1^k$ and U_{x_0} cover K, and, in addition,

$$\mathrm{Pr}_\xi H \cap \overline{S_{x_0}^{n-1}} = \varnothing, \qquad \mathrm{Pr}_x H \cap \overline{U}_{x_0} = \varnothing.$$

Summing up all the above, we conclude that there exists a finite open covering

$$\bigcup_{j=0}^{k} \left(S_{x_j}^{n-1} \times U_{x_j} \right) \supset S^{n-1} \times K,$$

possessing the following property: on any set $S_{x_j}^{n-1} \times U_{x_j}, j = 1, \ldots, k$, a factorization of the form (40) is satisfied (i.e., with a chosen direction $\xi_l, 1 \leqslant l \leqslant n$), and $|p_m(x, \xi)| \geqslant C_0 > 0$ on

$$\overline{S_{x_0}^{n-1}} \times \overline{U}_{x_0}. \tag{41}$$

We now construct a special partition of unity, subordinate to the covering $\{U_{x_j} \times S_{x_j}^{n-1}\}_0^k$. To do this we first construct a partition of unity on the unit sphere S^{n-1}, subordinate to $\{S_{x_j}^{n-1}\}_0^k$. If we denote the functions of this partition by $\psi_j(\xi)$, then they obviously belong to $C^\infty(S^{n-1})$. We extend them to C^∞-functions, defined for all $\xi \neq 0$, and homogeneous of degree zero. Next we construct a partition of unity on K, subordinate to $\{U_{x_j}\}_0^k$. We denote the functions of this partition by $\varphi_j'(x)$.

We now consider the functions

$$\varphi_j(x, \xi) = \frac{\varphi_j'(x)\psi_j(\xi)}{\Sigma_{j=0}^{k} \varphi_j'(x)\psi_j(\xi)}.$$

They are infinitely smooth for all x from some neighborhood of K and $|\xi| > 0$. Moreover, $\Sigma_0^k \varphi_j(x, \xi) \equiv 1$ in some neighborhood of K. Finally, we note that

$$\mathrm{supp}\, \varphi_j \subseteq \mathrm{supp}\, \varphi_j' \times \mathrm{supp}\, \psi_j \subseteq U_{x_j} \times S_{x_j}^{n-1}.$$

Editor's note. It appears that here and everywhere in this section the "x" in subscripts of S^{n-1} should be changed to "ξ".

Then we obtain the following identity:

$$\operatorname{Re}\big(P_{2m-1}(x, D)u, u\big) = \operatorname{Re} \sum_{j=0}^{k} \big(P_{2m-}(x, D) \circ \varphi_j(x, D)u, u\big)$$

$$+ \operatorname{Re}\left(P_{2m-1}(x, D) \circ \left(I - \sum_{j=0}^{k} \varphi_j(x, D) \right)u, u \right), \qquad u \in C_0^\infty(K). \quad (42)$$

Since the symbol of the operator $I - \Sigma_0^k \varphi_j(x, D)$ is identically zero in a neighborhood of K and the operator itself is pseudolocal, we find that the second term on the right side of (42) may be estimated in terms of $\|u\|_{m-1}^2$. In what follows we shall study the expression $(P_{2m-1}(x, D) \circ \varphi_j(x, D)u, u)$. It can be represented in the form

$$\big((P_{2m-1}\varphi_j)(x, D)u, u\big) + O\big(\|u\|_{m-1}^2\big).$$

We now consider a pseudodifferential operator with the symbol $p_{2m-1}(x, \xi)\varphi_j(x, \xi)$. Obviously, this symbol is concentrated on the set

$$\operatorname{supp} \varphi_j \subseteq U_{x_j} \times S_{x_j}^{n-1}.$$

Using Taylor's formula and the factorization (40), we obtain

$$p_{2m-1}(x, \xi)\varphi_j(x, \xi) = p_{2m-1}\big(x, a_j(x, \xi'), \xi'\big)\varphi_j(x, \xi)$$

$$+ \big(\xi_1 - a_j(x, \xi')\big)\varphi_j(x, \xi)A_{2m-2}^{(j)}(x, \xi),$$

where the symbol $A_{2m-2}^{(j)}$ is positively homogeneous of order $2m - 2$ in ξ. Consequently, the principal symbol of the operator $P_{2m-2} \circ \varphi_j$ has the form

$$\big(p_{2m-1}\varphi_j\big)(x, \xi) = p_{2m-1}\big(x, a_j(x, \xi'), \xi'\big)\varphi_j(x, \xi)$$

$$+ \varphi_j(x, \xi)\big(p_m(x, \xi)A_{2m-2}^{(j)}(x, \xi)\big)/q_{m-1}^{(j)}(x, \xi),$$

since $|q_{m-1}^{(j)}(x, \xi)| \geq c > 0$ on $\operatorname{supp} \varphi_j$. For $j = 0$ we find that

$$p_{2m-1}(x, \xi)\varphi_0(x, \xi) = p_{2m-1}(x, \xi)p_m(x, \xi)\varphi_0(x, \xi)/p_m(x, \xi),$$

since $|p_m(x, \xi)| \geq c_0 > 0$ on $\operatorname{supp} \varphi_0$.

Thus, the symbol of the operator $\Sigma_{j=0}^k(P_{2m-1}\varphi_j)(x, D)$ admits the representation

$$\sum_{j=1}^{k} p_{2m-1}\big(x, a_j(x, \xi'), \xi'\big)\varphi_j(x, \xi) + p_m(x, \xi)$$

$$\times \left[\frac{\varphi_0(x, \xi)}{p_m(x, \xi)}p_{2m-1}(x, \xi) + \sum_{j=1}^{k} \frac{\varphi_j(x, \xi)A_{2m-2}^{(j)}(x, \xi)}{q_{m-1}^{(j)}(x, \xi)} \right].$$

Using the fact that $\varphi_j(x, \xi)p_{2m-1}(x, a_j(x, \xi'), \xi') \geq 0$ in the whole space and also Gårding's inequality (see [4]), we obtain the desired inequality. We note that the symbol $p_{m-1}(x, \xi)$ has the form

$$-p_{m-1}(x, \xi) = \sum_{j=1}^{k} \frac{\varphi_j(x, \xi)A_{2m-2}^{(j)}(x, \xi)}{q_{m-1}^{(j)}(x, \xi)} + p_{2m-1}(x, \xi)\frac{\varphi_0(x, \xi)}{p_m(x, \xi)}.$$

COROLLARY. *Under the assumptions of Lemma 4, for any $\varepsilon > 0$ there exists a constant $C(\varepsilon)$ such that*

$$\mathrm{Re}\big(P_{2m-1}(x, D)u, u\big) + C(\varepsilon)\|u\|_{m-1}^2 + \varepsilon\|P_m u\|_0^2 \geq 0, \qquad \forall u \in C_0^\infty(K).$$

We come now to the proof of Theorem 5. We consider the equation

$$p_m^2(x, D)u + p_{2m-1}(x, D)u + p_{2m-2}(x, D)u = f, \tag{43}$$

where $p_{2m-2} \in S_{1,0}^{2m-2}$. We take the inner product of both sides of (43) with u. From the formula for the composition of pseudodifferential operators we find that

$$p_m \circ p_m = p_m^2 - i \sum_{j=1}^n \frac{\partial p_m}{\partial \xi_j} \frac{\partial p_m}{\partial x_j} - S_{2m-2}.$$

Consequently, (43) may be written as follows:

$$\big(p_m \circ p_m u, u\big) + i \sum_{j=1}^n \left(\frac{\partial p_m}{\partial \xi_j} \frac{\partial p_m}{\partial x_j} u, u\right) + \big(p_{2m-1}u, u\big) + \big(T_{2m-2}u, u\big) = (f, u),$$

where $T_{2m-2} \cap S_{1,0}^{2m-2}$. Hence

$$\big(p_m u, p_m^* u\big) + i \sum_{j=1}^n \left(\frac{\partial p_m}{\partial \xi_j} \frac{\partial p_m}{\partial x_j} u, u\right) + \big(p_{2m-1}u, u\big) + \big(T_{2m-2}u, u\big) = (f, u).$$

$$\tag{44}$$

The operator p_m^* has a symbol admitting the representation

$$p_m^* \sim p_m - i \sum_{j=1}^n \frac{\partial^2 p_m}{\partial x_j \partial \xi_j} + S'_{2m-2}.$$

Substituting the expression for p_m^* into (44) and transferring $\partial^2 p_m / \partial x_j \partial \xi_j$ over to the left side of the inner product $(p_m u, p_m^* u)$, we obtain

$$\|p_m u\|_0^2 + \big(p'_{2m-1}(x, D)u, u\big) + \big(R(x, D)u, u\big) = (f, u),$$

where $|(R(x, D)u, u)| \leq C\|u\|_{m-1}^2$. With the help of canonical transformations (see [14] or [16]), we readily find that

$$\|u\|_{m-1}^2 \leq C(\omega)\|p_m u\|_0^2, \qquad u \in C_0^\infty(\omega),$$

$$C(\omega) \to 0 \quad \text{as diam } \omega \to 0.$$

We now consider cases a), b), and c) separately.

Case a). From the relations we have obtained it follows, by virtue of the corollary to Lemma 4, that

$$\tfrac{1}{2}\|p_m u\|_0^2 + \mathrm{const}\|u\|_{m-1}^2 \leq \mathrm{Re}(f, u) \leq \|f\|_{1-m}\|u\|_{m-1}.$$

Since we can assume that the constant coefficient of $\|u\|_{m-1}^2$ on the left side of the inequality does not depend on ω, it follows that

$$\|u\|_{m-1}^2 \leq C'(\omega)\|f\|_{1-m}^2, \qquad \forall u \in C_0^\infty(\omega),$$

$$C'(\omega) \to 0, \quad \text{as diam } \omega \to 0.$$

Going over now to the adjoint operator $'P$ and using the fact that the real part of its subprincipal symbol is also nonnegative on the characteristic manifold of $p_m(x, -\xi)$, we obtain the theorem on local solvability.

Case b). In this case we employ the exact Gårding inequality from [2]. For this we introduce an operator of order $2m - 1$ with the symbol

$$A_{2m-1} = \operatorname{Re} p'_{2m-1} + p_m^2/\varepsilon|\xi|, \qquad \varepsilon > 0.$$

Let $x \in K$ and $|\xi| = 1$. Obviously, $\operatorname{Re} p'_{2m-1} > 0$ in some neighborhood of the characteristic manifold of the operator P_m. Consequently, choosing $\varepsilon > 0$ sufficiently small, we find that $A_{2m-1}(x, \xi) > 0$; that is, $A_{2m-1}(x, \xi)$ is a nondegenerate elliptic operator. Thus,

$$\tfrac{1}{2}\| p_m u\|_0^2 + \operatorname{Re}(A_{2m-1}(x, D)u, u) + O(\|u\|_{m-1}^2) \leq \operatorname{Re}(f, u)$$

or

$$\frac{1}{C(\omega)}\|u\|_{m-1}^2 + C_1\|u\|_{m-1/2}^2 - C_2\|u\|_{m-1}^2 \leq \|f\|_{1/2-m}\|u\|_{m-1/2},$$

where C_1 and C_2 are constants, $C_1 > 0$. Here C_1 and C_2 do not depend on ω, and $C(\omega) \to 0$ as $\operatorname{diam} \omega \to 0$. Taking a sufficiently small neighborhood ω, we find that

$$\|u\|_{m-1/2} \leq C\|f\|_{1/2-m}, \qquad \forall u \in C_0^\infty(\omega).$$

We take note here of the fact that the loss of smoothness is equal to one unit in this case (in comparison with elliptic operators).

Case c). Let $\operatorname{Im} p'_{2m-1}(x, \xi) > 0$ on $K, |\xi| > 0$. We may assume that $p'_{2m-1}(x, \xi)$ is pure imaginary. Indeed,

$$\operatorname{Im}(p'_{2m-1}u, u) = \operatorname{Im}(\operatorname{Re} p'_{2m-1}u, u) + \operatorname{Re}(\operatorname{Im} p'_{2m-1}u, u)$$

and since

$$(\operatorname{Re} p'_{2m-1}u, u) = (u, \operatorname{Re} p'_{2m-1}u) + O(\|u\|_{m-1}^2),$$

it follows that

$$|\operatorname{Im}(\operatorname{Re} p'_{2m-1}u, u)| \leq C\|u\|_{m-1}^2.$$

Taking the imaginary part of both sides of the identity

$$(p'_{2m-1}u, u) + \| p_m u\|_0^2 + (S_{2m-2}u, u) = (f, u),$$

we obtain

$$\operatorname{Im}(p'_{2m-1}u, u) + O(\|u\|_{m-1}^2) \leq \operatorname{Im}(f, u).$$

From the fact that $\operatorname{Im} p'_{2m-1}$ is a strictly positive symbol, it folows that constants $C_1 > 0$ and C_2 exist yielding the inequality

$$C_1\|u\|_{m-1/2}^2 + C_2\|u\|_{m-1}^2 \leq \|f\|_{1/2-m}\|u\|_{m-1/2}.$$

Since $m \geq 1$,

$$\|u\|_{m-1} \leq C(\omega)\|u\|_{m-1/2}, \qquad \forall u \in C_0^\infty(\omega)$$

(see [7]), where $C(\omega) \to 0$ as diam $\omega \to 0$. Taking a sufficiently small neighborhood ω, we obtain the required estimate.

The following corollary follows from Theorems 1 and 5.

COROLLARY. *Consider a pseudodifferential operator $P(x, D)$ with leading symbol $p_{2m}^0 = p_m^{0^2}$, where p_m^0 is a real symbol of principal type. Assume that Im $p_{2m-1}'(x, \xi)$ has along any zero bicharacteristic of p_m^0 a zero of finite odd order and that Re p_{2m-1}' preserves its sign, where $p_m^0 = 0$. Then the operator $P(x, D)$ is locally solvable if and only if Re $p_{2m-1}'(x, \xi) > 0$ at all the characteristic points of p_m^0.*

§9. On nonhypoellipticity of differential operators with double characteristics

In this section we shall show that there exist differential operators with characteristics of multiplicity identically two and with complex-valued coefficients, which are not hypoelliptic.

THEOREM 6. *Let $P(x, D)$ be a differential operator with infinitely smooth coefficients for the principal part of which the representation $p_{2m+r}^0 = p_m^2 q_r$ holds, where q_r is an elliptic symbol in some neighborhood of the point (x^0, ξ^0) while p_m is a symbol of principal type such that $p_m(x^0, \xi^0) = 0$. Assume that Im $p_m \equiv 0$ on some open arc with center at (x^0, ξ^0) of the zero bicharacteristic of Re p_m. In addition, assume that $\mathrm{Re\,grad\,} p_m$ and $\mathrm{Im\,grad\,} p_m$ are linearly dependent at any characteristic point of p_m. Then $P(x, D)$ is not hypoelliptic at x^0 if $p_{2m+r-1}' \equiv 0 \pmod{p_m}$ in a neighborhood of (x^0, ξ^0).*

For the proof we employ the following necessary condition for hypoellipticity of a differential operator (see [12]).

Let $P(x, D)$ be a differential operator, hypoelliptic in an open set $\Omega' \subset \Omega$. Then for any positive integer M and for each compactum $K_1 \subset \Omega'$ there exist a positive integer M_1, a compactum $K_1 \subset \Omega'$, and a constant $C > 0$ such that for any function $\varphi \in C^\infty(\Omega')$

$$\sup_{x \in K} \sum_{|\alpha| \leqslant M} |D^\alpha \varphi(x)| \leqslant C \sup_{x \in K_1} \left(|\varphi(x)| + \sum_{|\alpha| \leqslant M_1} |D^\alpha(P(x, D)\varphi)| \right).$$

Throughout the sequel we assume tht Ω' is a neighborhood of the point $x^0 = 0$ and that $0 \in \mathrm{int}\, K, K_1$. It follows from this that for any $\varphi \in C^\infty(\Omega')$

$$|\mathrm{grad}\,\varphi(0)| \leqslant C \sup_{x \in K_1} \left(|\varphi(x)| + \sum_{|\alpha| \leqslant M_1} |D^\alpha(P(x, D)\varphi)| \right). \tag{45}$$

We show that (45) is violated for some function $\varphi \in C^\infty(\Omega')$.

The first step in the proof consists in approximate solution of the characteristic equation $p_m(x, \mathrm{grad}\, w) = 0$. With the help of the implicit function theorem, a suitable change of variables, and a result from [9], we obtain the following factorization of the symbol $p_m(x, \xi)$ in a neighborhood of the characteristic point (x^0, ξ^0):

$$p_m(x, \xi) = (\xi_1 - a(x, \xi') - ib(x, \xi'))q_{m-1}(x, \xi), \tag{46}$$

where the symbol $q_{m-1}(x, \xi)$ is elliptic in a neighborhood of (x^0, ξ^0) and positively homogeneous of degree $m - 1$ in ξ', while the functions $a(x, \xi')$ and $b(x, \xi')$ are such that

$$a(x, \xi'^0) = \operatorname{grad}_{x, \xi'} a(x, \xi'^0) = 0; \qquad b(x_1, 0, \xi'^0) = 0 \quad \text{for } a < x_1 < b,$$

$$(a < 0 < b) \quad \text{and} \quad \operatorname{grad}_{x, \xi'} b(x_1, 0, \xi'^0) = 0 \quad \text{for all } x_1 \in (a, b).$$

We are required to find a function $w \in C^\infty$ which is a solution of the Cauchy problem for the equation

$$p_m(x, \operatorname{grad} w) = O(|y|^N), \quad y \to 0; \qquad \operatorname{Im} w(x_1, y) \sim C|y|^2, \quad C > 0.$$

In [9] Trèves outlined a scheme for determining w. In spite of this, we point out the main steps in the construction of $w(x_1, y)$.

To do this, we put

$$w(x_1, y) = (y, \xi'^0) + w_1(x_1, y) \Rightarrow p_m(x, \xi^0 + \operatorname{grad} w_1) = O(|y|^N), \quad y \to 0,$$
$$(47)$$

and we require that $w_1|_{x_1=0} = i|y|^2$. By virtue of the elliptic factorization (46), clearly it will be sufficient for us to solve the following Cauchy problem approximately:

$$\partial w_1/\partial x_1 - a(x, \xi'^0 + \operatorname{grad}_y w_1) - ib(x, \xi'^0 + \operatorname{grad}_y w_1) = O(|y|^N),$$

$$w_1(0, y) = i|y|^2.$$

Expanding $a(x, \xi'^0 + \operatorname{grad}_y w_1)$ and $b(x, \xi'^0 + \operatorname{grad}_y w_1)$ by Taylor's formula, we obtain for w_1 the equation

$$\frac{\partial w_1}{\partial x_1} = \sum_{|\alpha+\beta|<N} \frac{y^\beta}{\alpha!\beta!} \frac{\partial^{\alpha+\beta} a(x_1, 0, \xi'^0)}{\partial \xi'^\alpha \partial y^\beta} (\operatorname{grad}_y w_1)^\alpha$$

$$+ i \sum_{|\alpha+\beta|<N} \frac{1}{\alpha!\beta!} \frac{\partial^{\alpha+\beta} b(x_1, 0, \xi'^0)}{\partial \xi'^\alpha \partial y^\beta} y^\beta (\operatorname{grad}_y w_1)^\alpha, \qquad (48)$$

$$w_1(0, y) = i|y|^2.$$

We seek w_1 in the form

$$w_1(x_1, y) = i|y|^2 + \sum_{|\alpha|=2}^{N-1} w_\alpha(x_1) y^\alpha.$$

Substituting this expression for $w_1(x_1, y)$ into (48) and equating the coefficients of y^α, $|\alpha| < N$, to zero, we obtain the system of ordinary differential equations

$$w_\alpha'(x_1) = \Phi_\alpha(x_1, w_\beta(x_1)), \qquad |\alpha| < N, |\beta| < N,$$

$$w_\alpha(0) = 0,$$

for the determination of the coefficients $w_\alpha(x_1)$.

We need to show now that the domain of the w_α does not decrease as N increases.

Indeed, for $a \leqslant x_1 \leqslant b$ the function $\Phi_\alpha(x_1, w_\beta(x_1))$ does not depend on those $w_\beta(x_1)$ for which $|\beta| > |\alpha|$. In fact, the function $w_\beta(x_1)$ could only enter in Φ_α when $|\beta| = |\alpha| + 1$, in which case this dependence must be linear and with a coefficient independent of y. By virtue of (48) this coefficient must be equal to

$$(\partial a/\partial\xi_j)(x_1, 0, \xi'^0) + i(\partial b/\partial\xi_j)(x_1, 0, \xi'^0).$$

But these derivatives are equal to zero for $a \leqslant x_1 \leqslant b$ by virtue of (46). Thus, we can solve the equations in the system $w'_\alpha = \Phi_\alpha(x_1, w_\beta)$ successively.

Further, the same considerations show that for $|\alpha| \geqslant 3$ the functions Φ_α depend on $w_\alpha(x_1)$, $|\alpha| = j \geqslant 3$, linearly, and therefore the system of equations corresponding to these α has a solution over the whole interval $a \leqslant x_1 \leqslant b$. For $|\alpha| = 2$ the functions Φ_α depend on w_α nonlinearly, namely, they contain terms of the form $w_{ij} w_{ki} y_i y_j$. Therefore we first integrate the system with zero initial conditions for the function w_α, $|\alpha| = 2$, and determine w_α on the interval $(a + \varepsilon, b - \varepsilon)$, $\varepsilon > 0$. Then from the equations of the system we find w_α for $|\alpha| \geqslant 3$ over the whole interval $(a + \varepsilon, b - \varepsilon)$. Taking a smaller neighborhood $(a + \varepsilon, b - \varepsilon)$, we can assume that

$$\left| \sum_{|\alpha| \geqslant 2} w_\alpha(x_1) y^\alpha \right| \leqslant \frac{1}{4} |y|^2, \qquad \frac{\partial p_m(x_1, 0, \xi^0)}{\partial \xi_1} \neq 0.$$

Finally, from the inequality $|\operatorname{grad} w_1| \leqslant C|y|$ it follows that the solution of (48) is, in fact, a solution of (47).

We carry out all further considerations in some neighborhood Ω of the origin such that

$$\operatorname{Pr}_{x_1} \Omega \subseteq \{x_1 \mid a + \varepsilon < x_1 < b - \varepsilon\}.$$

We seek a function $\varphi(x_1, y)$, for which (45) is not satisfied, in the form

$$\varphi(x_1, y) = \sum_{j=0}^{L} \varphi_j(x_1, y) \lambda^{-j} e^{i\lambda w(x_1, y)}, \qquad \varphi_j \in C^\infty(\Omega).$$

If we succeed in determining φ_0 so that $\varphi_0(0) = 1$, then

$$\operatorname{grad} \varphi(0) = i\lambda \xi'^0 + o(1).$$

By means of a simple calculation we find that

$$P(x, D)\varphi = \sum_{j=0}^{2m+r+L-1} a_j(x_1, y) \lambda^{-j} e^{i\lambda w} \lambda^{2m+r},$$

where

$$a_0 = \varphi_0 p_{2m+r}(x, \mathrm{grad}\, w) = O(|y|^N), \qquad y \to 0,$$

$$a_1 = \varphi_1 p_{2m+r}(x, \mathrm{grad}\, w) + \sum_{j=1}^{n} \frac{\partial p_{2m+r}}{\partial \xi_j}(x, \mathrm{grad}\, w) D_{x_j}\varphi_0$$

$$+ \varphi_0\left(p_{2m+r-1}(x, \mathrm{grad}\, w) - \frac{i}{2} \sum_{j,k=1}^{n} \frac{\partial^2 p_{2m+r}}{\partial \xi_j \partial \xi_k}(x, \mathrm{grad}\, w)\frac{\partial^2 w}{\partial x_j \partial x_k} \right)$$

$$= \varphi_1 p_{2m+r}(x, \mathrm{grad}\, w) + \sum_{j=1}^{n} \frac{\partial p_{2m+r}}{\partial \xi_j}(x, \mathrm{grad}\, w) D_{x_j}\varphi_0$$

$$+ \varphi_0 p'_{2m+r-1}(x, \mathrm{grad}\, w) - \frac{i}{2}\varphi_0 \sum_{j=1}^{n} \frac{\partial}{\partial x_j}\left(\frac{\partial p_m}{\partial \xi_j}(x, \mathrm{grad}\, w) \right) = O(|y|^{N-1})$$

(since $p'_{2m+r-1} \equiv 0 \pmod{p_m}$).

We are not interested in all the terms in the coefficient of λ^{4m+r-2}. We merely note that from p_{2m+r-1} and p_{2m+r-2} there arise terms containing φ_0 or $\partial\varphi_0/\partial x_j$. The only expression containing $\partial^2\varphi_0/\partial x_j \partial\xi_k$ has the form

$$\frac{1}{2} \sum_{j,k=1}^{n} \frac{\partial^2 p_{2m+r}}{\partial \xi_j \partial \xi_k}(x, \mathrm{grad}\, w) D^2_{x_j x_k}\varphi_0,$$

and since, by hypothesis, $\partial p_m(x_1, 0, \xi^0)/\partial\xi_1 \neq 0$, it follows that the coefficient of $\partial^2\varphi_0/\partial x_1^2$ will be nonzero for $a + \varepsilon < x_1 < b - \varepsilon$, where $\varepsilon > 0$ is sufficiently small. We determine φ_0 as an approximate solution of some Cauchy problem for a second order differential operator with the initial data $\varphi_0(0, y) = 1$, $\partial\varphi_0(0, y)/\partial x_1 = 0$:

$$K\varphi_0 \equiv -\frac{1}{2} \sum_{j,k=1}^{n} \frac{\partial^2 p_{2m+r}}{\partial \xi_j \partial \xi_k}(x, \mathrm{grad}\, w)\frac{\partial^2\varphi_0}{\partial x_j \partial x_k}$$

$$+ \sum_{j=1}^{n} b_j \frac{\partial\varphi_0}{\partial x_j} + c\varphi_0 = O(|y|^{N-1}). \tag{49}$$

As in the cases considered earlier, we first expand the coefficients (49) up to order $N - 1$ in y and then replace these functions by segments of their Taylor series. We seek the function $\varphi_0(x_1, y)$ in the form

$$\varphi_0 = 1 + \sum_{|\alpha|=0}^{N-2} \tilde{\varphi}_j(x_1) y^\alpha.$$

Substituting this expression for φ_0 into (49) and equating to zero the coefficients of powers of y_j, we obtain for the determination of the $\tilde{\varphi}_j(x_1)$ a system of second order ordinary linear differential equations with zero initial conditions. Since the system is solved for $d^2\tilde{\varphi}_j/dx_1^2$, there exists a C^∞-solution "in the large" on the interval $(a + \varepsilon, b - \varepsilon)$.

We then multiply the function φ_0 by a cutoff function $g(y) \in C_0^\infty$ such that $g \equiv 1$ in some neighborhood of $y^0 = 0$. Obviously, $K(g\varphi_0) = O(|y|^{N-1})$ and, in addition, the domain of $g\varphi_0$ contains the domain $(a + \varepsilon, b - \varepsilon) \times \{|y| < \eta\}$, $\eta > 0$. Since

$$a_\mu(x_1, y) = O(|y|^{N-1}) - \frac{i}{2} \sum_{j,k=1}^{n} \frac{\partial^2 p_{2m+r}}{\partial \xi_j \partial \xi_k}(x, \text{grad } w) \frac{\partial^2 \varphi_{\mu-2}}{\partial x_j \partial x_k}$$

$$+ \sum_{j=1}^{n} b_j \frac{\partial \varphi_{\mu-2}}{\partial x_j} + C_\mu,$$

where C_μ is a linear combination of the functions $\varphi_0, \dots, \varphi_{\mu-3}$ and their derivatives, it is easy to find a function $\varphi_{\mu-2} \in C^\infty[(a + \varepsilon, b - \varepsilon) \times |y| < \eta]$ such that $a_\mu = O(|y|^{N-\mu})$, $y \to 0$.

We note, finally, that if supp g is sufficiently small, then in it we have the estimate

$$|e^{i\lambda w(x_1 y)}| \leqslant e^{-|y|^2/2}, \qquad a + \varepsilon \leqslant x_1 \leqslant b - \varepsilon.$$

We conclude the proof of the theorem with the help of Lemma 6.1.5 from [19].

REMARK. The condition on the lower terms $p'_{2m+r-1} \equiv 0 \pmod{p_m}$ cannot be discarded. Indeed, if we consider an operator with the leading symbol ξ_1^2, the condition of nonhypoellipticity on the lower terms reduces to the fact that the operator has the form $\xi_1^2 + j(x_1, x_2)\xi_1$. On the other hand, it is well known that the operator $\xi_1^2 + ia\xi_2$, where $a \neq 0$ is a real constant, is hypoelliptic.

§10. An application to a nonelliptic boundary value problem

1. As an example leading to a pseudodifferential equation with double characteristics, we consider a boundary value problem for the Laplace equation in a half-space.

Let $\mathbf{R}_+^{n+1}$ be the portion of $\mathbf{R}^{n+1}$ for whose points $X = (x_1, \dots, x_n, x_{n+1}) = (x, x_{n+1})$ the inequality $x_{n+1} \geqslant 0$ is satisfied.

In $\mathbf{R}_+^{n+1}$ we consider the equation

$$\Delta u = 0$$

with the boundary conditions

$$a\partial^2 u/\partial x_1^2 + \varphi(x_1)\partial u/\partial x_{n+1} = g(x) \quad \text{for } x_{n+1} = 0, \tag{50}$$

where $a = \alpha + i\beta$ is a constant, $\beta \neq 0$, $\varphi(x_1) \in C^\infty$, φ being a real function, and $g(x) \in C^\infty$.

Let

$$\tilde{u}(\xi, x_{n+1}) = \int u(x, x_{n+1}) e^{-i(x,\xi)} \, dx$$

be the partial Fourier transform with respect to x. By virtue of (50), for $x_{n+1} > 0$ the function $\tilde{u}(\xi, x_{n+1})$ satisfies

$$\partial^2 \tilde{u}/\partial x_{n+1}^2 - |\xi|^2 \tilde{u} = 0.$$

From this we find that

$$\tilde{u}(\xi, x_{n+1}) = v_1(\xi)e^{-x_{n+1}|\xi|} + v_2(\xi)e^{x_{n+1}|\xi|}.$$

Since we require a bounded solution for $x_{n+1} > 0$, we put $v_2 \equiv 0$. Substituting this expression for $\tilde{u}(\xi, x_{n+1})$ into the boundary condition (50), we find that

$$a\frac{\partial^2}{\partial x_1^2} \int v_1(\xi)e^{-x_{n+1}|\xi|+i(x,\xi)}\, d\xi + \varphi(x_1)\frac{\partial}{\partial x_{n+1}} \int v_1(\xi)e^{-x_{n+1}|\xi|+i(x,\xi)}\, d\xi$$

$$= (2\pi)^n g(x) \quad \text{for } x_{n+1} = 0.$$

Consequently,

$$\int \left(a\xi_1^2 + \varphi(x_1)|\xi|\right)v_1(\xi)e^{i(x,\xi)}\, d\xi = -(2\pi)^n g(x). \tag{51}$$

The expression (51) is a pseudodifferential equation with the symbol

$$p(x, \xi) = \xi_1^2 + \varphi(x_1)|\xi|/a.$$

Its leading symbol has the form ξ_1^2, i.e. $p_1(x, \xi) = \xi_1$. Thus,

$$p(x, \xi) = \xi_1^2 + (a - i\beta)\varphi(x_1)|\xi|/(a^2 + \beta^2).$$

From Theorem 5 we see that the boundary value problem (50) is solvable, at least locally, if one of the following requirements is satisfied:

a) $\alpha\varphi(x_1) \geqslant 0, \quad \forall x_1 \in (-\varepsilon, \varepsilon), \varepsilon > 0;$

b) $\beta\varphi(x_1) < 0, \quad \forall x_1 \in (-\varepsilon, \varepsilon).$

2. We now consider a boundary value problem of the following form for the Laplace equation:

$$\varphi(x_1)\partial u/\partial x_{n+1} + \partial^2 u/\partial x_1^2 = g(x) \quad \text{for } x_{n+1} = 0, \tag{52}$$

where $\varphi(x_1)$ is a smooth function of x_1, such that

$$\operatorname{Im}\varphi(x_1) = Cx_1^{2k+1} + O(x_1^{2k+2}), \quad C \neq 0, \operatorname{Re}\varphi(x_1) \neq 0.$$

Then for the determination of u we obtain a pseudodifferential equation with the symbol $p(x, \xi) = \xi_1^2 + \varphi(x_1)|\xi|$. Applying Theorems 1 and 5, we immediately find that problem (52) is solvable for any function g if and only if $\operatorname{Re}\varphi(x_1) > 0$ in some neighborhood of the origin.

Note added in proof. In a recent paper Cardoso and Treves [24] proved a theorem on the unsolvability of the operator $P(x, D)$ studied in Theorem 2, without any restrictions on the lower terms.

Received 19/SEPT/73

BIBLIOGRAPHY

1. Lars Hörmander, *Differential equations without solutions*, Math. Ann. **140** (1960), 169–173.

2. J. J. Kohn and L. Nirenberg, *An algebra of pseudo-differential operators*, Comm. Pure Appl. Math. **18** (1965), 269–305.

3. Lars Hörmander, *Pseudo-differential operators*, Comm. Pure Appl. Math. **18** (1965), 501–517.

4. ______, *Pseudo-differential operators and non-elliptic boundary problems*, Ann. of Math. (2) **83** (1966), 129–209.

5. ______, *Pseudo-differential operators and hypoelliptic equations*, Proc. Sympos. Pure Math., vol. 10, Amer. Math. Soc., Providence, R.I., 1967, pp. 138–183.

6. Louis Nirenberg and François Trèves, *On local solvability of linear partial differential equations. I: Necessary conditions*, Comm. Pure Appl. Math. **23** (1970), 1–38.

7. ______, *On local solvability of linear partial differential equations. II: Sufficient conditions*, Comm. Pure Appl. Math. **23** (1970), 459–510.

8. François Trèves, *Hypoelliptic partial differential equations of principal type with analytic coefficients*, Comm. Pure Appl. Math. **23** (1970), 637–651.

9. ______, *Hypoelliptic partial differential equations of principal type. Sufficient conditions and necessary conditions*, Comm. Pure Appl. Math. **24** (1971), 631–670.

10. J. J. Duistermaat and L. Hörmander, *Fourier integral operators. II*, Acta Math. **128** (1972), 183–269.

11. Anneli Lax, *On Cauchy's problem for partial differential equations with multiple characteristics*, Comm. Pure Appl. Math. **9** (1956), 135–169.

12. Lars Hörmander, *Hypoelliptic second order differential equations*, Acta Math. **119** (1967), 147–171.

13. Ju. V. Egorov, *On the local solvability of pseudo-differential equations*, Proc. Internat. Congr. Math. (Nice, 1970), Vol. 2, Gauthier-Villars, Paris, 1971, pp. 717–722.

14. ______, *Canonical transformations and pseudodifferential operators*, Trudy Moskov. Mat. Obšč. **24** (1971), 3–28; English transl. in Trans. Moscow Math. Soc. **24** (1971).

15. ______, *On necessary conditions for the solvability of pseudodifferential equations of principal type*, Trudy Moskov. Mat. Obšč. **24** (1971), 29–41; English transl. in Trans. Moscow Math. Soc. **24** (1971).

16. ______, *On the solvability of differential equations with simple characteristics*, Uspehi Mat. Nauk **26** (1971), no. 2(158), 183–198; English transl. in Russian Math. Surveys **26** (1971).

17. V. V. Grušin, *Singularities of solutions of a certain class of pseudodifferential and degenerating elliptic equations*, Uspehi Mat. Nauk **26** (1971), no. 1(157), 221–222. (Russian)

18. V. Ja. Ivriĭ, *Differential equations with multiple characteristics and with no solutions*, Dokl. Akad. Nauk SSSR **198** (1971), 279–282; English transl. in Soviet Math. Dokl. **12** (1971).

19. Lars Hörmander, *Linear partial differential operators*, Academic Press, New York, and Springer-Verlag, Berlin 1963.

20. O. A. Oleĭnik and E. V. Radkevič, *On the analyticity of solutions of linear partial differential equations*, Mat. Sb. **90(132)** (1973), 592–606; English transl. in Math. USSR Sb. **19** (1973).

21. Ju. V. Egorov and P. R. Popivanov, *Equations of principal type without solution*, Uspehi Mat. Nauk **29** (1974), no. 2(176), 172–189; English transl. in Russian Math. Surveys **29** (1974).

22. P. R. Popivanov, *The solvability and hypoellipticity of differential equations*, Uspehi Mat. Nauk **29** (1974), no. 1(175), 185–186. (Russian)

23. ______, *The local solvability of pseudodifferential equations with double characteristics*, Uspehi Mat. Nauk **29** (1974), no. 5(179), 233–234. (Russian)

24. Fernando Cardoso and François Treves, *A necessary condition of local solvability for pseudo-differential equations with double characteristics*, Ann. Inst. Fourier (Grenoble) **24** (1974), fasc. 1, 225–292.

Translated by J. F. HEYDA

Amer. Math. Soc. Transl.
(2) Vol. **118**, 1982

On Bifurcation of Topological Type
of Singular Points of Vector Fields
Depending on Parameters*

A. N. ŠOŠITAĬSVILI

§1. Formulation of results

This paper contains the proof of the results announced in the author's note [6].

A singular point of a system of ordinary differential equations is called *degenerate* if some of its eigenvalues lie on the imaginary axis. Degenerate singular points are encountered in the study of systems depending on parameters. In a k-parameter system points with degeneracy of codimension k occur in an inherent manner [1].

It is shown in this paper that both the study of the topological structure of a vector field in the neighborhood of a degenerate singular point and the study of its bifurcation reduce to a similar problem with the number of variables equal to the number k^0 of pure imaginary eigenvalues. The main theorem of the paper (Theorem 1) states that near a degenerate singular point and for values of the parameters close to critical, the family is topologically equivalent to the product of a family with a k^0-dimensional phase space and a standard multidimensional saddle.

We now introduce the terminology needed for a precise formulation of Theorem 1.

DEFINITION 1. By a C^k-*family of systems of ordinary differential equations* (in the future simply a C^k-*family*) we understand a C^k-system

$$\begin{cases} \dot{z} = v(z, \varepsilon), \\ \dot{\varepsilon} = 0, z \in R^n, \varepsilon \in R^l. \end{cases} \tag{1}$$

1980 *Mathematics Subject Classification.* Primary 57R25, 58F14.
* Translation of Trudy Sem. Petrovsk. **1** (1975), 279–309. MR **57** #17724.

DEFINITION 2. The family

$$\begin{cases} \dot{z} = v_2(z, \eta), \\ \dot{\eta} = 0, z \in R^n, \eta \in R^{l_2}, \end{cases} \tag{2}$$

is said to be *induced* from the family (1) if there exists a continuous mapping $h: R^n \times R^{l_2} \to R^n \times R^l$, the restriction of which to every plane of the form $\eta = c$ is a homeomorphism with the plane $\varepsilon = c_1(c)$ and which associates the system $\dot{z} = v_2(z, c)$ with the system $\dot{z} = v(z, c_1(c))$.

DEFINITION 3. The family (2) is said to be *homeomorphic* to the family (1) if the situation of Definition 2 holds and if the mapping $h: R^n \times R^{l_2} \to R^n \times R^l$ is a homeomorphism

DEFINITION 4. The family (1) is called *versal* if every family (2) such that $v_2(z, \eta)|_{\eta=0} = v(z, \varepsilon)|_{\varepsilon=0}$ is induced from the family (1). The family is called a *versal family of the system* $\dot{z} = w(z)$ if it is versal and if $v(z, 0) = w(z)$.

DEFINITION 5. By the C^k-*germ of the family* we understand the germ of the C^k-system (1) at the point ($z = 0, \varepsilon = 0$).

REMARK. Any assertion about germs should be understood as meaning that there exist representatives of the germs for which the given assertion holds.

THEOREM 1. *Let the C^k-germ, $2 \leqslant k < \infty$, of the family*

$$\begin{cases} \dot{z} = Bz + r(z, \varepsilon), \\ \dot{\varepsilon} = 0, z \in R^n, \varepsilon \in R^l, \end{cases} \tag{3}$$

be given, where $r \in C^k(R^n \times R^l)$, $r(0, 0) = 0$, $\partial_z r|_{(0,0)} = 0$, and $B: R^n \to R^n$ is a linear operator whose eigenvalues are divided into three groups:

$$\mathrm{I} = \left\{ \lambda_i, 1 \leqslant i \leqslant k^0 \,|\, \mathrm{Re}\, \lambda_i = 0 \right\},$$

$$\mathrm{II} = \left\{ \lambda_i, k^0 + 1 \leqslant i \leqslant k^0 + k^- \,|\, \mathrm{Re}\, \lambda_i < 0 \right\},$$

$$\mathrm{III} = \left\{ \lambda_i, k^0 + k^- + 1 \leqslant i \leqslant k^0 + k^- + k^+ \,|\, \mathrm{Re}\, \lambda_i > 0 \right\},$$

$$k^0 + k^- + k^+ = n.$$

Let the subspaces of R^n which are invariant with respect to B and which correspond to these groups be denoted by X, Y^-, and Y^+ respectively, and let $Y^- \times Y^+$ be denoted by Y.

Then the following assertions are true:

1) There exists a C^{k-1} manifold γ^0 that is invariant with respect to the germ (3), may be given by the graph of the mapping $\gamma^0: X \times R^l \to Y$, $y = \gamma^0(x, \varepsilon)$, and satisfies $\gamma^0(0, 0) = 0$ and $\partial_x \gamma^0(0, 0) = 0$.

2) The germ of the family (3) is homeomorphic to the product of the multidimensional saddle $\dot{y}^+ = y^+$, $\dot{y}^- = -y^-$, and the germ of the family

$$\begin{cases} \dot{x} = Bx + r_1(x, \varepsilon), \\ \dot{\varepsilon} = 0, \end{cases} \tag{4}$$

where $r_1(x, \varepsilon)$ is the x-component of the vector $r(z, \varepsilon)$, $z = (x, \gamma^0(x, \varepsilon))$, i.e. the germ of (3) is homeomorphic to the germ of the family

$$\begin{cases} \dot{y}^+ = y^+, & \dot{y}^- = -y^-, \\ \dot{x} = Bx + r_1(x, \varepsilon), & \dot{\varepsilon} = 0. \end{cases} \tag{5}$$

THEOREM 1'. *With the notation of Theorem 1 let the eigenvalues of B be divided into the three groups,* I, II, *and* III, *except that in group* I *the eigenvalues of B do not satisfy* $\operatorname{Re} \lambda_i = 0$. *Set*

$$\beta_1 = \sup_{\lambda_i \in \mathrm{I}} \operatorname{Re} \lambda_i, \qquad \beta_2 = \inf_{\lambda_i \in \mathrm{I}} \operatorname{Re} \lambda_i, \qquad \beta_3 = \sup_{\lambda_i \in \mathrm{II}} \operatorname{Re} \lambda_i, \qquad \beta_4 = \inf_{\lambda_i \in \mathrm{III}} \operatorname{Re} \lambda_i.$$

Let

$$\beta_3 < k\beta_2 < 0, \qquad k\beta_3 < \beta_2, \qquad 0 < \beta_1 < k\beta_4, \qquad k\beta_1 < \beta_4. \tag{6}$$

Then the conclusions of Theorem 1 also hold for such an operator B (recall that k is the smoothness of the germ of (3) under consideration).

REMARK. Theorem 1 can be applied to an individual system of differential equations. For example, the reduction principle proved by V. A. Pliss [2] follows from it. It states that if the stability of the zero solution of the system

$$\dot{z} = v(z), \qquad z \in R^n, \qquad v(0) = 0, \tag{7}$$

is not determined by the linear approximation, i.e. if the linear part B of the field v at zero has $n - k^0$ eigenvalues with negative real part and k^0 pure imaginary eigenvalues, then there exists a k^0-dimensional manifold γ_v^0 which passes through zero and is invariant with respect to the system (7), and the stability of the zero solution of (7) is the same as the stability of the zero solution of the system $(7)\big|_{\gamma_v^0}$.

Theorem 1 enables us to make a complete study of topological-type bifurcation of singular points of vector fields which depend in a general way on one parameter, i.e. bifurcation of codimension 1 (see [1]). These bifurcations are of two types. Corollaries 1 and 2, formulated below, give their normal forms.

We represent the number k^0 in the form $k^0 = k_1^0 + 2k_2^0$, where k_1^0 is the number of zero eigenvalues and k_2^0 is the number of conjugate pairs of pure imaginary nonzero eigenvalues.

COROLLARY 1. *Let $k_1^0 = 1$ and $k_2^0 = 0$. Consider the germ of the family*

$$\dot{w} = \beta + \rho w^2, \qquad \dot{y}^+ = y^+, \qquad \dot{y}^- = -y^-, \qquad \dot{\beta} = 0,$$
$$w \in R^1, \qquad \beta \in R^1, \qquad \rho = \pm 1, \qquad y^+ \in R^{k^+}, \qquad y^- \in R^{k^-}. \tag{8}$$

For the C^4-germ of the family (3) to be induced from the germ of the family (8) it is sufficient that $\partial_x^2 r_x \big|_{(0,0)} \neq 0$, where $r_x(z, \varepsilon)$ is the projection of $r(z, \varepsilon)$ onto the x-axis and $\rho = \operatorname{sgn} \partial_x^2 r \big|_{(0,0)}$.

COROLLARY 2. *Let $k_1^0 = 0$ and $k_2^0 = 1$. Consider the germ of the family*

$$\dot{w}_1 = -w_2 \cdot \mu(w, \varepsilon),$$

$$\dot{w}_2 = \left(w_1 + \beta w_2 + \rho w_2^3\right) \cdot \mu(w, \varepsilon),$$

$$\dot{y}^+ = y^+, \qquad \dot{y}^- = -y^-, \qquad \dot{\beta} = 0, \qquad \dot{\varepsilon} = 0, \tag{9}$$

where $(w_1, w_2) = w \in R^2$, $(\beta, \varepsilon) \in R^1 \times R^l$, $y^+ \in R^{k^+}$, $y^- \in R^{k^-}$, $\mu(w, \varepsilon) \in C(w, \varepsilon)$, $\mu(0, 0) \neq 0$ *and* $\rho = \pm 1$.

For the C^5-germ of the family (3) *to be induced from the germ of the family* (9) *it is sufficient that the third focal quantity d_3 of the system $\dot{x} = Bx + r_1(x, 0)$ (see (4)) be different from zero. The quantity d_3 can be expressed algebraically in terms of the coefficients of the first three powers of z in the Taylor series of the germ of the family* (3) *at the point* $(z = 0, \varepsilon = 0)$, $\rho = \operatorname{sgn} d_3$, *and the form of the function μ depends on the concrete form of the germ of the family* (3).

THEOREM 2. *The germ of the family* (3) *is versal if and only if the germ of the family* (4) *is versal.*

In this way Theorem 1 permits the reduction of various problems concerning the family (3), in particular the problem of finding versal families (Theorem 2), to problems concerning the lower-dimensional family (4).

However, Theorem 1 does not determine a natural mapping from the space of all systems of the form (3) to the space of systems of the form (4) obtained from (3) by retriction to the invariant manifold, since the invariant manifold itself is in general not unique. For example, the manifold $\gamma(c)$ in R^2, defined as the graph of $x_2 = c x_1^k(x_1, x_2) \in R^2$, $c \in R^1$, $k \in N$, is invariant for any c with respect to the system

$$\dot{x}_1 = x_1, \qquad \dot{x}_2 = kx_2. \tag{10}$$

The system (10) is homeomorphic to the Cartesian product of the different equations $\dot{x}_2 = kx_2$ and (10)$|_{\gamma(c)}$.

In the study of systems of the form (4) it is often necessary to exclude from consideration a certain "small" set M in the space of these systems. (For example, in the formulation of Corollary 2 we excluded the case where the first three focal quantities are zero.) Consider the set $\tilde{M}$, consisting of those systems of the form (3) the restriction of which to some invariant manifold proved to exist in Theorem 1 lies in the "small" set M in the space of systems of the form (4). A priori it could happen that simultaneously the set M is small while $\tilde{M}$ is large.

However, it follows from Theorem 3, formulated below, that if M is given by m equations for the values of the derivatives of the system (4) at the origin through order k, then $\tilde{M}$ is given by m equations for the values of the derivatives of (3) at the origin through order k.

In order to be able to formulate Theorem 3, we need several remarks.

Consider the space $\mathcal{L}(R^n)$ of linear operators $C: R^n \to R^n$, the manifold M^{k^0} of k^0-dimensional linear subspaces of R^n, the manifold M^{k^+} of k^+-dimensional subspaces of R^n and the manifold M^{k^-} of k^--dimensional subspaces of R^n, $k^+ + k^0 + k^- = n$.

Consider the linear operator $B: R^n + R^n$, and the linear subspaces X, Y^+, $Y^- \subset R^n$, where B, X, Y^+ and Y^- are defined in Theorem 1. For what follows, concerning the groups I, II, and III of eigenvalues of the operator B it will be sufficient to assume the inequalities (6) of Theorem 1′ in place of the conditions of Theorem 1.

Consider the neighborhoods $W(X) \subset M^{k^0}$, $W(Y^-) \subset M^{k^-}$ and $W(Y^+) \subset M^{k^+}$ of the points $X \in M^{k^0}$, $Y^- \in M^{k^-}$ and $Y^+ \in M^{k^+}$ respectively such that

$$R^n = \tilde{X} \times \tilde{Y}^- \times \tilde{Y}^+$$

$$\text{for any } \tilde{X} \in W(X), \ \tilde{Y}^+ \in W(Y^+) \text{ and } \tilde{Y}^- \in W(Y^-). \tag{11}$$

There exists a neighborhood $W_1(B) \subset \mathcal{L}(R^n)$ of the operator $B \in \mathcal{L}(R^n)$ such that for any operator $C \in W_1(B)$ there exists in each of the neighborhoods $W(X)$, $W(Y^+)$ and $W(Y^-)$ a unique subspace X_C, Y_C^+ and Y_C^- respectively which is invariant with respect to C. Denote the groups of eigenvalues of the operators $C|_{X_C}$, $C|_{Y_C^-}$ and $C|_{Y_C^+}$ by I_C, II_C and III_C respectively. There exists a neighborhood $W_2(B) \subset W_1(B)$ of B such that for any $C \in W_2(B)$ the inequalities (6) hold for the groups I_C, II_C and III_C.

Consider the set $U(R^n)$ of C^k-germs of vector fields on R^n which vanish at the origin and with linear part at the origin $C(v) \in W_2(B)$. From Theorem 1′, applied to an individual germ, it follows that there exists a manifold γ_v^0 which is invariant with respect to the germ $v \in U(R^n)$ and which is given by the graph of

$$y_{C(v)} = \gamma_v^0(x_{C(v)}), \qquad y_{C(v)} \in Y_{C(v)}, \qquad x_{C(v)} \in X_{C(v)}$$

where $Y_{C(v)} = Y_{C(v)}^+ \times Y_{C(v)}^-$. By virtue of (11) the manifold γ_v^0 can be given by the graphy of $\dot{y} = \gamma_v^0(x)$, $y \in Y$, $x \in X$, where Y and X are the subspaces invariant with respect to the operator B. In the sequel we shall use precisely this representation of γ_v^0.

Let $U(X)$ denote the set of C^{k-1}-germs at the origin of vector fields on X which vanish there, and let $I^{k-1}(X)$ and $I^{k-1}(R^n)$ denote the space of $(k-1)$-jets, at the origin, of the germs in $U(X)$ and $U(R^n)$ respectively.

We identify the manifold γ_v^0 with the subspace X by projecting γ_v^0 onto X along Y. The restriction of the germ v to γ_v^0 will become a germ in $U(X)$. If we associate with each germ $v \in U(R^n)$ its restriction to the manifold γ_v^0, and then identify γ_v^0 with X, we obtain a mapping $\pi: U(R^n) \to U(X)$.

The mapping π is ill-defined in the sense that it depends on the choice of γ_v^0. It can be shown that any manifold M which is invariant with respect to v is suitable as γ_v^0 if it is given by the graph of $y = f(x)$, i.e. that v is homeomorphic to the product of a multidimensional saddle and $v|_M$. However, because the inequalities (6) hold, it turns out that all manifolds M are tangent at the origin to order $k-1$. (See the example (10).)

In other words, the mapping π induces a mapping $\pi_{k-1}\colon I^{k-1}(R^n) \to I^{k-1}(X)$, which is well defined. Indeed, we have

THEOREM 3. 1. $\pi_{k-1}\colon I^{k-1}(R^n) \to I^{k-1}(X)$ *is a well-defined analytic mapping having maximal rank at every point.*

2. *Under* π_{k-1} *the preimage of any semialgebraic set in* $I^{k-1}(X)$ *is semialgebraic in* $I^{k-1}(R^n)$.

It follows from Theorem 3, for example, that there exist polynomials, in the coefficients of the Taylor series at the origin of the vector field that determines the system (3), of which the first $p + 1$ are zero if and only if the system which results from the restriction of (3) to some (and hence to any) two-dimensional invariant manifold γ^0 has its first p focal quantities equal to zero.

Theorems analogous to Theorems 1, 1′, 2 and 3 also hold for diffeomorphisms.

The principal method of proof of Theorems 1, 1′ and 2 is the construction of five invariant foliations in the neighborhood of a degenerate singular point of the vector field.

The study of singular points of vector fields and also of invariant manifolds has been the subject of a large number of papers (see, for example, the bibliography in [7]). Here we mention only those papers which are closest to the present paper.

In [4] results are formulated on existence of foliations for global diffeomorphisms that satisfy a condition more general than condition Y. In [5], under the same more general conditions, the existence of foliations for the system $\dot{x} = B(t)x + r(t, x)$ is proved (even in the nonautonomous case; in fact, the term "foliation" is not used there, and, more essentially, the smoothness of the leaves is not elucidated).

In [7] there is a theorem on the existence of an asymptotically stable invariant surface for a point mapping.

A theorem on an invariant surface, and its application to the problem of appearance of cycles, is contained in [10].

A number of assertions about trichotomic equations in Banach spaces are formulated without proof in [8].

The author expresses his deep gratitude to V. I. Arnol′d and A. G. Kušnirenko, without whom this work could not have been done.

§2. Proof of the existence of foliations

For a detailed definition of a foliation see [3]. For us it will be sufficient to use much simpler definitions. By a *foliation* of the space R^n we shall mean a representation of R^n in the form of a union of disjoint differentiable connected complete k-dimensional submanifolds such that there exists a homeomorphism $h\colon R^n \to R^n$ which sends this union into the union of parallel k-dimensional planes.

The submanifolds of a foliation are called *leaves*. A foliation is *invariant* with respect to a diffeomorphism $T\colon R^n \to R^n$ (a flow $e^{tv}\colon R^n \to R^n$) if each leaf of the foliation is sent by the diffeomorphism T (the mapping e^{tv} for each fixed t) into a leaf of the same foliation (but possibly a different leaf).

If the linear operator $B: R^n \to R^n$ (see (3)) has no eigenvalues on the imaginary axis, i.e. if $k^0 = 0$, equation (3) for fixed ε (sufficiently small) is an Anosov system (on a noncompact manifold), and the assertion that invariant foliations exist is a special case of Anosov's theorem on such systems ([3]).

The existence of foliations which are invariant with respect to the flow determined by the differential equation (3) in the case $k^0 \neq 0$ follows from Lemma 1, which we now proceed to formulate.

Consider an invertible linear operator $A: R^n \to R^n$. Let A have two invariant subspaces, X and Y, such that $R^n = X \times Y$.

The planes parallel to X form a foliation $\Gamma^1(A)$, and those parallel to Y form a foliation $\Gamma^2(A)$. The foliations $\Gamma^1(A)$ and $\Gamma^2(A)$ are invariant with respect to A.

Let the restriction $A|_X$ be denoted by P, and $A|_Y$ by Q. Let a norm on R^n be given such that

$$
\begin{aligned}
&|Px| < \lambda_1 |x|, \qquad \lambda_1 > 0, \forall x \in X, x \neq 0, \\
&|Qy| > \lambda_2 |y|, \qquad \lambda_2 > 0, \forall y \in Y, y \neq 0,
\end{aligned}
\tag{12}
$$

where λ_1 and λ_2 are constants which do not depend on $x \in X$ and $y \in Y$.

LEMMA 1. *For some integer* k, $2 \leqslant k < \infty$, *let*

$$
\lambda_1^k < \lambda_2, \qquad \lambda_1 < \lambda_2^k.
\tag{13}
$$

Then there exist a constant δ_0 and functions $c_i: (0, \delta_0) \to (0, \infty)$, $i = 1, 2$, depending only on λ_1 and λ_2, such that if a C^k-diffeomorphism $T: R^n \to R^n$ has derivatives through order k which are uniformly bounded in R^n and satisfies the condition

$$
\left|(T - A_*)|z|\right| < \delta, \qquad \forall z \in R^n
\tag{14}
$$

for some δ in the interval $(0, \delta_0)$, then

1) There exist unique foliations $\Gamma^1(T)$ and $\Gamma^2(T)$ which are invariant with respect to T and are such that

a) the leaves of $\Gamma^2(T)$ can be given by the graphs of C^{k-1}-mappings from Y into X, and the leaves of $\Gamma^1(T)$ can be given by the graphs of C^{k-1}-mappings from X into Y, and

b) the mappings which give the leaves of $\Gamma^1(T)$ and $\Gamma^2(T)$ satisfy a Lipschitz condition with constant $c_1(\delta)$, common to all of these mappings, where $c_1(\delta) \to \infty$ as $\delta \to 0$;

2) the mappings which give the leaves of $\Gamma^1(T)$ and $\Gamma^2(T)$ satisfy a Lipschitz condition with constant $c_2(\delta)$, common to all of these mappings, where $c_2(\delta) \to 0$ as $\delta \to 0$.

REMARK. For convenience in formulating Lemma 1 and the subsequent assertions we assume that $\delta \neq 0$. In the case $\delta = 0$ it follows from (14) that T differs from the linear operator A by a shift by a constant vector. Then $\Gamma^i(T) = \Gamma^i(A)$, $i = 1, 2$.

PROOF OF LEMMA 1. For the proof of Lemma 1 we first prove the existence of a "collection" which is invariant with respect to T. By a "collection" we mean a set of manifolds in R^n with a distinguished point on each manifold. Here each point of R^n is the distinguished point for exactly one manifold in the collection. The invariance of the collection with respect to T means that under the action of T the manifold with distinguished point z goes to the manifold in the same collection with distinguished point Tz.

In other words we consider the space Φ^c of continuous mappings $\varphi: X \to Y$ which satisfy a Lipschitz condition in x with constant c (common to all $\varphi \in \Phi^c$ and not depending on x):

$$\Phi^c = \{\varphi: X \to Y: |\varphi(x_1) - \varphi(x_2)| \leqslant c\,|x_1 - x_2|\,, \forall x_1, x_2 \in X\}.$$

We consider the space F^c of mappings $f: R^n \to \Phi^c$ such that $f_z(z_x) = z_y$, where $f_z \in \Phi^c$ is the value of f at the point $z \in R^n$, z_x is the x-component of z, and z_y is its y-component.

DEFINITION. $F^\infty = \bigcup_{c \geqslant 0} F^c$.

DEFINITION. A point of the space F^c is called a *c-collection*.

In the coordinates x and y let the diffeomorphism T have the form

$$T\binom{x}{y} = \binom{Px + p(x, y)}{Qy + q(x, y)}. \tag{15}$$

Condition (14) can be written in the form

$$\sup_{(x,\, y) \in R^n} \left(\left||\partial_x p\,|_{(x,\, y)}|,\, ||\partial_y p\,|_{(x,\, y)}|,\, ||\partial_x q\,|_{(x,\, y)}|,\, ||\partial_y q\,|_{(x,\, y)}| \right) < \delta. \tag{16}$$

The condition of invariance of the collection f with respect to T is written

$$Qf_z(x) + q(x, f_x(z)) = f_{Tz}(Px + p(x, f_z(x)))$$

or

$$f_z(x) = Q^{-1}f_{Tz}(Px + p(x, f_z(x))) - Q^{-1}q(x, f_z(x)). \tag{17}$$

For a T which satisfies (16) for some δ, consider the operator $\varphi_T: F^\infty \to F^\infty$ which associates with a collection f a collection g by means of the formula

$$g_z(x) = (\varphi_T f)_z(x) = Q^{-1}f_{Tz}(Px + p(x, f_z(x))) - Q^{-1}q(x, f_z(x)). \tag{18}$$

That φ_T indeed maps F^∞ into F^∞ follows from the following facts.

1) $g = \varphi_T f$ *is a collection, i.e.* $g_z(z_x) = z_y$. Indeed,

$$g_z(z_x) = Q^{-1}f_{Tz}((Tz)_x) - Q^{-1}q(z_x, f_z(z_x)) = Q^{-1}(Tz)_y - Q^{-1}q(z_x, z_y).$$

On the other hand, it follows from (15) that

$$(Tz)_y = Qz_y + q(z_x, z_y) \quad \text{or} \quad z_y = Q^{-1}(Tz)_y - Q^{-1}q(z_x, z_y).$$

2) *If* $f \in F^c \subset F^\infty$, *then* $g \in F^{c_1(c,\delta)} \subset F^\infty$. This follows from (16), $f \in F^c$, and the fact that the composition of mappings which satisfy Lipschitz conditions also satisfies a Lipschitz condition.

It follows from the definition of φ_T that a collection which is invariant with respect to T is a fixed point of the operator φ_T.

ASSERTION 1. *There exist a constant $\delta_1 > 0$ and functions $\beta_i: (0, \delta_1) \to (0, \infty)$, $i = 1, 2$, depending only on λ_1 and λ_2, such that if δ is chosen with $0 < \delta < \delta_1$ and c is chosen so that $\beta_2(\delta) \leqslant c \leqslant \beta_1(\delta)$, then φ_T, where T satisfies (16) for the given δ, takes a c-collection to a c-collection, i.e. $\varphi_T|_{F^c}: F^c \to F^c$. Also, $\beta_2(\delta) \to 0$ and $\beta_1(\delta) \to \infty$ as $\delta \to 0$.*

PROOF. We introduce convergence in F^∞ as follows: $f^n \to f^\infty$ as $n \to \infty \Leftrightarrow$ (for any $z \in R^n$, $f_z^n(x) \to f_z^\infty(x)$ as $n \to \infty$ uniformly in x in every compact set $K \subset X$). In the sense of this convergence the set D^c of c-collections with differentiable leaves

$$D^c = \left\{ f: \forall z \in R^n, \forall x_0 \in X \, \exists D_x(f_z)|_{x_0} \text{ and } |D_x(f_z)|_{x_0}| \leqslant c \right\}$$

is everywhere dense in $F^c \subset F^\infty$. To prove that φ_T maps F^c into F^c it is sufficient to prove that φ_T maps D^c into D^c and is continuous in F^c.

We shall prove that for sufficiently small $\delta > 0$ and a suitable c, the operator φ_T maps D^c into D^c. We have

$$D_x((\varphi_T f)_z) = Q^{-1} \cdot D_x(f_{Tz}) \cdot \left(P + \partial_x p + \partial_y p \cdot D_x(f_z) \right)$$
$$- Q^{-1} \cdot \partial_x q - Q^{-1} \cdot \partial_y q \cdot D_x(f_z).$$

(The points at which these derivatives are taken are not indicated, since the subsequent estimates are uniform in $x \in X$.) From this and from (12), (16), and the fact that $f \in D^c$ we obtain

$$|D_x((\varphi_T f)_z)| < c \cdot \left(\lambda_1 \cdot \lambda_2^{-1} + \lambda_2^{-1}\delta + \lambda_2^{-1}\delta \cdot c \right) + \lambda_2^{-1} \cdot \delta \cdot c + \lambda_2^{-1}\delta$$
$$= c \cdot \left(\lambda_1 \cdot \lambda_2^{-1} + \lambda_2^{-1}\delta \cdot c + 2\lambda_2^{-1} \cdot \delta \right) + \lambda_2^{-1} \cdot \delta. \tag{19}$$

Choose the constant $\delta > 0$ so that

$$\lambda_1 \cdot \lambda_2^{-1} + 2\lambda_2^{-1} \cdot \delta < 1. \tag{20}$$

This is possible since it follows from (13) that $\lambda_1 \cdot \lambda_2^{-1} < 1$. For example, we can take

$$0 < \delta < \delta_1' = \frac{1 - \lambda_1 \cdot \lambda_2^{-1}}{2\lambda_2^{-1}}. \tag{21}$$

Choose the nonnegative constant c so that

$$\lambda_1 \cdot \lambda_2^{-1} + 2\lambda_2^{-1} \cdot \delta + \lambda_2^{-1}\delta \cdot c < 1. \tag{22}$$

This is possible since (20) holds. For example, we take

$$0 \leqslant c \leqslant \beta_1(\delta) = \frac{1 - \lambda_1 \cdot \lambda_2^{-1} - 2\lambda_2^{-1} \cdot \delta}{2\lambda_2^{-1} \cdot \delta}. \tag{23}$$

Then for $0 < \delta < \delta_1'$ and $0 \leqslant c \leqslant \beta_1(\delta)$ we have from (12) that

$$|D_x((\varphi_T f)_z)| < c \cdot a + \lambda_2^{-1} \cdot \delta, \tag{24}$$

100 A. N. ŠOŠITAĬŠVILI

where

$$a = \lambda_1 \cdot \lambda_2^{-1} + 2\lambda_2^{-1} \cdot \delta_1' + \lambda_2^{-1} \cdot \delta_1' \cdot \beta_1(\delta) < 1.$$

We find c from the condition that $c \cdot a + \lambda_2^{-1} \cdot \delta \leq c$. For example, take

$$\beta_2(\delta) = \frac{2\lambda_2^{-1} \cdot \delta}{1 - a} \leq c. \tag{25}$$

Then from (24) we obtain that for $0 < \delta < \delta_1'$ and $\beta_2(\delta) \leq c \leq \beta_1(\delta)$ the operator φ_T maps D^c into D^c.

For the existence of a c such that $\beta_2(\delta) \leq c \leq \beta_1(\delta)$, it is necessary that $\beta_2(\delta) < \beta_1(\delta)$. From the definitions of β_2 and β_1 we see that $\beta_2(\delta) \to 0$ and $\beta_1(\delta) \to \infty$ as $\delta \to 0$. Therefore there is a sufficiently small $\delta_1'' > 0$ such that for $0 < \delta < \delta_1''$ we have $\beta_2(\delta) < \beta_1(\delta)$.

We set

$$\delta_1 = \min(\delta_1', \delta_1''). \tag{26}$$

We now show that φ_T is continuous on F^c. Let there be given a convergent sequence $f^m \in F^c$, $m = 1, 2, \ldots$, with $\lim_{m \to \infty} f^m = f$, i.e.

$$\forall z \in R^n, \qquad |f_z^m(x) - f_z(x)| \to 0 \quad \text{as } m \to \infty$$

uniformly in x in any compact set $K \subset X$. Then

$$|((\varphi_T f^m)_z)(x) - ((\varphi_T f)_z)(x)|$$

$$\leq |Q^{-1}f_{Tz}^m(Px + p(x, f_z^m(x))) - Q^{-1}q(x, f_z^m(x))$$

$$- Q^{-1}f_{Tz}(Px + p(x, f_z(x))) + Q^{-1}q(x, f_z(x))|$$

$$\leq |Q^{-1}f_{Tz}^m(Px + p(x, f_z^m(x))) - Q^{-1}f_{Tz}^m(Px + p(x, f_z(x)))|$$

$$+ |Q^{-1}f_{Tz}^m(Px + p(x, f_z(x))) - Q^{-1}f_{Tz}(Px, p(x, f_z(x)))|$$

$$+ |Q^{-1}q(x, f_z^m(x)) - Q^{-1}a(x, f_z(x))|. \tag{27}$$

The first term, by the mean-value theorem and by (12) and (16) ($f^m \in F^c$), is less than $\lambda_2^{-1} \cdot c \cdot \delta \cdot |f_z^m(x) - f_z(x)|$, so that for $x \in K \subset X$ it converges uniformly to zero as $m \to \infty$.

For $x \in K \subset X$ the second term converges uniformly to zero, since the mapping $x \to Px + p(x, f_z(x))$ maps the compact set K into some other compact set $K_1 \subset X$.

By virtue of the mean-value theorem, (12), and (16), the third term is less than $\lambda_2^{-1} \cdot \delta \cdot |f_z^m(x) - f_z(x)|$, and for $x \in K \subset X$ converges uniformly to zero as $m \to \infty$.

Assertion 1 is proved.

We now introduce a metric in the space F^∞. For any two collections $f^1, f^2 \in F^c$ we consider the number $\rho(f^1, f^2) \geq 0$, the minimum of all numbers $\gamma \geq 0$ such that

$$|f_z^1(x) - f_z^2(x)| \leq \gamma |x - z_x|, \qquad \forall x \in X, \forall z \in R^n. \tag{28}$$

The number $\rho(f^1, f^2)$ exists for any $f^1, f^2 \in F^\infty$. Indeed there is a $c > 0$ such that $f^1, f^2 \in F^c$. Then it follows from $f_z^1(z_x) = f_z^2(z_x) = z_y$ that

$$\left|f_z^1(x) - f_z^2(x)\right| \leq \left|f_z^1(x) - f_z^1(z_x) + f_z^2(z_x) - f_z^2(x)\right|$$

$$\leq \left|f_z^1(x) - f_z^1(x)\right| + \left|f_z^2(z_x) - f_z^2(x)\right| \leq 2c\,|x - z_x|. \quad (29)$$

LEMMA. $\rho: F^\infty \times F^\infty \to R$ *is a metric in* F^∞. *For any* $c > 0$, *the metric space* $(F^c, \rho|_{F^c}) \subset (F^\infty, \rho)$ *is complete.*

PROOF. 1. We show that the triangle inequality holds for ρ. Let there be given three collections, $f^1, f^2, f^3 \in F^c$. Assume that $x \neq z_x$, we set

$$\gamma_{i,j}(x, z) = \frac{\left|f_z^i(x) - f_z^j(x)\right|}{|x - z_x|} \qquad i, j = 1, 2, 3, i \neq j.$$

Clearly

$$\rho(f^i, f^j) = \sup_{x \in X, z \in R^n} \gamma_{i,j}(x, z).$$

We have

$$\left|f_z^1(x) - f_z^3(x)\right| \leq \left|f_z^1(x) - f_z^2(x)\right| + \left|f_z^2(x) - f_z^3(x)\right| \Rightarrow$$

$$\gamma_{1,3}(x, z)\,|x - z_x| \leq \gamma_{1,2}(x, z)\,|x - z_x| + \gamma_{2,3}(x, z)\,|x - z_x| \Rightarrow$$

$$\gamma_{1,3}(x, z) \leq \gamma_{1,2}(x, z) + \gamma_{2,3}(x, z) \Rightarrow$$

$$\sup_{\substack{x \in X \\ z \in R^n}} \gamma_{1,3}(x, z) = \sup_{\substack{x \in X \\ z \in R^n}} (\gamma_{1,2}(x, z) + \gamma_{2,3}(x, z))$$

$$\leq \sup_{\substack{x \in X \\ z \in R^n}} \gamma_{1,2}(x, z) + \sup_{\substack{x \in X \\ z \in R^n}} \gamma_{2,3}(x, z) \Rightarrow$$

$$\rho(f^1, f^3) \leq \rho(f^1, f^2) + \rho(f^2, f^3).$$

2. We now prove that the space F^c is complete in the metric ρ. Consider a fundamental sequence $f^i \in F^c$, $i = 1, 2, \cdots$. For each compact set $K \subset X$ we have for $x \in K$

$$\left|f_z^i(x) - f_z^j(x)\right| < \rho(f^i, f^j)\,|x - z_x| \leq \rho(f^i, f^j)d,$$

where $d = \sup_{x \in K} |x - z_x| < \infty$. From this and from the fact that $\rho(f^i, f^j) \to 0$ as $i, j \to \infty$ it follows that $\{f_z^i\}$, $i = 1, 2, \ldots$, converges for fixed z uniformly on each compact set $K \subset X$. Consequently the limit

$$\lim_{i \to \infty} f_z^i = f_z^\infty : X \to Y$$

exists. Clearly f_z^∞ satisfies a Lipschitz condition in x with constant c. The collection $f^\infty \in F^c$ with leaves f_z^∞ is the limit of the sequence f^i, $i = 1, 2, \cdots$. The lemma is proved.

ASSERTION 2. *There exist a constant $\delta_2 > 0$ and a function $\beta_3 \colon (0, \delta_2) \to (0, \infty)$, $\beta_3(\delta) \to \infty$ as $\delta \to 0$, which depend only on λ_1 and λ_2 and are such that if $0 < \delta < \delta_2$ and $0 \leqslant c \leqslant \beta_3(\delta)$, then for any $f^1, f^2 \in F^c$ and T which satisfies (16) for the given δ we have $\rho(\varphi_T f^1, \varphi_T f^2) < k\rho(f^1, f^2)$, where k is a constant, $0 < k < 1$, which does not depend on f^1 and f^2.*

PROOF. We have

$$\left(\varphi_T f^1\right)_z(x) - \left(\varphi_T f^2\right)_z(x) = Q^{-1}f^1_{Tz}\big(Px + p(x, f^1_z(x))\big) - Q^{-1}q\big(x, f^1_z(x)\big)$$
$$- Q^{-1}f^2_{Tz}\big(Px + p(x, f^2_z(x))\big) + Q^{-1}q\big(x, f^2_z(x)\big).$$

Set $\rho = \rho(f^1, f^2)$ and

$$\mathrm{I} = Q^{-1}f^1_{Tz}\big(Px + p(x, f^1_z(x))\big) - Q^{-1}f^1_{Tz}\big(Px + p(x, f^2_z(x))\big),$$
$$\mathrm{II} = Q^{-1}f^1_{Tz}\big(Px + p(x, f^2_z(x))\big) - Q^{-1}f^2_{Tz}\big(Px + p(x, f^2_z(x))\big),$$
$$\mathrm{III} = Q^{-1}q\big(x, f^1_z(x)\big) - Q^{-1}q\big(x, f^2_z(x)\big).$$

We have $(\varphi_T f^1)_z(x) - (\varphi_T f^2)_z(x) = \mathrm{I} + \mathrm{II} - \mathrm{III}$. From the mean-value theorem, (16), and the fact that f^1 and f^2 belong to F^c we have

$$|\mathrm{I}| \leqslant |Q^{-1}| \cdot c \cdot \delta \cdot \left|f^1_z(x) - f^2_z(x)\right|,$$

whence by virtue of (12) and (28) $|\mathrm{I}|$ is less than $\lambda_2^{-1} \cdot c \cdot \delta \cdot \rho \cdot |x - z_x|$.

Furthermore,

$$|\mathrm{II}| \leqslant |Q^{-1}| \cdot \rho \,|\, Px + p\big(x, f^2_z(x)\big) - (Tz)_x \,|$$
$$= |Q^{-1}| \cdot \rho \cdot |\, Px + p\big(x, f^2_z(x)\big) - Pz_x - p\big(x, f^2_z(z_x)\big) \,|.$$

From the mean-value theorem, (12), (16), and the fact that $f^2 \in F^c$ we obtain

$$\left|Px + p\big(x, f^2_z(x)\big) - Pz_x - p\big(x, f^2_z(z_x)\big)\right| < (\lambda_1 + \delta c)\,|x - z_x|.$$

Therefore

$$|\mathrm{II}| < \lambda_2^{-1} \cdot \rho \cdot (\lambda_1 + \delta c) \cdot |x - z_x|.$$

By the mean-value theorem, (12), (16), and (28) we obtain

$$|\mathrm{III}| \leqslant |Q^{-1}| \cdot |\partial_y q| \cdot |f^1_z(x) - f^2_z(x)| < \lambda_2^{-1} \cdot \delta \cdot \rho \cdot |x - z_x|.$$

Finally,

$$\left|\left(\varphi_T f^1\right)_z(x) - \left(\varphi_T f^2\right)_z(x)\right| \leqslant |\mathrm{I}| + |\mathrm{II}| + |\mathrm{III}|$$
$$< \rho \cdot \left(\lambda_2^{-1} \cdot \delta \cdot c + \lambda_1 \cdot \lambda_2^{-1} + \lambda_2^{-1} \cdot \delta \cdot c + \lambda_2^{-1} \cdot \delta\right) \cdot |x - z_x|$$

$$= \rho \cdot \left(2 \cdot \lambda_2^{-1} \cdot \delta \cdot c + \lambda_1 \cdot \lambda_2^{-1} + \lambda_2^{-1} \cdot \delta\right) \cdot |x - z_x|. \tag{30}$$

Choose $\delta > 0$ so that

$$\lambda_1 \cdot \lambda_2^{-1} + \lambda_2^{-1} \cdot \delta < 1. \tag{31}$$

This is possible, since it follows from (13) that $\lambda_1 \cdot \lambda_2^{-1} < 1$. For example, take

$$0 < \delta < \delta_2 = \frac{1 - \lambda_1 \cdot \lambda_2^{-1}}{2 \cdot \lambda_2^{-1}}. \tag{32}$$

Choose $c > 0$ so that

$$2\lambda_2^{-1} \cdot \delta \cdot c + \lambda_1 \cdot \lambda_2^{-1} + \lambda_2^{-1} \cdot \delta < 1. \tag{33}$$

This is possible since (31) holds. For example, take

$$0 \leqslant c \leqslant \beta_3(\delta) = \frac{1 - \lambda_1\lambda_2^{-1} - \lambda_2^{-1}\delta}{4\lambda_2^{-1}\delta}. \tag{34}$$

Set

$$k = 2\lambda_2^{-1}\delta_2\beta_3(\delta) + \lambda_1\lambda_2^{-1} + \lambda_2^{-1}\delta_2 \tag{35}$$

which, by the preceding, is less than 1. Then for $0 < \delta < \delta_2$ and $0 \leqslant c \leqslant \beta_3(\delta)$ it follows from (30), (33) and (35) that $|(\varphi_T f^1)_z(x) - (\varphi_T f^2)_z(x)| < k\rho |x - z_x|$, where $0 < k < 1$. Assertion 2 is proved.

Set

$$\beta_4(\delta) = \min(\beta_1(\delta), \beta_3(\delta)). \tag{36}$$

From (23), (25) and (34) we have $\beta_2(\delta) \to 0$ and $\beta_4(\delta) \to \infty$ as $\delta \to 0$. Choose $\delta_3' > 0$ such that if $0 < \delta < \delta_3'$, then $\beta_2(\delta) < \beta_4(\delta)$. Set

$$\delta_3 = \min(\delta_1, \delta_2, \delta_3') \tag{37}$$

(see Assertion 1 for the definition of δ_1).

ASSERTION 3. *If $0 < \delta < \delta_3$ and $\beta_2(\delta) \leqslant c \leqslant \beta_4(\delta)$, the operator φ_T (where T satisfies (16) with the given δ) maps F^c into F^c and has a unique fixed point in F^c.*

PROOF. From the definitions of δ_3, $\beta_2(\delta)$, and $\beta_4(\delta)$ it follows that Assertions 1 and 2 hold simultaneously for φ_T. It follows from Assertion 1 that φ_T maps F^c into F^c, and from Assertion 2 that $\varphi_T|_{F^c}$ is a contraction mapping. Assertion 3 is proved.

REMARK. In the proofs of Assertions 1–3 we have used only the facts that $\lambda_1 < \lambda_2$, that T is a C^1 diffeomorphism, and that (16) holds.

ASSERTION 4. *There exist a constant $\delta_4 > 0$ and a function $\beta_5: (0, \delta_4) \to (0, \infty)$, $\beta_5(\delta) \to \infty$ as $\delta \to 0$, which depend only on λ_1 and λ_2, such that if $0 < \delta < \delta_4$ and $c > 0$ satisfies $\beta_2(\delta) \leqslant c \leqslant \beta_5(\delta)$ (see Assertion 1 and (25) for the definition of $\beta_2(\delta)$). Then the fixed point $f^\infty \in F^c$ of the operator $\varphi_T|_{F^c}: F^c \to F^c$ (where T satisfies (16) with the given δ) is a c-collection with $(k - 1)$-times continuously differentiable leaves, i.e. $f_z^\infty: X \to Y$ is a C^{k-1}-mapping for any $z \in R^n$.*

PROOF. Consider a c-collection with k-times differentiable leaves. Reasoning inductively, we construct k constants c_l, $l = 1, \ldots, k$, such that

$$\left\{ \left| D_x^l(f_z)|_{x_0} \right| < c_l, \qquad l = 1, \ldots, k, \forall x_0 \in X, \forall z \in R^n \right\} \tag{38}$$

implies that

$$\left\{ \left| D_x^l(\varphi_T f_z) \big|_{x_0} \right| < c_l, \qquad l = 1,\dots,k, \, \forall x_0 \in X, \, \forall z \in R^n \right\}. \tag{39}$$

Assume that we have already constructed c_l, $l = 1,\dots,k_1 < k$. We can estimate the norm of $D_x^{k_1+1}((\varphi_T f)_z)\big|_{x_0}$ as follows:

$$\left| D_x^{k_1+1}((\varphi_T f)_z) \right| \leqslant |Q^{-1}| \cdot \left| D_x^{k_1+1}(f_{Tz}) \right| \left(\left| P + \partial_x p + \partial_y p \cdot D_x(f_z) \right| \right)^{k_1+1}$$

$$+ |Q^{-1}| \, \| D_x(f_z) \| \cdot |\partial_y p| \cdot \left| D_x^{k_1+1}(f_z) \right| + |Q^{-1}| \, |\partial_y q| \cdot \left| D_x^{k_1+1}(f_z) \right| + K, \tag{40}$$

where K is a constant that depends on $|Q^{-1}|$, $|P|$, $|\partial_x^r p|$, $|\partial_y^r p|$, $|\partial_x^r q|$ and $|\partial_y^r q|$, $r = 1,\dots,\, k_1 + 1 \leqslant k$, which by the conditions of Lemma 1 are uniformly bounded on all of R^n, and on $|D_x^l(f_z)|$ and $|D_x^l(f_{Tz})|$, $l = 1,\dots,k_1$, which by the induction hypothesis are uniformly bounded on all of R^n by the constants c_l, $l = 1,\dots,k_1$. Therefore we can take the constant K, $0 < K < \infty$, independent of $x \in X$ and $z \in R^n$. From (40), (12), (16), and the fact that $f \in F^c$ we have

$$\left| D_x^{k_1+1}((\varphi_T f)_z) \right| < \lambda_2^{-1} \left| D_x^{k_1+1}(f_{Tz}) \right| \cdot (\lambda_1 + \delta + \delta c)^{k_1+1}$$

$$+ \lambda_2^{-1} \cdot c \cdot \delta \cdot \left| D_x^{k_1+1}(f_z) \right| + \lambda_2^{-1} \cdot \delta \cdot \left| D_x^{k_1+1}(f_z) \right| + K. \tag{41}$$

We find the constant $c_{k_1+1} \geqslant 0$ from the condition

$$c_{k_1+1} \leqslant \lambda_2^{-1} c_{k_1+1}(\lambda_1 + \delta + \delta c)^{k_1+1} + \lambda_2^{-1} \cdot c \cdot \delta \cdot c_{k_1+1} + \lambda_2^{-1} \cdot \delta \cdot c_{k_1+1} + K$$

$$= c_{k_1+1}\left[\lambda_2^{-1}(\lambda_1 + \delta + \delta c)^{k_1+1} + \lambda_2^{-1} \cdot \delta \cdot c + \lambda_2^{-1}\delta \right] + K. \tag{42}$$

For this we note that (13) implies that

$$\lambda_2^{-1} \lambda_1^l < 1, \qquad l = 1,\dots,k. \tag{43}$$

Choose δ_4' so that when $0 < \delta < \delta_4'$ we have

$$\lambda_2^{-1}(\lambda_1 + \delta)^l + \lambda_2^{-1}\delta < 1, \qquad l = 1,2,\dots,k. \tag{44}$$

This is possible since the left member of (44) tends to $\lambda_2^{-1}\lambda_1^l < 1$ as $\delta \to 0$.

For any δ such that $0 < \delta < \delta_4'$ there exists a function $\beta_6(\delta) > 0$ such that if $0 \leqslant c \leqslant \beta_6(\delta)$, then

$$\lambda_2^{-1} \cdot (\lambda_1 + \delta + \delta \cdot c)^l + \lambda_2^{-1} \cdot \delta \cdot c + \lambda_2^{-1}\delta < 1, \qquad l = 1,\dots,k. \tag{45}$$

Indeed, as $c \to 0$ the left member of (45) tends to the left member of (44), which for this δ is less than 1. In addition, since for any fixed $c > 0$ the left member of (45) tends to $\lambda_2^{-1}\lambda_1^l < 1$ as $\delta \to 0$, $\beta_6(\delta)$ can be chosen so that $\beta_6(\delta) \to \infty$ as $\delta \to 0$. Set

$$\beta_5 = \min(\beta_4, \beta_6) \tag{46}$$

(see (36) for the definition of β_4).

From the definition of β_2 (see (25)) and β_5 it follows that $\beta_2(\delta) \to 0$ and $\beta_5(\delta) \to \infty$ as $\delta \to 0$. Choose δ_4'' so that for $0 < \delta < \delta_4''$ we have $\beta_2(\delta) < \beta_5(\delta)$.

Set

$$\delta_4 = \min(\delta_4', \delta_4'', \delta_3) \qquad (47)$$

(see (37) for the definition of δ_3).

It is clear from the preceding that if $0 < \delta < \delta_4$ and $\beta_2(\delta) \leqslant c \leqslant \beta_5(\delta)$, then (42) will hold if we take

$$c_{k_1+1} = \frac{CK}{1 - \lambda_2^{-1}(\lambda_1 + \delta_4 + \delta_4 \cdot \beta_6(\delta)) + \lambda_2^{-1} \cdot \delta_4 \cdot \beta_6(\delta) + \lambda_2^{-1} \cdot \delta_4} . \qquad (48)$$

As c_1 we choose a c such that $\beta_2(\delta) \leqslant c \leqslant \beta_5(\delta)$. Since $\beta_5(\delta) < \beta_4(\delta)$, it follows from Assertion 3 that φ_T maps F^c into F^c. This completes the induction.

Since $\varphi_T^n f \to f^\infty$ as $n \to \infty$ $\forall f \in F^c$, as the initial c-collection to which the iterations of φ_T are applied we may take $f \equiv 0$ ($f_z(x) = 0$, $\forall x \in X$, $\forall z \in R^n$). Since (38) holds for $f \equiv 0$, it also holds for $\varphi_T^n f$, $n = 1, 2, \cdots$. Consequently for any $l = 1,\ldots,k$ and any $z \in R^n$ the sequence $D_x^l((\varphi_T^n f)_z)|_{x_0}$, $n = 1, 2,\ldots$, is uniformly bounded in $x_0 \in X$. Hence for any $l = 1,\ldots,k-1$ and any $z \in R^n$ the sequence $D_x^l((\varphi_T^n f)_z)|_{x_0}$, $n = 1, 2,\ldots$, is uniformly bounded in $x_0 \in X$ and is equicontinuous in $x_0 \in X$. Hence, by Arzelà's lemma, for all $z \in R^n$ there exists a subsequence n_j, $j = 1, 2,\ldots$, such that $(\varphi_T^{n_j} f)_z$ along with $k-1$ derivatives converges uniformly to f_z^∞ as $n_j \to \infty$ on every compact set $K \subset X$. Consequently $f_z^\infty \in C^{k-1}(X, Y)$. Assertion 4 is proved.

ASSERTION 5. *Let $\delta > 0$ and $c \geqslant 0$ be such that Assertions 1–4 hold. Then the leaves of the collection f^∞, which is invariant with respect to $\varphi_T|_{F^c}: F^c \to F^c$, depend continuously on $z \in R^n$. That is, $f_{z_n}^\infty \to f_{z_0}^\infty$ as $z_n \to z_0$ along with $k-1$ derivatives uniformly on every compact set $K \subset X$.*

PROOF. We first show that the $f_{z_n}^\infty$, $n = 1, 2,\ldots$, are uniformly bounded on every compact set $K \subset X$. Set

$$r = \max_{x \in K} |x - (z_0)_x|, \qquad r_1 = \max_n (|(z_0)_x - (z_n)_x|, |(z_n)_y - (z_0)_y|).$$

For $x \in K$ we have

$$\left| f_{z_n}^\infty(x) - (z_0)_y \right| \leqslant \left| f_{z_n}^\infty(x) - (z_n)_y \right| + \left| (z_n)_y - (z_0)_y \right|$$
$$\leqslant c \cdot |x - (z_n)_x| + |(z_n)_y - (z_0)_y|$$
$$\leqslant c \cdot |x - (z_0)_x| + c \cdot |(z_0)_x - (z_n)_x| + |(z_n)_y - (z_0)_y|$$
$$\leqslant c \cdot r + r_1 \cdot (c + 1).$$

It follows from this that $|f_{z_n}^\infty(x)| < c \cdot r + r_1(c + 1) + |(z_0)_y|$, $n = 1, 2,\ldots$, for $x \in K$. Since $f^\infty \in F^c$, the $f_{z_n}^\infty$, $n = 1, 2,\ldots$, are equicontinuous in x.

It is clear from the proof of Assertion 4 that for any l, $1 \leqslant l \leqslant k-1$, the sequence $D_x^l(f_{z_n}^\infty)|_{x_0}$, $n = 1, 2,\ldots$, is uniformly bounded and equicontinuous in $x_0 \in X$.

In the same way we can show for any integer p that the sequence of leaves $f_{T^p z_n}^\infty$, $-\infty < n < \infty$, is uniformly bounded and equicontinuous on every compact set

$K \subset X$, and that for any $l = 1, \ldots, k - 1$ the sequence $D_x^l(f_{T^p z_n})$, $n = 1, 2, \ldots$, is uniformly bounded and equicontinuous on X.

By applying Arzelà's lemma and the diagonal process we see that we can choose a subsequence n_j such that for any integer p the leaves $f_{T^p z_{n_j}}$ converge, along with their first $k - 1$ derivatives, uniformly on every compact set $K \subset X$.

We denote the limit of the sequence $f_{T^p z_{n_j}}$ as $n_j \to \infty$ by $f^1_{T^p z_0}$.

Assume that $f^1_{z_0} \neq f^\infty_{z_0}$. Consider the collection f^1, of which all leaves except the leaves $f^\infty_{T^p z_0}$, $-\infty < p < \infty$, coincide with the leaves of the collection f^∞, while the leaves $f^\infty_{T^p z_0}$ are replaced by $f^1_{T^p z_0}$.

Let us show that f^1 is a c-collection. For this it is sufficient to show that $f^1_{T^p z_0}$ is a c-mapping and that

$$f^1_{T^p z_0}\big((T^p z_0)_x\big) = (T^p z_0)_y, \qquad -\infty < p < \infty.$$

These both follow from the fact that $f^1_{T^p z_0}$ is the uniform limit, on every compact set, of the c-mappings $f^\infty_{T^p z_{n_j}}$, which satisfy

$$f^\infty_{T^p z_{n_j}}\big((T^p z_{n_j})_x\big) = (T^p z_{n_j})_y.$$

It follows from the proof of the continuity of φ_T (carried out in (27)) that

$$\varphi_T\big(f^1_{T^p z_0}\big) = \varphi_T \lim_{n_j \to \infty} f^\infty_{T^p z_{n_j}} = \lim_{n_j \to \infty} \varphi_T f^\infty_{T^p z_{n_j}} = \lim_{n_j \to \infty} f^\infty_{T^{p+1} z_{n_j}} = f^1_{T^{p+1} z_0},$$

i.e. the collection f^1 is invariant with respect to φ_T.

By assumption $f^1 \neq f^\infty$. This contradicts the fact that f^∞ is the unique c-collection which is invariant with respect to φ_T.

It follows from the contradiction just obtained that $f^1_{z_0} = f^\infty_{z_0}$, i.e. that $f^\infty_{z_{n_j}} \to f^\infty_{z_0}$ in C^{k-1}. Since this is true for an arbitrary convergent subsequence $f^\infty_{z_{n_j}}$, and since we can choose a convergent subsequence from any subsequence $f^\infty_{z_n}$ (as is shown by the preceding considerations), it follows that $f^\infty_{z_n} \to f^\infty_{z_0}$ in C^{k-1} as $z_n \to z_0$. Assertion 5 is proved.

If we replace T by T^{-1}, and consider c-collections ψ with leaves $x = \psi_z(y)$, we can prove Assertions 1–5 for them as well. We shall use the same symbols for the constants arising in this situation as for the diffeomorphism T and the collections f with leaves $y = f_z(x)$.

Assume further that T is a C^2-diffeomorphism which satisfies (16) for δ with $0 < \delta < \delta_4$ (see (47) for the definition of δ_4). It then follows from Assertions 1–5 that there exist, uniquely, two $\beta_5(\delta)$-collections f^∞ and ψ^∞, invariant with respect to T, with C^1-leaves. In this connection f^∞ and ψ^∞ are automatically $\beta_2(\delta)$-collections. From the definitions of β_2 (see Assertion 1) and β_5 (see Assertion 4) it follows that $\beta_2 \to 0$ and $\beta_5 \to \infty$ as $\delta \to 0$. In what follows we shall assume f^∞ and ψ^∞ to be $\beta_2(\delta)$-collections.

We now turn to the proof that the collection f^∞ (ψ^∞) gives a partition of the space R^n, i.e. that if for the leaves $y = f^\infty_{z_1}(x)$ and $y = f^\infty_{z_2}(x)$ there exists a point x_0 such that $g_{z_1}(x_0) = g_{z_2}(x_0)$, then $g_{z_1}(x) = g_{z_2}(x) \ \forall x \in X$.

The proof of this is based on the fact that whether two points a and b belong to a single leaf is characterized by the speed with which their images move apart under iterations of T.

For what follows we shall need estimates for the norm of the restriction of T to tangent planes to the leaves of f^∞ and ψ^∞. From the invariance of f^∞ and ψ^∞ with respect to T it follows that the tangent planes to the leaves of f^∞ (ψ^∞) go into tangent planes to leaves of f^∞ (ψ^∞) under the action of T.

Let the vector v lie in the plane $\{dy = D_x(f_z^\infty)|_{x_0}dx\}$. Since $f^\infty \in F^{\beta_2(\delta)}$, we have

$$|v_x| \leqslant |v| \leqslant |v_x| + \beta_2(\delta)|v_x|,$$

$$\left|\left(T_*|_{z_0}v\right)\right| \leqslant \left|\left(T_*|_{z_0}v\right)_x\right| + \beta_2(\delta) \cdot \left|\left(T_*|_{z_0}v\right)_x\right|,$$

where $(\cdot)_x$ is the x-component of the vector $(\cdot)$, and $z_0 = (x_0, f_z^\infty(x_0))$. It follows from this and from (12) and (16) that

$$\left|\left(T_*|_{z_0}v\right)\right| \leqslant \left|\left(T_*|_{z_0}v\right)_x\right| \cdot (1 + \beta_2(\delta))$$

$$\leqslant (|P| + |\partial_x p| + |\partial_y p| \cdot \beta_2(\delta)) \cdot (1 + \beta_2(\delta)) \cdot |v| < \lambda_1(\delta) \cdot |v|, \quad (49)$$

where $\lambda_1(\delta) = (\lambda_1 + \delta + \delta \cdot \beta_2(\delta)) \cdot (1 + \beta_2(\delta))$.

Similarly, for $v \in \{dx = D_y(\psi_z^\infty)|_{y_0}dy\}$ we have

$$\left|\left(T_*|_{z_0}v\right)\right| > \lambda_2(\delta) \cdot |v|, \quad (50)$$

where

$$\lambda_2(\delta) = \frac{\lambda_2 - \delta - \delta \cdot \beta_2(\delta)}{1 + \beta_2(\delta)}, \qquad z_0 = (\psi_z^\infty(y_0), y_0).$$

Choose $\delta_5 > 0$ such that if $0 < \delta < \delta_5$, then

$$\lambda_1(\delta) < \lambda_2(\delta) \quad (51)$$

i.e. such that

$$(1 + \beta_2(\delta)) \cdot (\lambda_1 + \delta + \delta \cdot \beta_2(\delta)) < \frac{\lambda_2 - \delta - \delta \cdot \beta_2(\delta)}{1 + \beta_2(\delta)}. \quad (52)$$

This is possible since as $\delta \to 0$ (recall that $\beta_2(\delta) \to 0$) the left member of (52) converges to λ_1 and the right member of (52) converges to λ_2, and $\lambda_1 < \lambda_2$.

ASSERTION 6. *There exists a constant $\delta_6 > 0$ such that if $0 < \delta < \delta_6$, then for any $z \in R^n$ and $z_1 \in R^n$ the leaf $y = f_z^\infty(x)$, belonging to the T-invariant $\beta_2(\delta)$-collection f^∞, and the leaf $x = \psi_{z_1}^\infty(y)$, belonging to the T-invariant $\beta_2(\delta)$-collection ψ^∞, intersect at exactly one point.*

PROOF. Let the leaves f_z^∞ and $\psi_{z_1}^\infty$ intersect at $z_0 = (x_0, y_0)$, i.e. let $y_0 = f_z^\infty(x_0)$ and $x_0 = \psi_{z_1}^\infty(y_0)$. Then

$$x_0 = \psi_{z_1}^\infty(f_z^\infty(x_0)), \qquad y_0 = f_z^\infty(\psi_{z_1}^\infty(y_0)).$$

Assertion 6 will be proved if we can show that there exists a unique fixed point of the mapping $\psi_{z_1}^\infty \circ f_z^\infty \colon X \to X$. We have

$$\left| \psi_{z_1}^\infty(f_z^\infty(x_1)) - \psi_{z_1}^\infty(f_z^\infty(x_2)) \right| \leqslant \beta_2(\delta) \cdot \left| f_z^\infty(x_1) - f_z^\infty(x_2) \right| \leqslant \beta_2^2(\delta) \left| x_1 - x_2 \right|. \tag{53}$$

From the definition of $\beta_2(\delta)$ (see Assertion 1) it follows that $\beta_2(\delta) \to 0$ as $\delta \to 0$. Choose $\delta_6 > 0$ such that for $0 < \delta < \delta_6$

$$\beta_2(\delta) < 1. \tag{54}$$

It then follows from (53) that $\psi_{z_1}^\infty \circ f_z^\infty \colon X \to X$ is a contraction, so that there exists a unique fixed point x_0. Assertion 6 is proved.

ASSERTION 7. *If the points $a \in R^n$ and $b \in R^n$ lie on the leaf $y = f_z^\infty(x)$, then*

$$\frac{\rho(T^n a, T^n b)}{\rho(a, b)} < \sqrt{1 + \beta_2^2(\delta)} \cdot \lambda_1^n(\delta), \qquad 1 \leqslant n < \infty \tag{55}$$

(*see* (49) *for the definition of* $\lambda_1(\delta)$), *where $\rho(z_1, z_2)$ is the distance between the points $z_1 \in R^n$ and $z_2 \in R^n$.*

PROOF. Let $\sigma(a, b)$ denote the curve $x(t) = (1 - t)a_x + tb_x$, $y(t) = f_z^\infty(x(t))$, $0 \leqslant t \leqslant 1$, which joins the points a and b on the leaf f_z^∞. Let $l(\sigma(a, b))$ denote the length of this curve. We have

$$l(\sigma(a, b)) = \int_0^1 \sqrt{\left| \frac{dx}{dt} \right|^2 + \left| \frac{dy}{dt} \right|^2} \, dt$$

$$\leqslant \int_0^1 \left| \frac{dx}{dt} \right| \sqrt{1 + \left| \frac{dy}{dx} \right|^2} \, dt \leqslant \sqrt{1 + \beta_2^2(\delta)} \cdot \left| a_x - b_x \right| \tag{56}$$

$$\leqslant \sqrt{1 + \beta_2^2(\delta)} \cdot \rho(a, b).$$

Let $T^n \sigma(a, b)$ denote the image of $\sigma(a, b)$ under the mapping T^n. From the definition of $\lambda_1(\delta)$ (see (49)) we have

$$l(T^n \sigma(a, b)) = \int_a^b \left| T_*^n \right|_{z \in \sigma(a, b)} \overline{d\sigma} \, \left| < \lambda_1^n(\delta) \cdot l(\sigma(a, b)). \right.$$

Hence

$$\rho(T^n a, T^n b) \leqslant l(T^n \sigma(a, b)) < \lambda_1^n(\delta) \cdot l(\sigma(a, b)). \tag{57}$$

It follows from (56) and (57) that

$$\frac{\rho(T^n a, T^n b)}{\rho(a, b)} < \sqrt{1 + \beta_2^2(\delta)} \cdot \lambda_1^n(\delta).$$

Assertion 7 is proved.

ASSERTION 8. *If the points $a \in R^n$ and $b \in R^n$ lie on the leaf $x = \psi_z^\infty(y)$, then*

$$\frac{\rho(T^n a, T^n b)}{\rho(a, b)} > \frac{1}{\sqrt{1 + \beta_2^2(\delta)}} \cdot \lambda_2^n(\delta), \qquad 1 \leqslant n < \infty \tag{58}$$

(see (50) for the definition of $\lambda_2(\delta)$).

PROOF. Apply the reasoning of Assertion 7 to the diffeomorphism T^{-1} and the points $a' = T^n a$ and $b' = T^n b$.

ASSERTION 9. 1) *If the leaves $f_{z_1}^\infty \in f^\infty$ and $f_{z_2}^\infty \in f^\infty$ have a point in common, then they coincide. In other words, if there is a point $x_0 \in X$ such that $f_{z_1}^\infty(x_0) = f_{z_2}^\infty(x_0)$, then $f_{z_1}^\infty(x) = f_{z_2}^\infty(x) \; \forall x \in X$.*
2) *If the leaves $\psi_{z_1}^\infty \in \psi^\infty$ and $\psi_{z_2}^\infty \leftrightarrow \psi^\infty$ have a point in common, then they coincide.*

PROOF. Let $f_{z_1}^\infty$ and $f_{z_2}^\infty$ intersect at the point $b \in R^n$. Let there exist a point a belonging to $f_{z_1}^\infty$ but not to $f_{z_2}^\infty$. Consider the leaf ψ_a^∞ of the collection ψ^∞. According to Assertion 6 there exists a unique point of intersection of the leaves ψ_a^∞ and $f_{z_2}^\infty$. Let d denote this point. By the triangle inequality we have

$$\rho(T^n a, T^n d) \leqslant \rho(T^n a, T^n b) + \rho(T^n b, T^n d), \qquad n = 1, 2, \ldots. \tag{59}$$

It follows from Assertions 7 and 8 that

$$\rho(T^n a, T^n d) > \frac{\lambda_2^n(\delta)}{\sqrt{1 + \beta_2^2(\delta)}} \cdot \rho(a, d),$$

$$\rho(T^n a, T^n b) + \rho(T^n b, T^n d) < \lambda_1^n(\delta)\sqrt{1 + \beta_2^2(\delta)} \cdot (\rho(a, b) + \rho(b, d)). \tag{60}$$

It follows from (59) and (60) that

$$\lambda_2^n(\delta) \cdot \frac{\rho(a, d)}{\sqrt{1 + \beta_2^2(\delta)}} < \lambda_1^n(\delta) \cdot \sqrt{1 + \beta_2^2(\delta)} \cdot (\rho(a, b) + \rho(b, d)). \tag{61}$$

Since $\lambda_1(\delta) < \lambda_2(\delta)$ (see (51)), inequality (61) will not hold for sufficiently large n. The first part of Assertion 9 follows from this contradiction. If we consider T^{-1} instead of T, and apply the above reasoning to the collection ψ^∞, we obtain a proof of the second part. Assertion 9 is proved.

We now set

$$\delta_0 = \min(\delta_4, \delta_5, \delta_6), \qquad c_2(\varepsilon) = \beta_2(\delta), \qquad c_1(\delta) = \beta_5(\delta). \tag{62}$$

To see that the numbers δ_0, c_2, and c_1 thus constructed satisfy all the requirements of Lemma 1, it remains to show that there exist homeomorphisms of R^n into itself which transform f^∞ and ψ^∞ into families of parallel planes of appropriate dimension.

We construct a homeomorphism

$$h(f^\infty, \psi^\infty): R^n \to R^n, \tag{63}$$

which simultaneously takes f^∞ to the family of parallel planes $y = \text{const}$ and ψ^∞ to the family of parallel planes $x = \text{const}$.

Consider the mapping $h(f^\infty, \psi^\infty)\colon R^n \to R^n$ which takes the point $z_0 = (x_0, y_0) \in R^n$ to the point of intersection of the leaves $\psi^\infty_{z_1}$ and $f^\infty_{z_2}$, where $z_1 = (x_0, f_0^\infty(x_0))$ and $z_2 = (\psi_0^\infty(y_0), y_0)$. It follows from Assertion 6 that $h(f^\infty, \psi^\infty)z_0$ exists and is unique. It follows from Assertion 5 that $h(f^\infty, \psi^\infty)(z_0)$ depends continuously on z_0.

Consider the mapping $g\colon R^n \to R^n$ which takes the point $z_0 \in R^n$ to the point $g(z_0) = (x(z_0), y(z_0))$, where $x(z_0) = (\psi^\infty_{z_0} \cap f_0^\infty)_x$ and $y(z_0) = (f^\infty_{z_0} \cap \psi_0^\infty)_y$. It follows from Assertion 9 that the leaves $\psi^\infty_{z_1}$ and $f^\infty_{z_2}$ coincide with the leaves $\psi^\infty_{z_3}$ and $f^\infty_{z_3}$, where $z_3 = \psi^\infty_{z_1} \cap f^\infty_{z_2}$. Therefore $g \circ h(f^\infty, \psi^\infty)(z_0) = z_0$. Next, the point $g(z_0)$ depends continuously on z_0. It follows from this that g and $h(f^\infty, \psi^\infty)$ are mutually inverse continuous mappings, i.e. $h(f^\infty, \psi^\infty)\colon R^n \to R^n$ is a homeomorphism. It is easy to see that $h(f^\infty, \psi^\infty)$ takes f^∞ into a family of planes of the form $y = \text{const}$ and ψ^∞ into a family of planes of the form $x = \text{const}$.

Thus the collection f^∞ gives a foliation $\Gamma^1(T)$, and the collection ψ^∞ gives a foliation $\Gamma^2(T)$. Lemma 1 is proved.

§3. Proof of the theorems

PROOF OF THEOREM 1. Consider the linear operator $C\colon R^n \times R^l \to R^n \times R^l$, $C\colon (z, \varepsilon) \mapsto (Bz, 0)$, where $z \in R^n$ and $\varepsilon \in R^l$, and the linear operators

$$C^0 = C|_{X \times R^l}; \qquad C^- = C|_{Y^-}; \qquad C^+ = C|_{Y^+};$$
$$C^{0,-} = C|_{Y^- \times X \times R^l}; \qquad C^{+,0} = C|_{Y^+ \times X \times R^l}.$$

The spectrum $\sigma(e^C)$ of the linear operator e^C can be represented in the form $\sigma(e^C) = \sigma(e^{C^{-,0}}) \cup \sigma(e^{C^+})$. It is clear from the definition of $C^{-,0}$ that $\sigma(e^{C^{-,0}})$ is interior to the open ball of radius $1 + \rho$, for any constant $\rho > 0$. From the definition of C^+ it is clear that there exists a $\xi_1 > 0$ such that if $0 < \xi < \xi_1$, then (e^{C^+}) lies in the complement of the closed ball of radius $1 + \xi$. It is known that a norm can be chosen in $R^n \times R^l$ such that

$$\left| e^{C^{-,0}}(x, y^-, \varepsilon) \right| < (1 + \rho)\,|(x, y^-, \varepsilon)|\,,$$
$$\left| e^{C^+} y^+ \right| > (1 + \xi)\,|y^+|\,. \tag{64}$$

Choose $0 < \rho < \xi < \xi_1$ so that

$$(1 + \rho)^k < (1 + \xi), \qquad (1 + \rho) < (1 + \xi)^k. \tag{65}$$

In the conditions of Lemma 1 set

$$A = e^C, \qquad P = e^{C^{-,0}}, \qquad Q = e^{C^+}, \qquad \lambda_1 = (1 + \rho), \qquad \lambda_2 = (1 + \xi). \tag{66}$$

Then condition (13) of Lemma 1 holds.

Choose a representative of the germ (3) defined on all of $R^n \times R^l$ such that the mapping $T\colon R^n \times R^l \to R^n \times R^l$ of the flow per unit time for this representative satisfies condition (14) of Lemma 1. For this we prove

ASSERTION 10. *For any $\kappa > 0$ there exists a representative of the germ* (3)

$$\dot{z} = v'_\kappa(z, \varepsilon), \qquad \dot{\varepsilon} = 0, \qquad z \in R^n, \qquad \varepsilon \in R^l, \tag{67}$$

such that

$$v'_\kappa \in C^k(R^n \times R^l), \qquad |v'_\kappa(z, \varepsilon) - Bz| < \kappa,$$

$$|\partial_z v'_\kappa|_{(z, \varepsilon)} - B| < \kappa, \qquad |\partial_\varepsilon v'_\kappa|_{(z, \varepsilon)}| < d$$

for all $(z, \varepsilon) \in R^n \times R^l$; d is a nonnegative constant independent of (z, ε).

PROOF. Consider the differential equation

$$\dot{z} = Bz + f(|z|) \cdot r(z, \varepsilon \cdot \chi(|\varepsilon|)) = v'_\kappa(z, \varepsilon), \qquad \dot{\varepsilon} = 0,$$

where

$$(z, \varepsilon) \in R^n \times R^l; \qquad \chi, f \in C^\infty(R^1); \qquad 0 \leqslant f(t), \qquad \chi(t) \leqslant 1,$$

$$f(t) = \begin{cases} 1 & \text{if } t < \alpha_1, \\ 0 & \text{if } t > \alpha_2; \end{cases} \qquad \chi(t) = \begin{cases} 1 & \text{if } t < \alpha_3, \\ 0 & \text{if } t > \alpha_4; \end{cases}$$

with the constants $\alpha_i > 0$, $i = 1, \ldots, 4$, to be chosen later (see Theorem 1 for the definitions of B and $r(z, \varepsilon)$).

We have

$$v'_\kappa(z, \varepsilon) - Bz = f(|z|) \cdot r(z, \varepsilon \cdot \chi(|\varepsilon|)), \tag{68}$$

$$\partial_z v'_\kappa(z, \varepsilon) - B = \partial_z(f(|z|) \cdot r(z, \varepsilon \cdot \chi(|\varepsilon|))). \tag{69}$$

We estimate the norm of the operator $(\partial_z v'_\kappa(z, \varepsilon) - B)$ in terms of its matrix elements:

$$|\partial_z(f(|z|) \cdot r(z, \varepsilon \cdot \chi(|\varepsilon|)))|$$

$$\leqslant \sqrt{n} \sup_{i,j} \left| \frac{\partial f(|z|)}{\partial |z|} \cdot \frac{\partial |z|}{\partial z_i} \cdot r(z, \varepsilon\chi(|\varepsilon|))_j + f(|z|) \cdot \frac{\partial r(z, \varepsilon\chi(|\varepsilon|))_j}{\partial z_i} \right|. \tag{70}$$

It is clear that we can choose $f(t)$ so that

$$\frac{\partial f(t)}{\partial t} < \frac{2}{\alpha_2 - \alpha_1}, \qquad \forall t \in R^1. \tag{71}$$

Furthermore,

$$\left| \frac{\partial |z|}{\partial z_i} \right| \leqslant \frac{2|z_i|}{|z|} \leqslant \frac{2\alpha_2}{\alpha_1} \quad \text{for } \alpha_1 \leqslant |z| \leqslant \alpha_2. \tag{72}$$

From the conditions that $r(0, 0) = 0$ and $\partial_z r(0, 0) = 0$ we obtain for $|z| \leqslant \alpha_2$ and $|\varepsilon| \leqslant \alpha_4$ that

$$|r(z, \varepsilon \cdot \chi(|\varepsilon|))| \leqslant k_1(\alpha_4) + k_2(\alpha_4) \cdot \alpha_2 + k_3(\alpha_4, \alpha_2) \cdot \alpha_2^2, \tag{73}$$

$$\left| \frac{\partial r(z, \varepsilon)}{\partial z} \right| \leqslant k_2(\alpha_4) + k_3(\alpha_4, \alpha_2) \cdot \alpha_2, \tag{74}$$

where

$$k_1(\alpha_4) = \sup_{|\varepsilon| \leqslant \alpha_4} |r(0, \varepsilon)|, \qquad k_2(\alpha_4) = \sup_{|\varepsilon| \leqslant \alpha_4} |\partial_z r(0, \varepsilon)|,$$

$$k_3(\alpha_2, \alpha_4) = \sup_{\substack{|z| \leqslant \alpha_2 \\ |\varepsilon| \leqslant \alpha_4}} |\partial_z^2 r(z, \varepsilon)|;$$

for sufficiently small α_2 and α_4, $k_1(\alpha_4)$ and $k_2(\alpha_4)$ are arbitrarily close to zero, and $k_3(\alpha_2, \alpha_4)$ is arbitrarily close to $k_3(0, 0)$.

Set $\alpha_1 = \alpha_2/2$ and $\alpha_3 = \alpha_4/2$. From (71)–(74) we have

$$\left| \frac{\partial f(|z|)}{\partial |z|} \cdot \frac{\partial |z|}{\partial z_i} \cdot r(z, \varepsilon \cdot \chi(|\varepsilon|))_j \right|$$

$$< \frac{2\alpha_2}{\alpha_1} \cdot \frac{2}{\alpha_2 - \alpha_1} \cdot \left(k_1(\alpha_4) + k_2(\alpha_4) \cdot \alpha_2 + k_3(\alpha_2, \alpha_4) \cdot \alpha_2^2 \right)$$

$$= \frac{16}{\alpha_2} \cdot \left(k_1(\alpha_4) + k_2(\alpha_4) \cdot \alpha_2 + k_3(\alpha_2, \alpha_4) \cdot \alpha_2^2 \right), \tag{75}$$

$$\left| f(|z|) \cdot \frac{\partial r(z, \varepsilon \cdot \chi(|\varepsilon|))_j}{\partial z_i} \right| \leqslant k_2(\alpha_4) + k_3(\alpha_2, \alpha_4) \cdot \alpha_2. \tag{76}$$

Choose $\alpha_2, \alpha_4 > 0$ so that

$$k_3(\alpha_2, \alpha_4) < 2k_3(0, 0), \tag{77}$$

$$k_1(\alpha_4) + k_2(\alpha_4) \cdot \alpha_2 + 2k_3(0, 0) \cdot \alpha_2^2 < \kappa, \tag{78}$$

$$\frac{16}{\alpha_2} \cdot 2k_3(0, 0) \cdot \alpha_2^2 = 32k_3(0, 0) \cdot \alpha_2 < \frac{\kappa}{2\sqrt{n}}, \tag{79}$$

$$k_2(\alpha_4) + 2k_3(0, 0) \cdot \alpha_2 < \frac{\kappa}{2\sqrt{n}}. \tag{80}$$

It is clear that (77)–(80) hold if α_2 and α_4 are sufficiently small. Fix α_2. Choose $\alpha_4 > 0$ so that (77)–(80) hold and

$$\frac{16}{\alpha_2} \left(k_1(\alpha_4) + k_2(\alpha_4) \cdot \alpha_2 + 2k_3(0, 0)\alpha_2^2 \right)$$

$$= \frac{16}{\alpha_2} k_1(\alpha_4) + 16k_2(\alpha_4) + 32k_3(0, 0) \cdot \alpha_2 < \frac{\kappa}{2\sqrt{n}}. \tag{81}$$

This is possible, since $k_1(0) = 0$, $k_2(0) = 0$, and (79) holds. Fix α_4. From (68), (73), (77) and (78) we have $|v_\kappa'(z, \varepsilon) - Bz| < \kappa$; from (69), (70), (75), (76), (80) and (81) we have $|\partial_z v_\kappa'(z, \varepsilon) - B| < \kappa$. It only remains to note that

$$|\partial_\varepsilon v_\kappa'| \leqslant \sup_{|z| \leqslant \alpha_2, |\varepsilon| \leqslant \alpha_4} |\partial_\varepsilon v_\kappa'| < d < \infty. \tag{82}$$

Assertion 10 is proved.

Consider the vector field $v_{\kappa,l}$, $v_{\kappa,l}(z, \varepsilon) = v'_\kappa(z, l\varepsilon)$, where l is a constant with $0 < l < 1$. It follows from (82) that $|\partial_\varepsilon v_{\kappa,l}| < ld$, and hence for sufficiently small $l = l(\kappa)$ we have

$$|\partial_\varepsilon v_{\kappa, l(\kappa)}| < \kappa. \tag{83}$$

We denote $v_{\kappa, l(\kappa)}$ by v_κ. It follows from Lemma 8.3 of [3], p. 69, that for any $\delta > 0$ there exists $\kappa(\delta) > 0$ such that for the differential equation

$$\begin{cases} \dot{z} = v_{\kappa(\delta)}(z, \varepsilon), \\ \dot{\varepsilon} = 0 \end{cases} \tag{84}$$

we have

$$\left| e^{t(v_{\kappa(\delta)} \times 0)}(z, \varepsilon) - e^{tC}(z, \varepsilon) \right| < \delta,$$

$$\left| (e^{t(v_{\kappa(\delta)} \times 0)})_* |_{(z, \varepsilon)} - e^{tC} \right| < \delta,$$

$$\forall t : 0 \leqslant t \leqslant 1, \qquad \forall (z, \varepsilon) \in R^n \times R^l.$$

Recalling (66) and applying Lemma 1 to the diffeomorphism $e^{v_{\kappa(\delta)} \times 0}$, we obtain that there exist unique foliations Γ^+ with C^{k-1}-leaves $(x, y^-, \varepsilon) = \gamma^+_{(z,\bar{\varepsilon})}(y^+)$ and $\Gamma^{-,0}$ with C^{k-1}-leaves $y^+ = \gamma^-_{(z,\bar{\varepsilon})}(x, y^-, \varepsilon)$, which are invariant with respect to $e^{v_{\kappa(\delta)} \times 0}: R^n \times R^l \to R^n \times R^l$.

Setting $A = e^C$, $P = e^{C^-}$, $Q = e^{C^+ + D}$ and $T = e^{v_{\kappa(\delta)} \times 0}$ in Lemma 1, and reasoning similarly to the above, we obtain that there exist unique foliations $\Gamma^{+,0}$ with C^{k-1}-leaves $y^- = \gamma^{+,0}_{(z,\bar{\varepsilon})}(x, y^+, \varepsilon)$ and Γ^- with C^{k-1}-leaves $(x, y^+, \varepsilon) = \gamma^-_{(z,\bar{\varepsilon})}(y^-)$ which are invariant with respect to $e^{v_{\kappa(\delta)} \times 0}$.

We now prove that the foliations $\Gamma^{+,0}$ and $\Gamma^{-,0}$ intersect in a foliation Γ^0 which is invariant with respect to $e^{v_{\kappa(\delta)} \times 0}$. Namely, we have

ASSERTION 11. *For any $(z, \tilde{\varepsilon}) \in R^n \times R^l$ the leaves $y^- = \gamma^{+,0}_{(z,\bar{\varepsilon})}(x, y^+, \varepsilon)$ and $y^+ = \gamma^{-,0}_{(z,\bar{\varepsilon})}(x, y^-, \varepsilon)$ intersect in a C^{k-1}-manifold, which can be given by the graph of $(y^-, y^+) = \gamma^0_{(z,\bar{\varepsilon})}(x, \varepsilon)$, where $\gamma^0_{(z,\bar{\varepsilon})}: X \times R^l \to Y^- \times Y^+$ is a C^{k-1}-mapping which satisfies a Lipschitz condition with constant $c_3(\delta)$, where $c_3(\delta) \to 0$ as $\delta \to 0$.*

The collection of manifolds $\gamma^0_{(z,\bar{\varepsilon})}$, $(z, \tilde{\varepsilon}) \in R^n \times R^l$, gives a foliation Γ^0 which is invariant with respect to $e^{v_{\kappa(\delta)} \times 0}$.

PROOF. We write the desired graph $(y^-, y^+) = \gamma^0_{(z,\bar{\varepsilon})}(x, \varepsilon)$ in the form

$$y^+ = {}^1\gamma^0_{(z,\bar{\varepsilon})}(x, \varepsilon), \qquad y^- = {}^2\gamma^0_{(z,\bar{\varepsilon})}(x, \varepsilon). \tag{85}$$

The points of intersection of the manifolds $\gamma^{+,0}_{(z,\bar{\varepsilon})}$ and $\gamma^{-,0}_{(z,\bar{\varepsilon})}$ satisfy the equations

$$y^- - \gamma^{+,0}_{(z,\bar{\varepsilon})}(x, \gamma^{-,0}_{(z,\bar{\varepsilon})}(x, y^-, \varepsilon), \varepsilon) = 0, \qquad y^+ - \gamma^{-,0}_{(z,\bar{\varepsilon})}(x, \gamma^{+,0}_{(z,\bar{\varepsilon})}, \varepsilon), \varepsilon) = 0. \tag{86}$$

Since

$$\left| \partial_{y^+}\left(\gamma^{+,0}_{(z,\bar{\varepsilon})}\right) \cdot \partial_{y^-}\left(\gamma^{-,0}_{(z,\bar{\varepsilon})}\right) \right| \leqslant c_2^2(\delta) < 1,$$

$$\left| \partial_{y^-}\left(\gamma^{-,0}_{(z,\bar{\varepsilon})}\right) \cdot \partial_{y^+}\left(\gamma^{+,0}_{(z,\bar{\varepsilon})}\right) \right| \leqslant c_2^2(\delta) < 1,$$

(see (54) and (62)), by the implicit function theorem, (86) has a unique C^{k-1}-solution $y^-(x, \varepsilon)$, $y^+(x, \varepsilon)$. From (86) and the fact that $\Gamma^{+,0}$ and $\Gamma^{-,0}$ are $c_2(\delta)$-collections we obtain

$$\left| D_{(x,\varepsilon)} y^-(x, \varepsilon) \right| \leqslant \left(\left| \partial_{(x,\varepsilon)} \left(\gamma_{(z,\tilde{\varepsilon})}^{+,0} \right) \right| + \left| \partial_{y^+} \left(\gamma_{(z,\tilde{\varepsilon})}^{+,0} \right) \right| \cdot \left| \partial_{(x,\varepsilon)} \left(\gamma_{(z,\tilde{\varepsilon})}^{-,0} \right) \right| \right)$$

$$\times \left| \left(E - \partial_{y^+} \left(\gamma_{(z,\tilde{\varepsilon})}^{+,0} \right) \circ \partial_{y^-} \left(\gamma_{(z,\tilde{\varepsilon})}^{-,0} \right) \right) \right|^{-1}$$

$$\leqslant \frac{c_2(\delta) + c_2^2(\delta)}{1 - c_2^2(\delta)} \, .$$

Set $c_3(\delta)/2 = (c_2(\delta) + c_2^2(\delta))/(1 - c_2^2(\delta))$. In a similar way to the preceding we obtain $| D_{(x,\varepsilon)} y^+(x, \varepsilon) | \leqslant c_3(\delta)/2$.

Set

$$^1\gamma_{(z,\tilde{\varepsilon})}^0(x, \varepsilon) = y^+(x, \varepsilon), \qquad ^2\gamma_{(z,\tilde{\varepsilon})}^0(x, \varepsilon) = y^-(x, \varepsilon). \tag{87}$$

We have

$$\left| \gamma_{(z,\tilde{\varepsilon})}^0(x_2, \varepsilon_2) - \gamma_{(z,\tilde{\varepsilon})}^0(x_1, \varepsilon_1) \right| = \left| ^1\gamma_{(z,\tilde{\varepsilon})}^0(x_2, \varepsilon_2) - {}^1\gamma_{(z,\tilde{\varepsilon})}^0(x_1, \varepsilon_1) \right|$$

$$+ \left| ^2\gamma_{(z,\tilde{\varepsilon})}^0(x_2, \varepsilon_2) - {}^2\gamma_{(z,\tilde{\varepsilon})}^0(x_1, \varepsilon_1) \right|$$

$$\leqslant c_3(\delta) \cdot \left| (x_2, \varepsilon_2) - (x_1, \varepsilon_1) \right| \, .$$

Furthermore,

$$^1\gamma_{(z,\tilde{\varepsilon})}^0(z_x, \tilde{\varepsilon}) = y^+(z_x, \tilde{\varepsilon}) = z_{y^+}, \qquad ^2\gamma_{(z,\tilde{\varepsilon})}^0(z_x, \tilde{\varepsilon}) = z_{y^-} \, .$$

It follows from this that the manifolds $\gamma_{(z,\tilde{\varepsilon})}^0$, $(z, \tilde{\varepsilon}) \in R^n \times R^l$, give a $c_3(\delta)$-collection Γ^0. We see from the construction of Γ^0 that Γ^0 gives a partition of $R^n \times R^l$. The leaves of the collection Γ^0 depend continuously on the point $(z, \tilde{\varepsilon})$. Also, Γ^0 is invariant with respect to $e^{v_{\kappa(\delta)} \times 0}$.

In $R^n \times R^l$ consider the foliation Γ^Y of parallel planes $\{(y^+, y^-) = \text{const}\}$. Any leaf of Γ^0 intersects any leaf of Γ^Y at one point.

The homeomorphism $h(\Gamma^0, \Gamma^Y) \colon R^n \times R^l \to R^n \times R^l$ (see (63)) takes Γ^0 to the family of parallel planes $\{(x, \varepsilon) = \text{const}\}$. Therefore Γ^0 is a foliation. Assertion 11 is proved.

ASSERTION 12. *The foliations* $\Gamma^{+,0}$, $\Gamma^{-,0}$, Γ^+, Γ^- *and* Γ^0 *are invariant with respect to the flow* $e^{t(v_{\kappa(\delta)} \times 0)}$, $-\infty < t < \infty$.

PROOF. Every diffeomorphism $L \colon R^n \times R^l \to R^n \times R^l$ which commutes with $e^{(v_{\kappa(\delta)} \times 0)}$ carries a foliation Γ which is invariant with respect to $e^{(v_{\kappa(\delta)} \times 0)}$ to a foliation which is invariant with respect to $e^{(v_{\kappa(\delta)} \times 0)}$. Indeed,

$$e^{(v_{\kappa(\delta)} \times 0)}(L\Gamma) = Le^{(v_{\kappa(\delta)} \times 0)}\Gamma = L\Gamma.$$

Consider $e^{t(v_{\kappa(\delta)} \times 0)}\Gamma^{+,0}$, $|t| \leqslant t_0$, where t_0 is chosen so small that the leaves of foliation $e^{t(v_{\kappa(\delta)} \times 0)}\Gamma^{+,0}$ satisfy a Lipschitz condition with constant smaller than $c_1(\delta)$ (see Lemma 1 for the definition of $c_1(\delta)$). Since the mappings of the flow commute, $e^{t(v_{\kappa(\delta)}\times 0)}\Gamma^{+,0}$ is a $c_1(\delta)$-foliation invariant with respect to $e^{(v_{\kappa(\delta)}\times 0)}$. By virtue of the uniqueness of $\Gamma^{+,0}$ we have

$$e^{t(v_{\kappa(\delta)}\times 0)}\Gamma^{+,0} = \Gamma^{+,0}, \qquad |t| \leqslant t_0.$$

Since for any t, $-\infty < t < \infty$, the mapping $e^{t(v_{\kappa(\delta)}\times 0)}$ can be represented in the form of a composition of mappings $e^{t(v_{\kappa(\delta)}\times 0)}$, $|t| \leqslant t_0$, the foliation $\Gamma^{+,0}$ is invariant with respect to $e^{t(v_{\kappa(\delta)}\times 0)}$, $-\infty < t < \infty$.

The proof of the invariance of the foliations $\Gamma^{-,0}$, Γ^+ and Γ^- is similar. Since $\Gamma^0 = \Gamma^{+,0} \cap \Gamma^{-,0}$, Γ^0 is also invariant with respect to the flow $e^{t(v_{\kappa(\delta)}\times 0)}$, $-\infty < t < \infty$. Assertion 12 is proved.

Consider the leaves $\gamma_0^{+,0}$, $\gamma_0^{-,0}$, γ_0^+, γ_0^- and γ_0^0 of the foliations $\Gamma^{+,0}$, $\Gamma^{-,0}$, Γ^+, Γ^- and Γ^0 which pass through the origin ($z = 0$, $\varepsilon = 0$). Since the point ($z = 0$, $\varepsilon = 0$) these leaves are invariant manifolds. From the uniqueness of the ($k - 1$)-jet of the invariant manifolds (see Theorem 3) it follows easily that

$$D_{(y^+,x)}\left(\gamma_0^{+,0}\right)\Big|_{\substack{z=0 \\ \varepsilon=0}} = 0, \qquad D_{(y^-,x)}\left(\gamma_0^{-,0}\right)\Big|_{\substack{z=0 \\ \varepsilon=0}} = 0, \qquad D_{(y^+)}\left(\gamma_0^+\right)\Big|_{\substack{z=0 \\ \varepsilon=0}} = 0,$$

$$D_{(y^-)}\left(\gamma_0^-\right)\Big|_{\substack{z=0 \\ \varepsilon=0}} = 0, \qquad D_x\left(\gamma_0^0\right)\Big|_{\substack{z=0 \\ \varepsilon=0}} = 0. \tag{88}$$

Consider the homeomorphism $h(\Gamma^-, \Gamma^{+,0})$: $R^n \times R^l \to R^n \times R^l$ (see (63)). It follows from the invariance of the foliations Γ^- and $\Gamma^{+,0}$ that this homeomorphism conjugates the system

$$\begin{cases} \dot{z} = v_{\kappa(\delta)}(z, \varepsilon), \\ \dot{\varepsilon} = 0 \end{cases} \tag{89}$$

with the system

$$\begin{cases} \dot{y}^- = (v_{\kappa(\delta)})_{y^-}\left(y^-, \gamma_0^-\left(y^-\right)\right), \\ (\dot{x}, \dot{y}^+) = (v_{\kappa(\delta)})_{(x, y^+)}\left(x, y^+, \gamma_0^{+,0}(x, y^+, \varepsilon), \varepsilon\right), \\ \dot{\varepsilon} = 0, \end{cases} \tag{90}$$

where $(v_{\kappa(\delta)})_{y^-}$ is the y^--component of the vector $v_{\kappa(\delta)}$ and $(v_{\kappa(\delta)})_{(x, y^+)}$ is the (x, y^+)-component of this vector.

We note that each leaf of Γ^- lies in a plane $\{\varepsilon = \varepsilon_0\}$ (where ε_0 depends on the leaf in question). Indeed, for the field $(v_{\kappa(\delta)} \times 0)|_{\varepsilon=\varepsilon_0}$, given on the plane $\varepsilon = \varepsilon_0$, a foliation $\Gamma_{\varepsilon_0}^-$ is defined, which is invariant with respect to this field. It is clear that $\Gamma_{\varepsilon_0}^-$ is invariant with respect to the flow $e^{t(v_{\kappa(\delta)}\times 0)}$. It follows from the uniqueness of Γ^- that the leaves of $\Gamma_{\varepsilon_0}^-$ are the leaves of Γ^-.

It follows from the above and from the construction of the homeomorphism $h(\Gamma^-, \Gamma^{+,0})$ that the plane $\varepsilon = \varepsilon_0 \ \forall \varepsilon_0 \in R^l$ is carried into itself under the action of this homeomorphism. It follows from this that the families (89) and (90) are homeomorphic (see Definition 2).

Consider the C^{k-1}-family ($2 \leqslant k < \infty$)

$$(\dot{x}, \dot{y}^+) = v_1(x, y^+, \varepsilon), \qquad \dot{\varepsilon} = 0, \tag{91}$$

where

$$v_1(x, y^+, \varepsilon) = (v_{\kappa(\delta)})_{(x, y^+)}(x, y^+, \gamma^{+,0}(x, y^+, \varepsilon), \varepsilon). \tag{92}$$

We construct two foliations ${}^1\Gamma^+$ and ${}^1\Gamma^0$, with C^{k-1}-leaves, which are invariant with respect to $e^{t(v_1 \times 0)}: X \times Y^+ \times R^l \to X \times Y^+ \times R^l, -\infty < t < \infty$.

To do this we note that a leaf of Γ^+ or Γ^0 which has at least one point in common with the manifold $\gamma_0^{+,0}$ lies completely in this manifold. For Γ^+ this is proved in the same way as it is proved that a leaf of Γ^- lies in a plane of the form $\varepsilon = \varepsilon_0$. For Γ^0 it follows from the construction of Γ^0.

Denote the leaves of Γ^+ and Γ^0 which lie in $\gamma_0^{+,0}$ by $\Gamma^+ \cap \gamma_0^{+,0}$ and $\Gamma^0 \cap \gamma_0^{+,0}$. Project the manifold $\gamma_0^{+,0}$ along with the leaves $\Gamma^+ \cap \gamma_0^{+,0}$ and $\Gamma^0 \cap \gamma_0^{+,0}$ onto the plane $Y^+ \times X \times R^l$. It is easy to see the projection of $\Gamma^+ \cap \gamma_0^{+,0}$ yields the foliation ${}^1\Gamma^+$ with leaves $(x, \varepsilon) = {}^1\gamma^+_{(x_0, y_0^+, \varepsilon_0)}(y^+)$, and that the projection of $\Gamma^0 \cap \gamma_0^{+,0}$ yields the foliation ${}^1\Gamma^0$ with leaves $y^+ = {}^1\gamma^0_{(x_0, y_0^+, \varepsilon_0)}(x, \varepsilon)$, where ${}^1\gamma^+_{(x_0, y_0^+, \varepsilon_0)}$ is the (x, ε)-component of the mapping $\gamma^+_{(z_0, \varepsilon_0)}$ and ${}^1\gamma^0_{(x_0, y_0^+, \varepsilon_0)}$ is the y^+-component of the mapping $\gamma^0_{(z_0, \varepsilon_0)}$ (see (85)), and

$$z_0 = \left(x_0, y_0^- = \gamma_0^{+,0}(x_0, y_0^+, \varepsilon_0), y_0^+\right).$$

It follows from the construction of ${}^1\Gamma^+$ and ${}^1\Gamma^0$ that these foliations are invariant with respect to the flow $e^{t(v_1 \times 0)}$ as are the manifolds ${}^1\gamma_0^+$ and ${}^1\gamma_0^0$.

From arguments similar to those carried out for the families (89) and (90), we obtain that (91) is homeomorphic to the family

$$\begin{cases} \dot{y}^+ = (v_1)_{y^+}\left(y^+, {}^1\gamma_0^+(y^+)\right), \\ \dot{x} = (v_1)_x\left(x, {}^1\gamma_0^0(x, \varepsilon), \varepsilon\right), \\ \dot{\varepsilon} = 0. \end{cases} \tag{93}$$

Recalling the definition of v_1 (see (92)), we obtain

$$(v_1)_x\left(x, {}^1\gamma_0^0(x, \varepsilon), \varepsilon\right) = (v_{\kappa(\delta)})_x\left(x, {}^1\gamma_0^0(x, \varepsilon), \gamma_0^{+,0}\left(x, {}^1\gamma_0^0(x, \varepsilon), \varepsilon\right), \varepsilon\right). \tag{94}$$

It follows from the definitions of ${}^1\gamma_0^0$ and ${}^2\gamma_0^0$ (see (85), (87) and (86)) that

$$\gamma^{+,0}\left(x, {}^1\gamma_0^0(x, \varepsilon), \varepsilon\right) = {}^2\gamma_0^0(x, \varepsilon).$$

By taking this into account we may extend the equality (94) to

$$(v_{\kappa(\delta)})_x\left(x, {}^1\gamma_0^0(x, \varepsilon), \gamma_0^{+,0}\left(x, {}^1\gamma_0^0(x, \varepsilon), \varepsilon\right), \varepsilon\right)$$
$$= (v_{\kappa(\delta)})_x\left(x, {}^1\gamma_0^0(x, \varepsilon), {}^2\gamma_0^0(x, \varepsilon), \varepsilon\right)$$
$$= (v_{\kappa(\delta)})_x\left(x, \gamma_0^0(x, \varepsilon), \varepsilon\right).$$

Similarly,

$$(v_1)_{y^+}\left(y^+, {}^1\gamma_0^+(y^+)\right) = (v_{\kappa(\delta)})_{y^+}\left(y^+, \gamma_0^+(y^+)\right). \tag{95}$$

Therefore the family (91) is homeomorphic to the family

$$\begin{cases} \dot{y}^+ = (v_{\kappa(\delta)})_{y^+}(y^+, \gamma_0^+(y^+)), \\ \dot{x} = (v_{\kappa(\delta)})_x(x, \gamma_0^0(x, \varepsilon), \varepsilon), \\ \dot{\varepsilon} = 0. \end{cases}$$

Returning from the family (91) to (90) and recalling that the families (90) and (89) are homeomorphic, we obtain that (89) is homeomorphic to the family

$$\begin{aligned} \dot{y}^- = (v_{\kappa(\delta)})_{y^-}(y^-, \gamma_0^-(y^-)), \qquad \dot{y}^+ = (v_{\kappa(\delta)})_{y^+}(y^+, \gamma_0^+(y^+)), \\ \dot{x} = (v_{\kappa(\delta)})_x(x, \gamma_0^0(x, \varepsilon), \varepsilon), \qquad \dot{\varepsilon} = 0. \end{aligned} \tag{96}$$

Since (88) holds, the linear part of $(v_{\kappa(\delta)})_{y^-}(y^-, \gamma_0^-(y^-))$ at the origin is C^-, all eigenvalues of which have real part less than zero, and the linear part of the vector field $(v_{\kappa(\delta)})_{y^+}(y^+, \gamma_0^+(y^+))$ at the origin is C^+, all eigenvalues of which have real part greater than zero. By the theorem on the structural stability of a multidimensional node (see [9], for example), we obtain that (96) is homeomorphic to the family

$$\begin{aligned} \dot{y}^- = -y^-, \qquad \dot{y}^+ = y^+, \\ \dot{x} = (v_{\kappa(\delta)})_x(x, \gamma_0^0(x, \varepsilon), \varepsilon), \qquad \dot{\varepsilon} = 0. \end{aligned} \tag{97}$$

Make the substitution $z \mapsto z$, $\varepsilon \to l(\kappa(\delta)) \cdot \varepsilon$ (see (83) for the definition of $l(\kappa)$). Then the homeomorphic families (89) and (97) are taken to the homeomorphic families (67) and

$$\begin{aligned} \dot{y}^- = -y^-, \qquad \dot{y}^+ = y^+, \\ \dot{x} = (v_{\kappa(\delta)})_x(x, \gamma_0^0(x, \varepsilon/l(\kappa)), \varepsilon), \qquad \dot{\varepsilon} = 0. \end{aligned} \tag{98}$$

The family (67) is a representation of the germ (3) by construction. Denote $\gamma_0^0(x, \varepsilon/l(\kappa))$ by $\gamma^0(x, \varepsilon)$. Theorem 1 is proved.

PROOF OF THEOREM 2. Consider the germ of the family

$$\dot{z} = v_1(z, \eta), \qquad \dot{\eta} = 0, \qquad z \in R^n, \qquad \eta \in R^{l_2}, \tag{99}$$

such that

$$(99)\big|_{\eta=0} = (3)\big|_{\varepsilon=0}. \tag{100}$$

It follows from Theorem 1 that the germ of the family (99) is homeomorphic to the germ of the family

$$\begin{cases} \dot{y}^- = -y^-, \qquad \dot{y}^+ = y^+, \\ \dot{x} = (v_1)_x(x, \gamma_{v_1}^0(x, \eta), \eta), \qquad \dot{\eta} = 0. \end{cases} \tag{101}$$

Since (99) is homeomorphic to (101), and (3) is homeomorphic to (5), to prove Theorem 2 it is sufficient to prove that (101) is induced from (5). It follows from (100) that $\gamma_{v_1}^0$ can be chosen so that $\gamma_{v_1}^0\big|_{\eta=0} = \gamma^0\big|_{\varepsilon=0}$, i.e.

$$(v_1)_x(x, \gamma_{v_1}^0(x, \eta), \eta)\big|_{\eta=0} = (4)(x, \varepsilon)\big|_{\varepsilon=0}.$$

It follows from the fact that (4) is versal that (101) is induced from (5).

Theorem 2 is proved.

PROOF OF THEOREM 3. We shall show that the $(k-1)$-jet at zero $(\gamma_v^0)^{k-1}$ of the manifold γ_v^0 is uniquely determined by the $(k-1)$-jet at zero $(v)^{k-1}$ of a germ of the field v. Consider the extension $v_{\kappa(\delta)}$ of the germ of v (see (88)) and the diffeomorphism $T = e^{tv_{\kappa(\delta)}}$ at $t = 1$. Since $v(0) = 0$, it follows that $T(0) = 0$, and the $(k-1)$-jet $(T)^{k-1}$ at zero of the diffeomorphism T is completely determined by the jet $(v)^{k-1}$.

Since γ_v^0 is the intersection $\gamma_0^{+,0} \cap \gamma_0^{-,0}$ (see Assertion 11), it is sufficient to show that the $(k-1)$-jets at zero $(\gamma_0^{+,0})^{k-1}$ and $(\gamma_0^{-,0})^{k-1}$ of the manifolds $\gamma_0^{+,0}$ and $\gamma_0^{-,0}$ are completely determined by the jet $(T)^{k-1}$. To do this, consider the operator φ_T (see (19)). It follows from its definition that if $T(0) = 0$, then the $(k-1)$-jet at zero of the mapping $(\varphi_T f)_0$ from the collection $\varphi_T f$, where f is any collection with C^{k-1}-leaves, is completely determined by $(T)^{k-1}$ and the $(k-1)$-jet at zero $(f_0)^{k-1}$ of the mapping $f_0 \in f$. Consider the collection $f^\infty = \lim_{n \to \infty} \varphi_T^n f$, which is fixed under φ_T. The collection f^∞ does not depend on the choice of f. From this and the preceding we obtain that the $(k-1)$-jet at zero of the mapping f_0^∞ is completely determined by $(T)^{k-1}$.

The manifold $\gamma_0^{+,0}$, as well as $\gamma_0^{-,0}$, is the graph of a mapping f_0^∞, where f^∞ is a collection which is fixed with respect to the corresponding operator φ_T. Therefore $(\gamma_0^{+,0})^{k-1}$ and $(\gamma_0^{-,0})^{k-1}$ are completely determined by $(T)^{k-1}$.

We now establish the analyticity of the dependence of $(\gamma_v^0)^{k-1}$ on $(v)^{k-1}$.

Consider the set M of C^{k-1}-germs at zero of k^0-dimensional manifolds in R^n that pass through zero, which can be given by graphs $y = f(x)$ of mappings $f: X \to Y$. The correspondence associating with $m \in M$ the mapping $f(m): X \to Y$ whose graph yields m determines a one-to-one correspondence between M and the set $W(X)$ of C^{k-1}-mappings $f: X \to Y$ such that $f(0) = 0$. We identify $W(X)$ with M by means of this correspondence.

Consider the real analytic manifolds $I^{k-1}(R^n)$, $W^{k-1}(X)$ and $TW^{k-1}(X)$, where $I^{k-1}(R^n)$ is defined in the text preceding Theorem 3, $W^{k-1}(X)$ is the space of $(k-1)$-jets at zero of the mappings $f \in W(X)$, and $TW^{k-1}(X)$ is the tangent bundle over $W^{k-1}(X)$. Consider the analytic mapping

$$\psi^{k-1}: I^{k-1}(R^n) \times W^{k-1}(X) \to TW^{k-1}(X),$$
$$\psi^{k-1}: \left((v)^{k-1}, (f(x))^{k-1}\right) \to$$
$$\left((f(x))^{k-1}, \left(v_y(x, f(x)) - D_x f|_x \cdot v_x(x, f(x))\right)^{k-1}\right),$$

where $(\cdot)^{k-1}$ is the $(k-1)$-jet at zero of $(\cdot)$.

Note that $TW^{k-1}(X)$ is analytically isomorphic in a canonical way to $W^{k-1}(X) \times W^{k-1}(X)$, since $W^{k-1}(X)$ is furnished with a canonical structure of a linear space: mappings from X to Y and hence also their jets can be added. Let π denote the projection of $TW^{k-1}(X)$ onto the second factor of the product $W^{k-1}(X) \times W^{k-1}(X)$.

Consider the mapping

$$\varphi = \pi \circ \psi^{k-1}: I^{k-1}(R^n) \times W^{k-1}(X) \to W^{k-1}(X),$$

$$\varphi: \left((v)^{k-1}, (f(x))^{k-1}\right) \to \left(v_y(x, f(x)) - D_x f|_x v_x(x, f(x))\right)^{k-1}.$$

The $(k-1)$-jet at zero of the manifold γ_v^0, which is invariant with respect to the flow defined by the field v, lies in the inverse image of zero under the mapping $\varphi((v)^{k-1})$, where $\varphi((v)^{k-1}): W^{k-1}(X) \to W^{k-1}(X)$ is the mapping φ with fixed first argument $(v)^{k-1}$. Indeed, the invariance of γ_v^0 with respect to v is equivalent to the fact that $v(z) \in T_z \gamma_v^0 \ \forall z \in \gamma_v^0$. But the equation

$$v_y(x, f(x)) - D_x f|_x v_x(x, f(x)) = 0 \tag{102}$$

means exactly that $v(z) \in T_z \gamma_v^0 \ \forall z \in \gamma_v^0$.

The mapping φ is analytic.

To solve the equation

$$\varphi\left((v)^{k-1}, (\gamma_v^0)^{k-1}\right) = 0 \tag{103}$$

for $(\gamma_v^0)^{k-1}$ we show that in some neighborhood τ of the point $((v_0)^{k-1}, 0) \subset I^{k-1}(R^n) \times W^{k-1}(X)$, where $v_0(z) = Bz$ (see Theorem 1' for the definition of the linear operator B), the implicit function theorem holds for the mapping φ.

We see from the definition of φ that $\varphi((v_0)^{k-1}, 0) = 0$ and that $\varphi((v_0)^{k-1}): W^{k-1}(X) \to W^{k-1}(X)$ is linear. Therefore $\partial_w \varphi|_{(v_0)^{k-1}, 0} = \varphi(v_0)^{k-1}$. We assume that the operator B can be diagonalized, and let $(x_1, \ldots, x_{k^0}, y_1, \ldots, y_{n-k^0})$ be a basis in R^n in which B is diagonal. Let λ_i^0, $i = 1, \ldots, k^0$, be the eigenvalues of the operator $B|_X: X \to X$, and let $\lambda_j^{+,-}$, $j = 1, 2, \ldots, n - k^0$, be the eigenvalues of the operator $B|_Y: Y \to Y$. Then $\varphi((v_0)^{k-1})$ can be diagonalized, and its eigenvalues $\lambda_{i,j}^-$, where $i = i_1, \ldots, i_{k^0}$ runs through all sets of k^0 integers such that $0 \leqslant i_j, \ldots, i_{k^0}$ and $0 < i_1 + \cdots + i_{k^0} \leqslant k - 1$, $j = 1, \ldots, n - k^0$, are given by

$$\lambda_{i,j}^- = \lambda_j^{+,-} - \left(i_1 \cdot \lambda_1^0 + \cdots + i_{k^0} \cdot \lambda_{k^0}^0\right). \tag{104}$$

Indeed, it is easy to see that $(f_{i,j}^-(x))^{k-1}$, where

$$f_{i,j}^-(x) = \begin{cases} f_1(x) = 0 \\ \quad \vdots \\ f_j(x) = x_1^{i_1} \cdot \cdots \cdot x_{k^0}^{i_{k^0}} \\ \quad \vdots \\ f_{r-k^0}(x) = 0 \end{cases}$$

$$\bar{i} = i_1, \ldots, i_{k^0}, \qquad 0 \leqslant i_1, \ldots, i_{k^0},$$

$$0 < i_1 + \cdots + i_{k^0} \leqslant k - 1, \qquad j = 1, \ldots, n - k^0,$$

is a basis in $W^{k-1}(X)$ composed of eigenvectors of $\varphi((v_0)^{k-1})$ corresponding to the eigenvalues $\lambda_{i,j}^-$.

We have from (104) that $\operatorname{Re} \lambda_{i,j}^- = \operatorname{Re} \lambda_j^{+,-} - \operatorname{Re}(i_1\lambda_1^0 + \cdots + i_{k^0}\lambda_{k^0}^0)$. From (6) and from $\lambda_j^{+,-} \in \text{II} \cup \text{III}$, $\lambda_i^0 \in \text{I}$, we obtain that either

$$\left(\text{if } \lambda_j^{+,-} \in \text{II}\right) \qquad \operatorname{Re} \lambda_{i,j}^- < \beta_3 - (k-1)\beta_2 < 0,$$

or

$$\left(\text{if } \lambda_j^{+,-} \in \text{III}\right) \qquad \operatorname{Re} \lambda_{i,j}^- > \beta_4 - (k-1)\beta_1 > 0,$$

i.e. $\operatorname{Re} \lambda_{i,j}^- \neq 0 \; \forall i, j$. Hence $\lambda_{i,j}^- \neq 0 \; \forall i, j$. Thus $\varphi((v_0)^{k-1})$ is nonsingular. Since any linear operator in some (possibly complex) basis can be made as close to diagonal as desired, we have proved that for an operator B that satisfies the inequality (6), $\varphi((v_0)^{k-1})$ is nonsingular.

By the implicit function theorem the nonsingularity of $\partial_w\varphi|_{((v_0)^{k-1},0)} = \varphi((v_0)^{k-1})$ and $\varphi((v_0)^{k-1},0) = 0$ imply that there exists a neighborhood $\tau_1 \subset I^{k-1}(R^n)$ of the point $(v_0)^{k-1}$ such that for any $(v)^{k-1} \in \tau_1$ there exists a unique solution $(\gamma_v^0)^{k-1}$ of (103), and $(\gamma_v^0)^{k-1}$ depends analytically on $(v)^{k-1}$.

The above reasoning can be carried out not only in the neighborhood of the $(k-1)$-jet of the field Bz but also in the neighborhood of the $(k-1)$-jet of any linear field from $U(R^n)$, since by the definition of $U(R^n)$ such fields are in $W_2(B)$. (See before Theorem 3 for the definition of $W_2(B)$.) In other words, the inequality (6) of Theorem 1' holds for such fields. Therefore we can choose for $\tau_1 \subset I^{k-1}(R^n)$ some neighborhood of an n^2-dimensional domain $E \subset I^{k-1}(R^n)$, where E consists of the $(k-1)$-jets of all linear vector fields in $U(R^n)$.

We now note that for the contraction transformation $\lambda: R^n \to R^n$, $\lambda(z) = \lambda z$, $0 < \lambda < 1$, the linear part at zero of the germs from $U(R^n)$ is unchanged while the coefficients in the Taylor series of terms of degree l are "contracted" by a factor of λ^l. It follows from this that for any $(v)^{k-1} \in I^{k-1}(R^n)$ there is a λ, $0 < \lambda < 1$, such that as a result of the transformation $z \to \lambda z$ the jet $(v)^{k-1}$ is transformed into a jet contained in τ_1.

On performing the inverse transformation $z \to \lambda^{-1}z$ we obtain that (103) can be solved for any $(v)^{k-1} \in I^{k-1}(R^n)$, i.e. for any $v \in U(R^n)$ the $(k-1)$-jet at zero of a manifold which is invariant with respect to v is uniquely determined by $(v)^{k-1}$ and depends analytically on $(v)^{k-1}$.

Thus we have the analytic mapping

$$\varphi_1: I^{k-1}(R^n) \to I^{k-1}(R^n) \times W^{k-1}(X), \qquad \varphi_1: (v)^{k-1} \to \left((v)^{k-1}, (\gamma_v^0)^{k-1}\right).$$

$$(105)$$

Consider the mapping

$$\varphi_2: I^{k-1}(R^n) \times W^{k-1}(X) \to I^{k-1}(X),$$

$$\varphi_2\left((v)^{k-1}, (f(x))^{k-1}\right) = \left(v_x(f(x), x)\right)^{k-1},$$

where $v_x(f(x), x)$ is the x-component of the vector $v(z)$ at the point $z = (f(x), x)$. Clearly φ_2 is analytic.

Note that the mapping π_{k-1}, which is the subject matter of Theorem 3, is a composition: $\pi_{k-1} = \varphi_1 \cdot \varphi_2$. The analyticity of π_{k-1} is proved.

The mapping π_{k-1} has maximal rank at any point $(v)^{k-1} \in I^{k-1}(R^n)$. Let v be fixed. We may assume without loss of generality that the manifold γ_v^0, which is invariant with respect to the germ of v, coincides with the subspace X. Otherwise make the change of variables $x \to x, y \mapsto y - \gamma_v^0(x)$.

Consider the imbedding i of a sufficiently small neighborhood of zero $U \subset I^{k-1}(X)$ in $I^{k-1}(R^n)$ which associates $(r)^{k-1} \in I^{k-1}(X)$ with the $(k-1)$-jet of a germ of the form $\dot{x} = v_x(x, y) + r(x), \dot{y} = v_y(x, y)$.

Clearly $\pi_{k-1} \cdot i: U \to I^{k-1}(X)$ is a diffeomorphism of the neighborhood U with a neighborhood of the point $(r_1)^{k-1} \in I^{k-1}(X)$, where $r_1(x) = v_x(x, 0)$. Hence and from the fact that $i(0) = (v)^{k-}$, it follows that the rank of the mapping π_{k-1} at v is maximal.

It remains to prove that the preimage of a semialgebraic set under the mapping π_{k-1} is a semialgebraic set. For this it is sufficient to prove a similar assertion for φ_1 and φ_2. We shall prove this, say, for $\varphi_1: I^{k-1}(R^n) \to W^{k-1}(X)$ (the proof for φ_2 is analogous). The graph P of the mapping φ_1 is given by the algebraic equation (105), i.e. it is an algebraic subset of $\Delta = I^{k-1}(R^n) \times W^{k-1}(X)$. Let a semialgebraic subset M be given in $W^{k-1}(X)$. Consider the natural projections $p_1: \Delta \to I^{k-1}(R^n)$ and $p_2: \Delta \to W^{k-1}(X)$. Clearly the preimage of M under the mapping φ_1 is $p_1(P \cap p_2^{-1}(M))$. The set $P \cap p_2^{-1}(M)$ is semialgebraic in Δ, since it is the intersection of semialgebraic sets. By the Seidenberg-Tarski theorem a projection of a semialgebraic set is semialgebraic, so that $p_1(P \cap p_2^{-1}(M)) \subset I^{k-1}(R^n)$ is semialgebraic.

Theorem 3 is proved.

Received 5/JUNE/73

BIBLIOGRAPHY

1. V. I. Arnol'd, *Remarks on singularities of finite codimension in complex dynamical systems*, Funkcional. Anal. i Priložen. **3** (1969), no. 1, 1–6; English transl. in Functional Anal. Appl. **3** (1969).

2. V. A. Pliss, *A reduction principle in the theory of stability of motion*, Izv. Akad. Nauk SSSR Ser. Mat. **28** (1964), 1297–1324. (Russian)

3. D. V. Anosov, *Geodesic flows on closed Riemannian manifolds of negative curvature*, Trudy Mat. Inst. Steklov. **90** (1967); English transl. in Proc. Steklov Inst. Math. **90** (1967).

4. M. W. Hirsch, C. C. Pugh and M. Shub, *Invariant manifolds*, Bull. Amer. Math. Soc. **76** (1970), 1015–1019.

5. È. A. Tihonova, *Similarity and homeomorphism of a perturbed and an unperturbed system with block-triangular matrix*, Differencial'nye Uravnenija **6** (1970), 1221–1229; English transl. in Differential Equations **6** (1970).

6. A. N. Šošitaĭšvili, *Bifurcation of topological type of singular points of vector fields that depend on parameters*, Funkcional. Anal. i Priložen. **6** (1972), no. 2, 97–98; English transl. in Functional Anal. Appl. **6** (1972).

7. Ju. I Neĭmark, *The method of point transformation in the theory of nonlinear oscillations*, "Nauka", Moscow, 1972. (Russian)

8. M. G. Kreĭn and Ju. L. Daleckiĭ, *Certain results and problems in the theory of stability of solutions of differential equations in a Banach space*, Proc. Fifth Internat. Conf. Nonlinear Oscillations (Kiev, 1969), Vol. 1, Izdanie Inst. Mat. Akad. Nauk Ukrain. SSR, Kiev, 1970, pp. 332–347. (Russian)

9. V. I. Arnol'd, *Ordinary differential equations*, "Nauka", Moscow, 1971; English transl., M.I.T. Press, Cambridge, Mass., 1973.

10. G. Ja. Šafranov, *Generation of periodic motions from the equilibrium state*, Dokl. Akad. Nauk SSSR **193** (1970), 50–52; English transl. in Soviet Math. Dokl. **11** (1970).

Translated by W. S. LOUD

Amer. Math. Soc. Transl.
(2) Vol. **118**, 1982

Construction of the Solutions
of a System of Boundary-Layer Equations
by the Method of Lines with Respect to Time*

D. A. SILAEV

A system of boundary-layer equations for nonstationary flows of a viscous incompressible liquid near a stagnation point has been studied by O. A. Oleĭnik. In [1] approximate solutions of a Prandtl system were constructed as solutions of a certain boundary-value problem for a system of second-order ordinary differential equations, and their convergence was proved (the method of lines of Rothe type). In this article a different proof of the existence theorem is given: the method of lines with respect to the time variable is used to construct approximate solutions of the Prandtl system, and their convergence is proved; the approximations are obtained as solutions of a Cauchy problem for a certain system of ordinary differential equations [2]. A similar method was applied by Walter in the case of stationary flows [3].

For the case of an axially symmetric streamline flow of an incompressible liquid past a body we consider the system of boundary-layer equations for a nonstationary boundary layer:

$$u_t + uu_x + vu_y = -p_x + \nu u_{yy}, \qquad (ru)_x + (rv)_y = 0 \tag{1}$$

in a domain $D = \{0 \leqslant t < \infty, 0 \leqslant x \leqslant X, 0 \leqslant y < \infty\}$ under the conditions

$$u\big|_{t=0} = u_0(x, y), \qquad u\big|_{x=0} = 0, \qquad u\big|_{y=0} = 0, \qquad v\big|_{y=0} = v_0(t, x),$$

$$u(t, x, y) \to U(t, x) \quad \text{as } y \to \infty. \tag{2}$$

Here u and v are the velocity components, $U(t, x)$ is the longitudinal component of the velocity of the external flow, $U(t, 0) = 0$, $U(t, x) > 0$ when $x > 0$, $p_x = -(U_t + UU_x)$, ν is the coefficient of viscosity (the density $\rho \equiv 1$), $r(t, x)$ is the

---

1980 *Mathematics Subject Classification.* Primary 65M20, 65N40, 76D10.
* Translation of Trudy Sem. Petrovsk. **2** (1976), 223–241. MR **56** # 1947.

distance from the point x on the surface of the body to the axis of symmetry of the streamlined body, $r(t,0) = 0$ and $r(t, x) > 0$ when $x > 0$. We assume that $U_x > 0$, $r_x > 0$ and $p_x/U < 0$ in D.

By means of the substitution $\tau = t$, $\xi = x$, $\eta = u/U$, $w = u_y/U$, the problem is reduced to solution of the equation

$$\nu w^2 w_{\eta\eta} - w_\tau - \eta U w_\xi + A w_\eta + B w = 0 \tag{3}$$

in the domain $\Omega\{0 \leqslant \tau < \infty, 0 \leqslant \xi \leqslant X, 0 \leqslant \eta < 1\}$ under the conditions

$$w\,|_{\tau=0} = w_0 \equiv u_{0y}/U, \qquad (\nu w w_\eta - v_0 w + C)\,|_{\eta=0} = 0, \qquad w\,|_{\eta=1} = 0, \tag{4}$$

where $A = (\eta^2 - 1)U_x + (\eta - 1)U_t/U$, $B = \eta(r_x U/r - U) - U_t/U$ and $C = -p_x/U = U_x + U_t/U$ (for details see [1]).

We assume that A, B, C, U and v_0 are bounded and have bounded derivatives that are continuous with respect to τ in Ω, while w_0 is bounded together with its derivative with respect to ξ and is twice differentiable with respect to η when $\eta < 1$. Using the method of lines with respect to the time variable, we construct approximate solutions of problem (3), (4) and prove their convergence.

Let $f^{kj}(\tau) \equiv f(\tau, kh, jl)$ for any function $f(\tau, \xi, \eta)$, h, $l = \mathrm{const} > 0$. We replace equation (3) with conditions (4) by the system of first-order differential equations

$$\frac{dw^{kj}}{d\tau} = L_{kj}(\tau, w) \equiv \nu(w^{kj})^2 \delta_{\eta\eta} w^{kj} - jl U^k \delta_\xi w^{kj} + A^{kj} \delta_\eta w^{kj} + B^{kj} w^{kj} \tag{5}$$

$$(j = 1,\ldots,J - 1;\ k = 0, 1,\ldots,K;\ J = 1/l \text{ an integer}, K = [X/h]);$$

with the conditions

$$\nu w^{k,0} \delta_\eta w^{k,1} - v_0^k w^{k,0} + C^k = 0, \qquad w^{k,J} = 0, \tag{6}$$

$$w^{kj}(0) = w_0(kh, jl). \tag{7}$$

Here $\delta_\eta w^{kj} \equiv (w^{kj} - w^{k,j-1})/l$, $\delta_{\eta\eta} w^{kj} \equiv (w^{k,j-1} - 2w^{kj} + w^{k,j+1})/l^2$ and $\delta_\xi w^{kj} \equiv (w^{kj} - w^{k-1,j})/h$.

Let us consider the boundary condition (6) when $\eta = 0$. We rewrite this condition in such a way that it becomes a linear equation for $w^{k,0}$ by using (6) to express $w^{k,0}$. We have

$$w^{k,0} = \tfrac{1}{2}\left[\left(w^{k,1} - (l/\nu)v_0^k\right) \pm \sqrt{\left(w^{k,1} - (l/\nu)v_0^k\right)^2 + 4(l/\nu)C^k}\,\right].$$

We shall look for a solution of problem (5)–(7) that is positive when $\eta = 0$, i.e., such that $w^{k,0} > 0$. Since $C > 0$, we choose the plus sign before the root. Substituting the expression for $w^{k,0}$ into (6), we obtain

$$\lambda_k(w) \equiv \nu \delta_\eta w^{k,1} - v_0^k$$

$$+ 2C^k\left[\left(w^{k,1} - (l/\nu)v_0^k\right) + \sqrt{\left(w^{k,1} - (l/\nu)v_0^k\right)^2 + 4(l/\nu)C^k}\,\right]^{-1} = 0,$$

$$w^{k,J} = 0. \tag{8}$$

To investigate the system (5)–(7), we need certain information about differential inequalities. Let R^N be the space of vectors $y = (y_1,\ldots,y_N)$ with norm $\|y\| = \max |y_i|$. We introduce a partial ordering in R^N by writing $y \leqslant \bar{y}$ if $y_i \leqslant \bar{y}_i$ for all i. The vector-valued function $f(t, y) = (f_1(t, y),\ldots,f_N(t, y))$ is monotonically increasing in y if $f(t, y) \leqslant f(t, \bar{y})$ for $y \leqslant \bar{y}$. The basic concept is that of quasimonotonicity: the vector-valued function $f(t, y)$ is said to be *quasimonotonically increasing in y* if each component $f_i(t, y)$ is an increasing function of Y_j for $j \neq i$, i.e., $f_i(t, y) \leqslant f_i(t, \bar{y})$ if $y \leqslant \bar{y}$ and $y_i = \bar{y}_i$.

Suppose that in the domain $G\{0 \leqslant t < \infty, y \in R^N\}$ the system

$$y'(t) = f(t, y) \tag{9}$$

is defined with initial condition

$$y(0) = y_0. \tag{10}$$

Let $f(t, y)$ be continuous in t and satisfy a Lipschitz condition in y, i.e., $\|f(t, y) - f(t, \bar{y})\| \leqslant L\|y - \bar{y}\|$, where the constant L does not depend on t or y. Then, as we know, there exists for all $t \geqslant 0$ a unique continuously differentiable solution $u(t)$ of problem (9), (10) (see [14]). Moreover, let $f(t, y)$ be a vector-valued function quasimonotonically increasing in y. Then the following theorem holds.

THEOREM 1. *Let $u(t)$ be the solution of the Cauchy problem (9), (10), and suppose that the continuously differentiable function $v(t)$ satisfies the inequalities $v'(t) \leqslant f(t, v)$ and $v(0) \leqslant y_0$. Then $v(t) \leqslant u(t)$ for all $t \geqslant 0$. A similar assertion holds for the case in which the inequality signs in these three relations are reversed.*

The proof of this theorem is found in [3]. The generalization of Theorem 1 to Banach spaces was used by Walter to obtain bounds for approximate solutions of a Prandtl system in the plane stationary case in the absence of suction through the boundary of the body ($v_0 = 0$). Walter was forced to consider a system consisting of countably many first-order ordinary differential equations, since a von Mises transformation reduces the Prandtl system to a single quasilinear equation of parabolic type in an unbounded region. In our case, however, the method of lines for fixed h and l leads to a finite number of ordinary differential equations. This saves us from having to introduce any sort of supplementary boundary conditions at "infinity" in numerical computations, as in the case of the von Mises variables.

LEMMA 1. *Let $U \geqslant 0$, $C > 0$ and $U_x > 0$ in Ω; let A, B, C, U and v_0 be bounded functions continuous in τ; and let*

$$E_1(1 - \eta)\sigma \leqslant w_0 \leqslant E_2(1 - \eta)\sigma, \tag{11}$$

where $\sigma = \sqrt{-\ln \mu(1 - \eta)}$, and E_i and μ are positive constants, $\mu < 1$. Then problem (5), (8), (7) has a unique continuously differentiable solution $w^{kj}(\tau)$ such that for $0 \leqslant kh \leqslant X$

$$M_1(1 - jl)\sigma^j \leqslant w^{kj}(\tau) \leqslant M_2(1 - jl)\sigma^j, \tag{12}$$

where the positive constants M_1, M_2 and μ do not depend on h or l.

PROOF. Construct functions $V_1(\xi, \eta)$ and $V_2(\xi, \eta)$ such that $V_1 \leqslant w_0 \leqslant V_2$, $L_{kj}(V_1) \geqslant 0$, $\lambda_k(V_1) > 0$ and $L_{kj}(V_2) \leqslant 0$, $\lambda_k(V_2) < 0$. Consider the function $V_1 = M_3(1 - \eta)\sigma e^{-\alpha\xi}$. We calculate $L_{kj}(V_1)$ and $\lambda_k(V_1)$. We observe beforehand that for any smooth function $\varphi(\eta)$ we have

$$\delta_\eta \varphi^j = \varphi'^j + (l/2)\varphi''((j - \theta)l), \quad \delta_{\eta\eta}\varphi^j = \varphi''^j + (l^2/12)\varphi^{\mathrm{IV}}((j - \theta')l)$$
$$(0 < \theta < 1, |\theta'| < 1). \tag{13}$$

With these equalities taken into account, we have

$$L_{kj}(V_1) = V_1^{kj}\Bigg\{ -\nu M_3^2 e^{-2\alpha kh}\left(\tfrac{1}{2} + 1/4\sigma^{j_2}\right) - (1/2\sigma^{j_2})U_t^k/U^k$$

$$+ (1 + jl)U_x^k(1 - 1/2\sigma^{j_2}) + jl\left[\alpha U^k e^{\alpha h'} + \left(r_x^k U^k/r^k - U_x^k\right)\right]$$

$$-\nu M_3^2 e^{-2\alpha kh}\frac{(J - j)\sigma^j}{12(J - j + \theta')^3 \sigma^{j-\theta'}}\left[1 - \frac{1}{4(\sigma^{j-\theta'})^2} - \frac{3}{4(\sigma^{j-\theta'})^4} + \frac{39}{16(\sigma^{j-\theta'})^6}\right]$$

$$- \frac{l}{2\sigma^j}\left[(1 + jl)U_x^k + U_t^k/U^k\right]\frac{2 + (\sigma^{j-\theta})^{-2}}{4(J - j + \theta)l\sigma^{j-\theta}}\Bigg\}$$

$$(0 < \theta < 1, |\theta'| < 1, 0 < h' < h), \tag{14}$$

$$\lambda_k(V_1) = -\nu M_3 e^{-\alpha kh}\sigma^{1-\theta}\left(-\frac{1}{2(\sigma^{1-\theta})^2}\right) - \frac{v_0^k}{2}$$

$$+ \frac{2C^k + (l/2\nu)(v_0^k)^2 - M_3(1 - l)\sigma^l v_0^k e^{-\alpha kh}}{\sqrt{\left(M_3(1 - l)\sigma^l e^{-\alpha kh} - (l/\nu)v_0^k\right)^2 + 4(l/\nu)C^k} + M_3(1 - l)\sigma^l e^{-\alpha kh}}. \tag{15}$$

Since $U_x > 0$ and $C > 0$ in Ω, while $1/\sigma^j \leqslant 1/\sigma^0 \to 0$ as $\mu \to 0$, it is clear from (14) and (15) that the constants α, μ and M_3 can be chosen so that when $j \leqslant J - 2$ (when $j \leqslant J - 2$, $\tfrac{1}{2} \leqslant (J - j)/(J - j + \theta') \leqslant 2$) the inequalities $L_{kj}(V_1) > 0$ and $\lambda_k(V_1) > 0$ are satisfied. In the case $j = J - 1$ the term $\nu(V_1^{kj})^2\delta_{\eta\eta}V_1^{kj}$ in the expression for $L_{kj}(V_1)$ should be considered separately. We have

$$\nu\left(V_1^{k,J-1}\right)^2\delta_{\eta\eta}V_1^{k,J-1} = V_1^{k,J-1}\left\{-\nu M_3^2 e^{-2\alpha kh}\ln 2 \cdot 2\sigma^{J-1}/(\sigma^{J-1} + \sigma^{J-2})\right\}.$$

From this we see that $\nu V_1^{k,J-1}\delta_{\eta\eta}V_1^{k,J-1}$ is uniformly bounded with respect to l and can be made arbitrarily small in absolute value by choosing M_3 small. Decreasing M_3 if necessary, we now satisfy the inequality $V_1^{kj} \leqslant w_0^{kj}$.

Let $V_2 = M_4(1 - \eta)\sigma$. Expressions for $L_{kj}(V_2)$ and $\lambda_k(V_2)$ will be obtained if in (14) and (15) we replace V_1 by V_2, M_3 by M_4, and set $\alpha = 0$. After this it is not difficult to see that for sufficiently large M_4 and $\mu < 1/\sqrt{e}$ we have $L_{kj}(V_2) \leqslant 0$ and $\lambda_k(V_2) < 0$. Make M_4 so large that $V_2^{kj} \geqslant w_0^{kj}$. Thus we have constructed the V_1 and V_2 required.

Consider the problem (5), (8), (7). Using conditions (8), we exclude $w^{k,0}$ and $w^{k,J}$ from (5). We regularize the Cauchy problem obtained; namely, we consider the system

$$dw^{k,1}/d\tau = \tilde{L}_{k,1}(\tau, w)$$
$$\equiv \nu\Psi_1^2(w^{k,1})(1/l)\delta_\eta w^{k,2} - lU^k\delta_\xi w^{k,1} + a_l^k(\tau, \Psi_1(w^{k,1}))w^{k,1} + b_l^k(\tau, \Psi_1(w^{k,1})),$$

$$(16)$$

$$\frac{dw^{kj}}{d\tau} = \tilde{L}_{kj}(\tau, w) \equiv \nu\Psi_1^2(w^{kj})\delta_{\eta\eta}w^{kj} - jlU^k\delta_\xi w^{kj} + A^{kj}\delta_\eta w^{kj} + B^{kj}w^{kj},$$

$$(17)$$

$$j = 2,\ldots,J - 1; \quad w^{k,j} \equiv 0; \quad k = 0, 1,\ldots,K; \quad U^0 \equiv 0; \quad K = [X/h],$$

where

$$a_l^k(\tau, u) = -(1/l)v_0^k u$$
$$+ (2/l)C^k u\left[u - (l/\nu)v_0^k + \sqrt{\left(u - (l/\nu)v_0^k\right)^2 + 4(l/4)C^k}\,\right]^{-1} + B^{k,1},$$

$$b_l^k(\tau, u) = (1/\nu)A^{k,1}v_0^k$$
$$- (2/\nu)C^kA^{k,1}\left[u - (l/\nu)v_0^k + \sqrt{\left(u - (l/\nu)v_0^k\right)^2 + 4(l/\nu)C^k}\,\right]^{-1},$$

with initial condition (7). Here $\Psi_1(w^{kj})$ is a smooth function such that $\Psi_1(w^{kj}) \equiv w^{kj}$ when $V_1^{kj} \leqslant w^{kj} \leqslant V_2^{kj}$, $\Psi_1(w^{kj}) \equiv V_1^{kj}/2$ when $w^{kj} \leqslant V_1^{kj}/4$ and $\Psi_1(w^{kj}) \equiv 3V_2^{kj}/2$ when $w^{kj} \geqslant 2V_2^{kj}$, $0 \leqslant \Psi_1' \leqslant 1$. Since Ψ_1 is bounded, the $\tilde{L}_{kj}(\tau, w)$ satisfy a Lipschitz condition in w with a constant that does not depend on τ or w (although it depends on l). Moreover, all the coefficients are continuous in τ. Therefore there exists a unique continuously differentiable solution $w^{kj}(\tau)$ of problem (16), (17), (7) for all $\tau \geqslant 0$. The quantities $\tilde{L}_{kj}(\tau, w)$ satisfy the quasi-monotonicity condition. Since $\Psi_1(V_i^{kj}) = V_i^{kj}$ $(i = 1, 2)$, we have

$$\tilde{L}_{k,1}(\tau, V_i) = L_{k,1}(\tau, V_i) + \lambda_k(V_i)\left[(1/l)(V_i^{k,1})^2 - (1/\nu)A^{k,1}\right], \qquad (18)$$

$$\tilde{L}_{k,j}(\tau, V_i) = L_{k,j}(\tau, V_i) \qquad (j \geqslant 2),$$

from which it follows that $(V_1^{kj})_\tau \leqslant \tilde{L}_{kj}(\tau, V_1)$ and $(V_2^{kj})_\tau \geqslant \tilde{L}_{kj}(\tau, V_2)$. In addition, $V_1^{kj} \leqslant w_0^{kj} \leqslant V_2^{kj}$. Hence by virtue of Theorem 1 we obtain the bounds

$$V_1^{kj} \leqslant w^{kj}(\tau) \leqslant V_2^{kj}, \qquad (j \geqslant 1, 0 \leqslant kh \leqslant X). \qquad (19)$$

Let us evaluate $w^{k,0}$ and $w^{k,J}$ from conditions (8). We show that (19) also holds for $j = 0$. Consider the differences $y_1^{kj} = V_1^{kj} - w^{kj}$ and $y_2^{kj} = w^{kj} - V_2^{kj}$. From (8) we have

$$\nu(1/l)\left(y_i^{k,1} - y_i^{k,0}\right) - \Pi_l^k\left(V_i^{k,1}, w^{k,1}, \tau\right)y_i^{k,1} > 0,$$

where Π_l^k is a function uniformly bounded in l. Hence we get $y_i^{k,0} < y_i^{k,1}[1 - (l/\nu)\Pi_l^k] \leqslant 0$, provided that l is sufficiently small. This proves (19) for all j, and thus the $w^{kj}(\tau)$ so obtained constitute a solution of problem (5)–(7). Lemma 1 is proved.

LEMMA 2. *Let A, B, C, U and v_0 have bounded derivatives in Ω that are continuous in τ, and let*

$$v_0 \leqslant E_3\xi, \qquad v_{0\tau} \geqslant -E_4\xi, \qquad |U_\tau/U| \leqslant E_5\xi,$$
$$(U_\tau/U)_\tau \leqslant E_6\xi, \qquad U_{\xi\tau} \leqslant E_7\xi, \qquad |(r_xU/r - U_x)_\tau| \leqslant E_8\xi. \tag{20}$$

Let the hypothesis of Lemma 1 be satisfied with respect to w_0 and, in addition,

$$(1/h)\left|w_0^{kj} - w_0^{k-1,j}\right| \leqslant E_9(1-jl)\sigma^j, \tag{21}$$
$$-E_{10}\sigma^{j-1} \leqslant (1/l)\left(w_0^{kj} - w_0^{k,j-1}\right) \leqslant -E_{11}\sigma^{j-1}.$$

Suppose that the following compatibility conditions are satisfied:

$$-E_{12}(1-jl)\sigma^j \leqslant \nu\left(w_0^{kj}\right)^2\delta_{\eta\eta}w_0^{kj} + A(0, kh, jl)\delta_\eta w_0^{kj} \leqslant E_{13}kh(1-jl)\sigma^j - E_{14}l$$
$$\leqslant \nu w_0^{k,0}\delta_\eta w_0^{k,1} - v_0^k(0)w_0^{k,0} + C^k(0) \leqslant E_{15}lkh. \tag{22}$$

Then there is a positive number X_1, depending only U, r, v_0 and w_0, such that for $0 \leqslant kh \leqslant X_1$ the solution $w^{kj}(\tau)$ of problem (5)–(7) admits the following estimates:

$$|\delta_\xi w^{kj}| = (1/h)|w^{kj} - w^{k-1,j}| \leqslant M_5(1-jl)\sigma^j, \tag{23}$$
$$-M_6\sigma^{j-1} \leqslant \delta_\eta w^{kj} = (1/l)(w^{kj} - w^{k,j-1}) \leqslant -M_7\sigma^{j-1}, \tag{24}$$
$$-M_8(1-jl)\sigma^j \leqslant (d/d\tau)(w^{kj}) \leqslant M_9kh(1-jl)\sigma^j, \tag{25}$$
$$-M_{10} \leqslant w^{kj}\delta_{\eta\eta}w^{kj} \leqslant -M_{11}, \tag{26}$$

in which the positive constants M_i are independent of h and l.

The proof will proceed by induction on k. Let us set

$$r^{kj} = \delta_\xi w^{kj} = (1/h)(w^{kj} - w^{k-1,j}), \qquad z^{kj} = \delta_\eta w^{kj} = (1/l)(w^{kj} - w^{k,j-1}),$$

and $\rho^{kj} = dw^{kj}/d\tau$. Consider the functions $\Psi^{kj} = M_{12}w^{kj}$, $\Phi_1^{kj} = -M_{13}\sigma^{j-1}$, $\Phi_2^{kj} = -M_{14}\sigma^{j-1}$, $F_1^{kj} = -M_{15}w^{kj}$ and $F_2^{kj} = M_{16}khw^{kj}$. We note that the functions $r^{0,j}$, $\rho^{-1,j}$ and $z^{-1,j}$ are not defined, and their values, as we will see later, are irrelevant, since $U^0(\tau) \equiv 0$. Therefore for the proof of the lemma it is sufficient to show that if the estimates

$$|r^{k'j}| \leqslant \Psi^{k'j}, \qquad \Psi_1^{k'j} \leqslant z^{k'j} \leqslant \Phi_2^{k'j}, \qquad F_1^{k'j} \leqslant \rho^{k'j} \leqslant F_2^{k'j}$$

hold for $k' \leqslant k-1$, then they are valid also for $k' = k$. Indeed, these inequalities imply (23)–(25) by virtue of (12), and (26) is a consequence of them by virtue of (5).

We write down equations for r^{kj}. From (5) we get

$$\frac{dr^{kj}}{d\tau} = R_{kj}(\tau, r)$$

$$\equiv \nu(w^{kj})^2 \delta_{\eta\eta} r^{kj} - jlU^{k-1}\delta_\xi r^{kj} + A^{kj}\delta_\eta r^{kj} + B^{kj}r^{kj} - jl\delta_\xi U^k r^{kj}$$

$$+ \frac{w^{kj} + w^{k-1,j}}{(w^{k-1,j})^2}\left[\rho^{k-1,j} + jlU^{k-1}r^{k-1,j} - A^{k-1,j}z^{k-1,j} - B^{k-1,j}w^{k-1,j}\right]r^{kj}$$

$$+ \delta_\xi A^{kj}z^{k-1,j} + \delta_\xi B^{kj}w^{k-1,j},$$

$$k = 1,\ldots,K; \qquad j = 1,\ldots,J-1; \qquad U^0(\tau) \equiv 0; \quad \tau > 0. \tag{27}$$

We observe that the coefficient of r^{0j} is zero. It follows from (6) and (7) that when $k \geqslant 1$

$$\gamma_k(r) \equiv \left[\nu\delta_\eta r^{k,1} - \frac{C^k}{w^{k,0}w^{k-1,0}}r^{k,0} - \delta_\xi v_0^k + \delta_\xi C^k \frac{1}{w^{k-1,0}}\right] = 0,$$

$$r^{k,J} = 0, \tag{28}$$

$$r^{kj}(0) = \delta_\xi w_0^{kj} = (1/h)\left(w_0^{kj} - w_0^{k-1,j}\right). \tag{29}$$

We choose $M_{12} = M_{12}(M_1, M_2, M_{14})$ and $x^1 = x^1(M_{12}, M_{14}, M_{16})$ so that for $kh \leqslant x^1$

$$M_{12}\left[-C^k\left(\frac{1}{w^{k,0}} + \frac{1}{w^{k-1,0}}\right) + v_0^k\right] + \left|\delta_\xi v_0^k - \frac{1}{w^{k-1,0}}\delta_\xi C^k\right| \leqslant 0,$$

$$M_{12}\delta_\xi U^k \geqslant \left|\delta_\xi\left(r_x^k U^k/r^k - U_x^k\right)\right|,$$

$$\frac{1}{2}M_{12}w^{kj}\frac{w^{kj} + w^{k-1,j}}{(w^{k-1,j})^2}A^{k-1,j}z^{k-1,j} \geqslant \left|\delta_\xi A^{k-1,j}z^{k-1,j}\right| + \left|\delta_\xi\left(\frac{U_t}{U}\right)^k\right|w^{k-1,j},$$

$$\left(M_{16}(k-1)h + jlU^{k-1}\cdot M_{12} + B^{k-1,j}\right) \leqslant \tfrac{1}{2}A^{k-1,j}z^{k-1,j}/w^{k-1,j}$$

$$\leqslant \tfrac{1}{2}\left(C^{k-1} + jlU_x^{k-1}\right)M_{14}M_2^{-1}.$$

Then it is not difficult to show that for $kh \leqslant x^1$

$$d\Psi^{kj}/d\tau \geqslant R_{kj}(\tau, \Psi), \qquad d(-\Psi^{kj})/d\tau \leqslant R_{kj}(\tau, \Psi), \tag{30}$$

$$\gamma_k(\Psi) \leqslant 0, \qquad \gamma_k(-\Psi) \geqslant 0, \qquad \Psi^{k,J} = 0. \tag{31}$$

If necessary we increase the constant M_{12} so that

$$|r^{kj}(0)| \leqslant \Psi^{kj}(0). \tag{32}$$

Consider the system (27) with conditions (28) and (29). The functions $R_{kj}(\tau, r)$ are linear in r (and, consequently, they satisfy a Lipschitz condition uniformly in r), since according to the induction hypothesis $\rho^{k-1,j}$, $r^{k-1,j}$ and $z^{k-1,j}$ are considered to be known and continuous functions of τ. The functions $R_{kj}(\tau, r)$ are continuous in τ, and, moreover, they increase quasimonotonically in r.

Therefore, from (30)–(32) we get

$$|r^{kj}| \leq \Psi^{kj} = M_{12}w^{kj} \tag{33}$$

for $kh \leq x^1$ (see the proof of Lemma 2*). We can easily check this if we exclude $r^{k,0}$ and $r^{k,J}$ from (27) by using (28) and apply Theorem 1 to the system so obtained.

We now write equations for z^{kj}. From (5) we have for $k \geq 0$

$$dz^{kj}/d\tau = Z_{kj}(\tau, z)$$

$$\equiv \nu(w^{kj})^2\delta_{\eta\eta}z^{kj} + \nu(w^{kj} + w^{k,j-1})z^{kj}\delta_\eta z^{kj}$$

$$+A^{k,j-1}\delta_\eta z^{kj} + \left(B^{kj} + \delta_\eta A^{kj}\right)z^{kj} - jlU^k\delta_\xi z^{kj} - U^k r^{k,j-1} + \delta_\eta B^{kj}w^{k,j-1},$$

$$j = 2,\dots,J-1; \qquad U^0(\tau) \equiv 0. \tag{34}$$

From (5) with $j = J - 1$ and from the boundary condition $w^{k,J} = 0$ we obtain an equation for $z^{k,J}$ in the form

$$dz^{k,J}/d\tau = Z_{k,J}(\tau, z) \equiv \nu w^{k,J-1}z^{kJ}\delta_\eta z^{k,J} - (1 - l)U^k\delta_\xi z^{k,J}$$

$$+ \left[(2 - l)U_x^k + U_t^k/U^k\right]z^{k,J-1} + B^{k,J-1}z^{k,J}. \tag{35}$$

From (6) and (7) we get

$$z^{k,1} = (1/\nu)\left(v_0^k - C^k/w^{k,0}\right), \tag{36}$$

$$z^{kj}(0) = \delta_\eta w_0^{kj} = (1/l)\left(w_0^{kj} - w_0^{k,j-1}\right), \qquad j = 1,\dots,J. \tag{37}$$

Let us compute $\Phi_{1\tau}^{kj} - Z_{kj}(\tau, \Phi_1)$. We have

$$\Phi_{1\tau}^{kj} - Z_{kj}(\tau, \Phi_1)$$

$$= M_{13}\sigma^{j-1}\Bigg\{ \frac{\nu(w^{kj})^2}{2(1 - (j-1)l)^2(\sigma^{j-1})^2}\left(1 - \frac{1}{2(\sigma^{j-1})^2}\right)$$

$$+ \frac{\nu(w^{kj})^2}{4[J - j + 1 - \theta]^4 l^2\sigma^{j-1+\theta}\sigma^{j-1}}\left[1 - \frac{11}{12}(\sigma^{j-1+\theta})^{-2} + O\left((\sigma^{j-1+\theta})^{-4}\right)\right]$$

$$+ (2j - 1)lU_x^k - M_{13}\frac{\nu(w^{kj} + w^{k,j-1})}{2(1 - (j - 1 - \theta')l)\sigma^{j-1+\theta'}} + jl\left(r_x^k\frac{U^k}{r^k} - U_x^k\right)$$

$$- \frac{1 - (j-1)l}{2(1 - (j - 1 - \theta')l)\sigma^{j-1-\theta'}\sigma^{j-1}}\left[U_x^k(1 + (j-1)l) + U_t^k/U^k\right]\Bigg\}$$

$$+ U^k r^{k,j-1} - \left(r_x^k U^k/r^k - U_x^k\right)w^{k,j-1}, \tag{38}$$

$$j = 2,\dots,J-1; \qquad |\theta| < 1, \qquad 0 < \theta' < 1,$$

---

*Editor's note. This is a literal translation from the Russian.

$$\Phi_{1\tau}^{kj} - Z_{k,J}(\tau, \Phi_1)$$

$$= M_{13}\sigma^{J-1}\left\{-\nu M_{13}\ln 2\,\frac{w^{k,J-1}}{l(\sigma^{J-1}+\sigma^{J-2})} + (2-l)U_x^k\left[1 - \frac{\ln 2}{\sigma^{J-1}(\sigma^{J-1}+\sigma^{J-2})}\right]\right.$$

$$\left. + (1-l)(r_x^k U^k/r^k - U_x^k) - \frac{\ln 2}{\sigma^{J-1}(\sigma^{J-1}+\sigma^{J-2})}\,U_t^k/U^k\right\}. \quad (39)$$

We choose $M_{13}(M_{12})$ large enough to satisfy the inequalities

$$-M_{13}\sigma^{j-1} \leq -E_{10}\sigma^{j-1} \leq z^{kj}(0), \qquad -M_{13}\sigma^0 \leq z^{k,1} = \frac{1}{\nu}\left(v_0^k - C^k/w^{k,0}\right),$$

$$\Phi_{1\tau}^{kj} - Z_{kj}(\tau, \Phi_1) \leq 0 \qquad (2 \leq j \leq J). \quad (40)$$

Let us take $\Phi_2^{kj} = -M_{14}\sigma^{j-1}$. Expressions for $\Phi_{2\tau}^{kj} - Z_{kj}(\tau, \Phi_2)$ will be obtained if in (38) and (39) we replace M_{13} by M_{14} and Φ_1^{kj} by Φ_2^{kj}. We can make the expressions in braces positive when $kh \leq x^3$ by choosing the constants M_{14}, μ and x^3 sufficiently small. In fact the first three terms in (38) are positive:

$$\left(r_x^k U^k/r^k - U_x^k\right) \geq -N_1 kh, \qquad M_{14}\frac{\nu(w^{kj}+w^{k,j-1})}{2(1-(j-1-\theta')l)\sigma^{j-1-\theta'}} \to 0$$

$$\text{as } M_{14} \to 0,$$

$$\frac{1-(j-1)l}{2(1-(j-1-\theta')l)\sigma^{j-1-\theta'}\sigma^{j-1}} \to 0 \quad \text{as } \mu \to 0.$$

We choose M_{14} and μ from the conditions

$$-M_{14}\sigma^{j-1} \geq -E_{11}\sigma^{j-1} \geq z^{kj}(0), \qquad -M_{14}\sigma^{j-1} \geq z^{k,1} = (1/\nu)\left(v_0^k - C^k/w^{k,0}\right).$$

$$(41)$$

We choose $x^4(M_{12})$ so small that $\Phi_{2\tau}^{kj} - Z_{kj}(\tau, \Phi_2) \geq 0$ when $kh \leq x^4 \leq x^3$; this is possible because $U^k r^{k,j-1} \geq -N_2 khw^{k,j-1}$ and $r_x^k U^k/r^k - U_x^k \leq N_{kh}$. Here we should emphasize that M_{14} depends only on the data of the problem and the constants M_1 and M_2; it does not depend on M_{12}.

Consider the problem (34)–(37) (more precisely, the Cauchy problem obtained after substituting $z^{k,1}$ from (36) and (34)). We replace the product $z^{kj}\delta_\eta z^{kj}$ appearing in the expression for $Z_{kj}(\tau, z)$ by $\Psi_2(z^{kj})\delta_{\eta\eta}z^{kj}$. Here the smooth function $\Psi_2(z^{kj})$ is defined by the following relations:

$$\Psi_2(z^{kj}) = \tfrac{3}{2}\Phi_1^{kj} \quad \text{for } z^{kj} \leq 2\Phi_1^{kj},$$

$$\Psi_2(z^{kj}) = z^{kj} \quad \text{for } \Phi_1^{kj} \leq z^{kj} \leq \Phi_2^{kj},$$

$$\Psi_2(z^{kj}) = \tfrac{1}{2}\Phi_2^{kj} \quad \text{for } z^{kj} \geq \tfrac{1}{4}\Phi_2^{kj},\, 0 \leq \Psi_2' \leq 1.$$

The system regularized in this way satisfies a Lipschitz condition in z with constant independent of τ and z, and it also satisfies the quasimonotonicity condition (since $\Psi_2 \leq 0$). Therefore, by Theorem 1, for $\tilde{z}^{kj}(\tau)$—the solution of the regularized problem—we have the bound $\Phi_1^{kj} \leq \tilde{z}^{kj} \leq \Phi_2^{kj}$, and thus $\tilde{z}^{kj}(\tau)$ is a solution of the original problem (34)–(37). Since we are not interested in all of the solutions of this problem, but only those which are the left differences for $w^{kj}(\tau)$, i.e. $z^{kj} = \delta_\eta w^{kj}$, and the $w^{kj}(\tau)$ under investigation is unique, it follows that the solution $z^{kj}(\tau)$ sought is unique and coincides with the solution of the regularized problem. Thus the bound $\Phi_1^{kj} \leq z^{kj} \leq \Phi_2^{kj}$ is established for $kh \leq x^4$.

Now we write equations for $\rho^{kj} = dw^{kj}/d\tau$. To this end we differentiate (5) with respect to τ (the differentiation is legitimate since $w^{kj}(\tau)$ is a continuously differentiable function). We get

$$\frac{d\rho^{kj}}{d\tau} = T_{kj}(\tau, \rho)$$

$$\equiv \nu(w^{kj})^2 \delta_{\eta\eta}\rho^{kj} - jlU^k\delta_\xi\rho^{kj} + A^{kj}\delta_\eta\rho^{kj} + B^{kj}\rho^{kj}$$

$$+2\big[\rho^{kj} + jlU^k r^{kj} - A^{kj}z^{kj} - B^{kj}w^{kj}\big]\frac{\rho^{kj}}{w^{kj}} - jlU_\tau^k r^{kj} + A_\tau^{kj}z^{kj} + B_\tau^{kj}w^{kj}$$

$$(j = 1,\ldots,J-1,\ U^0 \equiv 0,\ k \geq 0). \tag{42}$$

Differentiating (6) with respect to τ, we find that

$$\Gamma_k(\tau, \rho) \equiv \nu\delta_\eta\rho^{k,1} - \frac{C^k}{(w^{k,0})^2}\rho^{k,0} - v_{0\tau}^k + \frac{C_\tau^k}{w^{k,0}} = 0, \qquad \rho^{k,J} = 0. \tag{43}$$

Set

$$\rho^{k,1}(0) = \tilde{L}_{k,1}(0, w_0) = L_{k,1}(0, w_0) + \lambda_k(w_0)\left[\frac{1}{l}(w_0^{k,1})^2 - \frac{1}{\nu}A^{k,1}(0)\right],$$

$$\rho^{kj}(0) = \tilde{L}_{kj}(0, w_0) = \nu(w_0^{kj})^2\delta_{\eta\eta}w_0^{kj} - jlU^k(0)\delta_\xi w_0^{kj}$$

$$+A^{kj}(0)\delta_\eta w_0^{kj} + B^{kj}(0)w_0^{kj} \qquad (j \geq 2), \tag{44}$$

where

$$\lambda_k(w_0) = \nu\delta_\eta w_0^{k,1} - v_0^k(0) + C^k(0)/w_0^{k,0}.$$

Since, by (5),

$$\rho^{kj} = \nu(w^{kj})^2(z^{kj} - z^{k,j-1})/l - jlU^k r^{kj} + A^{kj}z^{kj} + B^{kj}w^{kj},$$

we have for ρ^{kj} the bound $|\rho^{kj}| \leq N_4 w^{kj}/l$, $N_4 = \text{const} > 0$ (the quantities r^{kj} and z^{kj} have already been estimated). Therefore (42) is equivalent to

$$\frac{d\rho^{kj}}{d\tau} = \tilde{T}_{kj}(\tau, \rho)$$

$$\equiv \nu(w^{kj})^2\delta_{\eta\eta}\rho^{kj} - jlU^k\delta_\xi\rho^{kj} + A^{kj}\delta_\eta\rho^{kj} + B^{kj}\rho^{kj}$$

$$+2\frac{\Psi_3(\rho^{kj})}{w^{kj}}\big[\rho^{kj} + jlU^k r^{kj} - A^{kj}z^{kj} - B^{kj}w^{kj}\big] - jlU_\tau^k r^{kj} + A_\tau^{kj}z^{kj} + B_\tau^{kj}w^{kj}$$

$$(j = 1,\ldots,J-1;\ U^0 \equiv 0,\ k \geq 0). \tag{45}$$

Here

$$\Psi_3\big(\rho^{kj}\big) = \rho^{kj} \quad \text{for } |\rho^{kj}| \leqslant N_4 w^{kj}/l,$$
$$\Psi_3\big(\rho^{kj}\big) = -(1/2 + N_4/l)w^{kj} \quad \text{for } \rho^{kj} \leqslant -(1 + N_4/l)w^{kj}.$$
$$\Psi_3\big(\rho^{kj}\big) = (1/2 + N_4/l)w^{kj} \quad \text{for } \rho^{kj} \geqslant (1 + N_4/l)w^{kj}.$$

Since $\tilde{T}_{kj}(\tau, \rho)$ is continuous in τ and satisfies a Lipschitz condition with respect to ρ (with constant independent of τ and ρ but depending on l), problem (45), (43), (44) has a unique continuously differentiable solution $\rho^{kj}(\tau)$ for all $\tau \geqslant 0$.

We establish bounds for $\rho^{kj}(\tau)$, observing that $\tilde{T}^{kj}(\tau, \rho)$ is a quasimonotonically increasing function. Let $F_1^{kj} = -M_{15}w^{kj}$. We assume l to be so small that $\Psi_3(F_1^{kj}) = F_1^{kj}$. We have

$$F_{1\tau}^{kj} - \tilde{T}_{kj}(\tau, F_1) = -2M_{15}\big\{M_{15}w^{kj} - jlU^k r^{kj} + A^{kj}z^{kj} + B^{kj}w^{kj}\big\}$$
$$+ jlU_\tau r^{kj} - A_\tau^{kj}z^{kj} - B_\tau^{kj}w^{kj} \leqslant 0,$$

provided that M_{15} is sufficiently large; we observe that M_{15} does not depend on M_{13} since we are assuming $M_{15}A^{kj}z^{kj} \geqslant -A_\tau^{kj}z^{kj}$. Furthermore,

$$\Gamma_k(F_1) = M_{15}\big(-v_0^k + 2C^k/w^{k,0}\big) - v_{0\tau}^k + C_\tau^k/w^{k,0} > 0$$

for $kh \leqslant \xi^1$ if M_{15} is large enough. Choose M_{15} so large that $\rho^{kj}(0) \geqslant -M_{15}w_0^{kj}$. As we see from (44), this is possible because of the compatibility conditions (22). From these inequalities we get the estimate $\rho^{kj} \geqslant F_1^{kj} = -M_{15}w^{kj}$ (see Theorem 1).

We now establish an upper bound. Let $F_2^{kj} = M_{16}khw^{kj}$. Choose $M_{16} = M_{16}(M_{12}, M_{14})$ and $x^5 = x^5(x^1, M_{12}, M_{14}, M_{16})$ so that when $kh \leqslant x^5$ the following inequalities will be satisfied:

$$M_{16}kh\big(v_0^k - 2C^k/w^{k,0}\big) - v_{0\tau}^k + C_\tau^k/w^{k,0} \leqslant 0,$$
$$M_{16}U^k w^{k-1,j} \geqslant \big(r_x^k U^k/r^k - U_x^k\big)_\tau w^{kj} + M_{12}|U_\tau^k|w^{kj},$$
$$M_{16}A^{kj}z^{kj}kh \geqslant A_\tau^{kj}z^{kj} - \big(U_t^k/U^k\big)_\tau w^{kj},$$
$$\big(M_{16}kh + jlU^k M_{12} - B^{kj}\big)w^{kj} \leqslant \tfrac{1}{2}A^{kj}z^{kj}.$$

Fulfillment of these relations is a consequence of the conditions of the lemma, the smoothness of the coefficients and the fact that M_{12} is independent of M_{16}. If necessary, we shall make M_{16} so large that $\rho^{kj}(0) \leqslant F_2^{kj}(0)$ (see (22) and (44)). Then, as we can easily show, when $kh \leqslant x^5$ we shall have

$$\big(F_2^{kj}\big)_\tau \geqslant \tilde{T}_{kj}(\tau, F_2), \qquad \Gamma_k(F_2) < 0, \qquad F_2^{k,J} = 0,$$
$$\rho^{kj}(0) \leqslant F_2^{kj}(0). \tag{46}$$

From these relations, by applying Theorem 1, we obtain the bound $\rho^{kj}(\tau) \leqslant F_2^{kj}(\tau)$. Lemma 2 is proved.

THEOREM 2. *Suppose that the hypotheses of Lemma 2 pertaining to U, r and v_0 are satisfied, and let $w_0(\xi, \eta)$ be a function differentiable with respect to ξ when $\eta \leqslant 1$ and twice differentiable with respect to η when $\eta < 1$, such that*

$$E_1(1 - \eta)\sigma \leqslant w_0 \leqslant E_2(1 - \eta)\sigma,$$

$$|w_{0\xi}| \leqslant E_9(1 - \eta)\sigma, \qquad -E_{10}\sigma \leqslant w_{0\eta} \leqslant -E_{11}\sigma.$$

Assume in addition that w_0 satisfies the compatibility conditions (22). Then in $\Omega_{X_1}\{0 \leqslant \tau < \infty,\ 0 \leqslant \xi \leqslant X_1,\ 0 \leqslant \eta < 1\}$ (X_1 is a positive number depending on U, r, v_0 and w_0) there exists a solution of problem (3), (4) with the following properties: $w(\tau, \xi, \eta)$ is continuous in Ω_{X_1}, w_η is continuous with respect to η when $\eta < 1$, the generalized derivatives w_ξ, w_τ and $w_{\eta\eta}$ exist and

$$M_1(1 - \eta)\sigma \leqslant w \leqslant M_2(1 - \eta)\sigma, \qquad |w_\xi| \leqslant M_5(1 - \eta)\sigma,$$

$$-M_6\sigma \leqslant w_\eta \leqslant -M_7\sigma, \qquad -M_8(1 - \eta)\sigma \leqslant w_\tau \leqslant M_9\xi(1 - \eta)\sigma,$$

$$-M_{10} \leqslant ww_{\eta\eta} \leqslant -M_{11}.$$

The function w satisfies equation (3) almost everywhere as well as conditions (4). The solution of problem (3), (4) with the indicated properties is unique.

PROOF. Consider the family of functions $w^{h,l}(\tau, \xi, \eta)$ obtained from $w^{kj}(\tau)$, the solutions of the system (5)–(7), by linear interpolation with respect to η and ξ. From (12) of Lemma 1 it follows that $w^{h,l}$ satisfies the inequality

$$M_1(1 - \eta)\sigma - M_1\sigma^{j-1}l \leqslant w^{h,l}(\tau, \xi, \eta) \leqslant M_2(1 - \eta)\sigma \qquad (47)$$

for $(j - 1)l \leqslant \eta \leqslant jl$, where the constants M_1 and M_2 do not depend on h or l. We now show that the family in question is equicontinuous. Consider the functions $r^{kj}(\tau)$, $z^{kj}(\tau)$ and $\rho^{kj}(\tau)$, and, by using linear interpolation, form $r^{h,l}(\tau, \xi, \eta)$, $z^{h,l}(\tau, \xi, \eta)$ and $\rho^{h,l}(\tau, \xi, \eta)$. It is easy to see that the function $w^{h,l}(\tau, \xi, \eta)$ is piecewise continuously differentiable, and also

$$w_\tau^{h,l} = \rho^{h,l}, w_\xi^{h,l}(\tau, \xi, \eta) = r^{h,l}(\tau, kh, \eta) \quad \text{for } (k - 1)h < \xi < kh,$$

$$w_\eta^{h,l}(\tau, \xi, \eta) = z^{h,l}(\tau, \xi, jl) \quad \text{for } (j - 1)l < \eta < jl.$$

From Lemma 2 we obtain the following bounds:

$$-M_5(1 - \eta)\sigma \leqslant r^{h,l} \leqslant M_5(1 - \eta)\sigma, \qquad -M_8(1 - \eta)\sigma \leqslant \rho^{h,l} \leqslant M_9\xi(1 - \eta)\sigma,$$

$$-M_6\sigma(\eta) \leqslant z^{h,l}(\tau, \xi, \eta) \leqslant -M_7\sigma(\eta - l). \qquad (48)$$

Then for any $\varepsilon > 0$ we can choose δ_1 and δ_2 independent of h and l, such that

$$\left|w^{h,l}(\tau, \xi, \eta) - w^{h,l}(\bar{\tau}, \bar{\xi}, \bar{\eta})\right| \leqslant M_2(1 - \eta)\sigma < \varepsilon$$

$$\text{for } \eta \geqslant 1 - \delta_1 \text{ and } |\eta - \bar{\eta}| < \delta_1;$$

$$\left|w^{h,l}(\tau, \xi, \eta) - w^{h,l}(\bar{\tau}, \bar{\xi}, \bar{\eta})\right| \leqslant \varepsilon(|\eta - \bar{\eta}| + |\tau - \bar{\tau}| + |\xi - \bar{\xi}|)/\delta_2 < \varepsilon$$

$$\text{for } 0 \leqslant \eta \leqslant 1 - \delta_1 \text{ and } |\eta - \bar{\eta}| + |\tau - \bar{\tau}| + |\xi - \bar{\xi}| < \delta_2,$$

where

$$\delta_2 = \varepsilon\left\{\sup_{0<\eta<1-\tilde{\varepsilon}_1}\left(|z^{h,l}|, |\rho^{h,l}|, |r^{h,l}|\right)\right\}^{-1}.$$

Thus equicontinuity of the set $\{w^{h,l}\}$ has been proved. By Arzelà's theorem there exists a sequence w^{h_n,l_n} which converges uniformly as h_n, $l_n \to 0$ to some limit function $w(\tau, \xi, \eta)$. This function is continuous when $0 \leqslant \eta \leqslant 1$ and has generalized derivatives w_τ, w_ξ and w_η. As we see from (47) and (48), the functions w, w_τ, w_ξ and w_η satisfy the bounds required in the theorem.

Let us now prove that $w(\tau, \xi, \eta)$ is a generalized solution of problem (3), (4). Let $\varphi(\tau, \xi, \eta)$ be an arbitrary function twice continuously differentiable in the domain $\Omega_{X_1 T}\{0 \leqslant \tau \leqslant T, 0 \leqslant \xi \leqslant X_1, 0 \leqslant \eta \leqslant 1\}$ (T is an arbitrary positive number) and satisfying the conditions

$$\varphi|_{\xi=X_1} = 0, \qquad \varphi|_{\tau=T} = 0, \qquad \varphi_\eta|_{\eta=0} = 0, \qquad \varphi|_{\eta=1} = 0. \tag{49}$$

Multiply (5) by $hl\varphi^{kj}(\tau)[w^{kj}(\tau)]^{-2}$, sum on j from 1 to $J-1$ and on k from 0 to $K_1 = [X_1/h]$, and integrate with respect to τ from 0 to T. We have

$$\sum_{k=0}^{K_1}\sum_{j=1}^{J-1} hl\int_0^T\left[\frac{d}{d\tau}\left(\frac{1}{w^{kj}}\right) + \nu\delta_{\eta\eta}w^{kj} - jlU^k\delta_\xi w^{kj}\frac{1}{(w^{kj})^2}\right.$$
$$\left. + A^{kj}\delta_\eta w^{kj}\frac{1}{(w^{kj})^2} + B^{kj}\frac{1}{w^{kj}}\right]\varphi^{kj}\,d\tau = 0. \tag{50}$$

Transform this equality by integration and summation by parts. We have

$$-\sum_{k=0}^{K_1}\sum_{j=1}^{J-1} hl\frac{\varphi^{kj}(0)}{w_0^{kj}} + \sum_{k=0}^{K_1} h\int_0^T\varphi^{k,1}\left(-v_0^k + \frac{C^k}{w^{k,0}}\right)d\tau$$

$$\sum_{k=0}^{K_1-1}\sum_{j=1}^{J-1} hl\int_0^T\left\{-\frac{\varphi_\tau^{kj}}{w^{kj}} + \nu w^{kj}\delta_{\eta\eta}\varphi^{kj} - jlU^k\delta_\xi\varphi^{k+1,1}\frac{1}{w^{kj}}\right.$$

$$\left. -jl\delta_\xi U^{k+1}\frac{\varphi^{k+1,j}}{w^{kj}} + A^{kj}\delta_\eta\varphi^{k,j+1}\frac{1}{w^{kj}} + \delta_\eta A^{k,j+1}\frac{\varphi^{k,j+1}}{w^{kj}} + B^{kj}\frac{\varphi^{kj}}{w^{kj}}\right\}d\tau$$

$$+ \sum_{k=0}^{K_1} h\int_0^T\varphi^{k,1}\frac{A^{k,1}}{w^{k,0}}\,d\tau + \sum_{k=0}^{K_1} h\int_0^T\nu w^{k,1}\delta_\eta\varphi^{k,1}\,d\tau$$

$$+ \sum_{k=0}^{K_1}\sum_{j=1}^{J-1} h^2 l\int_0^T jlU^k\left(\frac{\delta_\xi w^{kj}}{w^{kj}}\right)^2\frac{\varphi^{kj}}{w^{k-1,j}}\,d\tau$$

$$+ \sum_{k=0}^{K_1}\sum_{j=1}^{J-1} hl^2\int_0^T(-A^{kj})\left(\frac{\delta_\eta w^{kj}}{w^{kj}}\right)^2\frac{\varphi^{kj}}{w^{k,j-1}}\,d\tau = 0. \tag{51}$$

Here we have used (6), (7) and (49), as well as the fact that $U^0(\tau) \equiv 0$. Consider the last three terms in (51). Since $\varphi_\eta \big|_{\eta=0} = 0$,

$$\sum_{k=0}^{K_1} h \int_0^T \nu w^{k,1} \delta_\eta \varphi^{k,1} \, d\tau = 0(l).$$

Furthermore,

$$\sum_{k=0}^{K_1} \sum_{j=1}^{J-1} h^2 l \int_0^T ji U^k \left(\frac{\delta_\xi w^{kj}}{w^{kj}} \right)^2 \frac{\varphi^{kj}}{w^{k-1,j}} \, d\tau = 0(h),$$

since according to the estimates of Lemma 2 the ratio $\delta_\xi w^{kj}/w^{kj}$ is bounded, while $\varphi^{kj}/w^{k-1,j}$ is bounded and is of order $O(1/\sigma^j)$ in a neighborhood of $\eta = 1$. This last follows from the lower bound for $w^{k-1,j}$ and the smoothness of φ: $\varphi(\tau, \xi, \eta) \sim (1 - \eta)$ in a neighborhood of $\eta = 1$. Similarly

$$\left| \sum_{k=0}^{K_1} \sum_{j=1}^{J-1} h l^2 \int_0^T A^{kj} \left(\frac{\delta_\eta w^{kj}}{w^{kj}} \right)^2 \frac{\varphi^{kj}}{w^{k-1,j}} \, d\tau \right|$$

$$\leqslant \text{const} \cdot X_1 T \sum_{j=1}^{J-1} l^2 (1 - jl) \frac{(\sigma^{j-1})^2}{(1 - jl)^2 (\sigma^j)^2} \cdot \frac{(1 - jl)}{(1 - (j-1)l)\sigma^{j-1}}$$

$$\leqslant \text{const} \cdot l \int_0^{1-l} \frac{d\eta}{(1 - \eta)\sigma} \leqslant \text{const} \cdot l\sigma^{J-1} \to 0 \quad \text{as } l \to 0.$$

Let $\tilde{w}^{h,l}(\tau, \xi, \eta)$ be a step function such that $\tilde{w}^{h,l}(\tau, \xi, \eta) = w^{kj}(\tau)$ for $kh \leqslant \xi < (k + 1)h$ and $jl \leqslant \eta < (j + 1)l$. Then (51) can be written as

$$-\int_0^{X_1} \int_l^{1-l} \frac{\tilde{\varphi}^{h,l}(0, \xi, \eta)}{\tilde{w}_0^{h,l}(\xi, \eta)} \, d\xi \, d\eta$$

$$+\int_0^{X_1} \int_0^T \tilde{\varphi}^{h,l}(\tau, \xi, l) \left[-\tilde{v}_0^{h,l} + \frac{\tilde{C}^{h,l}}{\tilde{w}^{h,l}(\tau, \xi, 0)} + \frac{\tilde{A}^{h,l}(\tau, \xi, l)}{\tilde{w}^{h,l}(\tau, \xi, 0)} \right] d\tau \, d\xi$$

$$+\int_0^T \int_0^{X_1 - h} \int_l^{1-l} \left[-\frac{\tilde{\varphi}_\tau^{h,l}}{\tilde{w}^{h,l}} + \nu \tilde{w}^{h,l} (\tilde{\varphi}_{\eta\eta})^{h,l} - \widetilde{\eta U}^{h,l} (\tilde{\varphi}_\eta)^{h,l} \frac{1}{\tilde{w}^{h,l}} \right.$$

$$- \widetilde{(\eta U_\xi)}^{h,l} \tilde{\varphi}^{h,l} \frac{1}{\tilde{w}^{h,l}} + \tilde{A}^{h,l} (\tilde{\varphi}_\eta)^{h,l} \frac{1}{\tilde{w}^{h,l}}$$

$$\left. + (\tilde{A}_\eta)^{h,l} \tilde{\varphi}^{h,l} \frac{1}{\tilde{w}^{h,l}} + \tilde{B}^{h,l} \tilde{\varphi}^{h,l} \frac{1}{\tilde{w}^{h,l}} \right] d\tau \, d\xi \, d\eta + \rho(h, l) = 0, \quad (52)$$

where $\rho(h, l) \to 0$ as $h + l \to 0$.

We note that

$$\left| \int_0^T \int_0^{X_1} \int_l^{1-l} (w^{h,l} - \tilde{w}^{h,l}) \Psi \, d\tau \, d\xi \, d\eta \right|$$

$$\leq \sup |\Psi| \cdot \left[\int_0^T \int_0^{X_1} \int_l^{1-l} (|r^{h,l}| \cdot h + |z^{h,l}| \cdot l) \, d\tau \, d\xi \, d\eta \right]$$

$$\leq \text{const} \cdot \sup |\Psi| \left(h + l \int_l^{1-l} \sigma \, d\eta \right) = O(h) + O(l),$$

$$(53)$$

$$\left| \int_0^T \int_0^{X_1} \int_l^{1-l} \left(\frac{1}{w^{h,l}} - \frac{1}{\tilde{w}^{h,l}} \right) \overbrace{(1 - \eta)}^{h,l} \Psi \, d\tau \, d\xi \, d\eta \right|$$

$$\leq \sup |\Psi| \int_0^T \int_0^{X_1} \int_l^{1-l} \frac{\overbrace{(1 - \eta)}^{h,l}}{w^{h,l} \tilde{w}^{h,l}} \left[h | r^{h,l} | + l | z^{h,l} | \right] d\tau \, d\xi \, d\eta$$

$$\leq \text{const} \cdot \sup |\Psi| \left[h + l \int_l^{1-l} \frac{d\eta}{(1 - \eta)\sigma} \right] \leq \text{const} \cdot [h + l\sigma(1 - l)] \sup |\Psi|.$$

$$(54)$$

Therefore (52) holds if we substitute $w^{h,l}$ for $\tilde{w}^{h,l}$ (along with a new function $\rho(h, l)$ which tends to zero as $h + l \to 0$). Passing now to the limit as h_n, $l_n \to 0$, we find that the limit function $w(\tau, \xi, \eta)$ (in the sense of uniform convergence) is a generalized solution of problem (3), (4) in the sense of the following integral identity:

$$\int_0^T \int_0^{X_1} \int_0^1 \left[\nu w \varphi_{\eta\eta} - \varphi_\tau \frac{1}{w} - \eta U \varphi_\xi \frac{1}{w} + A \varphi_\eta \frac{1}{w} + \eta r_x \frac{U}{r} \frac{\varphi}{w} \right] d\tau \, d\xi \, d\eta$$

$$- \iint_{\tau=0} \frac{\varphi}{w_0} d\xi \, d\eta - \iint_{\eta=0} \varphi v_0 \, d\tau \, d\xi = 0, \qquad (55)$$

which is valid for any function $\varphi \in C^2(\Omega_{X_1 T})$ satisfying (49). A similar integral identity was proposed by Oleĭnik [5] to define a stationary generalized solution in the Sobolev sense of problem (3), (4) for the case $r = 1$.

From (55) it follows that $w(\tau, \xi, \eta)$ has a generalized derivative w_η and satisfies (3) almost everywhere (for this one needs to take for φ a function with compact support and to transform some of the terms in (55) by integration by parts). From (26) of Lemma 2 it follows that for $w_{\eta\tau}$ we have

$$-M_{10} \leq w w_{\eta\eta} \leq -M_{11}. \qquad (56)$$

Consequently w_η is continuous in η when $\eta < 1$. The function $w(\tau, \xi, \eta)$, since it is continuous for $0 \leq \eta \leq 1$, satisfies the boundary condition in (4) for $\eta = 1$ and the initial condition in (4) for $\tau = 0$. From the boundary condition (6) for $\eta = 0$ we get

$$\frac{w^{h,l}(\tau, \xi, l) - w^{h,l}(\tau, \xi, 0)}{l} = \frac{1}{\nu} \left(v_0^h(\tau, \xi) - \frac{C^h(\tau, \xi)}{w^{h,l}(\tau, \xi, 0)} \right), \qquad (57)$$

where $w^{h,l}(\tau, \xi, \eta)$, $v_0^h(\tau, \xi)$ and $C^h(\tau, \xi)$ are functions obtained by linear interpolation from $w^{kj}(\tau)$, v_0^k and $C^k(\tau)$. Passing to the limit as $h_n \to 0$ in (57), we get

$$\frac{w^l(\tau, \xi, l) - w^l(\tau, \xi, 0)}{l} = \frac{1}{\nu}\left(v_0(\tau, \xi) - \frac{C(\tau, \xi)}{w^l(\tau, \xi, 0)} \right), \tag{58}$$

where $w^l(\tau, \xi, \eta) = \lim_{h_n \to 0} w^{h_n, l_n}(\tau, \xi, \eta)$. We assume that τ and ξ are fixed and that $\eta \in [0, 1 - \delta]$, where $\delta > 0$. Since

$$\frac{1}{l}\left(w^l(\tau, \xi, l) - w^l(\tau, \xi, 0) \right) = z^l(\tau, \xi, l) = \frac{1}{\varepsilon} \int_0^\varepsilon z^l(\tau, \xi, \eta) d\eta + \rho(\varepsilon, z^l),$$

where

$$\rho(\varepsilon, z) \leqslant \sup_{0 \leqslant \eta \leqslant \varepsilon} |z(\tau, \xi, \eta) - z(\tau, \xi, l)|,$$

we obtain, on passing to the limit as $l_n \to 0$ in (58),

$$\frac{1}{\varepsilon} \int_0^\varepsilon z(\tau, \xi, \eta) d\eta + \rho(\varepsilon, z) = \frac{1}{\nu}\left(v_0(\tau, \xi) - \frac{C(\tau, \xi)}{w(\tau, \xi, 0)} \right). \tag{59}$$

The function $z(\tau, \xi, \eta)$ coincides with w_η almost everywhere, since the sequence $z^{h_n, l_n} = \delta_\eta w^{h_n, l_n}$ converges weakly to w_η in $L_p(\Omega_{X_1 T}, p \geqslant 1)$. This proves that the function z in (59) can be replaced by w_η. Using the continuity of w_η in η and letting $\varepsilon \to 0$, we see that $\rho(\varepsilon, w_\eta) \to 0$ as $\varepsilon \to 0$, and consequently

$$w_\eta(\tau, \xi, 0) = \lim_{\varepsilon \to 0} \frac{1}{\varepsilon} \int_0^\varepsilon w_\eta(\tau, \xi, \eta) d\eta = \frac{1}{\nu}\left(v_0(\tau, \xi) - \frac{C(\tau, \xi)}{w(\tau, \xi, 0)} \right).$$

Thus, $w(\tau, \xi, \eta)$ also satisfies the boundary condition in (4) when $\eta = 0$.

It was proved in [1] that a solution of problem (3), (4) with the indicated properties is unique. From this it follows that any sequence $\{w^{h_n, l_n}\}$ converges to the same limit function $w(\tau, \xi, \eta)$ as $h_n, l_n \to 0$. Therefore it is appropriate to speak of $w(\tau, \xi, \eta)$ as the limit of the family of functions $\{w^{h,l}(\tau, \xi, \eta)\}$ as $h, l \to 0$.

The theorem is proved.

REMARK 1. It is interesting to compare the conditions of the theorem with those of the corresponding existence theorem for the solution of problem (3), (4), obtained by Oleĭnik [1]. The difference lies in the fact that instead of our conditions (22), the following compatibility conditions are given in Theorem 11 of [1]:

$$-E_{12}(1 - \eta)\sigma \leqslant \nu w_0^2 w_{0\eta\eta} + A(0, \xi, \eta) w_{0\eta} \leqslant E_{13}\xi(1 - \eta)\sigma, \tag{60}$$

$$\left(\nu w_0 w_{0\eta} - v_0 w_0 + C \right)\big|_{\eta=0, \tau=0} = 0. \tag{61}$$

It is obvious that every continuous functiion $w_0(\xi, \eta)$ that is twice continuously differentiable when $\eta < 1$ and is such that the compatibility conditions (22) are met satisfies (60) and (61). The converse, in general, is not true. Nevertheless, we can point out a sufficiently large class of initial functions w_0 for which the converse assertion also holds.

We have the following

ASSERTION. *Let $w_0(\xi, \eta)$ be continuous in ξ and η when $0 \leqslant \eta \leqslant 1$, differentiable with respect to ξ when $\eta \leqslant 1$, and twice continuously differentiable with respect to η when $\eta < 1$, and let*

$$E_1(1 - \eta)\sigma \leqslant w_0 \leqslant E_2(1 - \eta)\sigma, \qquad |w_{0\xi}| \leqslant E_9(1 - \eta)\sigma,$$

$$-E_{10}\sigma \leqslant w_{0\eta} \leqslant -E_{11}\sigma, \qquad -E_{16} \leqslant w_0 w_{0\eta\eta} \leqslant -E_{17}.$$

Suppose that w_0 satisfies the compatibility conditions (60) *and* (61). *Suppose, moreover, that $w_{0\eta\eta}(\xi, \eta)$ satisfies a Hölder condition in η when $0 \leqslant \eta \leqslant 1 - \delta$, $\delta > 0$, and let $w_{0\eta\eta}(\xi, \eta)$ be convex upwards in η when $1 - \delta \leqslant \eta < 1$, i.e.,*

$$|w_{0\eta\eta}(\xi, \eta_1) - w_{0\eta\eta}(\xi, \eta_2)| \leqslant L(\delta)|\eta_1 - \eta_2|^\gamma, \tag{62}$$

$$\tfrac{1}{2}\big[w_{0\eta\eta}(\xi, \eta_1) + w_{0\eta\eta}(\xi, \eta_2)\big] \leqslant w_{0\eta\eta}(\xi, \tfrac{1}{2}(\eta_1 + \eta_2)). \tag{63}$$

Then conditions (22) *are valid for the function w_0 when $h \geqslant O(l^\gamma)$. If w_0 satisfies* (63) *for all $\eta < 1$, none of the conditions on h and l occurs.*

We note that condition (63) is satisfied by the function $f(\xi)(1 - \eta)\sigma$, where $f(\xi) > 0$.

For the proof of the Assertion we need the following lemma.

LEMMA 3. *Let the function $y(x)$ be defined and continuous on the interval $[x_0 - \delta, x_0]$, $\delta = \mathrm{const} > 0$, and twice continuously differentiable in the interior of the interval. Then the following formulas for numerical differentiation hold:*

$$\frac{1}{h}\big[y(x_0 - h) - y(x_0 - 2h)\big] = y'(x_0 - h) - \frac{1}{h}\int_0^h ty''(x_0 - 2h + t)\, dt,$$

$$\tag{64}$$

$$\frac{1}{h}\big[y(x_0) - y(x_0 - h)\big] = y'(x_0 - h) + \frac{1}{h}\int_0^h ty''(x_0 - t)\, dt, \tag{65}$$

$$\frac{1}{h^2}\big[y(x_0) - 2y(x_0 - h) + y(x_0 - 2h)\big]$$

$$= \frac{1}{h^2}\int_0^h t\big[y''(x_0 - t) + y''(x_0 - 2h + t)\big]\, dt. \tag{66}$$

PROOF OF LEMMA 3. By using integration by parts it is easy to show that for any twice continuously differentiable function $\varphi(\xi)$ we have

$$\varphi(\xi) = \varphi(\xi_0) + (\xi - \xi_0)\varphi'(\xi_0) + \int_{\xi_0}^\xi (\xi - t)\varphi''(t)\, dt. \tag{67}$$

Applying this formula to the function $y(x)$ at the point $\xi = x_0 - 2h$, $\xi_0 = x_0 - h$ and $\xi = x_0$, $\xi_0 = x_0 - h$, we get

$$y(x_0 - h) - y(x_0 - 2h) = hy'(x_0 - h) - \int_{x_0 - h}^{x_0 - 2h}(x_0 - 2h - t_1)y''(t_1)\, dt_1,$$

$$y(x_0) - y(x_0 - h) = hy'(x_0 - h) + \int_{x_0 - h}^{x_0}(x_0 - t_2)y''(t_2)\, dt_2.$$

From this, after making the substitutions $t = t_1 - x_0 + sh$ and $t = x_0 - t_2$, we obtain (64) and (65). Formula (66) is a consequence of (64) and (65). Lemma 3 is proved.

PROOF OF THE ASSERTION. We have

$$\nu\delta_\eta w_0^{k,1} - v_0^k(0) + C^k(0)/w_0^{k,0} = \left(\nu w_{0\eta}^{k,0} - v_0^k(0) + C^k(0)/w_0^{k,0}\right)$$

$$+ \nu\frac{l}{2} w_{0\eta\eta}(kh, \theta l)$$

$$= l\frac{\nu}{2} w_{0\eta\eta}(kh, \theta l), \qquad 0 < \theta < 1,$$

from which the validity of the boundary compatibility condition (22) follows. Using the formulas of Lemma 3, we have

$$\nu\left(w_0^{kj}\right)^2 \delta_{\eta\eta}w_0^{kj} + A(0, kh, jl)\delta_\eta w_0^{kj}$$

$$= \left[\nu\left(w_0^{kj}\right)^2 w_{0\eta\eta}^{kj} + A(0, kh, jl)w_{0\eta}^{kj}\right]$$

$$+ \nu\left(w_0^{kj}\right)^2 \frac{1}{l^2}\int_0^l t\left[w_{0\eta\eta}(kh, (j+1)l - t)\right.$$

$$\left. - 2w_{0\eta\eta}(kh, jl) + w_{0\eta\eta}(kh, (j-1)l + t)\right]dt$$

$$+ \left[(C^k + jlU_x^k)(1 - jl)\frac{1}{l}\int_0^l tw_{0\eta\eta}(kh, (j-1)l + t)\,dt\right]\bigg|_{\tau=0}. \quad (68)$$

Since $C > 0$, $U_x > 0$, $w_{0\eta\eta} < 0$ and the function $w_{0\eta\eta}$ is convex upwards in a neighborhood of $\eta = 1$, (68) implies the validity of the upper bound in (22) for $1 - \delta \leqslant jl < 1$. Consider the last two terms in (68). By virtue of the assumptions made concerning w_0, we have the following estimates:

$$\left|\frac{1}{l^2}\int_0^l t\left[w_{0\eta\eta}(kh, (j+1)l - t) - 2w_{0\eta\eta}(kh, jl) + w_{0\eta\eta}(kh, (j-1)l + t)\right]dt\right|$$

$$\leqslant \frac{2L}{l^2}\int_0^l t\,|l - t|^\gamma dt = 2Ll^\gamma\int_0^1 t(1 - t)^\gamma dt \leqslant C(L)h$$

$$\text{for } h \geqslant O(l^\gamma) \text{ and } 0 \leqslant jl \leqslant 1 - \delta; \quad (69)$$

$$\left|\frac{1}{l^2}\int_0^l tw_{0\eta\eta}(kh, (j-1)l + t)\,dt\right| \leqslant \frac{E_{16}}{2E_1}\cdot\frac{1}{\left[(1 - \eta)\sigma\right]\big|_{\eta=(j-1)l+\theta l}}$$

$$\leqslant \frac{E_{16}}{2E_1}\cdot\frac{1}{(1 - jl)\sigma^j}, \qquad \text{since } 0 < \theta < 1; \quad (70)$$

$$\left|\frac{1}{l^2}\int_0^l tw_{0\eta\eta}(kh, (j+1)l - t)\,dt\right|$$

$$\leqslant \frac{1}{l^2}\int_0^l \frac{E_{16}}{E_1}\cdot\frac{1}{(1 - (j+1)l + t)\sqrt{-\ln\mu(1 - (j+1)l + t)}}$$

$$\leqslant \frac{2E_{16}}{E_1}\cdot\frac{1}{(1 - jl)\sigma^j}. \quad (71)$$

Here we have taken into account that the function $1/(1 - \eta)\sigma$ increases when $0 \leq \eta < 1$, the ratio $(1 - jl)\sigma^j/(1 - (j + 1)l)\sigma^{j+1}$ is bounded uniformly in l for $j \leq J - 2$, and

$$\frac{1}{l^2} \int_0^l \frac{t\,dt}{t\sqrt{-\ln \mu t}} \leq \frac{1}{l\sqrt{-\ln \mu l}} \, .$$

Then it can be seen from (68) that w_0 satisfies (22).

REMARK 2. The set of initial functions w_0 satisfying the conditions of the theorem is dense (in the metric of the space of continuous functions on $[0, X_1] \times [0, 1]$) in the set of initial functions $\tilde{w}_0$ which satisfy the hypothesis of the existence theorem found in [1]. In this event also the solutions w and $\tilde{w}$ themselves, as Lemma 10 of [1] shows, will likewise differ by an arbitrarily small amount in the metric of the space of continuous functions on $\Omega_{X_1 T}$.

In conclusion we mention that if $w(\tau, \xi, \eta)$ has the properties indicated in Theorem 2, then by means of the substitution

$$w(\tau, \xi, \eta) = w\!\left(t, x, \frac{u}{U}\right) = \frac{u_y}{U}, \qquad y = \int_0^{u(t,x,y)/U(t,x)} \frac{ds}{w(t, x, s)}$$

we can pass from the solution of problem (3), (4) to the solution u, v of the system (1), (2) and in this way obtain a theorem on the existence and uniqueness of the original problem (see [1], Theorems 12 and 2). The requirements on the function $U(t, x)$ (the longitudinal component of the external flow) formulated in Lemma 2 and in Theorem 2 are here somewhat weaker than the corresponding restrictions in [1].

I am deeply grateful to Professor O. A. Oleĭnik for her attention and help in this work.

BIBLIOGRAPHY

1. O. A. Oleĭnik, *Mathematical problems of boundary layer theory*, Uspehi Mat. Nauk **23** (1968), no. 3(141), 3–65; English transl. in Russian Math. Surveys **23** (1968).

2. D. A. Silaev, *Construction of the solution of a nonstationary Prandtl system by the method of lines with respect to time*, Uspehi Mat. Nauk **28** (1973), no. 2(1970), 243–244. (Russin)

3. Wolfgang Walter, *Existence and convergence theorems for the boundary layer equations based on the line method*, Arch. Rational Mech. Anal. **39** (1970), 169–188.

4. I. G. Petrovskiĭ, *Lectures on the theory of ordinary differential equations*, 5th ed., "Nauka", Moscow, 1964; English transl., Prentice-Hall, Englewood Cliffs, N. J., 1966.

5. O. A. Oleĭnik, *Weak solutions in the Sobolev sense for a system of boundary layer equations*, Partial Differential Equations (Proc. Sympos. Sixtieth Birthday S. L. Sobolev), "Nauka", Moscow, 1970, pp. 177–190; English transl. in Amer. Math. Soc. Transl. (2) **105** (1976).

Translated by R. N. GOSS

Amer. Math. Soc. Transl.
(2) Vol. **118**, 1982

On Prandtl's System of Equations
for the Boundary Layer
with a Surface of Discontinuity*

A. I. SUSLOV

In the flow of liquid or gas about a solid surface, in certain cases thin layers are formed with a sharply distinguished boundary on which the characteristics of the motion and state change in a jumpwise manner. For example, in the flow of steam about a cold surface (in condensing devices) a film of water is formed on the wall, which is carried along by the flow of steam and the force of gracity. Thin films with a sharply distinguished boundary are formed during the fusion of the surface of bodies moving with high supersonic velocities (the entry of a space ship into the dense layers of the atmosphere) and in a number of other cases.

The phenomenon occurs in a thin layer near the surface of the body, which makes it possible to investigate the problem by boundary-layer methods. In comparison with the usual theory the situation is complicated by the fact that there are one or several surfaces of discontinuity within the boundary layer which are unknown a priori and on which certain conditions must be satisfied depending on the character of the phenomenon studied.

The problem of a boundary layer with a surface of discontinuity for stationary plane-parallel flows of a compresible fluid or gas was posed in [1], where self-similar solutions for a constant velocity of the external flow were investigated. Self-similar solutions of the system of equations of a magnetohydrodynamic boundary layer with a surface of discontinuity are considered in [2] and [3].

In this paper we consider a system of boundary-layer equations with a surface of discontinuity for plane-parallel stationary flows of incompressible fluids. It is assumed that a fluid with properties distinct from the properties of the fluid of the main flow is injected into the boundary layer through the surface over which the flow occurs. The surface of discontinuity, which is unknown a priori,

---

1980 *Mathematics Subject Classification.* Primary 76D10; Secondary 76D15.

*Translation of Trudy Sem. Petrovsk. **2** (1976), 243–259. MR **56** #4398.

separates the immiscible fluids. In this case the dynamic problem (the determination of the velocity field) can be solved independently of the thermal problem (the determination of the temperature). A solution of the dynamic problem is constructed below, and its uniqueness is proved. A sufficient condition for the break-off of the boundary layer is also established.

§1. Formulation of the problem. Construction of a solution

We consider the case of stationary plane-parallel flow of an incompressible fluid about a porous surface through which another incompressible fluid with properties distinct from the properties of the fluid of the main flow is injected into the boundary layer. Suppose that the surface $y = y_*(x)$, where $y_*(x)$ is a function which remains to be defined, separates the immiscible fluids.

The velocity of the injected fluid u_1, v_1 in the domain $D_A^1\{0 < x < A, 0 < y < y_*(x)\}$ then satisfies the Prandtl system

$$u_1 \frac{\partial u_1}{\partial x} + v_1 \frac{\partial u_1}{\partial y} = \nu_1 \frac{\partial^2 u_1}{\partial y^2} - \frac{1}{\rho_1} \frac{dp}{dx},$$

$$\frac{\partial u_1}{\partial x} + \frac{\partial v_1}{\partial y} = 0, \tag{1}$$

where ν_1 and ρ_1 are respectively the viscosity and density of the injected fluid. In the domain $D_A^2\{0 < x < A, y_*(x) < y < \infty\}$

$$u_2 \frac{\partial u_2}{\partial x} + v_2 \frac{\partial u_2}{\partial y} = \nu_2 \frac{\partial^2 u_2}{\partial y^2} - \frac{1}{\rho_2} \frac{dp}{dx},$$

$$\frac{\partial u_2}{\partial x} + \frac{\partial v_2}{\partial y} = 0, \tag{2}$$

where u_2, v_2 is the velocity, ν_2 the viscosity, and ρ_2 the density of the main flow.

There are, moreover, the boundary conditions

$$u_1\big|_{x=0} = u_{10}(y), \quad 0 \leqslant y \leqslant y_*(0), \quad u_2\big|_{x=0} = u_{20}(y),$$

$$y_*(0) \leqslant y < \infty, \quad u_1\big|_{y=0} = 0, \quad v_1\big|_{y=0} = v_0(x) > 0, \quad u_2 \underset{y\to\infty}{\to} U(x), \tag{3}$$

where $y_*(0)$ is given quantity and $2p(x) + \rho_2 U^2(x) = \text{const}$ (Bernoulli's law).

We assume that the fluid does not flow through the surface of discontinuity. In this case for $y = y_*(x)$ (see [1]) we have the relations

$$\frac{dy_*(x)}{dx} = \frac{v_1\big(x, y_*(x)\big)}{u_1\big(x, y_*(x)\big)} = \frac{v_2\big(x, y_*(x)\big)}{u_2\big(x, y_*(x)\big)}. \tag{4}$$

For $y = y_*(x)$ there are the junction conditions

$$u_1\big(x, y_*(x)\big) = u_2\big(x, y_*(x)\big), \tag{5}$$

$$\nu_1 \rho_1 \frac{\partial u_1\big(x, y_*(x)\big)}{\partial y} = \nu_2 \rho_2 \frac{\partial u_2\big(x, y_*(x)\big)}{\partial y}. \tag{6}$$

It follows from (4) and (5) that $v_1(x, y_*(x)) = v_2(x, y_*(x))$.

With regard to the data in these problems we assume that $u_{10}(y) > 0$ and $u_{20}(y) > 0$ for $y > 0$, $u_{10}(0) = 0$, $u'_{10}(0) > 0$, $u_{20}(y) \to U(0)$ as $y \to \infty$, and $U(x) > 0$ for all x; dp/dx and $v_0(x)$ are continuously differentiable; $u_{10}(y)$, $u'_{10}(y)$, and $u''_{10}(y)$ are bounded for $0 \leqslant y \leqslant y_*(0)$ and satisfy a Hölder condition; $u_{20}(y)$, $u'_{20}(y)$, and $u''_{20}(y)$ are bounded for $y_*(0) \leqslant y < \infty$ and satisfy a Hölder condition. Moreover,

$$u_{10}\big(y_*(0)\big) = u_{20}\big(y_*(0)\big), \qquad \nu_1\rho_1 u'_{10}\big(y_*(0)\big) = \nu_2\rho_2 u'_{20}\big(y_*(0)\big).$$

We also assume that the consistency condition at the point $(0,0)$

$$\nu_1 u''_{10}(y) - p'(0)/\rho_1 - v_0(0)u'_{10}(y) = O(y^2),$$

is satisfied for small y. We denote by $u_0(y)$ a continuous function such that $u_0(y) = u_{10}(y)$ for $0 \leqslant y \leqslant y_*(0)$ and $u_0(y) = u_{20}(y)$ for $y_*(0) \leqslant y < \infty$.

We shall prove that a solution of problem $(1) - (6)$ exists. In the proof of the existence theorem we use the methods of [4] and [5].

THEOREM 1. *For some $A > 0$ there exists a solution u_1, v_1, u_2, v_2, $y_*(x)$ of problem (1)–(6) possessing the following properties: u_1 is continuous in $\overline{D}_A^1$, $u_1 > 0$ for $y > 0$, $\partial u_1/\partial y > m > 0$ for $0 < y \leqslant y_0$, where m and y_0 are constants; $\partial u_1/\partial y$ and $\partial^2 u_1/\partial y^2$ are continuous and bounded in D_A^1; $\partial u_1/\partial x$, v_1, and $\partial v_1/\partial y$ are continuous and bounded in D_A^1 for $y = 0$; u_2 is continuous and bounded in $\overline{D}_A^2$, $u_2 > 0$ in $\overline{D}_A^2$; $\partial y_2/\partial y$ and $\partial^2 u_2/\partial y^2$ are continuous and bounded in D_A^2; $\partial u_2/\partial x$, v_2, and $\partial v_2/\partial y$ are continuous and bounded in any finite part of the domain D_A^2; and $\partial u_i/\partial y$ and v_i are continuous in D_A^i $(i = 1,2)$ up to the curve $y = y_*(x)$. If $u'_{20}(y) \to 0$ as $y \to \infty$ sufficiently fast, then $\partial u_2/\partial x$ and $\partial v_2/\partial y$ are bounded in D_A^2. The function $y_*(x)$ is continuously differentiable on $[0, A]$. In the case where $dp/dx < 0$ such a solution of problem (1)–(6) exists in D_A for any $A > 0$.*

The next lemma reduces the proof of Theorem 1 to the investigation of a problem in von Mises form.

LEMMA 1. *Suppose that the following conditions are satisfied:*
1) *In the domain $G_A^1\{0 < x < A, 0 < \psi < \psi_*(x)\}$, where*

$$\psi_*(x) = \psi_*(0) + \int_0^x v_0(t)\, dt, \qquad \psi_*(0) = \int_0^{y_*(0)} u_{10}(s)\, ds, \qquad (7)$$

there exists a solution $w_1(x, \psi)$ of the equation

$$\frac{\partial w}{\partial x} + v_0(x)\frac{\partial w}{\partial \psi} = \nu_1\sqrt{w}\,\frac{\partial^2 w}{\partial \psi^2} - \frac{2}{\rho_1}\frac{dp}{dx}, \qquad (8)$$

which satisfies the boundary conditions

$$w_1\big|_{\psi=0} = 0, \qquad w_1\big|_{x=0} = w_{10}(\psi) \equiv u_{10}^2(y). \qquad (9)$$

2) *In the domain $G_A^2\{0 < x < A, \psi_*(x) < \psi < \infty\}$ there exists a solution $w_2(x, \psi)$ of the equation*

$$\frac{\partial w}{\partial x} + v_0(x)\frac{\partial w}{\partial \psi} = \nu_2\sqrt{w}\,\frac{\partial^2 w}{\partial \psi^2} - \frac{2}{\rho_2}\frac{dp}{dx}, \qquad (10)$$

which satisfies the boundary conditions

$$w_2\big|_{x=0} = w_{20}(\psi) \equiv u_{20}^2(y), \qquad w_2 \to U^2(x), \psi \to \infty. \tag{11}$$

3) *For $\psi = \psi_*(x)$ the following relations are satisfied*:

$$w_1\big(x, \psi_*\big) = w_2\big(x, \psi_*\big), \tag{12}$$

$$\nu_1\rho_1\frac{\partial w_1\big(x, \psi_*\big)}{\partial \psi} = \nu_2\rho_2\frac{\partial w_2\big(x, \psi_*\big)}{\partial \psi}. \tag{13}$$

4) *The functions $w_i(x, \psi)$ ($i = 1, 2$) possess the following properties: the w_i are continuous and bounded in $\overline{G}_A^i$ and $w_i(x, \psi) > 0$ in G_A^i; the $w_i(x, \psi)$ in G_A^i have the continuous derivatives contained in the respective equations (8) and (10), the $\partial w_i/\partial \psi$ are continuous up to the curve $\psi = \psi_*(x)$, and*

$$\left|\frac{\partial w_i}{\partial \psi}\right| \leqslant M, \qquad \left|\sqrt{w_1}\,\frac{\partial^2 w_1}{\partial \psi^2}\right| \leqslant M, \qquad \left|\frac{\partial^2 w_2}{\partial \psi^2}\right| \leqslant M;$$

moreover,

$$\left|\frac{\partial w_1}{\partial x}\right| \leqslant M\psi^{1-\beta}, \qquad \frac{\partial w_1}{\partial \psi} \geqslant m > 0$$

for $0 \leqslant \psi \leqslant \psi_1$ and $0 < \beta < 1/2$, where the constants M, m, and ψ_1 depend on A, $v_0(x)$, $p(x)$, and $u_0(y)$.

Then in the domain $D_A\{0 < x < A, \; 0 < y < \infty\}$ there exists a solution of problem (1)–(6) possessing the properties indicated in Theorem 1.

PROOF. We go over from the independent variables x, y to the von Mises variables

$$x = x, \qquad \psi = \psi(x, y), \tag{14}$$

where

$$\psi(x, y) = \begin{cases} \displaystyle\int_0^y u_1(x, s)\, ds & \text{for } 0 \leqslant y \leqslant y_*(x), \\[2ex] \displaystyle\int_0^{y_*(x)} u_1(x, s)\, ds + \int_{y_*(x)}^y u_2(x, t)\, dt & \text{for } y_*(x) \leqslant y < \infty. \end{cases}$$

With this change of variables the system (1) goes over into (8) if we set $w_1 = u_1^2$; the domain D_A^1 goes over into $\{0 < x < A, \; 0 < \psi < \psi_*(x)\}$, where $\psi_*(x) = \psi(x, y_*(x))$. Under the change of variables (14) the system (2) goes over into (10) if we set $w_2 = u_2^2$; the domain D_A^2 goes over into $\{0 < x < A, \; \psi_*(x) < \psi < \infty\}$. The boundary conditions (3) go over into (9) and (11), while (5) and (6) go over into (12) and (13).

Since the conditions (4) are satisfied, it follows that

$$\frac{d\psi_*}{dx} = \frac{\partial\psi_1(x, y_*)}{\partial x} + \frac{\partial\psi_1(x, y_*)}{\partial y}\frac{dy_*}{dx}$$

$$= v_0(x) - v_1(x, y_*) + u_1(x, y_*)\frac{v_1(x, y_*)}{u_1(x, y_*)} = v_0(x).$$

Hence $\psi_*(x) = \psi_*(0) + \int_0^x v_0(t)\,dt$. The equality (7) is proved.

If there exists a solution $w_1(x, \psi)$, $w_2(x, \psi)$ of problem (8)–(13) with the properties indicated in Lemma 1, then

$$u_i = \sqrt{w_i} \qquad (i = 1, 2),$$

$$y = \begin{cases} \displaystyle\int_0^\psi \frac{dt}{\sqrt{w_1(x, t)}}, & 0 \leqslant \psi \leqslant \psi_*(x), \\[3mm] \displaystyle\int_0^{\psi_*(x)} \frac{dt}{\sqrt{w_1(x, t)}} + \int_{\psi_*(x)}^\psi \frac{dt}{\sqrt{w_2(x, t)}}, & \psi_*(x) \leqslant \psi < \infty, \end{cases}$$

$$v_i - v_0(x) = -\frac{\partial\psi}{\partial x} \qquad (i = 1, 2).$$

The remainder of the proof of Lemma 1 is analogous to the proof of Lemma 1 of [4].

We set

$$w_0(\psi) = w_{10}(\psi) \quad \text{for } 0 \leqslant \psi \leqslant \psi_*(0),$$

$$w_0(\psi) = w_{20}(\psi) \quad \text{for } \psi_*(0) \leqslant \psi < \infty.$$

It follows from (6), (9), and (11) that $w_0'(\psi)$ is discontinuous at $\psi = \psi_*(0)$.

We write (8) and (10) as a single equation with discontinuous coefficients:

$$\sqrt{w}\,\frac{\partial}{\partial\psi}\left((\nu\rho)\frac{\partial w}{\partial\psi}\right) - \rho\frac{\partial w}{\partial x} - \rho v_0(x)\frac{\partial w}{\partial\psi} = 2\frac{dp}{dx}, \tag{15}$$

where $\nu\rho = \nu_i\rho_i$ and $\rho = \rho_i$ in the domain G_A^i $(i = 1, 2)$.

The construction of a solution of problem (8)–(13) with the required properties is carried out as in [4]. In order to apply the method of constructing a solution used in [4], it is necessary to smooth the coefficients of (15). The idea of smoothing discontinuous coefficients in elliptic and parabolic equations has previously been applied in [5].

We denote by $(\nu\rho)^\varepsilon(\eta)$ and $\rho^\varepsilon(\eta)$ smooth functions of the argument $\eta = \psi - \psi_*(x)$, which are uniformly bounded in ε, such that $(\nu\rho)^\varepsilon = \nu_1\rho_1$ and $\rho^\varepsilon = \rho_1$ for $\eta \leqslant -\varepsilon$, and $(\nu\rho)^\varepsilon = \nu_2\rho_2$ and $\rho^\varepsilon = \rho_2$ for $\eta \geqslant \varepsilon$.

In the domain G_A^ε $(\{0 < x < A, 0 < \psi < 1/\varepsilon\}, \varepsilon > 0)$, we consider the equation

$$\sqrt{w}\,\frac{\partial}{\partial\psi}\left((\nu\rho)^\varepsilon\frac{\partial w}{\partial\psi}\right) - \rho^\varepsilon\frac{\partial w}{\partial x} - \rho^\varepsilon v_0(x)\frac{\partial w}{\partial\psi} = 2\frac{dp}{dx} \tag{16}$$

with boundary conditions

$$w\,|_{\psi=0} = w_0^\varepsilon(\varepsilon)e^{\mu(\varepsilon)x/w_0^\varepsilon(\varepsilon)}, \qquad w\,|_{x=0} = w_0^\varepsilon(\varepsilon + \psi),$$

$$w\,|_{\psi=1/\varepsilon} = w_0^\varepsilon(\varepsilon + 1/\varepsilon)e^{\mu(\varepsilon+1/\varepsilon)w_0^\varepsilon(\varepsilon+1/\varepsilon)}. \tag{17}$$

As the function $w_0^\varepsilon(\psi)$ we choose a smooth function with derivatives $dw_0^\varepsilon/d\psi$ and $\sqrt{w_\varepsilon}\,d((\nu\rho)^\varepsilon\,dw_0^\varepsilon/d\psi)/d\psi$ uniformly bounded in ε which coincides with $w_0(\psi)$ for small ψ. We require that as $\varepsilon \to 0$ the sequence $w_0^\varepsilon(\psi)$ converge to $w_0(\psi)$ uniformly in ψ. The function $\mu(\psi)$ in (17) has the form

$$\mu(\psi) = \sqrt{w_0^\varepsilon(\psi)}\,\big((\nu\rho)^\varepsilon w_0^{\varepsilon'}(\psi)\big)' - 2p'(0) - p^\varepsilon(0,\psi)v_0(0)w_0^{\varepsilon'}(\psi).$$

From the fact that $w_0^\varepsilon(\psi) = w_0(\psi)$ for small ψ and the consistency condition at the point $(0,0)$ it follows that $\mu(\varepsilon) = O(\varepsilon)$ for small ε.

We shall show that for some $A > 0$ depending on dp/dx, $v_0(x)$, and $u_0(y)$ solutions $w_\varepsilon(x,\psi)$ of problem (16), (17) exist for small $\varepsilon > 0$, and as $\varepsilon \to 0$ they converge to a solution of problem (8)–(13).

Lemmas 2–6 formulated below are proved in the same way as the corresponding lemmas in [4].

LEMMA 2. *If a positive solution $w_\varepsilon(x,\psi)$ of problem (16), (17) exists in the domain G_A^ε, then for $x \leqslant A_0$ and sufficiently small ε the a priori estimate*

$$w_\varepsilon(x,\psi) \geqslant w_\varepsilon(x,0) + f(\psi)(1 + e^{-\alpha x})$$

holds, where $\alpha > 0$, $f(\psi) = A_1\psi^{4/3} + A_2$ for $\psi \leqslant \psi_0$, $A_1 > 0$, $A_2 > 0$, $f(\psi_0) \leqslant f(\psi)$ $\leqslant A_3$, $|f'(\psi)| \leqslant A_4$, $|f''(\psi)| \leqslant A_5$ for $\psi \geqslant \psi_0$ and $f(\psi) = \mathrm{const}$ for $\psi \geqslant \psi_*(0)/2$; $A_1, A_2, A_3, 1/\alpha$, and A_0 are sufficiently small and do not depend on ε; A_0 depends on dp/dx, $v_0(x)$, and $u_0(y)$. If $dp/dx \leqslant -\beta_0 < 0$, then the a priori estimate

$$w_\varepsilon(x,\psi) \geqslant w_\varepsilon(x,0) + f(\psi)$$

holds in G_A^ε for any fixed $A > 0$ and small $\varepsilon > 0$.

The proof is carried out by the maximum principle. Since

$$\partial(\nu\rho)^\varepsilon\big(x,\psi_*(x)\big)/\partial\psi \sim \varepsilon^{-1}$$

as $\varepsilon \to 0$, the additional restriction $f(\psi) \equiv \mathrm{const}$ for $\psi \geqslant \psi_*(0)/2$ is imposed on the function $f(\psi)$. Here $f'(\psi)\partial(\nu\rho)^\varepsilon/\partial\psi \equiv 0$ in G_A^ε.

LEMMA 3. *A solution $w_\varepsilon(x,\psi)$ of problem (16), (17) exists in the domain $G_{A_0}^\varepsilon$ for $0 < \varepsilon < \varepsilon_0$, where A_0 is the constant defined in Lemma 2 and ε_0 is a sufficiently small number. If $dp/dx \leqslant 0$, then a solution of problem (16), (17) exists in the domain G_A^ε for any $A > 0$. The solution $w_\varepsilon(x,\psi)$ belongs to the class $C^{2+\beta}(\overline{G}_{A_0}^\varepsilon)$. At the points of $G_{A_0}^\varepsilon$, equation (16), which $w_\varepsilon(x,\psi)$ satisfies, can be differentiated twice with respect to ψ and once with respect to x. The derivative $\partial^2 w_\varepsilon/\partial x\partial\psi$ belongs to the class $L_2(G_{A_0}^\varepsilon)$.*

The proof uses the existence theorem for a solution of the first boundary value problem for a quasilinear parabolic equation ([6], Theorem 6.1 of Chapter V).

We denote by M_i $(i = 0, 1, 2, \dots)$ positive constants not depending on ε.

LEMMA 4. *In the domain* $G_{A_0}^\varepsilon$, $0 < w_\varepsilon(x, \psi) \leqslant M_0$; *for* $0 \leqslant x \leqslant A_0$

$$M_1 \leqslant \frac{\partial w_\varepsilon(x, 0)}{\partial \psi} < M_2.$$

LEMMA 5. *In the domain* $P\{0 < x < A_0, \Delta < \psi < \psi_*(x) - \Delta, \psi_*(x) + \Delta < \psi < 1/\varepsilon\}$ *the functions* $w_\varepsilon(x, \psi)$ *have derivatives* $\partial w_\varepsilon/\partial \psi$, $\partial w_\varepsilon/\partial x$, *and* $\partial^2 w_\varepsilon/\partial \psi^2$ *which satisfy a Hölder condition, and the maxima of the moduli of these derivatives and the Hölder constants are bounded by constants not depending on ε but depending on Δ; the Hölder exponents are bounded by constants not depending on ε or Δ.*

For $\psi > \Delta$, (16) is a nondegenerate parabolic equation. From Lemmas 4 and 6 of [7] it follows that $w_\varepsilon(x, \psi)$ satisfies a Hölder condition in the domain P, with constant and exponent not depending on ε. Lemma 5 therefore follows from estimates of Schauder type for linear equations.

LEMMA 6. *In the domain* $\{0 < x < A_0, 0 < \psi < \Delta\}$, *for small* Δ,

$$0 < M_4 \leqslant \frac{\partial w_\varepsilon}{\partial \psi} \leqslant M_3, \qquad \left|\frac{\partial w_\varepsilon}{\partial x}\right| \leqslant M_5 \psi^{1-\beta}, \qquad \left|\sqrt{w_\varepsilon}\,\frac{\partial^2 w_\varepsilon}{\partial \psi^2}\right| \leqslant M_6.$$

We now consider (16) in the domain $P_1\{0 < x < A_0, \psi_*(x) - \Delta < \psi - \psi_*(x) + \Delta_1\}$. We make a change of independent variables of the form $x = x$, $\eta = \psi - \psi_*(x)$. Under this change of variables (16) goes over into the equation

$$\sqrt{w}\,\frac{\partial}{\partial \eta}\left((\nu\rho)^\varepsilon(\eta)\frac{\partial w}{\partial \eta}\right) - \rho^\varepsilon(\eta)\frac{dw}{dx} = 2\frac{dp}{dx}, \tag{18}$$

and P_1 goes over into the domain $Q_1\{0 < x < A_0, |\eta| < \Delta_1\}$.

We shall establish some integral estimates for the derivatives of w_ε in the domain $Q_k\{0 < x < A_0, |\eta| < \Delta_1/k\}$, where $k = 1, \dots, 4$. We note that in Q_k the inequality $w_\varepsilon \geqslant M_9 > 0$ holds for all k.

LEMMA 7. *In the domain* Q_3

$$\left\|\frac{\partial w_\varepsilon}{\partial \eta}\right\|_{L_4(Q_3)} \leqslant M_7, \qquad \left\|\frac{\partial}{\partial \eta}\left((\nu\rho)^\varepsilon\frac{\partial w_\varepsilon}{\partial \eta}\right)\right\|_{L_2(Q_3)} \leqslant M_8.$$

PROOF. Let $\zeta_k(\eta) = [\eta^2 - (\Delta_1/k)^2]^2$ be a function equal to zero on the lateral sides of the rectangle Q_k. Obviously $\zeta_k'^2/\zeta_k$ is bounded in Q_k.

We multiply (18) by $w_\varepsilon\zeta_1$, and integrate over Q_1:

$$\iint_{Q_1} \rho^\varepsilon \frac{\partial w_\varepsilon}{\partial x} w_\varepsilon \zeta_1\, dx\, d\eta = \iint_{Q_1} w_\varepsilon^{3/2}\zeta_1 \frac{\partial}{\partial \eta}\left((\nu\rho)^\varepsilon\frac{\partial w_\varepsilon}{\partial \eta}\right) dx\, d\eta - \iint_{Q_1} 2\frac{dp}{dx} w_\varepsilon\zeta_1\, dx\, d\eta$$

$$= I_1 + I_2. \tag{19}$$

Since ζ_1 and ρ^ε depend only on η, it follows that

$$\iint_{Q_1} \rho^\varepsilon \frac{\partial w_\varepsilon}{\partial x} w_\varepsilon \zeta_1 \, dx \, d\eta = -\int_{-\Delta_1}^{\Delta_1} \frac{\rho^\varepsilon \zeta_1}{2} w_\varepsilon^2 \, d\eta \, \Big|_{x=0}^{x=A}. \tag{20}$$

We transform the integral I_1 by integration by parts:

$$I_1 = -\iint_{Q_1} \frac{3}{2} w_\varepsilon^{1/2} \zeta_1 (\nu\rho)^\varepsilon \left(\frac{\partial w_\varepsilon}{\partial \eta} \right)^2 dx \, d\eta - \iint_{Q_1} (\nu\rho)^\varepsilon w_\varepsilon^{3/2} \zeta_1' \frac{\partial w_\varepsilon}{\partial \eta} dx \, d\eta. \tag{21}$$

Substituting (20) and (21) into (19), we obtain

$$\iint_{Q_1} \frac{3}{2} w_\varepsilon^{1/2} \zeta_1 (\nu\rho)^\varepsilon \left(\frac{\partial w_\varepsilon}{\partial \eta} \right)^2 dx \, d\eta = -\int_{-\Delta_1}^{\Delta_1} \frac{\rho_\varepsilon \zeta_1}{2} w_\varepsilon^2 \, d\eta \, \Big|_{x=0}^{x=A}$$

$$- \iint_{Q_1} (\nu\rho)^\varepsilon w_\varepsilon^{5/2} \zeta_1' \frac{\partial w_\varepsilon}{\partial \eta} dx \, d\eta - \iint_{Q_1} 2 \frac{dp}{dx} w_\varepsilon \zeta_1 \, dx \, d\eta. \tag{22}$$

From Lemma 4 it follows that the first and third integrals on the right side of (22) are uniformly bounded in ε. We estimate the second integral by means of the inequality $ab \leqslant \delta a^2/2 + b^2/2\delta$ for some δ:

$$\left| \iint_{Q_1} (\nu\rho)^\varepsilon w_\varepsilon^{3/2} \zeta_1' \frac{\partial w_\varepsilon}{\partial \eta} dx \, d\eta \right| \leqslant \iint_{Q_1} w_\varepsilon^{1/2} (\nu\rho)^\varepsilon \zeta_1 \left(\frac{\partial w_\varepsilon}{\partial \eta} \right)^2 dx \, d\eta$$

$$+ \frac{1}{4} \iint_{Q_1} (\nu\rho)^\varepsilon w_\varepsilon^{5/2} \frac{1}{\zeta_1} (\zeta_1')^2 \, dx \, d\eta. \tag{23}$$

It thus follows from (22) and (23) that

$$\iint_{Q_1} \zeta_1 \left(\frac{\partial w_\varepsilon}{\partial \eta} \right)^2 dx \, d\eta \leqslant M_{10},$$

whence

$$\left\| \frac{\partial w_\varepsilon}{\partial \eta} \right\|_{L_2(Q_2)} \leqslant M_{11}. \tag{24}$$

To prove the second assertion of Lemma 7 it obviously suffices to establish the inequality

$$\left\| \frac{\partial w_\varepsilon}{\partial x} \right\|_{L_2(Q_3)} \leqslant M_{12}. \tag{25}$$

We multiply (18) by $\zeta_2 (\partial w_\varepsilon/\partial x)/\sqrt{w_\varepsilon}$, and integrate over Q_2. We obtain

$$\iint_{Q_2} \frac{\rho^\varepsilon \zeta_2}{\sqrt{w_\varepsilon}} \left(\frac{\partial w_\varepsilon}{\partial x} \right)^2 dx \, d\eta = \iint_{Q_2} \zeta_2 \frac{\partial w_\varepsilon}{\partial x} \frac{\partial}{\partial \eta} \left((\nu\rho)^\varepsilon \frac{\partial w_\varepsilon}{\partial \eta} \right) dx \, d\eta$$

$$- \iint_{Q_2} 2 \frac{dp}{dx} \frac{\zeta_2}{\sqrt{w_\varepsilon}} \frac{\partial w_\varepsilon}{\partial x} dx \, d\eta$$

$$= I_3 + I_4.$$

Integrating by parts in I_3 (twice) and I_4, and using (24), we obtain (25).

We now multiply (18) by $\zeta_2 w_\varepsilon [(\nu\rho)^\varepsilon \partial w_\varepsilon / \partial \eta]^2$, and integrate over Q_2. Integrating certain terms of the resulting equality by parts and using (24) and (25), as above, we arrive at the estimate

$$\left\| \frac{\partial w_\varepsilon}{\partial \eta} \right\|_{L_4(Q_3)} \leqslant M_7.$$

Lemma 7 is proved.

To prove the boundedness of the derivatives of w_ε in Q_3 we need a result from [6].

In Chapter III of [6], in particular, linear equations of the form

$$\frac{\partial z}{\partial x} = \frac{\partial}{\partial \eta}\left(a(x, \eta)\frac{\partial x}{\partial \eta} \right) + b(x, \eta)\frac{\partial z}{\partial \eta} + c(x, \eta)z + f(x, \eta) \qquad (26)$$

with discontinuous unbounded coefficients are studied. For equations of the form (26) in [6] generalized solutions of class $V_2^{1,0}(P)$ are defined, where $V_2^{1,0}(P)$ is the Banach space of functions which are continuous in x in the L_2-norm and have a generalized derivative with respect to η in $L_2(P)$; P is the domain $\{0 < x < A, -\gamma < \eta < \gamma\}$. The following norm is defined in $V_2^{1,0}(P)$:

$$|z|_P = \max_{0 \leqslant x \leqslant A} \| z(x, \eta) \|_{L_2(-\gamma, \gamma)} + \left\| \frac{\partial z}{\partial \eta} \right\|_{L_2(P)}.$$

Theorem 7.1 of [6], Chapter III, asserts that if the coefficients of (26) possess the properties

$$0 < a_0 \leqslant a(x, \eta) \leqslant a_1, \qquad (27)$$

$$\| b^2(x, \eta), c(x, \eta), f(x, \eta) \|_{L_{q,r}(P)} \leqslant \mu, \qquad (28)$$

where

$$1/r + 1/2q = 1 - \kappa_1 \qquad (29)$$

$$\left(\text{here } q \in [1, \infty], r \in \left[\frac{1}{(1 - \kappa_1)}, \frac{2}{(1 - 2\kappa_1)} \right], 0 < \kappa_1 < \frac{1}{2} \right), \qquad (30)$$

then for any generalized solution $z(x, \eta)$ in $V_2^{1,0}(P)$ not exceeding k for $x = 0$ and $\eta = \pm\gamma$, the quantity $\operatorname{ess\,sup}_P z(x, \eta)$ is finite and is bounded above by a constant $c_1(k, a_0, \mu, q, r, \operatorname{mes} P)$. If conditions (27)–(30) are satisfied and $z(x, \eta) \geqslant -k$ on the boundary of P, then

$$\operatorname*{ess\,inf}_P z(x, \eta) \geqslant c_2(-k, a_0, \mu, q, r, \operatorname{mes} P).$$

We shall prove the following lemma.

LEMMA 8. *In the domain Q_3*

$$\left| \frac{\partial w_\varepsilon}{\partial x} \right| \leqslant M_{13}, \qquad \left| \frac{\partial w_\varepsilon}{\partial \eta} \right| \leqslant M_{14}, \qquad \left| \frac{\partial}{\partial \eta}\left((\nu\rho)^\varepsilon \frac{\partial w_\varepsilon}{\partial \eta} \right) \right| \leqslant M_{15}.$$

PROOF. We differentiate (18) with respect to x. The function $s_\varepsilon = \partial w_\varepsilon / \partial x$ satisfies the equation

$$\rho_\varepsilon \frac{\partial s_\varepsilon}{\partial x} = \frac{\partial}{\partial \eta}\left(\sqrt{w_\varepsilon}\,(\nu\rho)^\varepsilon \frac{\partial s_\varepsilon}{\partial \eta}\right) - \frac{(\nu\rho)^\varepsilon}{2\sqrt{w_\varepsilon}} \frac{\partial w_\varepsilon}{\partial \eta} \frac{\partial s_\varepsilon}{\partial \eta}$$

$$+ \frac{1}{2\sqrt{w_\varepsilon}} \frac{\partial}{\partial \eta}\left((\nu\rho)^\varepsilon \frac{\partial w_\varepsilon}{\partial \eta}\right) s_\varepsilon - 2\frac{d^2 p}{dx^2}. \tag{31}$$

We consider this equation as a linear equation for s_ε. It follows from Lemma 3 that the function s_ε belongs to $V_2^{1,0}(Q_3)$. By Lemmas 2 and 7 the coefficients of (31) satisfy the conditions (27)–(30) with $q = r = 2$ and $\kappa_1 = 1/4$. Equation (31) differs from equations of the form (26) by the presence of the coefficient $\rho^\varepsilon(\eta)$ of $\partial s_\varepsilon / \partial x$. Since ρ^ε does not depend on x, all the arguments of Theorem 7.1 in [6], Chapter III, remain valid for (31). The uniform boundedness of s_ε in ε on the boundary of Q_3 follows from Lemma 5 and the uniform boundedness of the function $d((\nu\rho)^\varepsilon\, dw_0^2/d\eta)/d\eta$ in ε. Lemma 8 is proved.

In investigating the smoothness of $(\nu\rho)^\varepsilon \partial w_\varepsilon / \partial \eta$ below we use the function classes $H_{(p_1,\dots,p_n)}^{(r_1,\dots,r_n)}(P; K)$ introduced in [8]. Here $r_i > 0$, $1 \leqslant p_i \leqslant \infty$ $(i = 1,\dots,n)$, and the domain $P \subset \mathbf{R}^n$, where $\mathbf{R}^n$ is n-dimensional Euclidean space; K is a constant bounding the norms of functions in $H_{(p_1,\dots,p_n)}^{(r_1,\dots,r_n)}(P; K)$. Theorem 2.4 of [8] asserts that if $f(x) \in H_{(p_1,\dots,p_n)}^{(r_1,\dots,r_n)}(K)$, then

$$f(x) \in H_q^{(\rho_1,\dots,\rho_n)}(\overline{K})$$

in $\mathbf{R}^n$ with some constant $\overline{K}$, while $p_i \leqslant q \leqslant \infty$ and $\rho_i = r_i\kappa/\kappa_i$, where

$$\kappa = 1 - \sum_{j=1}^{n} \frac{1}{r_j}\left(\frac{1}{p_j} - \frac{1}{q}\right), \qquad \kappa_i = 1 - \sum_{j=1}^{n} \frac{1}{r_j}\left(\frac{1}{p_j} - \frac{1}{p_i}\right).$$

It is further proved in [8] that Theorem 2.4 is also valid for funitons $f(x)$ defined in parallelepipeds $\{a_i \leqslant x_i \leqslant b_i,\ i = 1,\dots,n\}$.

LEMMA 9. *The function* $(\nu\rho)^\varepsilon \partial w_\varepsilon / \partial \eta$ *satisfies a Hölder condition in* Q_4 *with Hölder constant and exponent not depending on* ε.

PROOF. We differentiate (18) with respect to x, then multiply by $\zeta_3 \partial w_\varepsilon / \partial x$ and integrate over Q_3. We obtain

$$\iint_{Q_3} \rho^\varepsilon \zeta_3 \frac{\partial w_\varepsilon}{\partial x} \frac{\partial^2 w_\varepsilon}{\partial x^2}\, dx\, d\eta = \iint_{Q_3} \zeta_3 \frac{\partial w_\varepsilon}{\partial x} \sqrt{w_\varepsilon}\, \frac{\partial}{\partial \eta}\left((\nu\rho)^\varepsilon \frac{\partial^2 w_\varepsilon}{\partial x \partial \eta}\right) dx\, d\eta$$

$$+ \iint_{Q_3} \frac{\zeta_\varepsilon}{2\sqrt{w_\varepsilon}} \left(\frac{\partial w_\varepsilon}{\partial x}\right)^2 \frac{\partial}{\partial \eta}\left((\nu\rho)^\varepsilon \frac{\partial w_\varepsilon}{\partial \eta}\right) dx\, d\eta$$

$$- \iint_{Q_3} 2\frac{d^2 p}{dx^2} \zeta_3 \frac{\partial w_\varepsilon}{\partial x}\, dx\, d\eta$$

$$= I_5 + I_6 + I_7. \tag{32}$$

By Lemma 8 the integral on the left side of (32) and the integrals I_6 and I_7 are bounded uniformly in ε. We transform I_5 by integration by parts:

$$I_5 = -\iint_{Q_3} \zeta_3 \sqrt{w_\varepsilon}\, (\nu\rho)^\varepsilon \left(\frac{\partial^2 w_\varepsilon}{\partial x \partial \eta} \right)^2 dx\, d\eta$$

$$- \iint_{Q_3} \frac{\partial}{\partial \eta} \left(\zeta_3 \sqrt{w_\varepsilon} \right) \frac{\partial w_\varepsilon}{\partial x} (\nu\rho)^\varepsilon \frac{\partial^2 w_\varepsilon}{\partial x \partial \eta}\, dx\, d\eta. \tag{33}$$

Substituting (33) into (32) and using Lemma 8, we get

$$\left\| \frac{\partial^2 w_\varepsilon}{\partial x \partial \eta} \right\|_{L_2(Q_4)} \leqslant M_{16}. \tag{34}$$

It follows from Lemma 8 and (34) that $(\nu\rho)^\varepsilon \partial w_\varepsilon / \partial \eta \in H_{(2,\infty)}^{(1,1)}(Q_4; M_{17})$; that is, $n = 2$, $p_1 = 2$, $p_2 = \infty$ and $r_1 = r_2 = 1$. Therefore, $(\nu\rho)^\varepsilon \partial w_\varepsilon / \partial \eta$ belongs to the class $H_\infty^{(\rho_1, \rho_2)}(Q_4; M_{18})$ with $\rho_1 = 1/3$, $\rho_2 = 1$, and some constant M_{18} not depending on ε. Since $(\nu\rho)^\varepsilon \partial w_\varepsilon / \partial \eta$ has classical first derivatives in Q_4, it follows that $(\nu\rho)^\varepsilon \partial w_\varepsilon / \partial \eta \in C^\alpha(Q_4)$, $\alpha = 1/3$; here the Hölder constant does not depend on ε. Lemma 9 is proved.

THEOREM 2. *A solution of problem* (8)–(13) *possessing the properties indicated in Lemma 1 exists in the domain* $G_{A_0}\{0 < x < A_0,\ 0 < \psi < \infty\}$, *where* A_0 *is some constant. If* $dp/dx < 0$, *then a solution exists in* G_A *for any* $A > 0$.

PROOF. It follows from Lemmas 2–9 that we can extract a subsequence from the sequence w_ε that converges uniformly in G_{A_0} for $\psi \leqslant N$, where $N > 1$ is any number, and is such that the derivatives $\partial w_\varepsilon / \partial \psi$, $\partial w_\varepsilon / \partial x$, and $\partial^2 w_\varepsilon / \partial \psi^2$ converge uniformly for $1/N \leqslant \psi \leqslant \psi_*(x) - 1/N$, $\psi_*(x) + 1/N \leqslant \psi$, and the functions $(\nu\rho)^\varepsilon \partial w_\varepsilon / \partial \psi$ converge uniformly for $1/N \leqslant \psi \leqslant N$. It is obvious that the limit function $w(x, \psi)$ is continuous in $\overline{G}_{A_0}$; the functions w_i, which coincide with $w(x, \psi)$ in $\overline{G}_{A_0}^i$, are continuous in $\overline{G}_{A_0}^i$ $(i = 1, 2)$; w_1 satisfies (8) and (9) in $G_{A_0}^1$, and w_2 satisfies (10) and (11) in $G_{A_0}^2$. The junction conditions (12) and (13) are also satisfied. The remainder of the proof of Theorem 2 proceeds just as the proof of Theorem 2 in [4].

The validity of Theorem 1 follows from Lemma 1 and Theorem 2.

§2. The uniqueness theorems

We now establish uniqueness theorems for solutions of problems (1)–(6) and (8)–(13). We first prove a uniqueness theorem for the solution of problem (8)–(13) (in the von Mises plane x, ψ), and we obtain a uniqueness theorem for solutions of problem (1)–(6) as a corollary.

THEOREM 3. *Suppose that* $\{w_1(x, \psi),\ w_2(x, \psi)\}$ *is a solution of problem* (8)–(13) *in the domain* $G_A\{0 < x < A,\ 0 < \psi < \infty\}$ *such that the* w_i *are continuous in* G_A^i *and* $0 < w_i < C_1^2$ *for* $\psi > 0$ $(i = 1, 2)$, *and* $k_1\psi \leqslant w \leqslant k_2\psi$ *for* $0 \leqslant \psi \leqslant \psi_1$, *where* k_1, k_2, *and* ψ_1 *are constants; the derivatives* $\partial w_i / \partial \psi$ *and* $\partial w_i / \partial x$ *are bounded in* G_A^i $(i = 1, 2)$. *There can only be one solution of problem* (8)–(13) *possessing these properties.*

154 A. I. SUSLOV

PROOF. We suppose that in G_A there are two solutions of problem (8)–(13), $\{w_1, w_2\}$ and $\{\tilde{w}_1, \tilde{w}_2\}$. Then the difference $w_i - \tilde{w}_i$ in G_A^i satisfies the equation

$$\nu_i \rho_i \frac{\partial^2 (w_i - \tilde{w}_i)}{\partial \psi^2} - 2\rho_i \frac{\partial \left(\sqrt{w_i} - \sqrt{\tilde{w}_i} \right)}{\partial x} - 2\rho_i v_0(x) \frac{\partial \left(\sqrt{w_i} - \sqrt{\tilde{w}_i} \right)}{\partial \psi}$$

$$+ \frac{2}{\sqrt{w_i \tilde{w}_i}} \frac{dp}{dx} \left(\sqrt{w_i} - \sqrt{\tilde{w}_i} \right) = 0. \quad (35)$$

We multiply (35_1) by a smooth function $\Phi_1(x, \psi)$ which is zero for $\psi = \varepsilon$, $x = A - \varepsilon$, and we integrate over the domain $H_\varepsilon\{0 \leqslant x \leqslant A - \varepsilon, \varepsilon \leqslant \psi \leqslant \psi_*(x)\}$. We multiply (35_2) by a smooth function $\Phi_2(x, \psi)$ equal to zero for $\psi = 1/\varepsilon$, $x = A - \varepsilon$, and we integrate over the domain $H_\varepsilon^2\{0 \leqslant x \leqslant A - \varepsilon, \psi_*(x) \leqslant \psi \leqslant 1/\varepsilon\}$. We add the resulting equalities and transform the terms containing derivatives of the functions $(w_i - \tilde{w}_i)$ by integrating by parts. As a result,

$$\sum_{i=1}^{2} \left\{ \iint_{H_\varepsilon^i} \left[\nu_i \rho_i (w_i - \tilde{w}_i) \frac{\partial^2 \Phi_i}{\partial \psi^2} + 2\rho_i \left(\sqrt{w_i} - \sqrt{\tilde{w}_i} \right) \frac{\partial \Phi_i}{\partial x} \right. \right.$$

$$\left. + 2\rho_i v_0(x) \left(\sqrt{w_i} - \sqrt{\tilde{w}_i} \right) \frac{\partial \Phi_i}{\partial \psi} + \frac{2}{\sqrt{w_i \tilde{w}_i}} \frac{dp}{dx} \left(\sqrt{w_i} - \sqrt{\tilde{w}_i} \right) \Phi_i \right] dx\, d\psi$$

$$- \int_0^{A-\varepsilon} \nu_1 \rho_1 (w_1 - \tilde{w}_1) \frac{\partial \Phi_1}{\partial \psi} \Big|_{\psi=\varepsilon}^{\psi=\psi_*} dx - \int_0^{A-\varepsilon} \nu_2 \rho_2 (w_2 - \tilde{w}_2) \frac{\partial \Phi_2}{\partial \psi} \Big|_{\psi=\psi_*}^{\psi=1/\varepsilon} dx$$

$$+ \int_0^{A-\varepsilon} \left[\nu_1 \rho_1 \Phi_1 \frac{\partial (w_1 - \tilde{w}_1)}{\partial \psi} - \nu_2 \rho_2 \Phi_2 \frac{\partial (w_2 - \tilde{w}_2)}{\partial \psi} \right] \Big|_{\psi=\psi^*(x)} dx = 0. \quad (36)$$

Since $w_i(x, \psi)$ and $\tilde{w}_i(x, \psi)$ satisfy the junction conditions (12) and (13), for $\psi = \psi_*(x)$ we have

$$w_1 - \tilde{w}_1 = w_2 - \tilde{w}_2, \qquad \nu_1 \rho_1 \frac{\partial (w_1 - \tilde{w}_1)}{\partial \psi} = \nu_2 \rho_2 \frac{\partial (w_2 - \tilde{w}_2)}{\partial \psi}. \quad (37)$$

For the pair of functions Φ_1, Φ_2 we choose a solution of the problem

$$\nu_1 \rho_1 \left(\sqrt{w_1} - \sqrt{\tilde{w}_1} \right) \frac{\partial^2 \Phi_1}{\partial \psi^2} + 2\rho_1 \frac{\partial \Phi_1}{\partial x} + 2\rho_1 v_0(x) \frac{\partial \Phi_1}{\partial \psi}$$

$$+ \frac{2}{\sqrt{w_1 \tilde{w}_1}} \frac{dp}{dx} \Phi_1 = F \quad \text{in } H_\varepsilon^1, \quad (38)$$

$$\nu_2 \rho_2 \left(\sqrt{w_2} + \sqrt{\tilde{w}_2} \right) \frac{\partial^2 \Phi_2}{\partial \psi^2} + 2\rho_2 \frac{\partial \Phi_2}{\partial x} + 2\rho_2 v_0(x) \frac{\partial \Phi_2}{\partial \psi}$$

$$+ \frac{2}{\sqrt{w_2 \tilde{w}_2}} \frac{dp}{dx} \Phi_2 = F \quad \text{in } H_\varepsilon^2, \quad (39)$$

$$\Phi_1 \big|_{\omega=\varepsilon} = 0, \qquad \Phi_2 \big|_{\psi=1/\varepsilon} = 0, \qquad \Phi_i \big|_{x=A-\varepsilon} = 0 \quad (i = 1, 2), \quad (40)$$

$$\Phi_1\big|_{\psi=\psi_*}, \qquad \Phi_2\big|_{\psi=\psi_*}, \qquad \nu_1\rho_1\frac{\partial\Phi_1}{\partial\psi}\bigg|_{\psi=\psi_*} = \nu_2\rho_2\frac{\partial\Phi_2}{\partial\psi}\bigg|_{\psi=\psi_*}, \qquad (41)$$

where $F(x,\psi)$ is a smooth function defined in G_A and equal to zero near the lines $\psi = 0$, $x = 0$ and for $\psi \geq N$, where $N > 0$ is some number.

If a solution Φ_1^ε, Φ_2^ε of problem (38)–(41) exists, then from (36)–(41) we get

$$\iint_{H_\varepsilon^1 \cup H_\varepsilon^2}\left(\sqrt{w} - \sqrt{\tilde{w}}\right)F(x,\psi)\,dx\,d\psi + \int_{\psi=\varepsilon}\nu_1\rho_1(w_1 - \tilde{w}_1)\frac{\partial\Phi_1^\varepsilon}{\partial\psi}\,dx$$

$$-\int_{\psi=1/\varepsilon}\nu_2\rho_2(w_2 - \tilde{w}_2)\frac{\partial\Phi_2^\varepsilon}{\partial\psi}\,dx = 0, \quad (42)$$

where w and $\tilde{w}$ denote functions which coincide in $\overline{H}_\varepsilon^i$ with w_i and $\tilde{w}_i$ respectively.

If, moreover, the values of $\partial\Phi_1^\varepsilon/\partial\psi\big|_{\psi=\varepsilon}$ and $\partial\Phi_2^\varepsilon/\partial\psi\big|_{\psi=1/\varepsilon}$ are bounded uniformly in ε, then, since $w_1(x,\varepsilon) - \tilde{w}_1(x,\varepsilon) \to 0$ and $w_2(x,1/\varepsilon) - \tilde{w}_2(x,1/\varepsilon) \to 0$ as $\varepsilon \to 0$, in the limit (42) goes over into

$$\iint_{G_A}\left(\sqrt{w} - \sqrt{\tilde{w}}\right)F(x,\psi)\,dx\,d\psi = 0. \qquad (43)$$

Since $F(x,\psi)$ is an arbitrary compactly supported function in G_A, it follows from (43) that $w \equiv \tilde{w}$, and Theorem 3 is proved. It thus remains to show that a solution of problem (38)–(41) exists and possesses derivatives $\partial\Phi_1^\varepsilon/\partial\psi\big|_{\psi=2}$ and $\partial\Phi_2^\varepsilon/\partial\psi\big|_{\psi=1/\varepsilon}$ which are uniformly bounded in ε.

We make the change of independent variables $\tau = A - \varepsilon - x$, $\psi = \psi$ and consider the boundary value problem

$$L(\Phi) \equiv \left(\sqrt{w^h} + \sqrt{\tilde{w}^a}\right)\frac{\partial}{\partial\psi}\left((\nu\rho)^h\frac{\partial\Phi}{\partial\psi}\right)$$

$$+ 2\rho^h v_0(\tau)\frac{\partial\Phi}{\partial\psi} + \frac{2}{\sqrt{w^h\tilde{w}^h}}\frac{dp}{dx}\Phi = F \qquad (44)$$

in the domain $H_\varepsilon\{0 \leq \tau \leq A - \varepsilon, \varepsilon \leq \psi \leq 1/\varepsilon\}$ with boundary conditions

$$\Phi\big|_{\psi=\varepsilon} = 0, \qquad \Phi\big|_{\tau=0} = 0, \qquad \Phi\big|_{\psi=1/\varepsilon} = 0, \qquad (45)$$

where $(\nu\rho)^h(\eta)$ and $\rho^h(\eta)$ are smooth functions of the argument $\eta = \psi - \psi_*(\tau)$, uniformly bounded in h, such that $(\nu\rho)^h(\eta) = \nu_1\rho_1$ and $\rho^h(\eta) = \rho_1$ for $\eta \leq -h$, and $(\nu\rho)^h(\eta) = \nu_2\rho_2$ and $\rho^h(\eta) = \rho_2$ for $\eta \geq h$. We denote by w^h and $\tilde{w}^h$ the averages of the solutions $\{w_1, w_2\}$ and $\{\tilde{w}_1, \tilde{w}_2\}$ of problem (8)–(13) defined by the formula

$$w^h(\tau,\psi) = \frac{1}{h^2}\int_{\tau-h}^{\tau}\int_{\psi-h}^{\psi+h}\omega\left(\frac{|\psi - \xi|}{h}\right)w(\xi,t)\,d\xi\,dt,$$

where $\omega(|\xi|)$ is a smooth, nonnegative function which is equal to zero for $|\xi| \geq 1$ and such that

$$\int_{|\xi|\leq 1}\omega(|\xi|)\,d\xi = 1.$$

The coefficients of (44) for fixed h satisfy a Hölder condition in $\overline{H}_\varepsilon$. Since $F(\tau, \psi) = 0$ near the lines $\psi = \varepsilon$ and $\psi = 1/\varepsilon$, the consistency conditions of first order at the points $(0, \varepsilon)$ and $(0, 1/\varepsilon)$ are satisfied. Therefore there is a unique solution Φ_h^ε of problem (44), (45), and it belongs to $C^{2+\alpha}(\overline{H}_\varepsilon)$.

In Lemmas 10–12 proved below some properties of solutions Φ_h^ε of problem (44), (45) are established. These properties are such that from the sequence Φ_h^ε for $\varepsilon = \mathrm{const} > 0$ and $h \to 0$ it is possible to extract a subsequence which converges uniformly in $\overline{H}_\varepsilon$ to a continuous function Φ^ε. Moreover, the derivatives $\partial \Phi_h^\varepsilon / \partial \psi$, $\partial \Phi_h^\varepsilon / \partial \tau$, and $\partial^2 \Phi_h^\varepsilon / \partial \psi^2$ converge uniformly for $\varepsilon \leqslant \psi \leqslant \psi_*(x) - 1/N$, $\psi_*(x) + 1/N \leqslant \psi \leqslant N$, where N is any number greater than one. The function $(\nu\rho)^h \partial \Phi_h^\varepsilon / \partial \psi$ converges uniformly as $h \to 0$ for $\varepsilon \leqslant \psi \leqslant N$.

We set $\Phi_1^\varepsilon = \Phi^\varepsilon$ for $(x, \psi) \in H_\varepsilon^1$ and $\Phi_2^\varepsilon = \Phi^\varepsilon$ for $(x, \psi) \in H_\varepsilon^2$. The function Φ_1^ε satisfies (38) in H_ε^1, while Φ_2^ε satisfies (39) in H_ε^2. That the junction conditions are satisfied for $\psi = \psi_*(\tau)$ follows from the continuity of Φ^ε and Lemma 12. The derivatives $\partial \Phi_1^\varepsilon(x, \varepsilon)/\partial \psi$ and $\partial \Phi_2^\varepsilon(x, 1/\varepsilon)/\partial \psi$ are bounded uniformly in ε. Hence, Φ_1^ε and Φ_2^ε satisfy the conditions required for the proof of the uniqueness theorem.

LEMMA 10. *A solution Φ_h^ε of problem (44), (45) is bounded uniformly in ε and h in the region $\overline{H}_\varepsilon$. The derivatives $\partial \Phi_h^\varepsilon / \partial \psi$ for $\psi = \varepsilon$ and $\psi = 1/\varepsilon$ are bounded uniformly in ε and h.*

PROOF. In H_ε we consider the auxiliary function $r_\pm = \pm \Phi_h^\varepsilon - f(\tau, \psi)$, where $f(\tau, \psi) = \varphi(\psi)e^{\alpha(\tau+1)}$. We choose $\varphi(\psi)$ so that $\varphi = (\psi - \varepsilon) - (\psi - \varepsilon)^{4/3}$ for $\psi - \varepsilon \leqslant \delta$, $\varphi = \mathrm{const} > 0$ for $\psi \geqslant \psi_*(0)/2$, and $\varphi'(\psi) \leqslant 0$ for $\varepsilon \leqslant \psi \leqslant 1/\varepsilon$. We shall choose the positive constants α and δ later. We note that the condition $\varphi = \mathrm{const}$ for $\psi \geqslant \psi_*(0)/2$ implies the equality $\varphi'(\psi)\partial(\nu\rho)^h/\partial \psi \equiv 0$ in H_ε.

We consider $L(f)$, where L is the linear operator in (44). We have

$$L(f) = e^{\alpha(\tau+1)}\left[\left(\sqrt{w^h} + \sqrt{\tilde{w}^h}\right)(\nu\rho)^h \varphi'' - 2\rho^h \alpha \varphi\right.$$

$$\left. + 2\rho^h v_0(x)\varphi' + \frac{2}{\sqrt{w^h \tilde{w}^h}}\frac{dp}{dx}\varphi\right].$$

Here we have used the fact that $\varphi'\partial(\nu\rho)^h/\partial \psi \equiv 0$.

Let $0 \leqslant \psi - \varepsilon \leqslant \delta$. Since by hypothesis $w_1 \geqslant k_1\psi$ and $\tilde{w} \geqslant k_1\psi$ for small ψ, it follows that $w^h \geqslant k_1\psi$ and $\tilde{w}^h \geqslant k_1\psi$, and hence $(\sqrt{w^h} + \sqrt{\tilde{w}^h}) \geqslant 2\sqrt{k_1(\psi - \varepsilon)}$ for small $\psi \geqslant \varepsilon$. Thus for $0 \leqslant \psi - \varepsilon \leqslant \delta$

$$L(f) \leqslant e^{\alpha(\tau+1)}\left[-2k_1^{1/2}\frac{4}{9}(\nu\rho)^h(\psi - \varepsilon)^{-1/6} - 2\rho^h \alpha \varphi\right.$$

$$+ 2\rho^h v_0(x)\left(1 - \frac{4}{3}(\psi - \varepsilon)\right)$$

$$\left. + \frac{2}{k_1(\psi - \varepsilon)}\left|\frac{dp}{dx}\right|\left((\psi - \varepsilon) - (\psi - \varepsilon)^{4/3}\right)\right]. \quad (46)$$

Since the first term on the right side of (46) decreases without bound as $\psi - \varepsilon \to 0$ while the remaining terms are bounded, it follows that $L(f) < -\max |F(\tau, \psi)|$ for $0 \leqslant \psi - \varepsilon \leqslant \delta$ if δ is sufficiently small. Therefore, $L(r_\pm) = \pm L(\Phi) + L(f) = \pm F + L(f) < 0$ for $0 \leqslant \psi - \varepsilon \leqslant \delta$.

For $\psi - \varepsilon \geqslant \delta$ we have

$$L(f) \leqslant e^{\alpha(\tau+1)}\left[-2\rho^h \alpha \varphi + 2\rho^h v_0(x)\varphi' + \frac{2}{k_1\delta}\left|\frac{dp}{dx}\right||\varphi| \right]$$

$$< -\max |F(\tau, \psi)|,$$

if $\alpha = \alpha(\delta)$ is sufficiently large. Therefore, $L(r_\pm) < 0$ everywhere in H_ε.

Since $r_\pm \geqslant 0$ on the boundary of H_ε and the coefficient of $r_\pm$ in the expression for $L(r_\pm)$ is bounded above for each fixed ε, from the inequality $L(r_\pm) < 0$ in H_ε it follows that $r_\pm \geqslant 0$ in $\overline{H}_\varepsilon$. Therefore,

$$|\Phi_h^\varepsilon| \leqslant f(\tau, \psi) \quad \text{in } \overline{H}_\varepsilon. \tag{47}$$

We have proved that Φ_h^ε is uniformly bounded in ε and h.

From (47) and the fact that $\Phi_h^\varepsilon|_{\psi=\varepsilon} = 0$ it follows that $\partial\Phi_h^\varepsilon/\partial\psi$ is uniformly bounded in ε and h for $\psi = \varepsilon$. Since (44) is nondegenerate near the line $\psi = 1/\varepsilon$, from the fact that Φ_h^ε is uniformly bounded in ε and h it follows that $\partial\Phi_h^\varepsilon/\partial\psi$ for $\psi = 1/\varepsilon$ is bounded uniformly in ε and h. Lemma 10 is proved.

LEMMA 11. *In the domain* $\{0 \leqslant \tau \leqslant A - \varepsilon, \psi_*(\tau) - \Delta \geqslant \psi \geqslant \varepsilon, \psi_*(\tau) + \Delta \leqslant \psi \leqslant 1/\varepsilon\}$ *the solution* Φ_h^ε *of problem* (44), (45) *belongs to* $C^{2+\gamma}$ *and has the norm which does not depend on* h *but does depend on* ε *and* Δ.

This follows from well-known estimates of Schauder type for linear equations.

LEMMA 12. *In the domain* $\{0 \leqslant \tau \leqslant A - \varepsilon, \psi_*(\tau) - \Delta \leqslant \psi \leqslant \psi_*(\tau) + \Delta\}$ *the derivatives* $\partial\Phi_h^\varepsilon/\partial\psi$, $\partial\Phi_h^\varepsilon/\partial\tau$, *and* $\partial((v\rho)^\varepsilon\partial\Phi_h^\varepsilon/\partial\psi)/\partial\psi$ *are bounded by constants not depending on* h. *The function* $(v\rho)^h\partial\Phi_h^\varepsilon/\partial\psi$ *is Hölder continuous with a constant and exponent not depending on* h.

The assertions of Lemma 12 are proved in exactly the same way as Lemmas 7–9 of §2. The fact that the first derivatives of w^h and $\tilde{w}^h$ are uniformly bounded in h is used in the proof.

We obtain the uniqueness theorem for solutions of problem (1)–(6) as a corollary of Theorem 3.

THEOREM 4. *Let* u_1, v_1, u_2, v_2, $y_*(x)$ *be a solution of problem* (1)–(6) *such that* u_1 *and* v_1 *are continuous in* $\overline{D}_A^1$, u_2 *and* v_2 *are continuous in* $\overline{D}_A^2$, *and* $0 < u_1 < C_1$ *for* $y > 0$ ($i = 1, 2$); $k_1 y \leqslant u_1 \leqslant k_2 y$ *for* $0 \leqslant y \leqslant y_1$, *where* k_1, k_2, *and* y_1 *are constants; and the derivatives* $\partial u_i/\partial y$ *and* $\partial^2 u_i/\partial y^2$ *are bounded in* D_A^i ($i = 1, 2$). *There can only be one solution of problem* (1)–(6) *possessing these properties.*

§3. On a sufficient condition for the break-off of the boundary
layer of immiscible fluids

A sufficient condition for the break-off of the boundary layer of a homogeneous fluid was established in [9]. Now we establish a similar condition for a boundary layer with a surface of discontinuity.

We introduce a natural definition.

DEFINITION. Let A_0 be the supremum of A such that in the domain $D_A\{0 < x < A, 0 < y < \infty\}$ there exists a solution u_1, v_1, u_2, v_2, $y_*(x)$ of problem (1)–(6) possessing the properties $u_1 > 0$ for $y > 0$ and $\partial u_1(x,0)/\partial y > 0$. If $A_0 < \infty$, then we say that the boundary layer *breaks off*, and we call the point $x_s = A_0$ on the surface of the body around which the flow occurs the *break-off point*. If $A_0 = \infty$, we say that *break-free flow occurs*.

THEOREM 5. *Suppose that $\partial p/\partial x \geq 0$ and for some x_1 and x_2, $x_1 < x_2$,*

$$\int_{x_1}^{x_2} v_0(x)\, dx > \sqrt{6\nu_1 M_0(x_2 - x_1)},\tag{48}$$

where ν_1 is the viscosity of the injected fluid, and $M_0 = \sup u_0^2(y)$. Then the boundary layer breaks off in the domain D_A, where $A \leq x_2$.

PROOF. Problem (1)–(6) reduces to problem (8)–(13) by means of Lemma 1. It is easy to see that if in the domain $G_{x_2}\{0 < x < x_2, 0 < \psi < \infty\}$ there does not exist a positive solution of problem (8)–(13), then the reverse transition to the physical variables x, y is impossible, and hence the boundary layer breaks off in D_{x_2}.

We suppose that in G_{x_2} there exists a positive solution $\{w_1, w_2\}$ of problem (8)–(13). In the domain $G_{x_2}^1\{0 < x < x_2, 0 < \psi < \psi_*(x)\}$ the function w_1 satisfies the equation

$$\frac{\partial w_1}{\partial x} = \nu_1 \sqrt{w_1}\, \frac{\partial^2 w_1}{\partial \psi^2} - v_0(x)\frac{\partial w_1}{\partial \psi} - \frac{2}{\rho_1}\frac{dp}{dx},$$

and also the boundary conditions $w_1|_{\psi=0} = 0$ and $w_1|_{x=0} = w_{10}(\psi)$. We make a change of independent variables of the form $x = x$, $\xi = \psi - [\psi_*(x) - \psi_*(0)] = \psi - T(x)$. Obviously, $\xi(x, \psi_*(x)) = \psi_*(0) = $ const. With this change of variables the domain $G_{x_2}^1$ goes over into $\tilde{G}\{0 < x < x_2, -T(x) < \xi < \psi_*(0)\}$. The function $w_1(x, \xi)$ satisfies the equation

$$\frac{\partial w_1}{\partial x} = \nu_1 \sqrt{w_1}\, \frac{\partial^2 w_1}{\partial \xi^2} - \frac{2}{\rho_1}\frac{dp}{dx}\tag{49}$$

in $\tilde{G}$. Moreover, $w_1(x, -T(x)) = 0$ and $w_1(0, \xi) = w_{10}(\xi)$.

In the domain $Q_\delta\{0 < x < x_2 - \delta, \ -T(x) < \xi < -T(x_1)\}$ we consider the auxiliary function

$$\Phi_1(x, \xi) = \frac{(\xi + l)^2}{6\nu_1(x_2 - x)},$$

where

$$l = \int_0^{x_1} v_0(t)\, dt + \sqrt{6\nu M_0(x_2 - x_1)}\,.$$

In Q_δ the function $\Phi_1(x, \xi)$ satisfies the equation

$$\frac{\partial \Phi_1^2}{\partial x} = \nu_1 \Phi_1 \frac{\partial^2 \Phi_1^2}{\partial \xi^2}\,.$$

Since $Q_\delta \subset \tilde{G}$, it follows that $w_1(x, \xi)$ satisfies (49) everywhere in Q_δ. From the maximum principle it follows that in Q_δ

$$w_1(x, \xi) \leqslant \Phi_1^2(x, \xi) = \frac{(\xi + l)^4}{36 \nu_1^2 (x_2 - x)^2}\,. \tag{50}$$

By continuity, $w_1(x, \xi) \leqslant \Phi_1^2(x, \xi)$ in $Q_0\{0 < x < x_2,\ -T(x) < \xi < -T(x_1)\}$. If (48) is satisfied, then

$$\int_0^{x_2} v_0(t)\, dt = \int_0^{x_1} v_0(t)\, dt + \int_{x_1}^{x_2} v_0(t)\, dt > \int_0^{x_1} v_0(t)\, dt + \sqrt{6 \nu_1 M_0(x_2 - x_1)} = l.$$

This implies that some segment of the line $\xi = -l$ belongs to the domain $Q_0 \subset \tilde{G}$. Since $\Phi_1(x, -l) = 0$, it follows from (50) that $w_1(x, \psi) = 0$ at some points inside $\tilde{G}$, and hence $w_1(x, \psi) = 0$ at some points inside $G_{x_2}^1$. We have arrived at a contradiction, which proves the theorem.

In conclusion I wish to thank O. A. Oleĭnik for her assistance with the work.

Received 20/APR/74

Bibliography

1. G. G. Černyĭ, *Laminar motion of gas and liquid in a boundary layer with a surface of discontinuity*, Izv. Akad. Nauk SSSR Otdel. Tehn. Nauk **1954**, no. 12, 38–67. (Russian)

2. A. B. Vatažin, *On the injection of an electrically-conducting liquid or gas into a boundary layer in the presence of a magnetic field*, Prikl. Mat. Meh. **24** (1960), 909–911; English transl. in J. Appl. Math. Mech. **24** (1960).

3. A. V. Gotovcev, *Questions of separation of a magnetohydrodynamical boundary layer*, Author's Summary of Candidate's Dissertation, Izdat. Moskov. Gos. Univ., Moscow, 1973. (Russian)

4. O. A. Oleĭnik, *On a system of equations in boundary layer theory*, Ž. Vyčisl. Mat. i Mat. Fiz. **3** (1963), 489–507; English transl. in USSR Comput. Math. and Math. Phys. **3** (1963).

5. ______, *Boundary-value problems for linear elliptic and parabolic equations with discontinuous coefficients*, Izv. Akad. Nauk SSSR Ser. Math. **25** (1961), 3–20; English transl. in Amer. Math. Soc. Transl. (2) **42** (1964).

6. O. A. Ladyženskaja, V. A. Solonnikov and N. N. Ural'ceva, *Linear and quasilinear equations of parabolic type*, "Nauka", Moscow, 1967; English transl., Amer. Math. Soc., Providence, R.I., 1968.

7. O. A. Oleĭnik and S. N. Kružkov, *Quasilinear parabolic second-order equations with several independent variables*, Uspehi Math. Nauk **16** (1961), no. 5(101), 115–155; English transl. in Russian Math. Surveys **16** (1961).

8. S. M. Nikol'skiĭ, *Imbedding theorems for functions with partial derivatives considered in various metrics*, Izv. Akad. Nauk SSSR Ser. Mat. **22** (1958), 321–336; English transl. in Amer. Math. Soc. Transl. (2) **90** (1970).

9. A. I. Suslov, *On the effect of injection of boundary layer separation*, Prikl. Mat. Meh. **38** (1974), 166–169; English transl. in J. Appl. Math. Mech. **38** (1974).

Translated by J. R. SCHULENBERGER

Amer. Math. Soc. Transl.
(2) Vol. **118**, 1982

On the Solvability of the Cauchy Problem
for the Hopf Equation
Corresponding to
a Nonlinear Hyperbolic Equation*

M. I. VIŠIK AND A. I. KOMEČ

Contents

The nonlinear hyperbolic equation

$$\ddot{u}(x, t) = \Delta u(x, t) - f(u(x, t)), \qquad x \in T^n, t \in \mathbf{R}_t, \qquad (0.1)$$

where T^n is the n-dimensional torus, is equivalent to the Hamiltonian system

$$\begin{cases} \dot{u}(x, t) = \delta H/\delta p(x) \equiv p(x, t), \\ \dot{p}(x, t) = -\delta H/\delta u(x) \equiv \Delta u(x, t) - f(u(x, t)) \end{cases} \qquad (0.2)$$

with Hamiltonian

$$H(p, u) = \int_{T^n} \left[\frac{p^2(x)}{2} + \frac{(\nabla u(x))^2}{2} + F(u(x)) \right] dx, \qquad (0.3)$$

where $F'(u) = f(u)$. A *statistical solution* of the system (0.2) is a family of
measures $\mu_t(Q) = \mu_0(S_t^{-1}Q)$, where μ_0 is a given measure on the phase space $\mathfrak{M}$
of pairs of functions $(p(x), u(x)) = y(x)$ for which the energy integral (0.3) is
finite, $Q \subset \mathfrak{M}$, and $S_t y = S_t(p(x), u(x)) = (p(x, t), u(x, t))$ is the operator of

---

1980 *Mathematics Subject Classification*. Primary 35L15, 35R60; Secondary 35Q99.
* Translation of Trudy Sem. Petrovsk. **3** (1978), 19–42. MR **81** a 35087.

translation along the trajectories (solutions) of the system (0.2); $S_t^{-1}Q = \{y(x): S_t y \in Q\}$. This is one of the possible definitions of a statistical solution of the system (0.2), and it assumes the existence of a single-valued operator S_t of translation along the trajectories of this system.

The characteristic functional $\chi(t, \omega)$, $\omega = (\pi(x), \varphi(x))$ of the statistical solution $\mu_t(dy)$

$$\chi(t, \omega) = \int_{\mathfrak{M}} e^{i\langle \omega, y \rangle} \mu_t(dy), \qquad \langle \omega, y \rangle = \int_{T^n} \left[\pi(x)p(x) + \varphi(x)u(x) \right] dx$$

$$(0.4)$$

satisfies the Hopf equation

$$\frac{\partial \chi(t, \omega)}{\partial t} = i \int_{\mathfrak{M}} \left(\int_{T^n} \{ \pi(x)[\Delta u(x) - f(u(x))] + \varphi(x)p(x) \} \, dx \right) e^{i\langle \omega, y \rangle} \mu_t(dy)$$

$$= i \int_{T^n} \left\{ \pi(x) \left[\Delta_x \frac{\delta \chi(t, \omega)}{\delta(i\varphi(x))} - f \left(\frac{\delta}{\delta(i\varphi(x))} \right) \chi(t, \omega) \right] + \varphi(x) \frac{\delta \chi(t, \omega)}{\delta(i\pi(x))} \right\} dx,$$

$$(0.5)$$

where $\delta\chi/\delta\varphi(x)$ and $\delta\chi/\delta\pi(x)$ are the variational derivatives of the functional χ (see §2).

This equation was first derived and considered by Hopf [1] in the case of the system of Navier-Stokes equations.

In this paper (0.5) is studied directly. The solvability of the Cauchy problem for the Hopf equation (0.5) is proved in the case where the initial condition

$$\chi|_{t=0} = \chi(0, \omega) = \int_{\mathfrak{M}} e^{i\langle \omega, y \rangle} \mu_0(dy) \tag{0.6}$$

is the characteristic functional of a measure $\mu_0(dy)$ for which

$$\int_{\mathfrak{M}} H(y)\mu_0(dy) < +\infty. \tag{0.7}$$

The solution $\chi(t, \omega)$ of problem (0.5), (0.6) is for each t the characteristic functional of a measure $\mu_t(dy)$, i.e., (0.4) holds, and

$$\int_{\mathfrak{M}} H(y)\mu_t(dy) \leq C \int_{\mathfrak{M}} H(y)\mu_0(dy) + D. \tag{0.8}$$

From what has been said above it is natural to call the measure $\mu_t(dy)$ a *statistical solution* of the system (0.2) (cf. [2]).

We point out that the solution $\chi(t, \omega)$ of problem (0.5), (0.6) as well as the measure $\mu_t(dy)$ corresponding to it are constructed here without using the operator S_t of translation along the trajectories of (0.2). Moreover, under the conditions imposed here on the nonlinear function $f(u)$ it has so far not been established, in general, whether the solution of the Cauchy problem for (0.1) is unique, and it is therefore not known whether a single-valued operator S_t even exists [3].

The solution $\chi(t, \omega)$ of the Cauchy problem is constructed as the limit of solutions $\chi_m(t, \omega)$ of the Galerkin approximations of (0.5). It is proved that the family of functions $\{\chi_m(t, \omega)\}$, considered on the appropriate function space, is equicontinuous and uniformly bounded, and Arzelà's theorem can therefore be applied to it. It is established that the limit functional $\chi(t, \omega)$ is a solution of the Cauchy problem (0.5), (0.6) and possesses property (0.8). Here an essential role is played by Lemma 1.1 on the compactness of the imbedding of any bounded set in the space $\mathfrak{M}$ into a larger space $\tilde{\mathfrak{M}}$ introduced in §1 and Lemma 1.2 on the boundedness (uniformly in m) in $L_q(T^n)$, $q > 1$, of the operators P_m projecting functions $u(x) \in L_q(T^n)$ onto finite segments of their Fourier series.

The existence of a nontrivial stationary solution of the Hopf equation is proved in §4; that is, the existence of a functional $\chi(\omega)$ which satisfies (0.5) with $\partial\chi/\partial t \equiv 0$. $\chi(\omega)$ is the characteristic functional of a measure $\mu(dy)$ invariant with respect to the dynamical system (0.2).

§1. Phase spaces of nonlinear hyperbolic equations

We denote by T^n the n-dimensional torus with cyclic coordinates $x_1, \ldots, x_n$ defined modulo 2π. On T^n we consider the nonlinear equation (0.1) in the class of real measurable functions, where $f(u)$ is a real continuous function satisfying the conditions enumerated below. It is obvious that $x \in T^n$ means that $u(x, t)$ satisfies the periodicity conditions in x

$$u(x_1, \ldots, x_k + 2\pi, \ldots, x_n, t) = u(x_1, \ldots, x_k, \ldots, x_n, t)$$

for $1 \leqslant k \leqslant n$. Equation (0.1) is equivalent to the Hamiltonian system (0.2). The Hamiltonian (energy) H is preserved along trajectories of (0.2) which are twice continuously differentiable:

$$H(p(\cdot, t), u(\cdot, t)) = \text{const.} \tag{1.1}$$

As for $f(u)$ in (0.1), we assume the following. First of all, $f(u)$ is a real, continuous function of $u \in R$ which locally satisfies a Lipschitz condition and for some real C_0 and q, depending on f,

$$|f(u)| \leqslant C_0(|u| + 1)^{q-1}, \qquad u \in \mathbf{R}, q > 1. \tag{1.2}$$

The function $F(u) = \int_0^u f(z)\,dz + F(0)$ is then continuously differentiable and admits the upper bound

$$|F(u)| \leqslant C_1|u|^q + D_1, \qquad u \in \mathbf{R}. \tag{1.2'}$$

For brevity only, the case $0 < q \leqslant 1$ is not considered. The main results of the paper, Theorems 1 and 2, are also valid for these q. This can be proved by obvious modifications of the estimates.

Furthermore, it is assumed that $F(u)$ admits a lower bound analogous to (1.2') with the same exponent q:

$$F(u) \geqslant C_2|u|^q + D_2, \qquad u \in \mathbf{R}, \tag{1.3}$$

where $C_2 > 0$. The condition (1.3) ensures the boundedness below of the Hamiltonian (0.3), and because of (1.1) the solutions of (0.1) admit an a priori estimate.

A typical example of a function $f(u)$ satisfying the conditions formulated above is $f(u) = u|u|^{q-2}$ for $q \geqslant 2$ (it then satisfies a Lipschitz condition at $u = 0$). In this case $F(u) = |u|^q/q$. If q is a rational number with even numerator and odd denominator, then $u|u|^{q-2} = u^{q-1}$.

The main results of the paper are valid for $q < 4 + 6/(n-2)$ if $n \geqslant 3$ and for any q if $n = 1, 2$. We note that uniqueness of a weak solution for (0.1) has been proved (in our notation) only for $q \leqslant 2 + 2/(n-2)$ [3]. For these q and n the solvability of the Hopf equation and the existence of a statistical solution are obvious. It may be seen that the conditions we impose on $f(u)$ are considerably weaker. For example, if $n = 3$, here $q < 10$, and uniqueness is known [3] only for $q \leqslant 4$; if $n = 4$, here $q < 7$, while in [3] $q \leqslant 3$; as $n \to \infty$, here $q \leqslant 4$, while in [3] $q \leqslant 2$.

DEFINITION 1.1. The *phase space* $\mathfrak{M}$ of equation (0.1) (system (0.2)) is the set of pairs $(p(\cdot), u(\cdot))$ of real functions on the torus T^n with finite energy

$$\int_{T^n} \left[\frac{p^2(x)}{2} + \frac{(\nabla u(x))^2}{2} + F(u(x)) \right] dx = H(p, u) < \infty.$$

Here $\nabla u(x)$ is understood in the sense of generalized functions.

According to (1.2) and (1.3), a norm can be defined in $\mathfrak{M}$ by the following formula: for $y = (p, u)$

$$\|y\|_{\mathfrak{M}} = \|p\|_{L_2} + \|u\|_{\mathscr{H}_1} + \|u\|_{L_q} = \|p\|_0 + \|u\|_1 + \|u\|_{L_q}.$$

It is obvious that $\mathfrak{M} = L_2(T^n) \oplus [\mathscr{H}_1(T^n) \cap L_q(T^n)]$ is a linear, separable Banach space.[1] We observe that for $q \leqslant 2$ the term $\|u\|_{L_q} \leqslant C\|u\|_{L_2}$ is subordinate to the term $\|u\|_{\mathscr{H}_1}$: $\|u\|_{L_q} \leqslant C\|u\|_1$. In this case

$$\mathfrak{M} = L_2(T^n) \oplus \mathscr{H}_1(T^n).$$

It follows from (1.3) that

$$H(y) + H_0 \geqslant 0 \quad \text{for } y \in \mathfrak{M},$$

where

$$H_0 = - \inf_{y \in \mathfrak{M}} H(y) \leqslant - \int_{T^n} \left(\min_{u \in \mathbf{R}} F(u) \right) dx < \infty.$$

Moreover, the conditions on f imply the useful inequality

$$\|\nabla u\|_0^2 + \|p\|_0^2 + \|u\|_{L_q}^q \leqslant C(1 + H_0 + H(y)). \tag{1.3'}$$

---

[1] Here, as usual, $\mathscr{H}_s = \mathscr{H}_s(T^n)$ is the space of functions

$$u(x) = \sum_{\xi \in \mathbf{Z}^n} \tilde{u}(\xi) e^{i\langle \xi, x \rangle},$$

for which the following norm is finite:

$$\|u\|_s^2 = \sum_{\xi \in \mathbf{Z}^n} (1 + \xi \cdot \xi)^s |\tilde{u}(\xi)|^2 < \infty.$$

We shall also use the "weak" phase space $\tilde{\mathfrak{M}}$. We choose any $\delta > 0$.

DEFINITION 1.1'. We denote by $\tilde{\mathfrak{M}}$ the space

$$\mathcal{H}_{-\delta}(T^n) \oplus \left[\mathcal{H}_{1-\delta}(T^n) \cap L_{\tilde{q}}(T^n) \right], \quad \text{where } \tilde{q} = \max(q-1, 1),$$

$$\|y\|_{\tilde{\mathfrak{M}}} = \|p\|_{\mathcal{H}_{-\delta}} + \|u\|_{\mathcal{H}_{1-\delta}} + \|u\|_{L_{\tilde{q}}} = \|p\|_{-\delta} + \|u\|_{1-\delta} + \|u\|_{L_{\tilde{q}}}.$$

The space $\tilde{\mathfrak{M}}$, just as $\mathfrak{M}$, is a linear, separable Banach space consisting of real generalized functions on the torus T^n. The imbedding $\mathfrak{M} \subset \tilde{\mathfrak{M}}$ is continuous, and $\mathfrak{M}$ is dense in $\tilde{\mathfrak{M}}$. We equip $\mathfrak{M}$ and $\tilde{\mathfrak{M}}$ with Borel σ-algebras $B(\mathfrak{M})$ and $B(\tilde{\mathfrak{M}})$ of sets in $\mathfrak{M}$ and $\tilde{\mathfrak{M}}$ respectively. By definition, these σ-algebras are generated by open and closed subsets in $\mathfrak{M}$ and $\tilde{\mathfrak{M}}$ respectively. The following imbedding theorem plays an important role in the sequel.

LEMMA 1.1. *The imbedding* $\mathfrak{M} \subset \tilde{\mathfrak{M}}$ *is compact for* $q < 4 + 6/(n-2)$ *if* $n \geqslant 3$, *and for all* q *if* $n = 1, 2$.

PROOF. Suppose that the sequence $y_k \in \mathfrak{M}$ is bounded: $\|y_k\|_{\mathfrak{M}} \leqslant R < \infty$. It must be shown that from $\{y_k\}$ we can extract a subsequence $\{y_{\tilde{k}}\}$ which converges in $\tilde{\mathfrak{M}}$. This means that $y_{\tilde{k}} = (p_{\tilde{k}}, u_{\tilde{k}})$ where $\{p_{\tilde{k}}\}$ converges in $\mathcal{H}_{-\delta}(T^n)$ while $\{u_{\tilde{k}}\}$ converges in $\mathcal{H}_{1-\delta}(T^n)$ and in $L_{\tilde{q}}(T^n)$. Now $\|p_k\|_0 \leqslant R$, so that there exists a subsequence $\{p_{k'}\}$ converging in $\mathcal{H}_{-\delta}(T^n)$. Similarly, $\|u_{k'}\|_1 \leqslant R$, and from $\{u_{k'}\}$ it is therefore possible to extract a subsequence $\{u_{k''}\}$ converging in $\mathcal{H}_{1-\delta}(T^n)$. It remains to select from $\{u_{k''}\}$ a subsequence $\{u_{\tilde{k}}\}$ converging in $L_{\tilde{q}}(T^n)$.

If $q \leqslant 2$, then $\tilde{q} = 1$, and the existence of a subsequence $\{u_{\tilde{k}}\}$ converging in $L_1(T^n)$ follows from the compactness of the imbedding operator of $\mathcal{H}_1(T^n)$ in $L_1(T^n)$.

Suppose now that $q > 2$. It may be assumed with no loss of generality that the $y_k(x)$ are smooth functions, since such functions are dense in $\mathfrak{M}$. We choose γ to be rational with an odd denominator. Let $1 \leqslant \gamma \leqslant q/2 + 1$. Then $0 \leqslant 2\gamma - 2 \leqslant q$, and

$$\sum_{i=1}^{n} \int_{T^n} \left| \frac{\partial u_{k''}^{\gamma}}{\partial x_i} \right| dx \leqslant C \left[\int_{T^n} |u_{k''}|^{2\gamma - 2} dx + \sum_{i=1}^{n} \int_{T^n} \left| \frac{\partial u_{k''}}{\partial x_i} \right|^2 dx \right]$$

$$\leqslant C_1 [1 + R^q + R^2].$$

Therefore, by Sobolev's theorem the sequence $\{u_{k''}^{\gamma}\}$ is precompact in $L_{\beta}(T^n)$ for $1/\beta > 1 - 1/n$, i.e., from it a subsequence $\{u_{\tilde{k}}^{\gamma}\}$ can be selected which converges in $L_{\beta}(T^n)$. The sequence $\{u_{\tilde{k}}\}$ then converges in $L_{\gamma\beta}(T^n)$ if $\beta \geqslant 1$. Indeed, $\gamma \geqslant 1$, and so for real a and b

$$|a - b|^{\gamma} \leqslant C|a^{\gamma} - b^{\gamma}|.$$

This implies that

$$\int_{T^n} |u_{k_1}(x) - u_{k_2}(x)|^{\gamma\beta} dx \leqslant C \int_{T^n} |u_{k_1}^{\gamma}(x) - u_{k_2}^{\gamma}(x)|^{\beta} dx.$$

Therefore, $\{u_{\tilde{k}}\}$ converges in $L_{\gamma\beta}(T^n)$, while if $\gamma\beta \geqslant \tilde{q}$ it also converges in $L_{\tilde{q}}(T^n)$.

It thus remains to show that γ and β can be chosen so that $1 \leqslant \gamma \leqslant q/2 + 1$, $\beta \geqslant 1$, $1/\beta > 1 - 1/n$, and $\gamma\beta \geqslant \tilde{q}$.

For $n = 1$ this is obvious, since γ can be any number in $[1, q/2 + 1]$, while β may be taken so large that $\beta \geqslant 1$ and $\gamma\beta \geqslant \tilde{q}$.

Suppose that $n \neq 1$. We choose the extreme values allowed by the conditions on γ and β:

$$\gamma = q/2 + 1 \quad \text{and} \quad \beta = n/(n-1).$$

Then

$$\gamma\beta = (q/2 + 1)n/(n-1).$$

We shall prove that

$$(q/2 + 1)n/(n-1) > \tilde{q}.$$

Since $q \geqslant 2$, it follows that $\tilde{q} = q - 1$. However

$(q/2 + 1)n/(n-1) > q - 1$ is equivalent to $q < 4 + 6/(n-2)$.

The lemma is proved.

We shall use the Galerkin approximations of equation (0.1) To define them we introduce the projections P_m: $\tilde{\mathfrak{M}} \to \tilde{\mathfrak{M}}$ onto the eigensubspaces of the Laplace operator Δ, and we study the properties of these projections. The eigenfunctions of Δ are the exponential functions $e^{i\langle \xi, x\rangle}$ with "integral" frequencies $\xi \in \mathbf{Z}^n$ due to the periodic boundary conditions with period 2π in each coordinate:

$$\Delta e^{i\langle \xi, x\rangle} = -\xi^2 e^{i\langle \xi, x\rangle}, \qquad x \in T^n, \xi^2 = \sum_{i=1}^{n} \xi_i^2. \tag{1.4}$$

Any vector-valued generalized function

$$g(x) = (g_1(x), \ldots, g_\kappa(x)) \in D'(T^n) \otimes \mathbf{R}^\kappa$$

admits the Fourier series expansion

$$g(x) = \sum_{\xi \in \mathbf{Z}^n} \tilde{g}(\xi)e^{i\langle \xi, x\rangle}, \qquad x \in T^n,$$

where $\tilde{g}(\xi) \in \mathbf{C}^\kappa$. This series converges to $g(x)$ in the sense of generalized functions for any permutation of the terms. In particular, the partial sums

$$P_m g(x) = \sum_{|\xi| \leqslant m} \tilde{g}(\xi)e^{i\langle \xi, x\rangle},$$

where $|\xi| = \max_{1 \leqslant i \leqslant n} |\xi_i|$, converge to $g(x)$. We note that $\tilde{\mathfrak{M}} \subset D'(T^n) \otimes \mathbf{R}^2$.

DEFINITION 1.2. The operator P_m: $\tilde{\mathfrak{M}} \to \tilde{\mathfrak{M}}$ is defined by

$$P_m\left(\sum_\xi \tilde{y}(\xi)e^{i\langle \xi, x\rangle}\right) = \sum_{|\xi| \leqslant m} \tilde{y}(\xi)e^{i\langle \xi, x\rangle}.$$

We denote the space $P_m\tilde{\mathfrak{M}}$ by $\tilde{\mathfrak{M}}^m$.

For any fixed m the operator P_m is continuous in $\tilde{\mathfrak{M}}$ and in $\mathfrak{M}$, while $P_m\tilde{\mathfrak{M}} = \tilde{\mathfrak{M}}^m$. It follows from Lemma 1.2, proved below (see Corollary 1.2), that

$P_m y$ converges to y in the metric of $\mathfrak{M}$ for $y \in \mathfrak{M}$ and in the metric of $\tilde{\mathfrak{M}}$ for $y \in \tilde{\mathfrak{M}}$ as $m \to \infty$, and the norms of the operators P_m: $\mathfrak{M} \to \mathfrak{M}$ are bounded uniformly in m.

LEMMA 1.2. *If* $u \in L_q(T^n)$ *and* $q > 1$, *then for any* $n \geqslant 1$

$$\|P_m u\|_{L_q} \leqslant C_{q,n} \|u\|_q \tag{1.5}$$

and $P_m u \to u$ *in the metric of* $L_q(T^n)$ *as* $m \to \infty$. *The constant* $C_{q,n}$ *does not depend on* m.

PROOF. For $n = 1$ these assertions are proved in [4]. We assume that they hold for some $n \geqslant 1$ and demonstrate their validity for $n + 1$. For $x \in T^{n+1}$ we set $x' = (x_2, \ldots, x_{n+1})$, and for $\xi \in \mathbf{Z}^{n+1}$ we set $\xi' = (\xi_2, \ldots, \xi_{n+1})$. We define the operator $P_{1,m}$ in $C^\infty(T^{n+1})$ by

$$P_{1,m}\left(\sum_\xi \tilde{u}(\xi) e^{i\langle \xi, z \rangle} \right) = \sum_{\xi'} \sum_{|\xi_1| \leqslant m} \tilde{u}(\xi) e^{i\langle \xi, x \rangle}$$

and similarly the operator P'_m:

$$P'_m\left(\sum_\xi \tilde{u}(\xi) e^{i\langle \xi, x \rangle} \right) = \sum_{\xi_1} \sum_{|\xi'| \leqslant m} \tilde{u}(\xi) e^{i\langle \xi, x \rangle}.$$

Then $P_m = P'_m P_{1,m}$. By the induction hypothesis, for $u \in C^\infty(T^{n+1})$

$$\|P_m u\|_{L_q}^q = \int_{T^1} \left[\int_{T^n} |P'_m(P_{1,m} u(x_1, x'))|^q \, dx' \right] dx_1$$

$$\leqslant C \int_{T^1} \left[\int_{T^n} |P_{1,m} u(x_1, x')|^q \, dx' \right] dx_1 \leqslant C\|u\|_{L_q}^q. \, (^2)$$

Therefore, (1.5) holds for all n.

Further, for $u \in L_q(T^{n+1})$

$$\|P_m u - u\|_{L_q} \leqslant \|P_m u - P_{1,m} u\|_{L_q} + \|P_{1,m} u - u\|_{L_q}. \tag{1.6}$$

By the induction hypothesis, for almost every $x_1 \in T^1$

$$\|P'_m u(x_1, \cdot) - u(x_1, \cdot)\|_{L_q(T^n)} \underset{m \to \infty}{\to} 0.$$

Moreover, from (1.5) it follows that

$$\|P'_m u(x_1, \cdot) - u(x_1, \cdot)\|_{L_q} \leqslant \|P'_m u(x_1, \cdot)\|_{L_q} + \|u(x_1, \cdot)\|_{L_q}$$

$$\leqslant (1 + C)\|u(x_1, \cdot)\|_{L_q} = \Phi(x_1).$$

Since $\Phi(x_1) \in L_q(0, 2\pi)$, by Lebesgue's theorem

$$\int_{T^1} \|P'_m u(x_1, \cdot) - u(x_1, \cdot)\|_{L_q(T^n)}^q \, dx_1 \to 0.$$

$(^2)$ Here and henceforth different constants are sometimes denoted by the same letter C.

Hence

$$\|P_m u - P_{1,m} u\|_{L_q(T^{n+1})}^q = \int_{T^1} \|P_{1,m}[P_m' u(x_1, \cdot) - u(x_1, \cdot)]\|_{L_q(T^n)}^q \, dx_1$$

$$\leq C \int_{T_1} \|P_m' u(x_1, \cdot) - u(x_1, \cdot)\|_{L_q(T^n)}^q \, dx_1 \underset{m \to \infty}{\to} 0.$$

Thus the first term on the right side of (1.6) tends to zero as $m \to \infty$. Similarly, the second term also tends to zero. Lemma 1.2 is proved.

COROLLARY 1.1. *For $y \in \mathfrak{M}$ and all m*

$$H(P_m y) \leq CH(y) + D, \tag{1.7}$$

where C and D do not depend on m, and as $m \to \infty$

$$H(P_m y) \to H(y). \tag{1.8}$$

PROOF. By Parseval's equality

$$H(P_m y) = (2\pi)^n \sum_{|\xi| \leq m} \left[\frac{|\tilde{p}(\xi)|^2}{2} + \frac{\xi^2 |\tilde{u}(\xi)|^2}{2} \right] + \int_{T^n} F(P_m u(x)) \, dx. \tag{1.9}$$

From (1.2′) and (1.5) it follows that

$$\int_{T^n} F(P_m u(x)) \, dx \leq \int_{T^n} (C_1 |P_m u(x)|^q + D_1) \, dx \leq C' \int_{T^n} |u(x)|^q \, dx + D. \tag{1.10}$$

Recalling (1.3), from (1.9) and (1.10) we obtain (1.7)
We shall prove that

$$\int_{T^n} F(P_m u(x)) \, dx \to \int_{T^n} F(u(x)) \, dx \tag{1.11}$$

as $m \to \infty$, whence we obtain (1.8). Since $F'(u) = f(u)$, denoting $P_m u(x)$ by $u_m(x)$, we obtain

$$\left| \int_{T^n} [F(u_m(x)) - F(u(x))] \, dx \right| \leq \left| \int_{T^n} \left(\int_{u(x)}^{u_m(x)} f(z) \, dz \right) dx \right|$$

$$\leq \int_{T^n} |f(z')(u_m - u)| \, dx, \tag{1.12}$$

where $z' = u(x) + \theta[u_m(x) - u(x)]$, $0 < \theta < 1$. Because of (1.2), $|f(z')| \leq C(1 + |u|^{q-1} + |u_m|^{q-1})$. Therefore, according to the Hölder inequality, from (1.12) we obtain

$$\left| \int_{T^n} [F(u_m) - F(u)] \, dx \right| \leq \left(\int_{T^n} |f(z')|^{q/(q-1)} dx \right)^{(q-1)/q} \left(\int_{T^n} |u_m - u|^q \, dx \right)^{1/q}$$

$$\leq C_1 \left(\int_{T^n} (1 + |u|^q + |u_m|^q) \, dx \right)^{(q-1)/q} \cdot \left(\int_{T^n} |u_m - u|^q \, dx \right)^{1/q}. \tag{1.12′}$$

By Lemma 1.2 the first factor on the right in (1.12′) is bounded uniformly in m, while the second tends to zero as $m \to \infty$. This implies (1.11).

The next assertion is proved in an altogether analogous way.

COROLLARY 1.2. *For $y \in \mathfrak{M}$ and all m*

$$\|P_m y\|_{\mathfrak{M}} \leqslant C\|y\|_{\mathfrak{M}} \tag{1.13}$$

and $P_m y \to y$ in the metric of $\mathfrak{M}$ as $m \to \infty$.

Sets in $\mathfrak{M}$ of the form $P_m^{-1}Q$, where Q is a Borel set in $\mathfrak{M}^m$, are called *cylinder sets*. They form an algebra $A(\mathfrak{M})$. Similarly, $A(\tilde{\mathfrak{M}})$ is the algebra of cylinder sets in $\tilde{\mathfrak{M}}$.

LEMMA 1.3. *The space $\mathfrak{M}$ is Borel measurable in $\tilde{\mathfrak{M}}$, and every set that is a Borel set in $\mathfrak{M}$ is also a Borel set in $\tilde{\mathfrak{M}}$.*

PROOF. Obviously, $A(\mathfrak{M}) = \mathfrak{M} \cap A(\tilde{\mathfrak{M}})$. But the σ-algebra generated by $A(\mathfrak{M})$ coincides with $B(\mathfrak{M})$ (see [5]), since $\mathfrak{M}$ is a separable Banach space, and $\bigcup_1^\infty \mathfrak{M}^m = \mathfrak{M}^\infty$ is dense in $\mathfrak{M}$. Similarly, the σ-algebra spanned by $A(\tilde{\mathfrak{M}})$ coincides with $B(\tilde{\mathfrak{M}})$. It remains to prove that $\mathfrak{M} \in B(\tilde{\mathfrak{M}})$. Then $\mathfrak{M} \cap A(\tilde{\mathfrak{M}}) \subset B(\tilde{\mathfrak{M}})$, i.e. $A(\mathfrak{M}) \subset B(\tilde{\mathfrak{M}})$. Hence $B(\mathfrak{M}) \subset B(\tilde{\mathfrak{M}})$.

Let $y = (p, u) \in \tilde{\mathfrak{M}}$. We set $F^N(u) = F(u)$ for $F(u) \leqslant N$ and $F^N(u) = 0$ for $F(u) > N$. We note that $y \in \mathfrak{M}$ if and only if there exists an $R < \infty$ such that

$$H^N(y) = (2\pi)^n \sum_{|\xi| \leqslant N} \left[\frac{|\tilde{p}(\xi)|^2}{2} + \frac{\xi^2 |\tilde{u}(\xi)|^2}{2} \right] + \int_{T^n} F^N(u(x))\, dx \leqslant R$$

$$\tag{1.14}$$

for all $N > 0$. Therefore

$$\mathfrak{M} = \bigcup_{R=1}^{\infty} \bigcap_{N=1}^{\infty} \tilde{\mathfrak{M}}_N(R), \tag{1.14′}$$

where $\tilde{\mathfrak{M}}_N(R) = \{y \in \tilde{\mathfrak{M}} \,|\, H^N(y) \leqslant R\}$, and N and R are integers. Now $\tilde{\mathfrak{M}}_N(R) \in B(\tilde{\mathfrak{M}})$, since H^N is a continuous function on $\tilde{\mathfrak{M}}$. From (1.14′) it therefore follows that $\mathfrak{M} \in B(\tilde{\mathfrak{M}})$. The lemma is proved.

REMARK 1.1. The functions $H^N(y)$ are continuous on $\tilde{\mathfrak{M}}$, and

$$H^N(y) \xrightarrow[N \to \infty]{} \begin{cases} H(y) & \text{for } y \in \mathfrak{M}, \\ +\infty & \text{for } y \in \tilde{\mathfrak{M}} \setminus \mathfrak{M}. \end{cases} \tag{1.15}$$

We have thus defined the projections $P_m: \tilde{\mathfrak{M}} \to \mathfrak{M}^m$. We write (0.1) and (0.2) in the form

$$\dot{y} = V(y),$$

where $V(y) = (\Delta u - f(u), p)$ for $y = (p, u)$. The Galerkin approximations of (0.1) are then given by the equation on $\mathfrak{M}$

$$\dot{y} = P_m V(P_m y). \tag{1.16}$$

It is also a Hamiltonian system with Hamiltonian $H_m(y) = H(P_m y)$. Because of (1.4), in expanded form the Galerkin approximations have the form

$$\begin{cases} \dot{u} = P_m p, \\ \dot{p} = \Delta P_m u - P_m f(P_m u), \qquad (P_m \Delta P_m u = \Delta P_m u). \end{cases} \qquad (1.16')$$

They are equivalent to the equation

$$\ddot{u} = \Delta P_m u - P_m f(P_m u). \qquad (1.16'')$$

We denote the vector field $P_m V(P_m y)$ on $\mathfrak{M}$ by $V_m(y)$. It is tangent to the space $\mathfrak{M}^m$ (and is even parallel to it) and locally satisfies on this space a Lipschitz condition, since f satisfies a Lipschitz condition. We consider (1.16) as an equation on $\mathfrak{M}^m$. The Hamiltonian $H_m(y)$ on $\mathfrak{M}^m$ coincides with $H(y)$. It is preserved along the trajectories of (1.16).

PROPOSITION 1.1. *A solution of the Galerkin equations* (1.16) *exists and is unique for all t for any initial conditions in $\mathfrak{M}^m$.*

PROOF. The local (in t) existence and uniqueness of the trajectories follow from the Lipschitz condition. Global existence follows from the fact that the field $V_m(y)$ on $\mathfrak{M}^m$ has a first integral $H_m(y) \equiv H(y)$ for $y \in \mathfrak{M}^m$, and $H(y) \to \infty$ as $\|y\| \to \infty$, $y \in \mathfrak{M}^m$, where $\|y\|$ is any norm on $\mathfrak{M}^m$. Indeed, $\|y\| \to \infty$ implies that the Euclidean norm of the Fourier coefficients of $y = (p(x), u(x))$ (they are finite in number) tends to infinity. By (1.3'), then also $H(y) \to \infty$.

Thus, the Galerkin approximations (1.16) determine a phase flow $S_m(t)$ on $\mathfrak{M}^m$. We note that these equations also determine a phase flow on the entire space $\mathfrak{M}$ by means of parallel translation of the phase flow already constructed.

For $y_0 \in \mathfrak{M}^m$, all m, and all $t \in \mathbf{R}$

$$H(S_m(t)y_0) = H(y_0). \qquad (1.17)$$

§2. Statistical solutions of equation (0.1) and the Hopf equation corresponding to it

In this section we define a statistical solution and the Hopf equation for the nonlinear equation (0.1).

We suppose that some vector field $W(y)$ on $\mathfrak{M}$ defines in $\mathfrak{M}$ a continuous phase flow $S(t)$:

$$\frac{d(S(t)y)}{dt} \equiv W(S(t)y), \qquad S(0)y \equiv y, \qquad (2.0)$$

where $S(t)y$ depends continuously on $y \in \mathfrak{M}$ in the metric of $\mathfrak{M}$, and the derivative in (2.0) exists in $\mathfrak{M}$. For example, the Galerkin field $W(y) \equiv V_m(y)$ satisfies these assumptions.

Let μ_0 be some Borel probability measure on $\mathfrak{M}$: $\mu_0(\mathfrak{M}) = 1$. We set $\mu_t(Q) = \mu_0(S^{-1}(t)Q)$ for $Q \in B(\mathfrak{M})$. We denote by $\mathfrak{M}^*$ the dual space of $\mathfrak{M}$. The characteristic function of a measure μ_t is its Fourier transform: for $\omega \in \mathfrak{M}^*$

$$\chi(t, \omega) = \tilde{\mu}_t(\omega) \equiv \int_{\mathfrak{M}} e^{i\langle \omega, y \rangle} \mu_t(dy) = \int_{\mathfrak{M}} e^{i\langle \omega, S(t)y_0 \rangle} \mu_0(dy_0). \qquad (2.1)$$

Here

$$\langle \omega, y \rangle = \langle \pi(x), p(x) \rangle + \langle \varphi(x), u(x) \rangle,$$

$\langle \, , \, \rangle$ is the duality relation, and $\pi(x)$ and $\varphi(x)$ are generalized functions on T^n. By the Lebesgue theorem the characteristic functional $\chi(t, \omega)$ is continuous in the weak topology on $\mathfrak{M}^*$. Therefore, it is uniquely determined by its values on any subset $\mathfrak{M}' \subset \mathfrak{M}^*$ provided that $\mathfrak{M}'$ is dense in $\mathfrak{M}^*$ in the weak topology. Knowing $\chi(t, \omega)$ for $\omega \in \mathfrak{M}'$, we can obtain it on the entire space $\mathfrak{M}^*$ by extension by weak continuity. For example, as $\mathfrak{M}'$ it is possible to take $\mathfrak{M}^\infty = \bigcup_k \mathfrak{M}^k$, since $\mathfrak{M}^\infty$ is weakly dense in $\mathfrak{M}^*$. Indeed, if $\omega \in \mathfrak{M}^*$, then $P_k \omega \to \omega$ in $\mathfrak{M}^*$. The point is that for $y \in \mathfrak{M}$, according to Corollary 1.2, $P_k y \to^{\mathfrak{M}} y$ as $k \to \infty$. Therefore,

$$\langle P_k \omega, y \rangle = \langle \omega, P_k y \rangle \underset{k \to \infty}{\to} \langle \omega, y \rangle.$$

We derive an equation which the functional $\chi(t, \omega)$ satisfies on the basis of the equation of motion (2.0). For this we assume that the differentiations performed below are legitimate for $\omega \in \mathfrak{M}' \subset \mathfrak{M}^*$. Differentiating (2.1) with respect to t and using (2.0), we find that

$$\dot{\chi}(t, \omega) = \frac{\partial \chi(t, \omega)}{\partial t} = \int_{\mathfrak{M}} e^{i\langle \omega, S(t) y_0 \rangle} i \left\langle \omega, \frac{dS(t) y_0}{dt} \right\rangle \mu_0(dy_0)$$

$$= i \int_{\mathfrak{M}} e^{i\langle \omega, y \rangle} \langle \omega, W(y) \rangle \mu_t(dy), \qquad \omega \in \mathfrak{M}'. \tag{2.2}$$

The last integral is uniquely determined by χ, since the measure $\mu_t(dy)$ is uniquely determined by $\chi(t, \omega)$, $\omega \in \mathfrak{M}'$. We denote the last integral in (2.2) by $H_W \chi(t, \omega)$. The evolution equation thus obtained for the characteristic functional χ

$$\dot{\chi}(t, \omega) = i H_W \chi(t, \omega), \qquad \omega \in \mathfrak{M}', \tag{2.3}$$

is called the Hopf equation([3]) corresponding to the dynamical system (2.0) on $\mathfrak{M}$.

([3]) In analogy with the theory of pseudodifferential operators the Hopf equation (2.3) can conveniently be written in the form

$$\dot{\chi}(t, \omega) = i \left\langle \omega, W\left(\frac{\delta}{i\delta\omega} \right) \right\rangle \chi(t, \omega) = i \int_{T^n} \omega(x) W\left(\frac{\delta}{i\delta\omega(x)} \right) \chi(t, \omega) \, dx$$

and called a pseudodifferential equation, while H_W is called a pseudodifferential operator (cf. [6]). If, for example, $W(y) \equiv y$, then the Hopf equation (2.3) under the condition

$$\int_{\mathfrak{M}} \|y\|_{\mathfrak{M}} \mu_t(dy) \leqslant \text{const} < \infty$$

is equivalent to the equation in variational derivatives

$$\dot{\chi}(t, \omega) = \langle \omega, \delta_\omega \chi(t, \omega) \rangle = \int_{T^n} \left[\pi(x) \frac{\delta\chi(t, \pi, \varphi)}{\delta\pi(x)} + \varphi(x) \frac{\delta\chi(t, \pi, \varphi)}{\delta\varphi(x)} \right] dx,$$

where $\delta_\omega \chi(t, \omega)$ is, for example, the Gâteaux variation of χ at the point $\omega = (\pi, \varphi)$.

We shall study the Hopf equation in more detail for the case $W(y) \equiv V(y)$.

For example, for the Galerkin vector field $W(y) \equiv V_m(y)$ the measures μ_t^m are determined by the relation $\mu_t^m(Q) = \mu_0(S_m^{-1}(t)Q)$ (see §3), and the Hopf equation has the form

$$\dot{\chi}_m(t, \omega) = iH_{V_m}\chi_m(t, \omega) \equiv i\int_{\mathfrak{M}} e^{i\langle \omega, y\rangle} \langle \omega, V_m(y)\rangle \mu_t^m(dy). \qquad (2.4)$$

LEMMA 2.1. *For the Galerkin field* $W(y) \equiv V_m(y)$ *the Hopf equation* (2.4) *is valid for all* $\omega \in \mathfrak{M}^*$ *if the measure* $\mu_0 = \mu_0^m$ *is concentrated on* $\mathfrak{M}^m$ *and has finite mean energy*

$$\int_{\mathfrak{M}} H(y)\mu_0^m(dy) < \infty. \qquad (2.5)$$

PROOF. Like μ_0^m, the Galerkin measures $\mu_t^m(Q) = \mu_0^m(S_m^{-1}(t)Q)$ are also concentrated on $\mathfrak{M}^m$ and have finite mean energy:

$$\int_{\mathfrak{M}} H(y)\mu_t^m(dy) = \int_{\mathfrak{M}^m} H(S_m(t)y_0)\mu_0^m(dy_0) = \int_{\mathfrak{M}} H(y_0)\mu_0^m(dy_0) < \infty. \quad (2.6)$$

Here the energy integral (1.17) has been used. Next, the integrals in (2.2) for $W \equiv V_m$ extend over $\mathfrak{M}^m$, since the μ_t^m are concentrated on $\mathfrak{M}^m$. Now $V_m(y)$ for $y \in \mathfrak{M}^m$ is a finite-dimensional vector field on the finite-dimensional vector space $\mathfrak{M}^m$. Let $\| \cdot \|$ be any norm in $\mathfrak{M}^m$. Then, because of (1.16′), (1.2), and (1.3′), for $y = (p, u) \in \mathfrak{M}^m$ and any $\omega \in \mathfrak{M}^*$

$$|\langle \omega, W(y)\rangle| = |\langle P_m\omega, V_m(y)\rangle| \leqslant C_m(\omega)\left(\|\nabla u\| + \|u\|^{q-1} + 1 + \|p\|\right)$$

$$\leqslant C_m'(\omega)(1 + H_0 + H(y))^\rho, \qquad \rho < 1.$$

Therefore, the convergence of the last integral in (2.2) for $W = V_m$ and $\mu_0 = \mu_0^m$, $\omega \in \mathfrak{M}^*$, uniformly in t, follows from (2.6). The lemma is proved.

Under the hypotheses of Lemma 2.1 the measures μ_t^m are concentrated on $\mathfrak{M}^m$. Since $P_m y = y$ for $y \in \mathfrak{M}^m$, (2.4) can be rewritten in the form

$$\dot{\chi}_m(t, \omega) = i\int_{\mathfrak{M}} e^{i\langle \omega, y\rangle} \langle P_m\omega, V(y)\rangle \mu_t^m(dy), \qquad (2.4')$$

since the measures μ_t^m are concentrated on $\mathfrak{M}^m$. This equality is used in an essential way below.

In analogy with (2.3) we make the following definition:

DEFINITION 2.1. The *Hopf equation for the nonlinear equation* (0.1) is the evolution equation

$$\dot{\chi}(t, \omega) = iH_V\chi(t, \omega) \equiv i\int_{\mathfrak{M}} e^{i\langle \omega, y\rangle} \langle \omega, V(y)\rangle \mu_t(dy), \qquad \omega \in \mathfrak{M}^\infty. \,(^4)$$

$$(2.7)$$

$(^4)$ If, for example, $f(u) = u^k$, where k is a natural number, then the Hopf equation can be written as an equation in variational derivatives (here $(\pi, \varphi) = \omega$)

$$\dot{\chi}(t, \pi, \varphi) = i\int_{T^n} \left[\pi(x)\left(\Delta_x \frac{\delta\chi(t, \pi, \varphi)}{i\delta\varphi(x)} - \frac{\delta^k\chi(t, \pi, \varphi)}{(i\delta\varphi(x))^k}\right) + \varphi(x)\frac{\delta\chi(t, \pi, \varphi)}{i\delta\pi(x)}\right] dx.$$

Here we restrict ourselves to $\omega \in \mathfrak{M}^\infty$ in contrast to (2.4), because the right side of (2.7) is not meaningful for all $\omega \in \mathfrak{M}^*$. The situation is that, generally speaking, $V(y)$ does not belong to $\mathfrak{M}$ if $y \in \mathfrak{M}$. Therefore, the symbol $\langle \omega, V(y) \rangle$ occurring in (2.7) is not defined for all $(\omega, y) \in \mathfrak{M}^* \times \mathfrak{M}$, and the right side of the Hopf equation (2.7) is not defined for all $\omega \in \mathfrak{M}^*$.

We shall found conditions on ω under which the symbol $\langle \omega, V(y) \rangle$ is defined and continuous for all $y \in \mathfrak{M}$. If $\omega = (\pi, \varphi)$ and $y = (p, u)$, then

$$\langle \omega, V(y) \rangle = \langle \pi, \Delta u - f(u) \rangle + \langle \varphi, p \rangle. \tag{2.8}$$

For the first term in (2.8) to be defined and continuous in y on $\mathfrak{M}$, it suffices that $\pi \in \mathcal{H}_1 \cap L_q$, since $f(u(x)) \in L_{q/(q-1)}$ because of (1.2) and the continuity of $f(u)$. For the continuity of the second term in y on $\mathfrak{M}$ it is necessary and sufficient that $\varphi \in L_2$. Thus, $\langle \omega, V(\cdot) \rangle$ is a continuous function on $\mathfrak{M}$ for $\omega \in [\mathcal{H}_1 \cap L_q] \oplus L_2$. The space of such ω is naturally denoted by $\mathfrak{M}^{\mathrm{tr}}$ ($\mathfrak{M}$ transposed), since

$$\mathfrak{M} = L_2 \oplus [\mathcal{H}_1 \cap L_q].$$

It is obvious that $\mathfrak{M}^{\mathrm{tr}} \subset \mathfrak{M}^*$. In $\mathfrak{M}^{\mathrm{tr}}$ we introduce the norm

$$\|\omega\|_{\mathfrak{M}^{\mathrm{tr}}} = \|\pi\|_1 + \|\pi\|_{L_q} + \|\varphi\|_{L_2}.$$

PROPOSITION 2.1. *The symbol* $\langle \omega, V(y) \rangle$ *is continuous separately in* ω *and* y *on* $\mathfrak{M}^{\mathrm{tr}} \times \mathfrak{M}$. *Moreover,*

$$|\langle \omega, V(y) \rangle| \leqslant C \|\omega\|_{\mathfrak{M}^{\mathrm{tr}}} \left(1 + \|\nabla u\|_0 + \|u\|_{L_q}^{q-1} + \|p\|_0 \right)$$
$$\leqslant C_0 \|\omega\|_{\mathfrak{M}^{\mathrm{tr}}} (1 + H_0 + H(y)). \tag{2.9}$$

PROPOSITION 2.2. *For* $\omega \in \mathfrak{M}^{\mathrm{tr}}$ *the projections* $P_k \omega$ *as* $k \to \infty$ *converge to* ω *in the metric of* $\mathfrak{M}^{\mathrm{tr}}$.

This fact is established in the same way as Corollary 1.2.

We shall investigate the solvability of the Cauchy problem for the Hopf equation with the initial condition $\chi(0, \omega) = \tilde{\mu}_0(\omega)$ (see (2.1)), $\omega \in \mathfrak{M}^*$. In integral form this Cauchy problem for (2.7) has the form

$$\chi(t, \omega) - \chi(0, \omega) = i \int_0^t H_V \chi(\tau, \omega) d\tau = i \int_0^t \left(\int_{\mathfrak{M}} e^{i \langle \omega, y \rangle} \langle \omega, V(y) \rangle \mu_\tau(dy) \right) d\tau. \tag{2.10}$$

Thus, suppose that the function $f(u)$ in (0.1) locally satisfies a Lipschitz condition and also satisfies (1.2) and (1.3). The next result is the main result of the paper.

THEOREM 1. *Let μ_0 be a Borel probability measure on $\mathfrak{M}$ which has finite mean energy*

$$\int_{\mathfrak{M}} H(y)\mu_0(dy) = E < \infty. \tag{2.11}$$

Then the following assertions are true:

(I) *The Cauchy problem for the Hopf equation* (2.7) *with the initial condition* $\chi(0, \omega) = \tilde{\mu}_0(\omega)$ *is solvable for* $q < 4 + 6/(n-2)$ *if* $n \geqslant 3$, *and for any* q *if* $n = 1$ *or* 2.

(II) *The Hopf equation* (2.7) *is satisfied for all* $\omega \in \mathfrak{M}^{\mathrm{tr}}$.

(III) *The measures* μ_t *on* $\mathfrak{M}$ *corresponding to the characteristic functionals* $\chi(t, \cdot)$ *have finite mean energy*

$$E_t = \int_{\mathfrak{M}} H(y)\mu_t(dy) \leqslant CE + D. \tag{2.12}$$

We shall prove this theorem in the next section.

DEFINITION 2.1' (cf. [2]). If $\chi(t, \omega)$ is a solution of the Hopf (2.7), then the collection of measures μ_t in (2.7) for which $\chi(t, \omega)$ are characteristic functionals is called a *statistical solution* of (0.1).

§3. Proof of the solvability of the Cauchy problem for the Hopf equation

The proof of Theorem 1 is given in this section. The solution $\chi(t, \omega)$ of the Cauchy problem for the Hopf equation is obtained as the limit of the Galerkin approximations $\chi_m(t, \omega)$. Here we make use of Arzelà's theorem. The Galerkin approximations are constructed as follows.

We project the measure μ_0 onto $\mathfrak{M}^m$: for $Q_m \in B(\mathfrak{M}^m) = \mathfrak{M}^m \cap B(\mathfrak{M})$ we set

$$\bar{\mu}_0^m(Q_m) = \mu_0(P_m^{-1}Q_m). \tag{3.1}$$

We define the measures $\bar{\mu}_t^m$ on $\mathfrak{M}^m$ by

$$\bar{\mu}_t^m(Q_m) = \bar{\mu}_0^m(S_m^{-1}(t)Q_m). \tag{3.2}$$

The Galerkin measures μ_t^m on $\mathfrak{M}$ are defined as follows: for $Q \in B(\mathfrak{M})$

$$\mu_t^m(Q) = \bar{\mu}_t^m(Q \cap \mathfrak{M}^m). \tag{3.3}$$

We may write symbolically

$$\mu_t^m = \bar{\mu}_t^m \times \delta(y - P_m y).$$

The measures μ_t^m are concentrated on $\mathfrak{M}^m$, and by (3.1)–(3.3) we have, as in §2,

$$\mu_t^m(Q) = \mu_0^m(S_m^{-1}(t)Q). \tag{3.4}$$

By Lemma 1.3 the measure μ_t^m is a Borel measure in $\tilde{\mathfrak{M}}$ as well, and (3.3) and (3.4) are also valid for $Q \subset B(\tilde{\mathfrak{M}})$. As in §2, for $\omega \in \mathfrak{M}*$

$$\chi_m(t, \omega) = \tilde{\mu}_t^m(\omega) \equiv \int_{\mathfrak{M}^m} e^{i\langle \omega, y \rangle}\mu_t^m(dy).$$

Equality (2.4') holds for $\omega \in \mathfrak{M}*$, since the measure μ_0^m has finite mean energy.

LEMMA 3.1. *The Galerkin measures μ_t^m have finite and bounded mean energy. For all m and t*

$$E_{m,t} = \int_{\mathfrak{M}} H(y)\mu_t^m(dy) \leqslant CE + D. \tag{3.5}$$

PROOF. According to (1.17) and (3.1)–(3.3),

$$E_{m,t} = \int_{\mathfrak{M}^m} H(S_m(t)y_0)\mu_0^m(dy_0) = \int_{\mathfrak{M}^m} H(y_0)\mu_0^m(dy_0)$$

$$= \int_{\mathfrak{M}} H(P_m y_0)\mu_0(dy_0).$$

It therefore follows from (1.7) that $E_{m,t} \leqslant CE + D$. The lemma is proved.

Let $\omega \in \mathfrak{M}^k$ and $k < m$. Then $P_m\omega = \omega$, and (2.4′) takes the form

$$\dot\chi_m(t, \omega) = i\int_{\mathfrak{M}} e^{i\langle\omega, y\rangle}\langle\omega, V(y)\rangle\mu_t^m(dy) = iH_V\chi_m(t, \omega), \tag{3.6}$$

or, in integral form,

$$\chi_m(t, \omega) - \dot\chi_m(0, \omega) = i\int_0^t H_V\chi_m(\tau, \omega)\,d\tau$$

$$= i\int_0^t \left(\int_{\mathfrak{M}} e^{i\langle\omega, y\rangle}\langle\omega, V(y)\rangle\mu_\tau^m(dy)\right)d\tau. \tag{3.7}$$

It remains for us to pass to the limit as $m \to \infty$ in this equality in order to obtain (2.7). We first prove that the family $\{\chi_m(t, \omega)\}_{m=1,2,\ldots}$ of characteristic functionals of the Galerkin measures is equicontinuous on each bounded set in $\mathbf{R}_t \times \mathfrak{M}^{\mathrm{tr}}$. We recall that $\mathfrak{M}^{\mathrm{tr}} \subset \mathfrak{M}^*$.

We note that the $\chi_m(t, \omega)$ are uniformly bounded: $|\chi_m(t, \omega)| \leqslant 1$ for $\omega \in \mathfrak{M}^*$ and all $t \in \mathbf{R}_t = \mathbf{R}$. To prove equicontinuity of χ_m we need two lemmas.

LEMMA 3.2. *For any $R > 0$ there is a constant α_R depending only on R such that for $\omega \in \mathfrak{M}^{\mathrm{tr}}$ and $\|\omega\|_{\mathfrak{M}^{\mathrm{tr}}} \leqslant R$*

$$|\partial\chi_m(t, \omega)/\partial t| \leqslant \alpha_R$$

for all t and m.

PROOF. From (2.4′), (2.9), and Proposition 2.1 we conclude that

$$|\dot\chi_m(t, \omega)| \leqslant C\int_{\mathfrak{M}} \|P_m\omega\|_{\mathfrak{M}^{\mathrm{tr}}}(1 + H_0 + H(y))\mu_t^m(dy) \leqslant C'\|\omega\|_{\mathfrak{M}^{\mathrm{tr}}}.$$

This proves the lemma.

LEMMA 3.3. *The functionals $\chi_m(t, \omega)$ for all t are equicontinuous in ω on the space $\mathfrak{M}^*$:*

$$|\chi_m(t, \omega_1) - \chi_m(t, \omega_2)| \leqslant L\|\omega_1 - \omega_2\|_{\mathfrak{M}^*}, \tag{3.8}$$

where L does not depend on m, t, or $\omega_1, \omega_2 \in \mathfrak{M}^*$.

PROOF. From the inequality $|e^{i\varphi} - 1| \leq |\varphi|$ for real φ it follows that, because of (1.3′).

$$|\chi_m(t, \omega_1) - \chi_m(t, \omega_2)| \leq \int_{\mathfrak{M}} |e^{i\langle \omega_1 - \omega_2, y \rangle} - 1| \mu_t^m(dy)$$

$$\leq C \int_{\mathfrak{M}} \|\omega_1 - \omega_2\|_{\mathfrak{M}*} \|y\|_{\mathfrak{M}} \mu_t^m(dy)$$

$$\leq C_1 \|\omega_1 - \omega_2\|_{\mathfrak{M}*} \int_{\mathfrak{M}} (1 + H_0 + H(y)) \mu_t^m(dy).$$

This and (3.5) prove the lemma.

The last two lemmas imply the equicontinuity of the functionals $\chi_m(t, \omega)$ on any set in $\mathbf{R}_t \times \mathfrak{M}^{\mathrm{tr}}$ bounded in ω.

LEMMA 3.4. *For any $R > 0$ there is a constant $\mathcal{C}_R$ depending only on R such that*

$$|\chi_m(t_1, \omega_1) - \chi_m(t_2, \omega_2)| \leq \mathcal{C}_R(|t_1 - t_2| + \|\omega_1 - \omega_2\|_{\mathfrak{M}*}) \qquad (3.9)$$

for $\|\omega_i\|_{\mathfrak{M}^{\mathrm{tr}}} \leq R$, $i = 1, 2$, and all $t_1, t_2 \in \mathbf{R}_t$.

PROOF. Obviously, $\mathcal{C}_R \leq \max(\alpha_R, L)$. Thus, on the set $\|\omega\|_{\mathfrak{M}^{\mathrm{tr}}} \leq R$ in $\mathbf{R}_t \times \mathfrak{M}^{\mathrm{tr}}$ the functionals $\chi_m(t, \omega)$ are equicontinuous and uniformly bounded. Since $\mathbf{R}_t \times \mathfrak{M}^{\mathrm{tr}}$ is a separable space, by Arzelà's theorem there exists a subsequence $\chi_{m'}$ convergent for $\|\omega\|_{\mathfrak{M}^{\mathrm{tr}}} \leq R$. Since R is arbitrary, by a diagonal process we can select a subsequence $\chi_{\tilde{m}}$ which converges on $\mathbf{R}_t \times \mathfrak{M}^{\mathrm{tr}}$. We denote the limit functional by $\chi(t, \omega)$:

$$\chi(t, \omega) = \lim_{\tilde{m} \to \infty} \chi_{\tilde{m}}(t, \omega), \qquad \omega \in \mathfrak{M}^{\mathrm{tr}}, t \in \mathbf{R}_t. \qquad (3.10)$$

We henceforth write m in place of $\tilde{m}$, $\chi_m(t, \omega)$ in place of $\chi_{\tilde{m}}(t, \omega)$, and μ_t^m in place of $\mu_t^{\tilde{m}}$. The functional $\chi(t, \cdot)$ is defined on $\mathfrak{M}^{\mathrm{tr}}$ for all t and satisfies the Lipschitz inequality (3.8) with the same constant L as the functional χ_m.

LEMMA 3.5. *For any t the sequence of measures μ_t^m on $\tilde{\mathfrak{M}}$ converges weakly. The weak limit μ_t of the measures μ_t^m has the Fourier transform*

$$\tilde{\mu}_t(\omega) = \chi(t, \omega), \qquad \omega \in \mathfrak{M}^{\mathrm{tr}}, \qquad (3.11)$$

where $\chi(t, \omega)$ is defined in (3.10).

PROOF. It is proved below that the set of measures μ_t^m in $\tilde{\mathfrak{M}}$ is weakly sequentially compact. Any weak limit point μ_t of this set then has the Fourier transform

$$\tilde{\mu}_t(\omega) = \int_{\tilde{\mathfrak{M}}} e^{i\langle \omega, y \rangle} \mu_t(dy)$$

$$= \lim_{m'' \to \infty} \int_{\tilde{\mathfrak{M}}} e^{i\langle \omega, y \rangle} \mu_t^{m''}(dy) = \lim_{m \to \infty} \chi_{m''}(t, \omega), \qquad \omega \in \tilde{\mathfrak{M}}*,$$

$$(3.11')$$

where $\{\mu_t^{m''}\}$ is a subsequence of the sequence of measures $\{\mu_t^m\}$ which converges weakly to the measure μ_t on $\tilde{\mathfrak{M}}$. On the other hand, by (3.10)

$$\lim_{m'' \to \infty} \chi_{m''}(t, \omega) = \chi(t, \omega), \qquad \omega \in \mathfrak{M}^{\mathrm{tr}}. \tag{3.11''}$$

From (3.11') and (3.11'') it follows that

$$\tilde{\mu}_t(\omega) = \chi(t, \omega) \quad \text{for } \omega \in \mathfrak{M}^* \cap \mathfrak{M}^{\mathrm{tr}}. \tag{3.12}$$

In particular, this is valid for

$$\omega \in \mathfrak{M}^\infty = \bigcup_{k=1}^\infty \mathfrak{M}^k. \tag{3.13}$$

This means that for $\omega \in \mathfrak{M}^\infty$ the functional $\chi(t, \omega)$ is uniquely defined. It is therefore also uniquely defined on $\mathfrak{M}^*$, since $\mathfrak{M}^\infty$ is dense in $\mathfrak{M}^*$ in the weak topology. Thus the measure μ_t is uniquely defined, i.e. the set of measures μ_t^m on $\tilde{\mathfrak{M}}$ has only one weak limit point. Therefore the sequence of measures μ_t^m converges weakly on $\tilde{\mathfrak{M}}$.

It remains to prove the weak sequential compactness of the sequence of measures μ_t^m on $\tilde{\mathfrak{M}}$. We shall now prove a stronger assertion. We denote by $\tilde{\Lambda}(y)$ the function

$$\tilde{\Lambda}(y) = 1 + \|\nabla u\|_{-\delta} + \|u\|_{L_{\tilde{q}}}^{\tilde{q}-1} + \|p\|_{-\delta} \tag{3.14}$$

on $\tilde{\mathfrak{M}}$. it is continuous and positive on $\tilde{\mathfrak{M}}$, since $\tilde{q} \geqslant 1$. For $y \in \mathfrak{M}$, because of (1.3').

$$\tilde{\Lambda}(y) \leqslant \tilde{C}(1 + H_0 + H(y))^\rho, \tag{3.15}$$

where

$$\rho = \max(1/2, (\tilde{q} - 1)/q) < 1.$$

It follows from (3.5) that for all m and t

$$\tilde{C} \int_{\mathfrak{M}} (1 + H_0 + H(y))\mu_t^m(dy) \leqslant \mathcal{E} < \infty. \tag{3.16}$$

The weak sequential compactness of $\{\mu_t^m\}_m$ follows from the next lemma.

LEMMA 3.6. *The sequence of measures* $\tilde{\Lambda}(y)\mu_t^m(dy)$ *on* $\tilde{\mathfrak{M}}$ *is weakly sequentially compact.*

PROOF. We choose any $R > 0$ and define

$$\tilde{K}_R = \overline{\{y \in \mathfrak{M} \mid H(y) \leqslant R\}}, \tag{3.17}$$

where the bar denotes closure in $\tilde{\mathfrak{M}}$. It follows from Lemma 1.1 that $\tilde{K}_R$ is compact in $\tilde{\mathfrak{M}}$. We set $\tilde{\Delta}_R = \tilde{\mathfrak{M}} \setminus \tilde{K}_R$. In analogy to the derivation of the Tchebycheff inequality, from (3.15) and (3.16) we obtain

$$\tilde{\Lambda}\mu_t^m(\tilde{\Delta}_R) \equiv \int_{\tilde{\Delta}_R} \tilde{\Lambda}(y)\mu_t^m(dy)$$

$$\leqslant \tilde{C} \int_{\tilde{\Delta}_R} (1 + H_0 + H(y))^{\rho - 1 + 1} \mu_t^m(dy) \leqslant (1 + H_0 + R)^{\rho - 1} \cdot \mathcal{E},$$

where $\rho - 1 < 0$. Therefore, for any t the measures $\tilde{\Lambda}\mu_t^m$ are (1) uniformly bounded:

$$\left(\tilde{\Lambda}\mu_t^m\right)\left(\tilde{\mathfrak{M}}\right) \leqslant \mathcal{E},$$

and (2) for any $\varepsilon > 0$ there exists a compact set $\tilde{K}_R \subset \tilde{\mathfrak{M}}$, $R = R(\varepsilon)$, such that

$$\left(\tilde{\Lambda}\mu_t^m\right)\left(\tilde{\mathfrak{M}} \smallsetminus \tilde{K}_R\right) = \left(\tilde{\Lambda}\mu_t^m\right)\left(\tilde{\Delta}_R\right) \leqslant \varepsilon. \tag{3.18}$$

It follows from (1) and (2) that the family of measures $\tilde{\Lambda}\mu_t^m$ is weakly sequentially compact on $\tilde{\mathfrak{M}}$ ([5], Chapter VI, §1, Theorem 1). Since $\tilde{\Lambda}(y) \geqslant 1$ and $\tilde{\Lambda}(y)$ is a continuous functional on $\tilde{\mathfrak{M}}$, the measures μ_t^m are also weakly sequentially compact on $\tilde{\mathfrak{M}}$.

Thus, Lemmas 3.5 and 3.6 are proved. Actually, the following result also holds.

LEMMA 3.5. *For any t the sequence of measures $\tilde{\Lambda}(y)\mu_t^m(dy)$ on $\tilde{\mathfrak{M}}$ converges weakly to $\tilde{\Lambda}(y)\mu_t(dy)$.*

PROOF. By Lemma 3.6 the sequence of measures $\{\tilde{\Lambda}(y)\mu_t^m(dy)\}$ on $\tilde{\mathfrak{M}}$ is weakly sequentially compact. Let $M_t(dy)$ be one of its limit points. It must be shown that

$$M_t(dy) = \tilde{\Lambda}(y)\mu_t(dy),$$

or, equivalently,

$$M_t(dy)/\tilde{\Lambda}(y) = \mu_t(dy).$$

But if $\psi(y)$ is a bounded continuous function on $\tilde{\mathfrak{M}}$, then

$$\int_{\tilde{\mathfrak{M}}} \psi(y) \frac{M_t(dy)}{\tilde{\Lambda}(y)} = \int_{\tilde{\mathfrak{M}}} \frac{\psi(y)}{\tilde{\Lambda}(y)} M_t(dy)$$

$$= \lim_{m \to \infty} \int_{\tilde{\mathfrak{M}}} \frac{\psi(y)}{\tilde{\Lambda}(y)} \tilde{\Lambda}(y)\mu_t^m(dy) = \int_{\tilde{\mathfrak{M}}} \psi(y)\mu_t(dy),$$

since $\psi(y)/\tilde{\Lambda}(y)$ is a bounded continuous function on $\tilde{\mathfrak{M}}$. The lemma is proved.

We shall show that the limit measure $\mu_t(dy)$ is actually concentrated on $\mathfrak{M}$.

LEMMA 3.7. *For any $t \in R$*

$$\mu_t\left(\tilde{\mathfrak{M}} \smallsetminus \mathfrak{M}\right) = 0.$$

PROOF. In (1.14) we defined functions $H^N(y)$ on $\tilde{\mathfrak{M}}$. For $R > 0$ we set

$$H_R^N(y) = \min(R, H^N(y)), \qquad y \in \tilde{\mathfrak{M}}.$$

According to Remark 1.1, the functions $H_R^N(y)$ are then continuous on $\tilde{\mathfrak{M}}$. They are uniformly bounded in $N = 1, 2, \ldots$, and for $y \in \tilde{\mathfrak{M}}$

$$H_R^N(y) \underset{N \to \infty}{\to} H_R(y) = \begin{cases} \min(R, H(y)), & y \in \mathfrak{M}, \\ R, & y \in \tilde{\mathfrak{M}} \smallsetminus \mathfrak{M}, \end{cases} \tag{3.19}$$

and $\{H_R^N(y)\}$ is a bounded sequence in N: $H_R^N(y) \leq R$. It therefore follows from Lemma 3.5 and (3.5) that

$$\int_{\mathfrak{M}} H_R(y)\mu_t(dy) = \lim_{N \to \infty} \int_{\mathfrak{M}} H_R^N(y)\mu_t(dy)$$

$$= \lim_{N \to \infty} \lim_{m \to \infty} \int_{\mathfrak{M}} H_R^N(y)\mu_t^m(dy) \leq CE + D. \qquad (3.20)$$

On the other hand,

$$R \cdot \mu_t(\tilde{\mathfrak{M}} \setminus \mathfrak{M}) \leq \int_{\tilde{\mathfrak{M}}} H_R(y)\mu_t(dy), \qquad (3.20')$$

since $H_R(y) \geq 0$ and $H_R(y) = R$ for $y \in \tilde{\mathfrak{M}} \setminus \mathfrak{M}$. From this and (3.20) it follows that

$$\mu_t(\tilde{\mathfrak{M}} \setminus \mathfrak{M}) \leq (CE + D)/R. \qquad (3.20'')$$

Since R is arbitrary, the lemma is proved.

The lemma implies that $\chi(t, \cdot)$ admits a unique extension by weak continuity from $\mathfrak{M}^{\mathrm{tr}}$ to $\mathfrak{M}^*$. This still does not imply that $\chi(t, \cdot)$ satisfies the Lipschitz condition (3.8) on $\mathfrak{M}^*$, since we know only that $\mathfrak{M}^{\mathrm{tr}}$ is dense in $\mathfrak{M}^*$ in the weak topology.

We shall now prove that the mean energy of the measure $\mu_t(dy)$ is finite. We note that this implies that the integral $\int_{\mathfrak{M}} \|y\|_{\mathfrak{M}} \mu_t(dy)$ is finite, and hence $\chi(t, \cdot)$ satisfies the Lipschitz condition (3.8) on the entire space $\mathfrak{M}^*$.

The sequence $\{H_R(y)\}_{R=1,2,\dots}$ increases monotonically as $R \to \infty$, and it tends to $H(y)$ for $y \in \mathfrak{M}$. From (3.20) it therefore follows that

$$E_t = \int_{\mathfrak{M}} H(y)\mu_t(dy) = \lim_{R \to \infty} \int_{\mathfrak{M}} H_R(y)\mu_t(dy) \leq CE + D. \qquad (3.20''')$$

Thus, the functional $\chi(t, \cdot)$ is defined on $\mathfrak{M}^*$ and satisfies the Lipschitz condition (3.8). Theorem 1 is almost proved: the measures μ_t on $\mathfrak{M}$ have been constructed, and the finiteness of the energy (2.12) has been demonstrated. It remains to verify the most important point—the Hopf equation (2.7) for $\omega \in \mathfrak{M}^{\mathrm{tr}}$. We shall prove that χ satisfies the equivalent equation (2.10). This we do by passing to the limit as $m \to \infty$ in (3.7). We recall that in (3.7) $\omega \in \mathfrak{M}^k$ and $k < m$.

LEMMA 3.8. *The function $\langle \omega, V(y) \rangle$ for $\omega \in \mathfrak{M}^k$ is continuous on $\tilde{\mathfrak{M}}$ and satisfies the estimate*

$$|\langle \omega, V(y) \rangle| \leq C_k \tilde{\Lambda}(y), \qquad y \in \tilde{\mathfrak{M}}. \qquad (3.21)$$

PROOF. Since $P_k \omega = \omega$, it follows from (2.8) that

$$\langle \omega, V(y) \rangle = \langle \pi, \Delta P_k u - P_k f(u) \rangle + \langle \varphi, P_k p \rangle, \qquad \omega = (\pi, \varphi).$$

But

$$|\Delta P_k u(x)| \leq C_k \|\nabla u\|_{-\delta} \quad \text{and} \quad |P_k p(x)| \leq C_k \|p\|_{-\delta},$$

and

$$|P_k f(u(x))| \leqslant C_k \int_{T^n} \left(1 + |u(x)|^{q-1}\right) dx$$

because of (1.2). This implies (3.21).

Everything is now ready for the passage to the limit in (3.7). We do this in two stages. We first pass to the limit in the inner integral over $\mathfrak{M}$.

LEMMA 3.9. *For $\omega \in \mathfrak{M}^\infty$ the sequence of continuous functions of $\tau \in \mathbf{R}$*

$$H_V \chi_m(\tau, \omega) = \int_{\mathfrak{M}} e^{i\langle \omega, y\rangle} \langle \omega, V(y)\rangle \mu_\tau^m(dy)$$

is uniformly bounded, and for any τ it converges as $m \to \infty$ to

$$H_V \chi(\tau, \omega) = \int_{\mathfrak{M}} e^{i\langle \omega, y\rangle} \langle \omega, V(y)\rangle \mu_\tau(dy).$$

PROOF. According to Lemma 3.8, the function $\langle \omega, V(y)\rangle / \tilde{\Lambda}(y)$ for $\omega \in \mathfrak{M}^\infty$ is bounded and continuous on $\mathfrak{M}$, and by Lemma 3.5' the measures $\tilde{\Lambda}(y)\mu_\tau^m(dy)$ on $\mathfrak{M}$ converge weakly to $\tilde{\Lambda}(y)\mu_\tau(dy)$. Hence

$$\lim_{m \to \infty} H_V \chi_m(\tau, \omega) = \lim_{m \to \infty} \int_{\mathfrak{M}} e^{i\langle \omega, y\rangle} \frac{\langle \omega, V(y)\rangle}{\tilde{\Lambda}(y)} \tilde{\Lambda}(y)\mu_\tau^m(dy)$$

$$= \int_{\mathfrak{M}} e^{i\langle \omega, y\rangle} \langle \omega, V(y)\rangle \mu_\tau(dy).$$

Finally, we pass to the limit as $m \to \infty$ in the relation

$$\chi_m(t, \omega) - \chi_m(0, \omega) = i\int_0^t H_V \chi_m(\tau, \omega)\, d\tau. \tag{3.22}$$

Here for $\omega \in \mathfrak{M}^\infty$ the left side converges as $m \to \infty$ to the left side of (2.10) because of (3.10). The right side of (3.22) converges to the right side of (2.10) by the Lebesgue theorem as a consequence of Lemma 3.9. Thus, letting $m \to \infty$ in (3.22), we obtain (3.10):

$$\chi(t, \omega) - \chi(0, \omega) = i\int_0^t \left(\int_{\mathfrak{M}} e^{i\langle \omega, y\rangle} \langle \omega, V(y)\rangle \mu_\tau(dy)\right) d\tau$$

for $\omega \in \mathfrak{M}^\infty = \bigcup_1^\infty \mathfrak{M}^k$. We shall pass that this equation also holds for $\omega \in \mathfrak{M}^{tr}$. By Proposition 2.2, $\mathfrak{M}^\infty$ is dense in $\mathfrak{M}^{tr}$. Therefore, by passing to the limit $\omega = \omega_k \to \omega$ as $k \to \infty$ in $\mathfrak{M}^{tr}$ it is easy to see that (2.10) is also valid for $\omega \in \mathfrak{M}^{tr}$. Indeed, the integrands then converge pointwise; by Proposition 2.1

$$\langle \omega_k, V(y)\rangle \underset{k \to \infty}{\to} \langle \omega, V(y)\rangle$$

for $y \in \mathfrak{M}$, and they are majorized uniformly in k, by summable functions because of (2.9) and (2.12). Hence, the Lebesgue theorem on passing to the limit under the integral sign is applicable.

The Hopf equation (2.7) is thus valid for all $\omega \in \mathfrak{M}^{tr}$. Theorem 1 is proved.

§4. Stationary solutions of the Hopf equation

In this section we construct a nontrivial stationary solution $\chi(t, \omega) \equiv \chi(\omega)$ of the Hopf equation (2.7). The function $\chi(\omega)$ satisfies the stationary Hopf equation

$$0 = H_V \chi(\omega) \equiv \int_{\mathfrak{M}} e^{i\langle \omega, y \rangle} \langle \omega, V(y) \rangle \mu(dy), \qquad \omega \in \mathfrak{M}^{\infty}. \qquad (4.1)$$

The measure $\mu(dy)$ corresponding to $\chi(\omega)$ is called an *invariant measure* with respect to the original system of equations (0.2).

Suppose that a probability measure is given on $\mathfrak{M}$, satisfying the condition (2.11) of finiteness of the mean energy, and let μ_t be the statistical solution corresponding to it, i.e. the family of probability measures for which the characteristic functionals $\chi(t, \omega)$ satisfy (2.10). With appropriate restrictions on $f(u)$ the existence of μ_t has been proved in §3. We recall that $\mu_t(dy)$ are Borel measures concentrated on $\mathfrak{M}$. In Theorem 2 below it is shown that stationary solutions of the Hopf equation and the invariant measures corresponding to them can be obtained as is usually done in the theory of dynamical systems by averaging the time-dependent solutions and their corresponding statistical solutions $\mu_t(dy)$ with respect to time (see, for example, [7] and [8]). The existence of averaged measures is based on the following result.

LEMMA 4.1. *For any* $t \in \mathbf{R}$ *there exists a probability measure* ν^t *on* $\mathfrak{M}$ *satisfying relation*

$$\int_{\mathfrak{M}} \psi(y) \nu^t(dy) = \frac{1}{t} \int_0^t \left(\int_{\mathfrak{M}} \psi(y) \mu_\tau(dy) \right) d\tau, \qquad (4.2)$$

where $\psi(y)$ *is an arbitrary bounded continuous function on* $\tilde{\mathfrak{M}}$.

PROOF. We prove, first of all, that the right side of (4.2) is defined for the indicated functions $\psi(y)$.

Indeed, the integral

$$\int_{\mathfrak{M}} \psi(y) \mu_t(dy) = \lim_{m \to \infty} \int_{\mathfrak{M}^m} \psi(y) \mu_t^m(dy) = \lim_{m \to \infty} \int_{\mathfrak{M}^m} \psi(S_m(t)y_0) \mu_0^m(dy_0)$$

$$(4.3)$$

is a bounded measurable function of $t \in \mathbf{R}$, since the integral on the right side of (4.3) is a continuous function of t which is uniformly bounded in m. In (4.3) we have used the weak convergence of the measures μ_t^m to μ_t on $\tilde{\mathfrak{M}}$ established in §3 and the fact that the measure $\mu_t(dy)$ is concentrated on $\mathfrak{M}$. Therefore, the integral on the right in (4.2) is defined for any functions $\psi(y)$ which are continuous and bounded on $\tilde{\mathfrak{M}}$.

We introduce the linear functional

$$J_t \psi = \frac{1}{t} \int_0^t \left(\int_{\mathfrak{M}} \psi(y) \mu_\tau(dy) \right) d\tau, \qquad \psi \in C(\tilde{\mathfrak{M}}). \qquad (4.4)$$

It is obvious that this functional is bounded:

$$|J_t \psi| \leqslant \|\psi\|_{C(\tilde{\mathfrak{M}})}; \qquad J_t \psi \geqslant 0, \quad \text{if } \psi(y) \geqslant 0. \qquad (4.5)$$

We shall prove that there exists a measure $\nu'(dy)$ defined on $\mathfrak{M}$ such that

$$J_t\psi = \int_{\mathfrak{M}}\psi(y)\nu'(dy).\tag{4.6}$$

For this we verify that the functional J_t satisfies the conditions of Theorem 1 of [5], Chapter V, §2. Indeed, according to (3.18) (where $\tilde\Lambda(y) \geqslant 1$ for $y \in \mathfrak{M}$), for any $\varepsilon > 0$ there exists a compact set $\tilde K_R$ in $\mathfrak{M}$, where $R = R(\varepsilon)$, such that if $\psi \in C(\mathfrak{M})$ and $\psi(y) = 0$ for $y \in \tilde K_R$, then

$$J_t\psi = \frac{1}{t}\int_0^t\left(\int_{\mathfrak{M}\setminus\tilde K_R}\psi(y)\mu_t(dy)\right)dt$$

$$\leqslant \frac{1}{t}\int_0^t\varepsilon\|\psi\|_{C(\mathfrak{M})}\,dt \leqslant \varepsilon\|\psi\|_{C(\mathfrak{M})},\quad\text{since }\mu_t\left(\mathfrak{M}\setminus\tilde K_R\right)\leqslant\varepsilon.\tag{4.7}$$

Because of the theorem mentioned, there exists a measure $\nu'(dy)$ for which (4.6) holds. From (4.4) and (4.6) it follows that

$$\int_{\mathfrak{M}}\psi(y)\nu'(dy) = \frac{1}{t}\int_0^t\left(\int_{\mathfrak{M}}\psi(y)\mu_\tau(dy)\right)d\tau.\tag{4.8}$$

It remains to prove that the measure ν' is concentrated on $\mathfrak{M}$, i.e., $\nu'(\mathfrak{M}\setminus\mathfrak{M}) = 0$, and hence on the left side of (4.8) the integral can be taken over $\mathfrak{M}$. The proof is altogether analogous to the proof of Lemma 3.7. Indeed, setting $\psi(y) = H_R^N(y)$ in (4.8) and letting $N \to \infty$, we find as in (3.20) that

$$\int_{\mathfrak{M}}H_R(y)\nu'(dy) = \lim_{N\to\infty}\ \int_{\mathfrak{M}}H_R^N(y)\nu'(dy)$$

$$= \lim_{N\to\infty}\frac{1}{t}\int_0^t\left(\int_{\mathfrak{M}}H_R^N(y)\mu_\tau(dy)\right)d\tau \leqslant CE + D.\tag{4.9}$$

As in (3.20′) and (3.20″), we deduce that

$$\nu'(\mathfrak{M}\setminus\mathfrak{M}) \leqslant (CE + D)/R,$$

and hence $\nu'(\mathfrak{M}\setminus\mathfrak{M}) = 0$. Formula (4.2) has been established.

From (4.9) it follows as in (3.20‴) that

$$\int_{\mathfrak{M}}H(y)\nu'(dy) = \lim_{R\to\infty}\int_{\mathfrak{M}}H_R(y)\nu'(dy) \leqslant CE + D\tag{4.10}$$

for all $t \in \mathbf{R}$.

We note further that (4.2) is valid not only for bounded functions $\psi(y)$ but also for any function of the form $\tilde\Lambda(y)\psi(y)$, where $\psi(y) \in C_b(\mathfrak{M})$, i.e.

$$\int_{\mathfrak{M}}\psi(y)\tilde\Lambda(y)\nu'(dy) = \frac{1}{t}\int_0^t\left(\int_{\mathfrak{M}}\psi(y)\tilde\Lambda(y)\mu_\tau(dy)\right)d\tau.\tag{4.11}$$

Indeed, let

$$\tilde\Lambda_R(y) = \min(\tilde\Lambda(y), R),\qquad y \in \mathfrak{M}.$$

Since $\tilde{\Lambda}_R(y)$ is a bounded, continuous functional on $\tilde{\mathfrak{M}}$, (4.11) holds for it if $\tilde{\Lambda}$ is replaced by $\tilde{\Lambda}_R$. But $\tilde{\Lambda}_R(y) \to \tilde{\Lambda}(y)$ as $R \to \infty$, $y \in \tilde{\mathfrak{M}}$, and

$$|\psi(y)\tilde{\Lambda}_R(y)| \leq C\|\psi\|_{C(\tilde{\mathfrak{M}})}(1 + H_0 + H(y)).$$

Therefore, because of (4.10), by the Lebesgue theorem we obtain (4.11) as $R \to \infty$.

It is now possible to formulate the theorem mentioned above on the convergence of the averaged measures. As in §§2 and 3, we assume that the function $f(u)$ in (0.1) satisfies the conditions formulated in §1, and $1 < q < 4 + 6/(n-2)$ if $n \geq 3$ and $q > 1$ if $n = 1, 2$.

THEOREM 2. *Suppose that the measure μ_0 in $\mathfrak{M}$ has finite mean energy. Then there exists a sequence $t_s \to \infty$ for which the measures ν^{t_s} defined in Lemma 4.1 converge weakly in $\mathfrak{M}$ to an invariant measure of the system* (0.2).

We first show that the family of measures $\nu^t(dy)$ in $\tilde{\mathfrak{M}}$ is weakly sequentially compact. In fact, a lemma similar to Lemma 3.6 holds.

LEMMA 4.2. *The family of measures $\{\tilde{\Lambda}(y)\nu^t(dy)\}_{t \in \mathbf{R}}$ is weakly sequentially compact.*

PROOF. By (4.10) this family is uniformly bounded:

$$(\tilde{\Lambda}\nu^t)(\tilde{\mathfrak{M}}) \equiv \int_{\tilde{\mathfrak{M}}} \tilde{\Lambda}(y)\nu^t(dy) = \int_{\mathfrak{M}} \tilde{\Lambda}(y)\nu^t(dy) \leq \text{const}, \qquad t \in \mathbf{R}. \qquad (4.12)$$

Moreover, for the component $\tilde{K}_R$ introduced in the proof of the preceding lemma, according to (4.10) and (3.15),

$$(\tilde{\Lambda}\nu^t)(\tilde{\mathfrak{M}} \setminus \tilde{K}_R) \equiv \int_{\tilde{\mathfrak{M}} \setminus \tilde{K}_R} \tilde{\Lambda}(y)\nu^t(dy) \leq \int_{\tilde{\mathfrak{M}} \setminus \tilde{K}_R} (1 + H_0 + H(y))^{\rho - 1 + 1}\nu^t(dy)$$

$$\leq (1 + H_0 + R)^{\rho - 1}(1 + H_0 + CE + D), \qquad \rho < 1.$$

Since $\rho - 1 < 0$, for any $\varepsilon > 0$ there is an $R = R(\varepsilon)$ such that

$$(\tilde{\Lambda}\nu^t)(\tilde{\mathfrak{M}} \setminus \tilde{K}_R) \leq \varepsilon. \qquad (4.13)$$

The lemma now follows from (4.12) and (4.13) by Theorem 1 of [5], Chapter VI, §1.

We now complete the proof of Theorem 2. Suppose that some sequence $t_s \to +\infty$ and

$$\tilde{\Lambda}\nu^{t_s} \to \tilde{\Lambda}\mu \quad \text{in } C^*(\tilde{\mathfrak{M}}) \quad \text{as} \quad s \to \infty, \qquad (4.14)$$

where μ is a probability measure on $\tilde{\mathfrak{M}}$; this is established as in Lemma 3.5'. As in Lemma 3.7, it can be shown that μ is concentrated on $\mathfrak{M}$ and the mean energy is finite:

$$\int_{\mathfrak{M}} H(y)\mu(dy) \leq CE + D.$$

We verify that the measure μ satisfies (4.1). For this we divide both sides of the Hopf equation (2.10) by t_s and let $s \to \infty$. Using (4.11) and (4.14), we then obtain

$$0 = \lim_{s \to \infty} \frac{1}{t_s} \big[\chi(t_s, \omega) - \chi(0, \omega) \big]$$

$$= \lim_{s \to \infty} \frac{1}{t_s} \int_0^{t_s} \left(\int_{\mathfrak{M}} e^{i\langle \omega, y \rangle} \langle \omega, V(y) \rangle \mu_\tau(dy) \right) d\tau$$

$$= \lim_{s \to \infty} \int_{\mathfrak{M}} e^{i\langle \omega, y \rangle} \frac{\langle \omega, V(y) \rangle}{\tilde{\Lambda}(y)} \tilde{\Lambda}(y) \nu^{t_s}(dy)$$

$$= \int_{\mathfrak{M}} e^{i\langle \omega, y \rangle} \frac{\langle \omega, V(y) \rangle}{\tilde{\Lambda}(y)} \tilde{\Lambda}(y) \mu(dy) = \int_{\mathfrak{M}} e^{i\langle \omega, y \rangle} \langle \omega, V(y) \rangle \mu(dy),$$

since $\langle \omega, V(y) \rangle / \tilde{\Lambda}(y) \in C_b(\mathfrak{M})$ for $\omega \in \mathfrak{M}^\infty$ because of (3.21). Theorem 2 is proved.

REMARK 4.1. The Hopf equation (4.1) is also valid for $\omega \in \mathfrak{M}^{\mathrm{tr}}$. This is established as in §3 by passing to the limit on ω.

REMARK ADDED IN PROOF. Lemma 1.1 and hence all the other results are valid for any q: the convergence of u_k'' in $H_1 - \delta$ implies the convergence of the sequence $\{u_k^-\}$ almost everywhere; the boundedness of $\{u_k^-\}$ then implies the convergence of $\{u_k^-\}$ in L_{q-1}.

BIBLIOGRAPHY

1. Eberhard Hopf, *Statistical hydromechanics and functional calculus*, J. Rational Mech. Anal. **1** (1952), 87–123.

2. C. Foiaş, *Solutions statistiques des équations d'évolutions non linéaires*, Problems in Non-Linear Analysis (C.I.M.E., IV Ciclo, Varenna, 1970), Edizioni Cremonese, Rome, 1971, pp. 129–188.

3. Jacques-Louis Lions, *Quelques méthodes de résolution des problèmes aux limites non linéaires*, Dunod and Gauthier-Villars, Paris, 1969.

4. A. Zygmund, *Trigonometrical series*, PWN, Warsaw, 1935; corrected reprint, Chelsea, New York, 1952.

5. I. I. Gihman and A. V. Skorohod, *The theory of stochastic processes*. I, "Nauka", Moscow, 1971; English transl., Springer-Verlag, 1974.

6. P. M. Bleher and M. I. Višik, *On a class of pseudodifferential operators with an infinite number of variables, and applications*, Mat. Sb. **86** (**128**) (1971), 446–494; English transl. in Math. USSR Sb. **15** (1971).

7. V. V. Nemyckiĭ and V. V. Stepanov, *Qualitative theory of differential equations*, OGIZ, Moscow, 1947; English transl., Princeton Univ. Press, Princeton, N. J., 1960.

8. C. Foiaş, *Statistical study of Navier-Stokes equations*. II, Rend. Sem. Mat. Univ. Padova **49** (1973), 9–123.

Translated by J. R. SCHULENBERGER

Amer. Math. Soc. Transl.
(2) Vol. **118**, 1982

On a Class of Pseudodifferential Equations
with Multiple Characteristics*

JU. V. EGOROV AND C. V. RANGELOV

In this paper we discuss the conditions of local solvability for an equation of the form

$$P(x, D)u(x) = f(x), \qquad x \in \Omega \subset \mathbf{R}^n, n > 1, \tag{1}$$

where $P(x, D)$ is a pseudodifferential operator with a smooth symbol $p(x, \xi)$ and $f(x)$ is an infinitely differentiable function with compact support. We recall that the operator $P(x, D)$ is said to be *locally solvable* at a point $x_0 \in \Omega$ if there exists a neighborhood $\omega \subset \Omega$ of x_0 in which, for any function $f \in C_0^\infty(\omega)$, we can find a distribution $u(x) \in D'(\Omega)$ such that $P(x, D)u(x) = f(x)$ in ω.

If $P(x, D)$ is an operator with simple real characteristics, conditions of local solvability for (1) are now quite well known (see, for example, [1]).

In this paper we consider the case where $P(x, D)$ is such that there exists a point $(x_0, \xi^0) \in T^*(\Omega)$ at which

$$p^0(x_0, \xi^0) = 0, \qquad \operatorname{grad}_{x,\xi} p^0(x_0, \xi^0) = 0, \tag{2}$$

where $p^0(x, \xi)$ is the principal part of the symbol $p(x, \xi)$. Various conditions of solvability for such operators were obtained in [2]–[7].

A striking feature of the class of pseudodifferential operators examined here is that the conditions of local solvability do not depend on the lower terms but are determined solely by the behavior of the principal symbol in a neighborhood of the point (x_0, ξ^0). No assumptions are made here regarding the structure of the characteristic set in the neighborhood of this point. As V. Ja. Ivriĭ has pointed out, the operator $P(x, D)$ is hypoelliptic microlocally in a neighborhood of (x_0, ξ^0).

1980 *Mathematics Subject Classification.* Primary 35S05, 35A07, 35A30.
* Translation of Trudy Sem. Petrovsk. **3** (1978), 43–48. MR **58** #12047.

If $P^*(x, D)$ is a pseudodifferential operator such that

$$(P(x, D)u(x), v(x))_{L_2(\Omega)} = (u(x), P^*(x, D)v(x))_{L_2(\Omega)}$$

for any functions $u(x), v(x) \in C_0^\infty(\Omega)$, then $P^*(x, D)$ is said to be *formally adjoint* to $P(x, D)$, and the principal part of its symbol is $p^0(x, \xi)$. The principal part of the symbol of the commutator $C_1 = [P^*, P] \equiv P^*P - PP^*$ is

$$c_1^0(x, \xi) = 2 \operatorname{Im} \sum_{j=1}^n \frac{\partial p^0(x, \xi)}{\partial x_j} \frac{\partial \overline{p^0(x, \xi)}}{\partial \xi_j}.$$

Assume that

$$c_1^0(x, \xi) \geqslant m_0\left(|x - x_0|^2 + |\xi - \xi^0|^2\right), \qquad m_0 > 0, \tag{3}$$

in some neighborhood ω of the point (x_0, ξ^0).

The main result of this paper is

THEOREM 1. *Assume that the conditions* (2) *and* (3) *are satisfied for the operator* $P(x, D)$. *Then for any neighborhood* ω *of the point* x_0 *there is a function* $f(x) \in C_0^\infty(\omega)$ *such that* (1) *has no solution in* ω *belonging to the class* $D'(\Omega)$.

The proof of Theorem 1 is based on several lemmas below and the following theorems.

THEOREM 2 (HÖRMANDER [8]). *Assume that* $p^0(\bar{x}, \bar{\xi}) = 0$ *and* $c_1^0(\bar{x}, \bar{\xi}) > 0$ *for some* $\bar{x} \in \Omega$ *and* $0 \neq \bar{\xi} \in T_{\bar{x}}^*$. *Then for any neighborhood* ω *of* $\bar{x}$ *there is a function* $f(x) \in C_0^\infty(\omega)$ *such that there is no distribution* $u(x)$ *in* $D'(\Omega)$ *for which* $Pu = f$ *in* Ω.

THEOREM 3. *Let* $p_2(x, \xi)$ *be a complex-valued quadratic form and* $p_2(x, \xi) \neq 0$ *for* $0 \neq (x, \xi) \in \mathbf{R}^n \oplus \mathbf{R}^n$, $n > 1$. *Let* $P_2(x, D)$ *be a differential operator obtained by changing* ξ *to* $D = (D_1, \ldots, D_n)$ *in* $p_2(x, \xi)$. *Then the following assertions are true*:

1. *There exists a number* $z \in \mathbf{C}$, $|z| = 1$, *such that the form* $\operatorname{Re} z p_2(x, \xi)$ *is positive definite.*

2. *The spectrum of the operator* $P_2(x, D)$: $L_2(\mathbf{R}^n) \to L_2(\mathbf{R}^n)$ *is a nonempty set.*

The proof is, in essence, contained in Sjöstrand's paper [7].

THEOREM 4. *Suppose that for the differential operator*

$$q_2(x, D) = \sum_{|\alpha+\beta|=2} q_{\alpha\beta} x^\beta D^\alpha,$$

where the $q_{\alpha\beta}$ *are real numbers and* $q_2(x, \xi) > 0$ *for* $0 \neq (x, \xi) \in \mathbf{R}^n \oplus \mathbf{R}^n$. *Then there is a constant* $k > 0$ *such that*

$$\operatorname{Re}(q_2(x, D)u(x), u(x))_{L_2(\mathbf{R}^n)} \geqslant k\|u(x)\|_{L_2(\mathbf{R}^n)}^2$$

for any $u(x) \in C_0^\infty(\mathbf{R}^n)$.

Theorem 4 is a particular case of Theorem 2.4 of Melin [9].

LEMMA 1. *If the conditions (2) and (3) are satisfied, then there exist positive constants C_1 and C_2 such that*

$$C_1 r \geq |\operatorname{grad} \operatorname{Re} zp^0(x, \xi)|, \tag{4}$$

$$C_2 r \leq |\operatorname{grad} \operatorname{Re} zp^0(x, \xi)| \tag{5}$$

everywhere in $\bar{\omega}$ for any $z \in \mathbf{C}$, $|z| = 1$, $|x - x_0|^2 + |\xi - \xi_0|^2 = r^2$.

PROOF. Note that (2) and (3) are invariant under the multiplication of $P(x, D)$ by $z \in \mathbf{C}$, $|z| = 1$, because

$$c_1^0(x, \xi) = 2 \operatorname{Im} \sum_{j=1}^n \frac{\partial zp^0(x, \xi)}{\partial x_j} \frac{\partial \overline{zp^0(x, \xi)}}{\partial \xi_j}$$

$$= 2 \operatorname{Im} \sum_{j=1}^n \frac{\partial p^0(x, \xi)}{\partial x_j} \frac{\partial \overline{p^0(x, \xi)}}{\partial \xi_j}$$

for any $z \in \mathbf{C}$, $|z| = 1$.

From (2) follows inequality (4) with

$$C_1 = \sup_{(x, \xi) \in \bar{\omega}} \max_{|\alpha| = 2} |D^\alpha p^0(x, \xi)|.$$

From (3) and (4) we have

$$m_0 r^2 \leq \left| \sum_{j=1}^n \left[\frac{\partial \operatorname{Re} zp^0(x, \xi)}{\partial x_j} \frac{\partial \operatorname{Im} zp^0(x, \xi)}{\partial \xi_j} - \frac{\partial \operatorname{Re} zp^0(x, \xi)}{\partial \xi_j} \frac{\partial \operatorname{Im} zp^0(x, \xi)}{\partial x_j} \right] \right|$$

$$\leq C_1 r |\operatorname{grad} \operatorname{Re} zp^0(x, \xi)|,$$

which implies (5) with $C_2 = m_0/C_1$. Lemma 1 is proved.

LEMMA 2. *If the conditions (2) and (3) are satisfied, then the form $\operatorname{Re} zp_2(y, \eta)$ cannot be definite for any $z \in \mathbf{C}$, $|z| = 1$. Here*

$$p_2(y, \eta) = \sum_{|\alpha + \beta| = 2} \frac{1}{\alpha! \beta!} p_{(\beta)}^{0(\alpha)}(x_0, \xi^0) \eta^\alpha y^\beta, \qquad (y, \eta) \in \mathbf{R}^n \oplus \mathbf{R}^n.$$

PROOF. We assume the contrary; that is, assume that there is a $z \in \mathbf{C}$, $|z| = 1$, for which the form $\operatorname{Re} zp_2(y, \eta)$ is definite. Since (2) and (3) are invariant under the multiplication of $P(x, D)$ by constants $z \in \mathbf{C}$, $|z| = 1$, we may suppose that $z = 1$ and the quadratic form $\operatorname{Re} p_2(y, \eta)$ is positive definite. The hypotheses of Theorem 3 are fulfilled for the form $p_2(y, \eta)$, and so the spectrum of the operator $P_2(y, D)$: $L_2(\mathbf{R}^n) \to L_2(\mathbf{R}^n)$ is a nonempty set. This implies that there exist a constant $\kappa_1 \in \mathbf{C}$ and a function $u_1(y) \in L_2(\mathbf{R}^n)$ such that

$$(P_2(y, D) + \kappa_1)u_1(y) = 0. \tag{6}$$

On the other hand, it follows from (3) that the symbol $q_2(y, \eta)$ of the commutator $q_2 = [P_2^*, P_2]$ is a positive definite quadratic form. Note that $q_2 = [P_2^* + \bar{\kappa}, P_2 + \kappa]$ for any $\kappa \in \mathbf{C}$. It is easy to verify that

$$\|(P_2(y, D) + \kappa)u(y)\|_{L_2(\mathbf{R}^n)}^2$$

$$= \|(P_2^*(y, D) + \bar{\kappa})u(y)\|_{L_2(\mathbf{R}^n)}^2 + (q_2(y, D)u(y, u(y)))_{L_2(\mathbf{R}^n)} \qquad (7)$$

for any function $u(y) \in C_0^\infty(\mathbf{R}^n)$. The operator $q_2(y, D)$ satisfies all the conditions of Theorem 4; consequently there exists a constant $K > 0$ such that

$$(q_2(y, D)u(y), u(y))_{L_2(\mathbf{R}^n)} \geqslant K\|u\|_{L_2(\mathbf{R}^n)}^2 \qquad (8)$$

for any function $u(y) \in C_0^\infty(\mathbf{R}^n)$. From (7) and (8) we see that

$$\|(P_2(y, D) + \kappa)u\|_{L_2(\mathbf{R}^n)}^2 \geqslant K\|u\|_{L_2(\mathbf{R}^n)}^2 \qquad (9)$$

for any function $u(y) \in C_0^\infty(\mathbf{R}^n)$.

Since (9) contradicts (6), our initial hypothesis is contradicted. Lemma 2 is proved.

REMARK. Lemma 2 can be established alternatively as follows. Assuming $\operatorname{Re} p_2(y, \eta)$ to be positive definite, we apply canonical transformations to reduce it to its normal form. To do this, it suffices to use only symplectic transformations of the variables (y, η). Expressing the commutator $q_2(y, \eta)$ in terms of the new coordinates, one can verify by a straightforward computation that the form $q_2(y, \eta)$ cannot be definite.

LEMMA 3. *If conditions (2) and (3) are fulfilled, then there exists a number r_0 such that for any $r \leqslant r_0$ a point (x, ξ) can be found where $p^0(x, \xi) = 0$ and $|x - x_0|^2 + |\xi - \xi^0|^2 = r^2$.*

PROOF. We consider two surfaces:

$$\Gamma_1 = \{(x, \xi), \operatorname{Re} p_2(x - x_0, \xi - \xi^0) = 0\},$$

$$\Gamma_2 = \{(x, \xi), \operatorname{Im} p_2(x - x_0, \xi - \xi^0) = 0\}.$$

Set $M_1^r = \Gamma_1 \cap S_{2n-1}^r$ and $M_2^r = \Gamma_2 \cap S_{2n-1}^r$, where S_{2n-1}^r is the sphere of radius r with center at (x_0, ξ^0). It follows from Lemma 2 and Theorem 3 that M_1^r and M_2^r have a nonempty intersection. Condition (3) implies that the vectors $\operatorname{Re} \operatorname{grad} p_2$ and $\operatorname{Im} \operatorname{grad} p_2$ are not collinear; therefore the surfaces M_1^r and M_2^r intersect transversally. Since the form p_2 is homogeneous, there is a $c_0 > 0$ such that the internal diameters of connected components of the sets $\operatorname{Re} p_2 \neq 0$ and $\operatorname{Im} p_2 \neq 0$ on the sphere S_{2n-1}^r are not less than $c_0 r$.

We take an arbitrary point $(x, \xi) \in M_1^r$. Assume that the curve $\zeta(t)$ given by $\zeta: [-r^{3/2}, r^{3/2}] \to S_{2n-1}^r$ is a portion of the geodesic arc on the sphere such that $\zeta(0) = (x, \xi)$ and the velocity vector is directed along the normal to the set M_1^r at the point (x, ξ).

We assume that the geodesic $\zeta(t) = (x(t), \xi(t))$ is parametrized by means of the arc length.

Consider the function $\operatorname{Re} p^0$ along the geodesic $\zeta(t)$. We have $f(t) = \operatorname{Re} p^0(x(t), \xi(t)) = \alpha + \beta t + \gamma(t)$, where $\alpha = \operatorname{Re} p^0(x, \xi)$, $\beta = f'(0)$ and $\gamma(t) = o(t^2)$, $t \in [-r^{3/2}, r^{3/2}]$. We remark that, in view of (2), $\operatorname{Re} p^0(x, \xi) = \operatorname{Re} p_2(x - x_0, \xi - \xi^0) + o(r^3)$, and so there exists a number $r' > 0$ such that for $r \leqslant r'$ the functions α and $\gamma(t)$ are uniformly bounded; that is, there are constants C_α and C_γ such that for any $r \leqslant r'$

$$|\alpha| \leqslant C_\alpha r^3, \qquad |\gamma(t)| \leqslant C_\gamma t^2, \quad t \in [-r^{3/2}, r^{3/2}].$$

On the other hand, $|\beta| > C_2 r$ by Lemma 1. These inequalities imply that there is a number r_1, $0 < r_1 \leqslant r'$, such that for any $r \leqslant r_1$ and any $(x, \xi) \in M_1^r$ the functions $f(t)$ and βt are of the same sign at the endpoints of $[-r^{3/2}, r^{3/2}]$; namely, both of them are positive at one endpoint and negative at the other. Thus there is a number $t' \in [-r^{3/2}, r^{3/2}]$ such that $f(t') = 0$ and the length of the geodesic $\zeta(t)$ joining the points (x, ξ) and $(x(t'), \xi(t'))$ does not exceed $r^{3/2}$. This length is in fact, less than $C^{(1)} r^2$, where $C^{(1)}$ does not depend on r. Indeed, equality $\alpha + \beta t' + \gamma(t') = 0$ yields

$$|t'| \leqslant \left|\frac{\alpha}{\beta}\right| + \left|\frac{\gamma(t')}{\beta}\right| \leqslant \frac{C_\alpha}{C_2} r^2 + \frac{C_\gamma}{C_2} r^2 = C^{(1)} r^2,$$

where $C^{(1)} = (C_\alpha + C_\gamma)/C_2$.

Denote by N_1^r the set of points of the form $(x(t'), \xi(t'))$ with $f(t') = 0$, where the point (x, ξ) runs through M_1^r. It follows from (3) that N_1^r is a smooth submanifold of S_{2n-1}^r and has the same dimension $2n - 1$ as the submanifold M_1^r.

In an analogous manner it can be proved that there exist quantities $r_2 > 0$ and $C^{(2)} > 0$ such that, for any $r \leqslant r_2$, on the geodesic passing through $(x, \xi) \in M_1^r$ a point (x_1, ξ^1) can be found at a distance not exceeding $C^{(2)} r^2$, where $\operatorname{Im} p^0(x_1, \xi^1) = 0$; denote the set of such points by N_2^r. Now choose a number $r_0 > 0$ such that

$$r_0 < \min(r_1, r_2), \qquad \max\left(C^{(1)} r_0^2, C^{(2)} r_0^2\right) < c_0 r_0.$$

Then for any $r \leqslant r_0$ every point of N_1^r is situated at a distance not exceeding $C^{(1)} r^2$ from some point of M_1^r, and correspondingly every point of N_2^r is situated at a distance not exceeding $C^{(2)} r^2$ from some point of M_2^r. Since on S_{2n-1}^r the internal diameters of the sets $\operatorname{Re} p_2 \neq 0$ and $\operatorname{Im} p_2 \neq 0$ are not less than $c_0 r$, it follows that $N_1^r \cap N_2^r$ is nonempty. Lemma 3 is proved.

PROOF OF THEOREM 1. By Lemma 3, we can construct a sequence of points $\{(x_j, \xi^j)\}$, $j = 1, 2, \ldots$, that converges to the point (x_0, ξ^0) and is such that $p^0(x_j, \xi^j) = 0$ and $(x_j, \xi^j) \neq (x_0, \xi^0)$. By Theorem 2, the operator $P(x, D)$ has no solution in some neighborhood of the point x_j, and hence it has no solution in any neighborhood of x_0. Theorem 1 is proved.

Received 17/SEPT/75

BIBLIOGRAPHY

1. Ju. V. Egorov, *On the solvability of differential equations with a simple characteristics*, Uspehi Mat. Nauk **26** (1971), no. 2 (158), 183–198; English transl. in Russian Math. Surveys **26** (1971).

2. V. V. Grušin, *On a class of hypoelliptic operators*, Mat. Sb. **83** (**125**) (1970), 456–473; English transl. in Math. USSR Sb. **12** (1970).

3. P. R. Popivanov, *On the local solvability of a class of pseudodifferential equations with double characteristics*, Trudy Sem. Petrovsk. **1** (1975), 237–278; English transl. in this volume.

4. Louis Boutet de Monvel and François Trèves, *On a class of pseudodifferential operators with double characteristics*, Invent. Math. **24** (1974), 1–34.

5. Fernando Cardoso and François Trèves, *A necessary condition of local solvability for pseudo-differential equations with double characteristics*, Ann. Inst. Fourier (Grenoble) **24** (1974), fasc. 1, 225–292.

6. Lars Hörmander, *A class of hypoelliptic pseudodifferential operators with double characteristics*, Math. Ann. **217** (1975), 165–188.

7. Johannes Sjöstrand, *Parametrices for pseudodifferential operators with multiple characteristics*, Ark. Mat. **12** (1974), 85–130.

8. Lars Hörmander, *Pseudo-differential operators and non-elliptic boundary problems*, Ann. of Math. (2) **83** (1966), 122–209.

9. Anders Melin, *Lower bounds for pseudo-differential operators*, Ark. Mat. **9** (1971), 117–140.

Translated by P. C. SINHA

Amer. Math. Soc. Transl.
(2) Vol. **118**, 1982

The Multiplicity of Limit Cycles
Arising from Perturbations
of the Form $w' = P_2/Q_1$
of a Hamilton Equation
in the Real and Complex Domain*

JU. S. IL'JAŠENKO

Contents

Introduction

In this paper we study perturbations of the equation

$$dw/dz = (3 - 3z^2)/2w \tag{1}$$

in the class of equations

$$dw/dz = P_2/Q_1, \tag{2}$$

where P_2 and Q_1 are polynomials of degree two and one, respectively. The solutions of (1) are the level curves $\varphi_c = \{H = c\} \subset \mathbf{C}^2$ of the polynomial

$$H = w^2 + 3z - z^3. \tag{3}$$

For real $c \in (-2, 2)$ the intersection $\varphi_c \cap \mathbf{R}^2$ has a closed component which circles the point $(0, -1)$. This component, denoted by $\delta_1(c)$, is called a vanishing cycle. (The requisite definitions are given in §1.) The vanishing cycle can be continued into the complex domain as a class of freely homotopic loops $\delta_1(c)$ on the Riemann surface φ_c, $c \in \mathbf{C}_1 \setminus \{ \pm 2 \}$. We prove the following.

1980 *Mathematics Subject Classification.* Primary 34C05; Secondary 58F21, 70K10.
*Translation of Trudy Sem. Petrovsk. **3** (1978), 49–60. MR **58** #13180.

THEOREM. *Under perturbations in the class* (2), *vanishing cycles* $\delta_1(c)$ *on the solutions of* (1) *can generate limit cycles with the following multiplicities*:

1) *in the real domain* ($c \in (-2, 2)$, $\delta_1(c) \in \mathbf{R}^2$), *real limit cycles of multiplicity at most* 1;

2) *in the complex domain, complex limit cycles of multiplicity at most* 2; *and*

3) *there exist vanishing cycles* $\delta_1(c)$ *which generate complex limit cycles of multiplicity* 2 *under special perturbations in the class* (2).

Statement (1) of the theorem was proved by R. I. Bogdanov [1] using statement $1°$ of the main lemma of the present paper, and also by G. E. Popov [2] (whose proof was based on approximation of integrals). Statement $1°$ of our main lemma was proved independently by V. I. Arnol'd, A. F. Filippov (unpublished), G. E. Popov [2] and the author.

The problem we have addressed is the simplest among those involving the multiplicity of limit cycles arising from closed algebraic curves which are solutions to Hamilton's equation $dH = 0$ under perturbations in the class $w' = P/Q$ where H, P and Q are polynomials of degree $n + 1$, n and n, respectively.

The choice of equation (1) is not accidental. A real Hamilton equation of class (2) with simple singular points and only one closed solution reduces to the form (1) by an affine change of coordinates. In fact, if $Q_1 = az + c$, then the line $z = -c/a$ is a solution of equation (2), and (2) does not have closed solutions. If $Q_1 = az + bw + c$, $b \neq 0$, then the change of coordinates $z_1 = z$, $w_1 = Q_1$ is nonsingular and brings equation (2) to the form

$$w_1' = \tilde{P}_2/w_1. \tag{2'}$$

Since equation (2') is Hamilton, the polynomial $\tilde{P}_2$ depends only on z_1. A suitable affine change of coordinates preserving the z_1 axis and carrying the roots of the polynomial $\tilde{P}_2$ (which are simple by assumption) to the points 1 and -1 converts equation (2') into equation (1).

All three assertions of the theorem, including even the purely real assertion 1), are proved using the same complex methods.

§1. Definitions and preliminary results

A. *Vanishing cycles.* The polynomial $H = w^2 + 3z - z^3$ has two critical points $\mathcal{P}_1 = (-1, 0)$ and $\mathcal{P}_2 = (1, 0)$ in $\mathbf{C}^2$ with critical values -2 and 2, respectively.

The level curves φ_{-2} and φ_2 have singularities (simple self-intersections) at the points $\mathcal{P}_1$ and $\mathcal{P}_2$ respectively. The remaining leaves φ_c are nonsingular.

The critical points $\mathcal{P}_1$ and $\mathcal{P}_2$ are nondegenerate (that is, the second differential of H at these points is a nondegenerate quadratic form). According to the Morse lemma, we can find an analytic change of coordinates in a neighborhood of each critical point which brings H to the form $H(\mathcal{P}_j) + z_1 w_1$, $j = 1, 2$. On the leaves φ_c where c is close to -2 there are classes of freely homotopic loops, called *local vanishing cycles* (at the point $\mathcal{P}_1$) and denoted by $\delta_1(c) \subset \varphi_c$. A representative of the class $\delta_1(c)$ is given in the coordinates z_1, w_1 by the equations $z_1 = \rho e^{i\varphi}$ and $w_1 = (c + 2)/\rho e^{i\varphi}$, where ρ is small and $\varphi \in [0, 2\pi]$. Similarly we define local vanishing cycles $\delta_2(c)$ at the point $\mathcal{P}_2$.

Let σ denote the interval $(-2, 2)$. We will define vanishing cycles $\delta_j(c)$ for all $c \in \sigma$. If $c \in \sigma$ is near -2, then representatives of $\delta_1(c)$ are the closed curves defined above. For any $c \in \sigma$ we define a representative of the class $\delta_1(c)$ to be the closed component of the intersection of the leaf φ_c with the plane $(\mathrm{Re}\, z, \mathrm{Re}\, w)$ and a representative of $\delta_2(c)$ to be the closed component of the intersection of φ_c with the plane $(\mathrm{Re}\, z, \mathrm{Im}\, w)$.

We now define vanishing cycles $\delta_1(c)$ and $\delta_2(c)$ for an arbitrary leaf φ_c. Let B denote the line $\mathbf{C}^1 = \{c\}$ with the points ± 2 deleted, and let $\hat{B}$ be the universal cover of B with the base point 0 and projection π. Let $\hat{\sigma} \subset \hat{B}$ denote the interval such that $\pi\hat{\sigma} = \sigma$ and $0 \in \hat{\sigma}$. Finally, let E be the space $\mathbf{C}^2 \setminus \varphi_{-2} \setminus \varphi_2$. Then, the map $H\colon E \to B$ is the projection map of a fibration with total space E, base space B, and fiber φ_c. To each point $\hat{c} \in \hat{B}$ there corresponds a class of homotopic (in B) curves $\{\lambda_{0,c}\}$ with initial point 0 and endpoint c, where $\pi\hat{c} = c$. To each such class of curves there corresponds a class of homotopic maps $\Delta_{\hat{c}}\colon \varphi_0 \to \varphi_c$. The class of freely homotopic loops $\{\Delta_{\hat{c}}\delta_j(0)\}$ is called a vanishing cycle $\delta_j(\hat{c}) \subset \varphi_c$, $j = 1, 2$.

B. *Complex limit cycles and their multiplicity.* By a complex cycle on a solution φ of the analytic equation $dw/dz = R(z, w)$, where R is the rational function (3), we mean an element of the fundamental group $\pi_1(\varphi)$. For each complex line Γ, transverse to the solution φ at the base point p of the fundamental group $\pi_1(\varphi)$, and each cycle l, there is a complex successor* function $\Delta_l\colon (\Gamma, p) \to (\Gamma, p)$. Δ_l is the germ of a conformal map $(\Gamma, p) \to (\Gamma, p)$ (for more details, see [3]). If ζ is a chart on Γ, then $\Delta_l(\zeta)$ is an analytic function. We define the multiplicity of the cycle l to be the multiplicity of the fixed point p of the map Δ_l or, equivalently, the multiplicity of the zero of the function $\Delta_l(\zeta) - \zeta$ at the point p. If the multiplicity is finite, l is called a limit cycle, and if it is infinite (that is, $\Delta_l = \mathrm{id}$), a constant cycle.

REMARK. The $\delta_j(\hat{c})$ are constant cycles of equation (1).

A constant cycle l of an equation α of the form (2) with a perturbation $\alpha(\tau)$ continuously dependent on a multidimensional parameter τ, $\alpha(0) = \alpha_0$, is said to generate a complex cycle of multiplicity k if $\alpha(\tau)$ has a complex limit cycle $l(\tau)$ for $\tau \neq 0$ (also continuously dependent on τ) of multiplicity k and such that $l(\tau) \to l$ as $\tau \to 0$.

We have now defined all the concepts used in the formulation of the theorem.

C. *Integrals along vanishing cycles.* Let ω be a holomorphic 1-form whose residues are zero on the leaves φ_c. Then the integral

$$I_j(\hat{c}, \omega) = \int_{\delta_j(\hat{c})} \omega, \qquad j = 1, 2, \tag{4}$$

is well defined as a single-valued analytic function on $\hat{B}$: the Cauchy theorem guarantees that the integral does not depend on the choice of representative of the class $\delta_j(\hat{c})$. The integral (4) can also be regarded as a many-valued analytic

---

Editor's note. The term "continuation" is used instead of "successor" in the translation of [3].

function on B. Now let $\mu_1(c)$ denote the loop with initial point $c \in \sigma$ which circles the point -2 in a counterclockwise direction along a circle of radius $c + 2$, and let $\mu_2(c)$ be a similar loop around the point 2. We let $\Delta_k I_j(c)$, $k = 1, 2$, denote the result of analytically continuing the integral I_j from the segment σ over the loop μ_k. It is known (see, for example, [3]) that, for $c \in \sigma$,

$$\Delta_1 \begin{pmatrix} I_1(c) \\ I_2(c) \end{pmatrix} = \begin{pmatrix} I_1(c) \\ I_2(c) - I_1(c) \end{pmatrix}, \tag{5}$$

$$\Delta_2 \begin{pmatrix} I_1(c) \\ I_2(c) \end{pmatrix} = \begin{pmatrix} I_1(c) + I_2(c) \\ I_2(c) \end{pmatrix}. \tag{6}$$

Let U_1, $U_2 \subset \mathbf{C}^1$ be neighborhoods of the points -2 and 2, respectively. We see that under continuation from the segment σ the integral I_1 is single-valued in U_1 and the integral I_2 has a logarithmic branch point. Therefore, in U_1,

$$I_2(c) = h_2(c) - \frac{\ln(c + 2)}{2\pi i} I_1(c), \tag{7}$$

where h_2 is a single-valued analytic function in the punctured neighborhood U_1. Similarly, in U_2,

$$I_1(c) = h_1(c) + \frac{\ln(c - 2)}{2\pi i} I_2(c), \tag{8}$$

where h_2 is a single-valued analytic function in the punctured neighborhood U_2.

§2. Reduction to the study of an abelian integral

Throughout this section we shall use the restriction $c = H \mid_\Gamma$ as a local parameter on a transversal Γ to a solution of equation (1).

A. *Transition from an arbitrary continuous perturbation to a one-parameter analytic perturbation.*

LEMMA 1. *Suppose that a cycle $l \subset \varphi_c$ of equation (1) generates a limit cycle $l(\varphi)$ of multiplicity k under some perturbation in class (2) which depends continuously on a multidimensional parameter τ ranging over the domain U. Then there exists an analytic one-parameter family $\alpha(t)$ of equations of class (2) such that $\alpha(0)$ coincides with (1) and the cycle l generates a limit cycle $l(t)$ of multiplicity k of the equation $\alpha(t)$.*

PROOF. Let α be a point in the space of coefficients of the polynomials P_2 and Q_1. Each such point corresponds to an equation of class (2), and conversely. Let α_0 be the point which corresponds to equation (1). According to the conditions of the lemma, there exist continuous functions $c(\tau)$ and $\alpha(\tau)$ such that at the point $c(\tau)$ the difference $\Delta_1(c, \alpha(\tau)) - c$ has a zero of multiplicity k with respect to c for all $\tau \neq 0$. In other words, the set of points $\{c(\tau), \alpha(\tau)\}$ satisfies the system of equations

$$\Delta_l(c, \alpha) - c = 0, \quad (\partial^m / \partial c^m)\Delta_l(c, \alpha) - c = 0, \quad m = 1, \ldots, k - 1.$$

However, the above system describes an analytic set M in a neighborhood of the point (c_0, α_0) in the space $\{(c, \alpha)\}$. M contains the image of the domain U under the continuous map $\tau \to (c(\tau), \alpha(\tau))$, and this image lies on the line $\alpha = \alpha_0$. Lemma 1 now follows from the following proposition:

Let $\mathbf{C}^n = \mathbf{C}^1 \oplus \mathbf{C}^{n-1} = \{(c, \alpha)\}$ *and suppose that an analytic subset* $M \subset \mathbf{C}^n$ *contains the point* (c_0, α_0) *and does not lie on the line* $\alpha = \alpha_0$. *Then* M *contains an analytic curve* $\{c(t), \alpha(t)\}$, $(c(0), \alpha(0)) = (c_0, \alpha_0)$, *which does not lie on the line* $\alpha = \alpha_0$.

The above proposition follows from the theorem concerning the local description of an analytic set [4]. We will not dwell any further on this point.

B. *The principal part of the successor function.* Every analytic one-parameter perturbation of (1) in class (2) can be written, after a change of variables $z_1 = z$, $w_1 = Q_1$, in the form

$$dH + \omega_1(t) = dH + \alpha(t)w\,dz + \beta(t)zw\,dz + \rho(t)w^2\,dz + dH_t(z), \qquad (9)$$

where α, β, ρ, and the polynomial H_1 depend analytically on t and vanish at $t = 0$. Define a form ω as follows. Let $\gamma = \lim_{t \to 0} \beta(t)/\alpha(t)$. If $\gamma \neq \infty$, then $\omega = wdz + \gamma zwdz$; if $\gamma = \infty$ then $\omega = zwdz$.

LEMMA 2. *If the cycle* $\delta_1(\hat{c}_0)$ *generates a limit cycle of multiplicity* k *under the perturbation* (9), *then* $I_1(\hat{c}, \omega)$ (*defined in* (4)) *has a zero of multiplicity not less than* k *at the point* $\hat{c}_0$. *For* $k = 2$ *the converse is true: if* $I_1(\hat{c}, \omega)$ *has a zero of multiplicity* 2 *at the point* $\hat{c}_0$, *then there exist analytic functions* $\alpha(t)$ *and* $\beta(t)$, *vanishing at* $t = 0$, *such that under the perturbation*

$$dH + \alpha(t)w\,dz + \beta(t)zw\,dz = 0$$

of (1) *the cycle* $\delta_1(\hat{c}_0)$ *generates a limit cycle of multiplicity* 2.

PROOF. Suppose first that $\gamma \neq \infty$, and let m be a natural number such that $\lim_{t \to 0} \alpha(t)/t^m \neq 0, \infty$. Note that $\int_{\delta_1(\hat{c})} \omega_1(t) = t^m[I_1(\hat{c}, \omega) + O(t)]$ (since $dH_t(z)$ is an exact form and $\int_{\delta_1(\hat{c})} w^2\,dz \equiv 0$ because both the cycle and the form are symmetric with respect to the z-axis). Standard arguments similar to those in [3] now allow us to compute the principal term (with respect to t) of the successor function:

$$\Delta_{\delta_1(\hat{c}_0)}(c, t) = c + t^m\left[I_1(\hat{c}, \omega) + O(t)\right], \qquad \pi\hat{c} = c. \qquad (10)$$

Here, the point $\hat{c}$ runs over a neighborhood of $\hat{c}_0$ small enough so that $\hat{c}$ is uniquely determined by its projection $c \in B$.

We conclude that if the function $(\Delta - c)/t^m$ has a zero $c(t)$, $t \neq 0$, of multiplicity k with respect to c, and if $c(t) \to c_0$ as $t \to 0$, then the limit $I_1(c, \omega)$ of this function as $t \to 0$ has a zero at $\hat{c}_0$ of multiplicity not less than k.

The case $\gamma = \infty$ is handled similarly.

Conversely, suppose that $\hat{c}_0$ is a double zero of $I_1(\hat{c}, \omega)$. Consider the perturbation

$$dH + t_1 w\,dz + t_2 zw\,dz = 0$$

of equation (1) which depends on two complex parameters t_1 and t_2. Inside the cone $|\gamma - t_2/t_1| < \varepsilon$ set $s = t_2/t_1$. Proceeding as in (9), we find that

$$\Delta_{\delta_1(\hat{c}_0)}(c, t_1, t_2) = c + \int_{\delta_1(\hat{c})} (t_1 w\, dz + t_2 zw\, dz) + O(|t_1|^2 + |t_2|^2).$$

The function

$$f(c, t_1, s) \stackrel{\text{def}}{=} \frac{\Delta(c, t_1, st_1) - c}{t_1} = \int_{\delta_1(\hat{c})} w\, dz + swz\, dz + O(t_1)$$

is holomorphic in a neighborhood of the point $(c, t_1, s) = (c_0, 0, \gamma)$. As $t_1 \to 0$, f tends to the limit $\int_{\delta_1(\hat{c})} w\, dz + szw\, dz$ which, by assumption, has a double zero with respect to c at the point $(c_0, 0, \gamma)$. It follows easily from the Weierstrass preparation theorem that there exists an analytic curve $t_1 = \alpha(t)$, $s = \tilde{\beta}(t)$, $(\alpha(0), \beta(0)) = (0, \gamma)$, such that the function $f(c, \alpha(t), \tilde{\beta}(t))$ has a double zero with respect to c for all sufficiently small t. The same holds for the function $\Delta(c, \alpha(t), \beta(t)) - c$, where $\beta(t) = \alpha(t)\tilde{\beta}(t)$. The case $\gamma = \infty$ is handled similarly. Lemma 2 is proved.

   C. *The main integral.* Set $\omega_1 = w\, dz$ and $\omega_2 = zw\, dz$. Henceforth the integral

$$I_1(\hat{c}, \alpha, \beta) = \int_{\delta_1(\hat{c})} \alpha\omega_1 + \beta\omega_2, \qquad (\alpha, \beta) \neq (0, 0) \tag{11}$$

will be called the *main integral.* Lemmas 1 and 2 reduce our theorem to the following lemma.

   Main Lemma. *The integral* (11) *is not identically zero and, as a function of $\hat{c}$, can have zeros of the following multiplicities:*
   $1°$. *For $\hat{c} \in \sigma = (-2, 2)$ and $\delta_1(\hat{c}) \subset \mathbf{R}^2$, the integral I_1 can have only simple zeros.*
   $2°$. *For arbitrary $\hat{c} \in \hat{B}$, the integral I_1 cannot have zeros of multiplicity greater than 2.*
   $3°$. *There exists a countable set of numbers $\gamma_n \in \mathbf{C}^1$ and points $\hat{c}_n \in \hat{B}$ such that the integral*

$$I_1(\hat{c}, \gamma) = \int_{\delta_1(\hat{c})} \omega_1 + \gamma\omega_2$$

with $\gamma = \gamma_n$ has a double zero with respect to c at c_n.

§3. Investigation of the main integral

   A. *Heuristic considerations.* Let ε be a holomorphic 1-form or a meromorphic 1-form whose residues on the solutions φ_c are zero. Then, the integral $I_1(\hat{c}, \omega)$ in (4) is well defined and, like every abelian integral, satisfies a linear differential equation of Fuchsian type. The order of this equation equals the dimension of the linear space generated by all branches of the function $I_1(\hat{c}, \omega)$. Formulas (5) and (6) show that this dimension is 2 and the integrals $I_1(\hat{c}, \omega)$ and $I_2(c, \omega)$ form a fundamental system of solutions. The singular points of the differential equation are the points ± 2, ∞ and those points at which the Wronskian determinant

$$W(\hat{c}, \omega) = \begin{vmatrix} I_1(\hat{c}, \omega) & I_2(\hat{c}, \omega) \\ \dfrac{d}{d\hat{c}} I_1(\hat{c}, \omega) & \dfrac{d}{d\hat{c}} I_2(\hat{c}, \omega) \end{vmatrix} \tag{12}$$

vanishes. At nonsingular points of the equation the functions I_1 and I_2 cannot have double zeros. Formulas (5) and (6) show that, under continuation from σ over the loops μ_1 and μ_2, the matrix (12) is multiplied on the right by the matrices $T_1 = \left(\begin{smallmatrix} 1 & -1 \\ 0 & 1 \end{smallmatrix}\right)$ and $T_2 = \left(\begin{smallmatrix} 1 & 0 \\ 1 & 1 \end{smallmatrix}\right)$, respectively. We note that $\det T_1 = \det T_2 = 1$ and, thus, the determinant (12) is single-valued as a function of c. Our major technique consists in studying this determinant.

   B. *Notation.* We put

$$\omega_1 = w\,dz, \quad \omega_2 = zw\,dz, \quad \omega_1' = \frac{dz}{w}, \quad \omega_2' = \frac{z\,dz}{w}, \tag{13}$$

$$I_{ij}(\hat{c}) = I_i(\hat{c}, \omega_j), \qquad I_{ij}'(\hat{c}) = I_i(\hat{c}, \omega_j'), \qquad i, j = 1, 2; \tag{14}$$

$$\omega = \alpha\omega_1 + \beta\omega_2, \qquad \omega' = \alpha\omega_1' + \beta\omega_2', \tag{15}$$

$$I_i(\hat{c}) = I_i(\hat{c}, \omega), \qquad J_i(\hat{c}) = I_i(\hat{c}, \omega'), \tag{16}$$

$$W_1 = \begin{vmatrix} I_1 & I_2 \\ I_1' & I_2' \end{vmatrix}, \qquad W_2 = \begin{vmatrix} J_1 & J_2 \\ J_1' & J_2' \end{vmatrix}. \tag{17}$$

   The last six functions above depend linearly or quadratically on α and β. Where relevant, we indicate this dependence by writing, for example, $I_1(\hat{c}, \alpha, \beta)$ instead of $I_1(\hat{c})$.

   C. *The derivative* $I_i'(\hat{c})$. Let $\gamma(c)$ be a cycle which varies continuously with respect to c on the nonsingular level sets of the polynomial H, and let $\tilde{\omega}$ be a holomorphic 1-form. Let $\tilde{\omega}'$ be a meromorphic 1-form such that $dH \wedge \tilde{\omega}' = d\tilde{\omega}$. It is well known that then

$$\frac{d}{dc} \int_{\gamma(c)} \tilde{\omega} = \int_{\gamma(c)} \tilde{\omega}'.$$

This relation shows that the residue of the form $\tilde{\omega}'$ on the nonsingular leaves φ_c are zero. If the curve $\gamma(c)$ is contractible in φ_c, then the right-hand side is incidentally equal to zero. Note that $dH \wedge \omega' = 2d\omega$. Consequently, the integrals J_1 and J_2 are well defined:

$$J_i(c) = 2I_i'(\hat{c}), \qquad i = 1, 2, \tag{18}$$

and the determinant W_2 is a single-valued function of c.

   D. *Reduction.* In this subsection and the next we will prove statement 1° of the Main Lemma.

   LEMMA 3. *The function $J_1(\hat{c})$ does not have a multiple zero on the interval σ.*

   We first show that the above lemma implies statement 1° of the main lemma. We shall prove the lemma later.

Suppose the contrary. That is, suppose $I_1(\hat{c})$ has a multiple zero at some $c_0 \in \sigma$. We note that $I_1(-2) = 0$ because ω is holomorphic and $\delta_1(c)$ vanishes (contracts to a point as $c \to -2$). Thus, the derivative I_1' vanishes at some point on $(0, c_0)$. In addition, $I_1'(c_0) = 0$, by assumption. Thus, the linear combination $\alpha J_{11} + \beta J_{12}$ has two zeros on σ. It easily follows that some other linear combination of J_{11} and J_{12} has a double zero on σ. In fact, consider the curve $J = \{J(c)\} = \{(J_{11}(c), J_{12}(c))\}$ on the (ξ, η)-plane. The line $\alpha\xi + \beta\eta = 0$ intersects the curve J in two points. Then there exists a line $\alpha_0\xi + \beta_0\eta = 0$ tangent to the curve J at some point $J(c_1)$. This means that the linear combination $\alpha_0 I_{11} + \beta_0 I_{12}$ has a zero of multiplicity 2 at the point $c_1 \in \sigma$. This contradicts Lemma 3.

E. *Symmetry properties of the foliation $H = c$.* We note that the map

$$j: \mathbf{C}^2 \to \mathbf{C}^2, \quad (z, w) \mapsto (-z, iw)$$

has the effect of multiplying the polynomial H by -1, $H \circ j = -H$, and thus preserves the foliation φ_c. In particular, $j\varphi_c = \varphi_{-c}$. Note, also, that

$$j\delta_1(c) = \delta_2(-c), \quad j\delta_2(c) = -\delta_1(c),$$
$$j^*\omega_1 = i\omega_1, \quad j^*\omega_2 = -i\omega_2, \quad j^*\omega_1' = -i\omega_1', \quad j^*\omega_2' = i\omega_2'.$$

Therefore,

$$I_{12}(c) = iI_{11}(-c), \quad I_{22}(c) = -iI_{21}(-c), \tag{19}$$
$$J_{12}(c) = -iJ_{11}(-c), \quad J_{22}(c) = iJ_{22}(-c). \tag{20}$$

These formulas reduce integration on $\delta_2(c)$ to integration on $\delta_1(c)$.

F. *The proof of Lemma 3.* To prove Lemma 3 we explicitly compute the determinant W_2 (in (17)) and show that it does not vanish on the interval σ. We shall take advantage of the representations (7) and (8). In a neighborhood U_2 of the point 2 the continuation of J_1 from σ equals

$$J_1(c) = \frac{\ln(c - 2)}{2\pi i} J_2(c) + h_2(c), \tag{21}$$

where h_2 is holomorphic in the punctured neighborhood U_2. By using an explicit estimate, it is easy to show that $J_1(c)$ grows no faster than $-\ln|c - 2|$ as $c \to 2$. As indicated above, $I_2(c)$, and hence also $J_2(c) = 2I_1'(c)$, is holomorphic in U_2 under continuation from σ. Consequently, h_2 can grow no faster than $-\ln|c - 2|$ and, thus, by the removable singularity theorem, is holomorphic in U_2. Therefore, in a neighborhood of 2,

$$W_2(c) = \begin{vmatrix} \dfrac{\ln(c - 2)}{2\pi i} J_2(c) + h_2(c) & J_2(c) \\[3mm] \dfrac{\ln(c - 2)}{2\pi i} J_2'(c) + h_2'(c) + \dfrac{J_2(c)}{2\pi i(c - 2)} & J_2'(c) \end{vmatrix}$$

$$= -\frac{J_2^2(2)}{2\pi i(c - 2)} + O(1). \tag{22}$$

Similarly, in a neighborhood U_1 of -2

$$W_2(c) = -\frac{J_1^2(-2)}{2\pi i(c+2)} + O(1). \tag{23}$$

Finally, as $c \to \infty$ along an arbitrary ray, $J_{11}(c) = O(c^{-1/6})$ (the length of the path of integration grows as $c^{1/3}$ and the integrand grows as $c^{-1/2}$). Similarly, $J_{12}(c) = O(c^{1/6})$. Therefore, $J_1(c) = O(c^{1/6})$ and $J_1'(c) = O(c^{-5/6})$. By (20), $J_2(c) = O(c^{1/6})$ and $J_2'(c) = O(c^{-5/6})$. Therefore, $W_2(c) = O(c^{-2/3})$ as $c \to \infty$ and, using the fact that $W_2(c)$ is single-valued, we conclude that $W_2(\infty) = 0$. Thus, $W_2(c)$ is a rational function with poles at ± 2. Its principal terms are described by (21) and (23), and $W_2(\infty) = 0$. Thus,

$$-2\pi i W_2(c) = J_1^2(-2)/(c+2) + J_2^2(2)/(c-2). \tag{24}$$

The numerators in (24) are easy to compute. Set $J_{11}(-2) = a$ and $J_{12}(-2) = b$. Then $J_1(-2) = \alpha a + \beta b$ and, by (20), $J_2(2) = -i\alpha a + i\beta b$. Computing the residues of $\omega_1'|_{\varphi_{-2}}$ and $\omega_2'|_{\varphi_2}$ at the point $\mathcal{P}_1$, we obtain

$$J_{11}(-2) = a = 2\pi/3, \qquad J_{12}(-2) = b = -a = -2\pi/3. \tag{25}$$

It is important for us that a and b be real. Thus,

$$-2\pi i W_2(c) = \frac{(4\alpha\beta ab)c - 4(\alpha^2 a^2 + \beta^2 b^2)}{c^2 - 4}.$$

The determinant $W_2(c)$ vanishes at the single point

$$c_0 = (\alpha^2 a^2 + \beta^2 b^2)/\alpha\beta ab.$$

For real α, β, a and b the modulus of this expression is not less than 2. Thus $W_2(c)|_o \neq 0$. Lemma 3 and, thus, statement 1° of the Main Lemma and statement 1) of the theorem are proved.

G. *Reduction.*

LEMMA 4. *The determinant $W_1(c)$ (in (17)) is not identically zero and is a polynomial of degree no higher than one.*

Statement 2° of the Main Lemma follows from Lemma 4. In fact, suppose that $I_1(\hat{c})$ had a zero of multiplicity greater than 2 at some point $\hat{c}_0 \in \hat{B}$, $\pi\hat{c}_0 = c_0$. Then the first column of the determinant $W_1(c)$, and thus also the determinant itself, would be divisible by $(c - c_0)^2$, and this contradicts Lemma 4.

H. *The proof of Lemma 4.* Since the proof is similar to the argument in subsection F, we will omit a number of details. In a neighborhood U_2 of the point 2

$$W_1(c) = \begin{vmatrix} \dfrac{\ln(c-2)}{2\pi i} I_2(c) + g_1(c) & I_2(c) \\[2ex] \dfrac{\ln(c-2)}{2\pi i} I_2'(c) + g_1'(c) + \dfrac{I_2(c)}{2\pi i(c-2)} & I_2'(c) \end{vmatrix},$$

and the function g_1 is single-valued in the punctured neighborhood U_2. The form ω is holomorphic, and so the continuation of $I_2(c)$ from σ is holomorphic in U_2, and $I_2(2) = 0$. The integral $I_1(c)$ is bounded as $c \to 2$, $c \in \sigma$. Thus, g_1 is bounded in $U_2 \setminus 2$ and, hence, holomorphic in U_2. Also, $g_1(2) = I_1(2)$. Thus the determinant $W_1(c)$ is holomorphic in the neighborhood U_2 of 2, and

$$W_1(2) = I_1(2)I_2'(2). \tag{26}$$

Similarly, in a neighborhood U_1 of -2 the determinant W_1 is a holomorphic function and

$$W_1(-2) = -I_2(-2)I_1'(-2). \tag{27}$$

Finally, as $c \to \infty$ along any ray we have

$$I_1(c) = O(c^{7/6}), \qquad I_2(c) = O(c^{7/6}),$$
$$I_1'(c) = O(c^{1/6}), \qquad I_2'(c) = O(c^{1/6}),$$

which implies that $W_1(c) = O(c^2)$. Since $W_1(c)$ is single-valued it must be a polynomial of degree no greater than one:

$$W_1(c) = \frac{W_1(2) - W_1(-2)}{4} c + \frac{W_1(2) + W_1(-2)}{2}.$$

The values $W_1(2)$ and $W_1(-2)$ given by (26) and (27) are easy to compute. By (18), $I_1'(-2) = \frac{1}{2}(\alpha a + \beta b)$, where $a \neq 0$ and $b \neq 0$ are given by (25). Furthermore, it follows from (20) that

$$I_2'(2) = -i(\alpha a - \beta b)/2. \tag{28}$$

Finally, the leaf φ_2: $w = (z - 2)\sqrt{z + 4}$ is uniformized by the parameter t: $z = t^2 - 4$, $w = (t^2 - 6)t$, and the integrals $I_1(2) \stackrel{\text{def}}{=} \alpha d + \beta e$ and $I_2(-2) = i(\alpha d - \beta e)$ (the latter equality follows from (20)) are calculated explicitly: $d = 16\sqrt{2}/15$ and $e = -d/7$. Thus

$$W_1(c) = iae\big[(3\alpha\beta)c + (7\alpha^2 - \beta^2)\big].$$

Lemma 4, together with statement 2° of the Main Lemma and statement 2) of the theorem, is proved.

J. REMARK. Put $\gamma = \beta/\alpha$. The determinant $W_1(c)$ vanishes at the single point

$$c_0(\gamma) = (\beta^2 - 7\alpha^2)/3\alpha\beta = (\gamma^2 - 7)/3\gamma. \tag{29}$$

K. *The many-valued functions* $k(\hat{\gamma})$ *and zeros of multiplicity two of the main integral.* Let Γ denote the punctured plane $\mathbf{C}^1 = \{\gamma\}$ with the points γ_i at which the function $c_0(\gamma)$ takes the values ± 2 and ∞ removed. Let $\hat{\Gamma}$ be the universal cover of Γ with base point $q = -\sqrt{7}$ $(c_0(q) = 0)$ and projection $\pi'\colon \hat{\gamma} \to \gamma$.

The map $c_0\colon \Gamma \to B$ lifts to a unique map $\hat{c}_0\colon \hat{\Gamma} \to \hat{B}$ if we require that the base point be carried to the base point: $\hat{c}_0(q) = 0$. Let $\sigma' \subset \hat{\Gamma}$ denote an interval such that $\hat{c}_0(\sigma') = \sigma$ and $\sigma' \ni q$. The interval σ' projects onto the interval $(-7, -1)$, and we will identify them. Let $A(\hat{\gamma})$ denote the matrix

$$A(\hat{\gamma}) = \begin{pmatrix} a_{11}(\hat{\gamma}) & a_{12}(\gamma) \\ a_{21}(\hat{\gamma}) & a_{22}(\hat{\gamma}) \end{pmatrix} = \begin{vmatrix} I_1(1, \gamma, \hat{c}_0(\hat{\gamma})) & I_2(1, \gamma, \hat{c}_0(\hat{\gamma})) \\ \dfrac{\partial I_1}{\partial c}(1, \gamma, \hat{c}_0(\gamma)) & \dfrac{\partial I_2}{\partial c}(1, \gamma, \hat{c}_0(\gamma)) \end{vmatrix}.$$

By definition of $c_0(\gamma)$ (see subsection J), $\det A(\hat{\gamma}) \equiv 0$. The main integral can have a double zero only at a point $\hat{c}$, $\pi\hat{c} = c$, for which the determinant $W_1(c)$ vanishes. In this case, we must have $c = c_0(\gamma)$. We now define a function $k(\hat{\gamma})$ on $\hat{\Gamma}$ by the expression

$$k(\hat{\gamma}) = \frac{a_{11}(\hat{\gamma})}{a_{12}(\hat{\gamma})} = \frac{I_1(1, \gamma, \hat{c}_0(\hat{\gamma}))}{I_2(1, \gamma, \hat{c}_0(\hat{\gamma}))}.$$

LEMMA 5. *The main integral*

$$I_1(\alpha, \gamma\alpha, \hat{c}) = \alpha I_1(1, \gamma, \hat{c}) \tag{30}$$

has a zero of multiplicity greater than 1 *at the point* $\hat{c}$ *if and only if* $c = \hat{c}_0(\hat{\gamma})$, $\pi'\hat{\gamma} = \gamma$ *and* $k(\hat{\gamma}) = 0$.

PROOF. Suppose the integral (30) has a double zero at the point $\hat{c}$, $c = \hat{c}_0(\hat{\gamma})$. Then $I_2(1, \gamma, \hat{c}) \neq 0$, because otherwise the determinant $W_1(1, \gamma, \hat{c})$ would have a double zero at $c = \pi\hat{c}$, contradicting Lemma 4. Consequently, $a_{11}(\hat{\gamma}) = 0 \neq a_{12}(\hat{\gamma})$ and $k(\hat{\gamma}) = 0$.

Conversely, suppose that $k(\hat{\gamma}) = 0$. We assume first that $a_{12}(\hat{\gamma}) \neq 0$. Then $a_{11}(\hat{\gamma}) = 0$, and so the condition $\det A(\hat{\gamma}) = a_{11}(\hat{\gamma}) = 0$ implies that $a_{21}(\hat{\gamma}) = 0$. Now suppose that $a_{12}(\hat{\gamma}) = 0$. In this case the equality $k(\hat{\gamma}) = 0$ implies that a_{11} has a zero of multiplicity greater than 1 at $\hat{\gamma}$. The lemma is proved.

L. *The character of the many-valuedness and the zeros of the function* $k(\hat{\gamma})$.

LEMMA 6. *The function* $k(\hat{\gamma})$ *has a countable number of zeros* $\{\hat{\gamma}_n\}$ *with distinct projections* γ_n.

PROOF. The function $k(\hat{\gamma})$ can be interpreted as a many-valued function on Γ. Abusing notation somewhat, we shall denote it by $k(\gamma)$. It is easy to investigate the many-valuedness of $k(\gamma)$ knowing (5) and (6). In what follows we use the notation introduced in §1. Let $U_2' \subset \Gamma$ be a neighborhood of -1 $(c_0(-1) = 2)$ such that $c_0(U_2') = U_2$. Note that $c_0'(-1) = 8/3$ and, therefore, the map $c_0|_{U_2'}$ may be assumed to be conformal. Let $\mu_2'(\gamma) \subset U_2'$ be the loop satisfying $c_0\mu_2'(\gamma) = \mu_2(c)$, $c_0(\gamma) = c$, and let $\Delta_2' k(\gamma)$ be the continuation of the function $k(\gamma)$ over the loop $\mu_2'(\gamma)$. It follows from (6) that

$$\Delta_2' k(\gamma) = k(\gamma) + 1, \qquad \gamma \in \sigma'.$$

Therefore, if one of the branches of the function k in the neighborhood U_2' takes an integer value at γ, then $k(\gamma)$ vanishes at some point lying over γ.

PROPOSITION. *The function* $k(\hat{\gamma})$ *takes a countable number of integer values at the points* $\hat{\gamma}_n$, *and the projections* $\pi\hat{\gamma}_n \in U_2'$ *are distinct.*

PROOF. We claim that, in U_2'',

$$k(\gamma) = \frac{\nu}{\gamma + 1} + \frac{\ln(\gamma + 1)}{2\pi i} + f(\gamma), \tag{31}$$

where f is a holomorphic function in U_2' and $\nu \neq 0$. Our proposition will follow immediately, since the principal term $\nu/(\gamma + 1)$ of the decomposition (31) maps U_2'' onto a full neighborhood of infinity. We now establish (31). From (7) it follows that

$$I_1(1, \gamma, c_0(\gamma)) = \frac{\ln(\gamma + 1)}{2\pi i} I_2(1, \gamma, c_0(\gamma)) + f_1(\gamma),$$

where f_1 is a single-valued analytic function in U_2' and

$$f_1(-1) = I_1(1, -1, 2) = I_{11}(2) - I_{12}(2) = 8e \neq 0$$

(the constants e and a, which appear below, were computed in subsections F and H). The integral $I_2(1, \gamma, c_0(\gamma))$ has a simple zero at $\gamma = -1$. In fact,

$$I_2(1, \gamma, c_0(\gamma)) = \left[I_{21}'(2) - I_{22}'(2) \right] c_0'(-1)\gamma + o(\gamma).$$

Using (28) and the relation $b = -a$, we conclude that $\nu = 3ei/a \neq 0$. Formula (31) follows immediately. Lemma 6, together with statement $3°$ of the Main Lemma and statement 3) of the theorem, is proved.

Received 12/SEPT/75

BIBLIOGRAPHY

1. R. I. Bodganov, *Bifurcations of a limit cycle of a family of vector fields on the plane*, Trudy Sem. Petrovsk. **2** (1976), 23–35; English transl. in this volume.

2. G. E. Popov, *Approximate determination of a limit cycle of a certain equation with a small parameter*, Izv. Vysš. Učebn. Zaved. Matematika **1973**, no. 11, 66–70. (Russian)

3. Ju. S. Il'jašenko, *The origin of limit cycles under a perturbation of the equation $dw/dz = -R_z/R_w$, where $R(z, w)$ is a polynomial*, Mat. Sb. **78 (120)** (1969), 360–373; English transl. in Math. USSR Sb. **7** (1969).*

4. M. Hervé, *Several complex variables. Local theory*, Tata Inst. Fund. Res., Bombay, and Oxford Univ. Press, London, 1963.

Translated by DONAL O'SHEA

Translator's note. See also I. G. Petrovskiĭ and E. M. Landis, *On the number of limit cycles of the equation $dy/dx = P(x, y)/Q(x, y)$, where P and Q are polynomials of the second degree*, Mat. Sb. **37 (79)** (1955), 209–250; English transl., Amer. Math. Soc. Transl. (2) **10** (1958), 177–221; and ________, *On the number of limit cycles of the equation $dy/dx = P(x, y)/Q(x, y)$, where P and Q are polynomials*, Mat. Sb. **43 (85)** (1957), 149–168; English transl., Amer. Math. Soc. Transl. (2) **14** (1960), 181–199.

Amer. Math. Soc. Transl.
(2) Vol. **118**, 1982

A Priori Estimates for Equations
of Parabolic Type*

T. F. KALUGINA

Contents

The main result of this paper consists in obtaining sharp estimates for equations of parabolic type in refined scales of spaces (§§3 and 4). For this we investigate classes of spaces with functional parameter of Hölder type $C[t^m\omega(t)]$ (§1) and of Besov type $B_p[t^m\omega(t)]$ (§2). In the case when $\omega(t) \equiv t^\alpha$, where $\alpha \in [0, 1)$, the corresponding estimates were obtained by Solonnikov [1], [2] for noncritical values of α (for any integer $m \geq 0$ and $\alpha \neq 0$ in a C-scale, and for $\alpha \neq 1/p$ in a B-scale). In this paper, main attention is given to estimates in spaces whose smoothness indices are critical or differ from critical by a function of less than power growth.

§1. On the extension of functions from $C[t^m\omega(t)](Q)$

Let E be a Banach space, $\{G(t), t \geq 0\}$ an equicontinuous semigroup of class (C^0), and Λ its infinitesimal generator [3]. A description of the set of interpolation functions f and averaged spaces $S_{f,p}(D(\Lambda^m), E)$ $(D(\Lambda^m) \subset S_{f,p}(D(\Lambda^m), E)$ $\subset E$; here $m \geq 1$ is an integer and $p \in [1, \infty])$ is contained in [4] and [5].

Below η is a positive number; the function $\omega(t)$ is assumed given for $t > 0$, continuously differentiable and nondecreasing, with $\omega(t) = o(\max\{1, t\})$. We set $(\Psi\omega)(t) = t\omega'(t)/\omega(t)$.

1980 *Mathematics Subject Classification.* Primary 35K10, 35B45.
*Translation of Trudy Sem. Petrovsk. **3** (1978), 61–80. MR **58** #1640.

DEFINITION 1. Let the function $\omega(t)$ be such that

$$\overline{\lim_{t \to 0}} \, (\Psi\omega)(t) < 1. \tag{1}$$

Then $C[t\omega(t)]$ is the linear space

$$C[t\omega(t)](E; \Lambda) = C[t\omega(t)] = \left\{ a \in E, \; \sup_{(0,\eta)} \|(I - G(t))^2 a\|_E / t\omega(t) < \infty \right\}.$$

The norm

$$\|a\|_{C[t\omega(t)]} = \|a\|_E + \|a\|_{\pi C[t\omega(t)]} = \|a\|_E + \sup_{(0,\eta)} \|(I - G(t))^2 a\|_E / t\omega(t)$$

makes $C[t\omega(t)]$ a Banach space [5].

PROPOSITION 1. *Let* $\omega_1(t) = \int_0^t s^{-1}\omega(s)\,ds < \infty$. *Then* Λa *exists whenever* $a \in C[t\omega(t)]$.

PROOF. According to [6] (Lemma 1.2 of Chapter VIII, §2) it suffices to establish the convergence of the integral $\int_0^\infty t^{-1}(I - G(t))^2 a t^{-1}\,dt$. Since

$$\int_0^\infty t^{-1}\|(I - G(t))^2 a\|_E \frac{dt}{t} \leq \int_0^1 t^{-1}\|(I - G(t))^2 a\|_E \frac{dt}{t} + \text{const} \int_1^\infty \|a\|_E \frac{dt}{t^2}$$

$$\leq \sup_{(0,1)} \frac{\|(I - G(t))^2 a\|_E}{t\omega(t)} \int_0^1 \omega(t) \frac{dt}{t} + \text{const}\|a\|_E \leq \text{const}\|a\|_{C[t\omega(t)]},$$

it follows that $a \in D(\Lambda)$ and $\Lambda a = \text{const} \int_0^\infty t^{-1}(I - G(t))^2 a t^{-1}\,dt$.

DEFINITION 2. Assume $\omega(t)$ satisfies (1) and m is a natural number. Then $C[t^m\omega(t)]$ is the linear space

$$C[t^m\omega(t)](E; \Lambda) = C[t^m\omega(t)] = \left\{ a \in E, \; \sup_{(0,\eta)} \frac{\|(I - G(t))^2 \Lambda^{m-1} a\|_E}{t\omega(t)} < \infty \right\}.$$

DEFINITION 3. Let $\omega(t)$ satisfy the condition

$$\lim_{t \to 0} \, (\Psi\omega)(t) > 0, \tag{2}$$

with m a natural number on 0. Then $C[t^m\omega(t)]$ is the linear space

$$C[t^m\omega(t)](E; \Lambda) = C[t^m\omega(t)] = \left\{ a \in E, \; \sup_{(0,\eta)} \frac{\|(I - G(t))^2 \Lambda^m a\|_E}{\omega(t)} < \infty \right\}.$$

After introduction of the appropriate norms

$$\|a\|_{C[t^m\omega(t)]} = \|a\|_E + \|a\|_{\pi C[t^m\omega(t)]} = \|a\|_E + \sup_{(0,\eta)} \frac{\|(I - G(t))^2 \Lambda^{m-1} a\|_E}{t\omega(t)}$$

$$\tag{3}$$

or

$$\|a\|_{C[t^m\omega(t)]} = \|a\|_E + \|a\|_{\textrm{п}C[t^m\omega(t)]} = \|a\|_E + \sup_{(0,\eta)} \frac{\|(I - G(t))^2\Lambda^m a\|_E}{\omega(t)}$$

$$(4)$$

$C[t^m\omega(t)]$ becomes a Banach space.

DEFINITION 4. $C^{m,\omega}(E; \Lambda)$, for $m \geq 0$, is a Banach space with norm

$$\|a\|_{C^{m,\omega}(E;\Lambda)} = \|a\|_E + \sup_{(0,\eta)} \frac{\|(I - G(t))\Lambda^m a\|_E}{\omega(t)}. \qquad (5)$$

REMARK 1. If the function $\omega(t)$ satisfies neither (1) nor (2), it suffices to go to the averaged space generated by a difference of order N, where N is such that $N^{-1}\overline{\lim}_{t\to 0}(\Psi\omega)(t) < 1$.

Further investigation of these spaces yields nothing new beyond the study of spaces generated by second order differences. Therefore we assume that at least one of conditions (1) or (2) holds.

REMARK 2. Equivalent moduli of continuity generate one and the same space, with equivalent norms, and the smoothness of the functions in a given class is determined only by the behavior of $\omega(t)$ for small t.

C-spaces may also be described as averaged spaces [5]; namely (below, the conditions to be satisfied by the function $\omega(t)$ appear above the equal sign):

$$C[t^m\omega(t)] \overset{(1)}{=} S_{t^{1/2}\omega(t^{1/2}),\infty}(D(\Lambda^{m+1}), D(\Lambda^{m-1}))$$

$$\|(1), (2)$$

$$C^{m,\omega} \overset{(1),(2)}{=} S_{\omega(t),\infty}(D(\Lambda^{m+1}), D(\Lambda^m))$$

$$\|(1), (2)$$

$$C[t^m\omega(t)] \overset{(2)}{=} S_{\omega(t^{1/2}),\infty}(D(\Lambda^{m+2}), D(\Lambda^m)).$$

Therefore for $\omega(t)$ satisfying (1) and (2) we can write

$$C^{m,\omega} = C[t^m\omega(t)].$$

EXAMPLE 1. Let $\alpha_0 \in (0, 1)$, and let $\alpha_1,\ldots,\alpha_k$ be arbitrary real numbers. Then for

$$a \in C\left[t^{m+\alpha_0}|\ln t|^{\alpha_1}\cdots\underbrace{|\ln \cdots |\ln t|}_{k \text{ times}}\cdots|^{\alpha_k}\right]$$

we have

$$\|a\|_{\textrm{п}C[t^{m+\alpha_0}|\ln t|^{\alpha_1}\cdots|\ln\cdots|\ln t|\cdots^{\alpha_k}]} = \sup_{(0,\tau_i)} \frac{\|(I - G(t))\Lambda^m a\|_E}{t^{\alpha_0}|\ln t|^{\alpha_1}\cdots|\ln \cdots |\ln t|\cdots|^{\alpha_k}}.$$

In the space $C[t^m |\ln t|^{\alpha_1} \cdots |\ln \cdots |\ln t| \cdots |^{\alpha_k}]$, where the α_i $(i = 1, \ldots, k)$ are arbitrary real numbers, we have

$$\|a\|_{\Pi C[t^m|\ln t|^{\alpha_1}\cdots|\ln\cdots|\ln t|\cdots|^{\alpha_k}]} = \sup_{(0,\eta)} \frac{\|(I - G(t))^2 \Lambda^{m-1}a\|_E}{t |\ln t|^{\alpha_1} \cdots |\ln \cdots |\ln t| \cdots |^{\alpha_k}}.$$

We introduce the notation

$$(D\omega)(t) = \int_0^t \omega(s)\frac{ds}{s}, \qquad (T_1\omega)(t) = t\int_t^\eta \omega(s)\frac{ds}{s},$$

$$(T_2\omega)(t) = t\int_t^\eta \frac{\omega(s)}{s}\frac{ds}{s}.$$

LEMMA 1. *If the function $\omega(t)$ satisfies* (1) *and* $(D\omega)(t) < \infty$, *then* $(T_2\omega)(t) \leq$ const$(D\omega)(t)$.

PROOF. (1) is equivalent to the existence of a constant $c > 1$ for which $c\omega(s) \leq (D\omega)(s)$ ([4], §1). Therefore

$$\frac{(c - 1)\omega(s)}{s^2} \leq \frac{(D\omega)(s) - \omega(s)}{s^2} = -\left(\frac{(D\omega)(s)}{s}\right)'.$$

Integrating with respect to s from $t > 0$ to η, we obtain

$$\int_t^\eta \frac{\omega(s)}{s}\frac{ds}{s} \leq \frac{1}{c - 1}\frac{(D\omega)(t)}{t} + \text{const},$$

from which the assertion of the lemma follows.

PROPOSITION 2. *If the function $\omega(t)$ satisfies* (1) *and* $(D\omega)(t) < \infty$, *then* $C[t^m\omega(t)] \subset C^{m,D\omega}$. *However, if $\int_0^\eta \omega(s)s^{-1}\,ds$ diverges at the origin, then* $C[t^m\omega(t)] \subset C^{m-1,T_1\omega}$.
If $\omega(t)$ satisfies (2), *then* $C[t^m\omega(t)] \subset C^{m,T_2\omega}$.

PROOF. Suppose that for $t \in (0, \eta)$ we have $(D\omega)(t) < \infty$ and $a \in C[t^m\omega(t)]$. Then, by assumption 1, $a \in D(\Lambda^m)$. It now suffices to use a particular case of an integral inequality ([8], inequality (6)) which follows from an integral representation of V. P. Il'in [7]. Namely, for any function $\sigma(h)$ continuous on $[0, \eta]$ such that $\sigma(0) = 0$ and $\sigma'(h) > 0$ (we may take $\sigma(h) \equiv h$), we have

$$\|(I - G(t))\Lambda^m a\|_E$$

$$\leq \text{const}\left(t\|\Lambda^{m-1}a\|_E + \left\{\int_0^t + \int_t^\eta \frac{t}{\sigma(h)}\right\}\frac{1}{\sigma(h)}\right.$$

$$\left. \times \int_0^1 \|(I - G(sh))^2 \Lambda^{m-1}a\|_E\, ds\, \frac{\sigma'(h)}{\sigma(h)}\, dh\right).$$

Setting $\sigma(h) \equiv h$, and using the facts that $s \leq 1$ and $\omega(t)$ is nondecreasing, we get

$$\|(I - G(sh))^2 \Lambda^{m-1}a\|_E \leq \|a\|_{C[t^m\omega(t)]} h\omega(h).$$

Therefore

$$\|(I - G(t))\Lambda^m a\|_F$$

$$\leqslant \text{const}\left(t\|\Lambda^{m-1}a\|_E + \|a\|_{C[t^m\omega(t)]}\left\{ \int_0^t \omega(h)\frac{dh}{h} + t\int_t^\eta \frac{\omega(h)}{h}\frac{dh}{h}\right\}\right).$$

It only remains to use Lemma 1. The remaining assertions of the proposition also reduce to the application of inequality (6) from [8].

REMARK 3. It is clear that the imbedding $C^{m,\omega} \subset C[t^m\omega(t)]$ holds for any m and $\omega(t)$ from Definition 2 or Definition 3.

Let $E = C^0((0, \infty))$. The operator $\Lambda = d/dx$ generated by the semigroup of translations defines the spaces

$$C^{m,\omega}\big(C^0((0, \infty)); d/dx\big) \quad \text{and} \quad C[t^m\omega(t)]\big(C^0((0, \infty)); d/dx\big).$$

We assume that the function of one variable $a(x)$ is defined on the positive semiaxis and vanishes together with some of its derivatives at $x = 0$. We define it to be zero on the negative semiaxis:

$$a^0(x) = \begin{cases} a(x) & \text{for } x \geqslant 0, \\ 0 & \text{for } x < 0. \end{cases}$$

THEOREM 1. *Let* $a(x) \in C[t^m\omega(t)]((0, \infty))$, *and let* $\omega(t)$ *satisfy* (1).

I. *If* $(D\omega)(t) < \infty$, *assume that* $a^{(k)}(0) = 0$ *for all integers* k *for which* $0 \leqslant k \leqslant k_{\max} = m_1 = m$. *If* $\omega_1(t) \sim (D\omega)(t)$, *then* $a^0(x) \in C[t^m\omega_1(t)](R)$ *and*

$$\|a^0\|_{C[t^m\omega_1(t)](R)} \leqslant \text{const}\|a\|_{C[t^m\omega(t)]((0,\infty))}. \tag{6.I}$$

II. *If* $(D\omega)(t)$ *does not exist, let* $a^{(k)}(0) = 0$ *for all integers* k *for which* $0 \leqslant k \leqslant k_{\max} = m_1 = m - 1$. *If* $\omega_1(t) \sim (T_1\omega)(t)$, *then* $a^0(x) \in C[t^{m-1}\omega_1(t)](R)$ *and*

$$\|a^0\|_{C[t^{m-1}\omega_1(t)](R)} \leqslant \text{const}\|a\|_{C[t^m\omega(t)]((0,\infty))}. \tag{6.II}$$

III. *Let* $a(x) \in C[t^m\omega(t)]((0, \infty))$, *and let* $\omega(t)$ *satisfy* (2). *If* $a^{(k)}(0) = 0$ *for* $0 \leqslant k \leqslant k_{\max} = m_1 = m$, *then for* $\omega_1(t) \sim (T_2\omega)(t)$ *we have* $a^{(0)}(x) \in C[t^m\omega_1(t)](R)$ *and*

$$\|a^0\|_{C[t^m\omega_1(t)](R)} \leqslant \text{const}\|a\|_{C[t^m\omega(t)]((0,\infty))}. \tag{6.III}$$

REMARK 4. If the modulus of continuity $\omega(t)$ satisfies conditions (1) and (2), then $D(\omega)(t)$, $t^{-1}(T_1\omega)(t)$, $(T_2\omega)(t)$ and $\omega(t)$ are equivalent [4], and the smoothness of the functions when continued by zero is preserved.

PROOF OF THE THEOREM. I. We must estimate

$$\sup_{(0,\eta)} \frac{\|(I - G(t))^2\Lambda^{m-1}a^0\|_E}{t(D\omega)(t)} = \sup_{(0,\eta)} \sup_{x\in R} \frac{|(I - G(t))^2\Lambda^{m-1}a^0(x)|}{t(D\omega)(t)}.$$

But

$$\sup_{x \in R} \left| (I - G(t))^2 \Lambda^{m-1} a^0(x) \right|$$

$$\leqslant \sup_{x > 0} \left| (I - G(t))^2 \Lambda^{m-1} a(x) \right|$$

$$+ \sup_{(-t,0)} \left| \Lambda^{m-1} a(x + 2t) - 2\Lambda^{m-1} a(x + t) \right| + \sup_{(-2t,-t)} \left| \Lambda^{m-1} a(x + 2t) \right|.$$

Since $C[t^m \omega(t)] \subset C^{m, D\omega}$, it follows that

$$\sup_{(-2t,-t)} \left| \Lambda^{m-1} a(x + 2t) \right| \leqslant t \sup_{(0,t)} \left| \Lambda^m a(z) \right|$$

$$= t \sup_{(0,t)} \left| \Lambda^m a(z) - \Lambda^m a(0) \right| \leqslant \| a \|_{{}_\Pi C^{m, D\omega}} \cdot t(D\omega)(t).$$

Furthermore,

$$\sup_{(-t,0)} \left| \Lambda^{m-1} a(x + 2t) - 2\Lambda^{m-1} a(x + t) \right|$$

$$\leqslant \sup_{(t,2t)} \left| \Lambda^{m-1} a(z) \right| + 2 \sup_{(0,t)} \left| \Lambda^{m-1} a(z) \right|$$

$$\leqslant \text{const} \| a \|_{{}_\Pi C^{m, D\omega}} \cdot t(D\omega)(t).$$

The first term obviously does not exceed $ct\omega(t)$, and $\omega(t) \leqslant \text{const}(D\omega)(t)$ (see Lemma 1).

Assertions II and III are proved in a completely analogous manner by reducing a second order difference to a first, and using the imbeddings of Proposition 2.

EXAMPLE 2. If the function $a(x) \in C[t^m |\ln t|^r]((0, \infty))$, where r is any real number, then $a^0(x) \in C[t^m |\ln t|^{r+1}](R)$ since

a) if $r < -1$, then

$$(D |\ln \cdot|^r)(t) = \int_0^t |\ln s|^r \frac{ds}{s} \sim |\ln t|^{r+1},$$

b) if $-1 \leqslant r \leqslant 0$, then

$$(T_1 |\ln \cdot|^r)(t) = t \int_t^\eta |\ln s|^r \frac{ds}{s} \sim t |\ln t|^{r+1},$$

c) if $0 \leqslant r < \infty$, then

$$(T_2 |\ln \cdot|^r)(t) = t \int_t^\eta \frac{s |\ln s|^r}{s} \frac{ds}{s} \sim t |\ln t|^{r+1}.$$

DEFINITION 5. Suppose the domain Q depends on l variables $x_1, \ldots, x_l$ and is of the form $Q = \{x_{i_1} \geqslant 0, \ldots, x_{i_k} \geqslant 0\}$ for $0 \leqslant k \leqslant l$. Let $E = C^0(Q)$ and $\Lambda_i = \partial/\partial x_i$, $i = 1, \ldots, l$. We denote

$$\Omega_k(h) = \begin{cases} \sqrt{h}\, \omega_k(\sqrt{h})(1), \\ \omega_k(h)(1),\,(2), \\ \omega_k(\sqrt{h})(2), \end{cases} \qquad M_k = \begin{cases} m_k - 1(1), \\ m_k(1),\,(2), \\ m_k(2), \end{cases} \qquad \delta_k = \begin{cases} 2(1), \\ 1(1),\,(2), \\ 2(2), \end{cases}$$

where the numbers of the conditions that must be satisfied by the function $\omega_k(h)$ are shown in parentheses. We let

$$C\left[h^{m_1}\omega_1(h),\ldots,h^{m_l}\omega_l(h)\right](Q)$$

$$= \bigcap_{k=1}^{l} S_{\Omega_k,\infty}\left(D(\Lambda_k^{M_k+\delta_k}) \cap C^{M_1,\ldots,M_l}(Q), C^{M_1,\ldots,M_l}(Q)\right).$$

In the case when $l = n + 1$, $m_1 = \cdots = m_n$ and $\omega_1(h) \sim \cdots \sim \omega_n(h)$ we separate out the last variable (denoting points of the domain Q by (x, t)) and, for short, simply write $C[h^{m_x}\omega_x(h), h^{m_t}\omega_t(h)](Q)$.

COROLLARY 1 TO THEOREM 1. *Let the function $a(x, t)$ be defined in the half-space $(t \geq 0)$, and let $a \in C[h^{m_x}\omega_x(h), h^{m_t}\omega_t(h)]$ $(t \geq 0)$. Let $(m_t)_1$ and $(\omega_t)_1(h)$ be defined from m_t and $\omega_t(h)$ in accordance with Theorem 1. If $a_t^{(k)}(x, 0) = 0$ for integer values of k, where $0 \leq k \leq k_{\max} = (m_t)_1$, then the function*

$$a^0(x, t) = \begin{cases} a(x, t) & \text{for } t \geq 0, \\ 0 & \text{for } t < 0 \end{cases}$$

belongs to $C[h^{m_x}(\omega_x)(h), h^{(m_t)_1}(\omega_t)_1(h)](R^{n+1})$ and satisfies the appropriate norm estimate.

§2. On the extension of functions from $B_p[t^m\omega(t)](Q)$

Let positive numbers η and $p \in (1, \infty)$ be given, as well as a Banach space E and an infinitesimal generator Λ.

DEFINITION 6. Let the function $\omega(t)$ and number m be the same as in Definition 2. Then $B_p[t^m\omega(t)]$ is the linear space

$$B_p[t^m\omega(t)](E; \Lambda) = B_p[t^m\omega(t)]$$

$$= \left\{a \in E, \left(\int_0^{\eta}\left(\frac{\|(I - G(t))^2\Lambda^{m-1}a\|_E}{t\omega(t)}\right)^p \frac{dt}{t}\right)^{1/p} = \|a\|_{\Pi B_p[t^m\omega(t)]} < \infty\right\}.$$

$$(7)$$

DEFINITION 7. Let the function $\omega(t)$ and number m be the same as in Definition 3. Then $B_p[t^m\omega(t)]$ is the linear space

$$B_p[t^m\omega(t)](E; \Lambda) = B_p[t^m\omega(t)]$$

$$= \left\{a \in E, \left(\int_0^{\eta}\left(\frac{\|(I - G(t))^2\Lambda^m a\|_E}{\omega(t)}\right)^p \frac{dt}{t}\right)^{1/p} = \|a\|_{\Pi B_p[t^m\omega(t)]} < \infty\right\}. \quad (8)$$

DEFINITION 8. Let m be a nonnegative integer, and let the function $\omega(t)$ satisfy (1) and (2). Then

$$B_p[t^m\omega(t)](E; \Lambda) = B_p[t^m\omega(t)]$$

$$= \left\{ a \in E, \left(\int_0^\eta \left(\frac{\|(I - G(t))^2 \Lambda^m a\|_E}{\omega(t)} \right)^p \frac{dt}{t} \right)^{1/p} = \|a\|_{\amalg B_p[t^m\omega(t)]} < \infty \right\}. \quad (9)$$

Introducing the norm

$$\|a\|_{B_p[t^m\omega(t)]} = \|a\|_E + \|a\|_{\amalg B_p[t^m\omega(t)]}$$

makes $B_p[t^m\omega(t)]$ a Banach space [5].

Definitions 6–8 of B-spaces are compatible. In fact (attached to the equals signs are the conditions the function $\omega(t)$ should satisfy),

$$B_p[t^m\omega(t)] \overset{(1)}{=} S_{t^{1/2}\omega(t^{1/2}),p}(D(\Lambda^{m+1}), D(\Lambda^{m-1})),$$

$$\|(1), (2)$$

$$B_p[t^m\omega(t)] \overset{(1),(2)}{=} S_{\omega(t),p}(D(\Lambda^{m+1}), D(\Lambda^m)),$$

$$\|(1), (2)$$

$$B_p[t^m\omega(t)] \overset{(2)}{=} S_{\omega(t^{1/2}),p}(D(\Lambda^{m+2}), D(\Lambda^m)).$$

REMARK 5. It is clear that the smoothness of functions of class $B_p[t^m(t)]$ is determined by the behavior of $\omega(t)$ only for small t. If $\omega_1(t) \sim \omega_2(t)$ the spaces generated by them coincide and their norms are equivalent.

EXAMPLE 3. Let $\eta < 1$. We denote the function

$$t^{\alpha_0} |\ln t|^{\alpha_1} \cdots |\underbrace{\ln \cdots |\ln t|}_{k \text{ times}} \cdots|^{\alpha_k}$$

by $\mathrm{Ln}(\alpha_0, \alpha_1, \ldots, \alpha_k)$. If $a \in B_p[t^l \mathrm{Ln}(\alpha_0, \alpha_1, \ldots, \alpha_k)]$, where $\alpha_0 \in (0, 1)$ and the α_i $(i = 1, \ldots, k)$ are arbitrary real numbers, then

$$\|a\|_{\amalg B_p[t^l \mathrm{Ln}(\alpha_0, \alpha_1, \ldots, \alpha_k)]} = \left(\int_0^\eta \left(\frac{\|(I - G(t))\Lambda^l a\|_E}{\mathrm{Ln}(\alpha_0, \alpha_1, \ldots, \alpha_k)} \right)^p \frac{dt}{t} \right)^{1/p}.$$

In the space $B_p[t^l \mathrm{Ln}(0, \alpha_1, \ldots, \alpha_k)]$, where l is an integer, we have

$$\|a\|_{\amalg B_p[t^l \mathrm{Ln}(0, \alpha_1, \ldots, \alpha_k)]} = \left(\int_0^\eta \left(\frac{\|(I - G(t))^2 \Lambda^{l-1} a\|_E}{\mathrm{Ln}(1, \alpha_1, \ldots, \alpha_k)} \right)^p \frac{dt}{t} \right)^{1/p}.$$

Below, let $E = L_p(0, \infty)$ and $\Lambda = d/dx$. The imbedding theorems yielding the existence of derivatives of functions in $B_p[t^l\omega(t)]$ are well known (see, for example, [7], [9] or [10]). In particular, if $a \in B_p[t^l\omega(t)]$, then $\Lambda^l a$ exists when the integral $\int_0^\eta (t^{-1}\omega(t))^q \, dt$ converges, where $q^{-1} = 1 - p^{-1}$.

We introduce the notation

$$(D_p\omega)(t) = t^{1/p}\left(\int_0^t \left(\frac{\omega(s)}{s}\right)^q ds\right)^{1/q} = t^{1/p}\left(\int_0^t \left(\frac{\omega(s)}{s^{1/p}}\right)^q \frac{ds}{s}\right)^{1/q},$$

$$(T_p\omega)(t) = t^{1/p}\left(\int_t^\eta \left(\frac{\omega(s)}{s^{1/p}}\right)^q \frac{ds}{s}\right)^{1/q}.$$

PROPOSITION 3. *Let* $a(x) \in B_p[t^l\omega(t)]$.

I. *If* $l = 0$ *and the segment* $[\underline{\lim}_{t\to 0}(\Psi\omega)(t), \overline{\lim}_{t\to 0}(\Psi\omega)(t)]$ *does not contain the point* $1/p$, *or* $l = 1$ *and* $\overline{\lim}_{t\to 0}(\Psi\omega)(t) < 1/p$, *then for any* ω_1 *with* $\omega_1(t) \sim \omega(t)$ *the inequality*

$$\left(\int_0^\infty \left(\frac{|a(x) - a(0)|}{x^l\omega_1(x)}\right)^p dx\right)^{1/p} \leqslant \text{const}\|a\|_{B_p[t^l\omega(t)]} \tag{10.a}$$

holds if $l + (\Psi\omega)(t) > 1/p$, *and the inequality*

$$\left(\int_0^\infty \left(\frac{|a(x)|}{\omega_1(x)}\right)^p dx\right)^{1/p} \leqslant \text{const}\|a\|_{B_p[\omega(t)]} \tag{10.b}$$

holds if $l + (\Psi\omega)(t) < 1/p$.

II. *If* $l = 0$ *and* $(D_p\omega)(\eta)$ *exists, then for any function* $\omega_1(t)$ *for which*

$$\int_0^\eta \left(\frac{(D_p\omega)(t)}{\omega_1(t)}\right)^p \frac{dt}{t} < \infty, \tag{11.a}$$

inequality (10.a) *holds.*

III. *If* $l = 0$ *and* $(D_p\omega)(\eta)$ *does not exist, then for any function* $\omega_1(t)$ *for which*

$$\int_0^\eta \left(\frac{(T_p\omega)(t)}{\omega_1(t)}\right)^p \frac{dt}{t} < \infty, \tag{11.b}$$

inequality (10.b) *holds.*

IV. *If* $l = 1$ *and* $(D_p\omega)(\eta)$ *does not exist, then for a function* $\omega_1(t)$ *satisfying* (11.b), *inequality* (10.a) *holds.*

PROOF. It suffices to consider only smooth functions vanishing for large x.

I. The proof of this part consists in a literal repetition of the proof of Lemma 2 of [1], §2, with the use of Theorem 2 of [4].

II. We use a particular case of Il'in's integral inequality ([7], inequality (54); for integral representations and inequalities see also [10], Chapter IV, §4.4):

$$|a(x) - a(0)| \leqslant \text{const}\left\{\int_0^x + \int_x^\eta \frac{x}{h}\right\} \int_0^h |a(y + h) - a(y)|\, dy\, \frac{dh}{h^2}.$$

Applying Hölder's inequality, we obtain

$$|a(x) - a(0)| \leqslant \text{const}\left(\int_0^x \left(\frac{\|(I - G(h))a\|_E}{\omega(h)}\right)^p \frac{dh}{h}\right)^{1/p}\left(\int_0^x \left(\frac{\omega(h)}{h^{1/p}}\right)^q \frac{dh}{h}\right)^{1/q};$$

hence

$$\left(\int_0^\eta \left(\frac{|a(x) - a(0)|}{\omega_1(x)} \right)^p dx \right)^{1/p}$$

$$\leqslant \left(\int_0^\eta \left(\frac{\|(I - G(h))a\|_E}{\omega(h)} \right)^p \left\{ \int_h^\eta \left(\frac{(D_p\omega)(x)}{\omega_1(x)} \right)^p \frac{dx}{x} \right\} \frac{dh}{h} \right)^{1/p}.$$

From this and (11.a), we obtain

$$\left(\int_0^\eta \left(\frac{|a(x) - a(0)|}{\omega_1(x)} \right)^p dx \right)^{1/p} \leqslant \mathrm{const} \|a\|_{\Pi B_p[\omega(t)]}.$$

Assertions III and IV are proved in the same manner with the help of [7], (38) and (54).

THEOREM 2. *Let* $a(x) \in B_p[t^l\omega(t)]$, *and set* $k_{\max} = l$ *if* $(D_p\omega)(t)$ *exists, and* $k_{\max} = l - 1$ *if it does not. Assume that* $a^{(k)} = 0$ *for all integral* k *satisfying* $0 \leqslant k \leqslant k_{\max}$. *Let*

$$a^0(x) = \begin{cases} a(x) & \text{for } x > 0, \\ 0 & \text{for } x < 0. \end{cases}$$

Then $a^0 \in B_p[t^l\omega(t)](R)$, *and*

$$\|a^0\|_{B_p[t^l\omega_1(t)](R)} \leqslant \mathrm{const} \|a\|_{B_p[t^l\omega(t)]((0,\infty))}, \tag{12}$$

where the function $\omega_1(t)$ *is defined in Proposition 3.*

REMARK 6. Thus, extending a function by zero does not preserve smoothness when $\omega(t)$ satisfies

$$1/p \in \left[\varliminf_{t\to 0} (\Psi\omega)(t), \ \varlimsup_{t\to 0} (\Psi\omega)(t) \right].$$

In this case, deterioration of smoothness occurs even in the class of compactly supported functions equal to zero in a neighborhood of the origin. In the case when $\omega(t) \sim t^{1/p}$, Solonnikov ([1], §3) constructed a sequence $\{\xi_n(x)\}$, $n = 0, 1, 2, \ldots$, of functions for which the seminorms $\|\xi_n\|_{\Pi B_p'[t^l\omega(t)]((0,\infty))}$ remain bounded, but $\|\xi_n^0\|_{\Pi B_p[t^l\omega(t)](R)} \to \infty$ as $n \to \infty$.

PROOF OF THEOREM 2. First, suppose that l and $\omega(t)$ satisfy condition I of Proposition 3. We have

$$\int_{-\infty}^{\infty} dx \int_0^{\infty} \left(\frac{|(I - G(h))^2 a^0(x)|}{h^l f(h)} \right)^p \frac{dh}{h}$$

$$= \int_0^{\infty} dx \int_0^{\infty} \left(\frac{|(I - G(h))^2 a(x)|}{h^l f(h)} \right)^p \frac{dh}{h} + \int_{-\infty}^0 dx \int_{-x/2}^{-x} \left(\frac{|a(x + 2h)|}{h^l f(h)} \right)^p \frac{dh}{h}$$

$$+ \int_{-\infty}^0 dx \int_{-x}^{\infty} \left(\frac{|a(x + 2h) - 2a(x + h)|}{h^l f(h)} \right)^p \frac{dh}{h}.$$

But

$$\int_{-\infty}^{0} dx \int_{-x/2}^{-x} \left(\frac{|a(x+2h)|}{h^l f(h)} \right)^p \frac{dh}{h} \leq \text{const} \int_{0}^{\infty} \left(\frac{|a(y) - a(0)|}{y^l f(y)} \right)^p dy,$$

since $a(0) = 0$ and ([4], §11)

$$\int_{y}^{\infty} |h^l f(h)|^{-p} \frac{dh}{h} \leq \text{const}(y^l f(y))^{-p}.$$

Furthermore,

$$\int_{-\infty}^{0} dx \int_{-x}^{\infty} \left(\frac{|a(x+2h) - 2a(x+h)|}{h^l f(h)} \right)^p \frac{dh}{h}$$

$$\leq \text{const} \left(\int_{-\infty}^{0} dx \int_{-x}^{\infty} \left(\frac{|a(x+2h)|}{h^l f(h)} \right)^p \frac{dh}{h} + \int_{-\infty}^{0} dx \int_{-x}^{\infty} \left(\frac{|a(x+h)|}{h^l f(h)} \right)^p \frac{dh}{h} \right)$$

$$\leq \text{const} \int_{0}^{\infty} \left(\frac{|a(y) - a(0)|}{y^l f(y)} \right)^p dy.$$

Thus (12) follows from Proposition 3, if we set $f(t) \sim \omega_1(t)$.

Now suppose l and $\omega(t)$ satisfy condition II, III, or IV of Proposition 3. Then

$$\int_{0}^{\infty} (f(h))^{-p} \int_{-\infty}^{\infty} |a^0(x+h) - a^0(x)|^p \, dx \, \frac{dh}{h}$$

$$= \|a\|_{B_p[f(h)]((0,\infty))}^p + \int_{0}^{\infty} (f(h))^{-p} \int_{-h}^{0} |a(x+h)|^p \, dx \, \frac{dh}{h}$$

$$\leq \text{const} \left(\|a\|_{B_p[f(h)]((0,\infty))}^p + \int_{0}^{\infty} \left(\frac{|a(y)|}{f(y)} \right)^p dy \right),$$

which means that the establishment of (12) again reduces to an application of Proposition 3 with $f(t) \sim \omega_1(t)$.

If $l = 1$ and $(D_p \omega)(t) < \infty$ or $l \geq 2$, we obtain the desired estimate by considering the function $a^{(k_{\max})}(x)$ for which one of the solutions I, II, or IV of Proposition 3 is already fulfilled.

DEFINITION 9. Let the domain Q be the same as in Definition 5, with $E = L_p(Q)$ and

$$\Lambda_i = \partial/\partial x_i \qquad (i = 1, \ldots, m).$$

We introduce the notation

$$\Omega_k(h) = \begin{cases} \sqrt{h}\,\omega_k(\sqrt{h})(1), \\ \omega_k(h) \quad (1), (2), \\ \omega_k(\sqrt{h}) \quad (2), \end{cases} \qquad L_k = \begin{cases} l_k - 1 \quad (1), \\ l_k \quad (1), (2), \\ l_k \quad (2), \end{cases} \qquad \delta_k = \begin{cases} 2 \ (1), \\ 1 \ (1), (2), \\ 2 \quad (2), \end{cases}$$

where the conditions that must be satisfied by $\omega_k(h)$ are shown in parentheses. We set

$$B_p\big[h^{l_1}\omega_1(h),\ldots,h^{l_m}\omega_m(h)\big](Q)$$

$$= \bigcap_{k=1}^{m} S_{\Omega_k,p}\big(D(\Lambda_k^{L_k+\delta}) \cap W_p^{L_1,\ldots,L_m}(Q),\, W_p^{L_1,\ldots,L_m}(Q)\big)$$

$$\overset{*}{=} \bigcap_{l=1}^{m} B_p\big[h^{l_i}\omega_i(h)\big](Q \cap R_{x_i}),$$

where $R_{x_i} = \{-\infty < x_i < \infty\}$, and the equality marked by an asterisk is satisfied for $1 < p < \infty$ ([10], §9.2).

In the case when $m = n + 1$, $l_1 = \cdots = l_n$ and $\omega_1(h) \sim \cdots \sim \omega_n(h)$, we separate out the last variable (denoting points of the domain Q by (x, t)) and shorten the notation to $B_p[h^{l_x}\omega_x(h),\, h^{l_t}\omega_t(h)](Q)$.

COROLLARY 1 OF THEOREM 2. *Let the function $a(x, t)$ be given in the half-space $t > 0$, and let $a \in B_p[h^{l_x}\omega_x(h),\, h^{l_t}\omega_t(h)]$ $(t > 0)$. Let $k_{\max}$ and $(\omega_t)_1(h)$ be defined from l_t and $\omega_t(h)$ according to Theorem 2. If $a_t^{(k)}(x, 0) = 0$ for integral k for which $0 \leqslant k \leqslant k_{\max}$, then the extension $a^0(x, t)$ of the function $a(x, t)$ by zero for negative t belongs to $B_p[h^{l_x}\omega_x(h),\, h^{l_t}(\omega_t)_1(h)](R^{n+1})$.*

Thus the smoothness $\omega_t(h)$ is in general not preserved, even in the class of compactly supported functions equal to zero for small t.

In B-spaces with functional parameter, all theorems having to do with restrictions to a subspace, or extension to a domain of higher dimension remain valid ([5], §2); moreover the change in smoothness does not depend on the functional parameter, and coincides with the corresponding change for spaces with numerical parameter. Below, these theorems will be assumed known.

COROLLARY 2 OF THEOREM 2. *Let nonnegative integers m_x and l_t be given, as well as smoothness functions $\varphi_x(h)$ and $\omega_t(h)$ satisfying the following conditions:*

$$l_t + \lim_{h \to 0} (\Psi\omega_t)(h) > 1/p, \tag{13}$$

$$2l_t + 2(\Psi\omega_t)(h) = l_x + (\Psi\omega_x)(h) = m_x + (\Psi\varphi_x)(h) + 2/p, \tag{14}$$

$$1/p \in \left[\varliminf_{h \to 0} (\Psi\varphi_x)(h),\, \varlimsup_{h \to 0} (\Psi\varphi_x)(h)\right], \tag{15}$$

where the points $1/p$ and $1 + 1/p$ do not lie in the interval

$$\left[2 \varliminf_{h \to 0} (\Psi\omega_t)(h),\, 2 \varlimsup_{h \to 0} (\Psi\omega_t)(h)\right]. \tag{16}$$

Let the function $a(x) \in B_p[h^{m_x}\varphi_x(h)]$ $(x_n > 0,\, t = 0)$ vanish for small x_n. Then there does not in general exist a function $b(x, t)$ on $(t > 0,\, x_n > 0)$ such that

$b(x, 0) = a(x)$ and $b_{x_n}^{(j)}|_{x_n=0} = 0$ *for all integers* $j \geq 0$ *with the following property that for any integer* k,

$$j \leq 2l_t + 2 \varliminf_{h \to 0} (\Psi \omega_t)(h) - \frac{1}{p} \leq 2l_t + 2(\Psi \omega_t)(h) - \frac{1}{p}$$

$$= k + (\Psi f)(h) - \frac{1}{p},$$

the equality

$$j = 2l_t + 2 \varliminf_{h \to 0} (\Psi \omega_t)h - 1/p$$

holding only when $(D_p f)(h) < \infty$; *and*

$$\| b(x, t) \|_{B_p[h^l x \omega_x(h), h^l t \omega_t(h)] \, (t>0, x_n>0)} \leq \mathrm{const} \| a \|_{B_p'[h^m x \varphi_x(h)] \, (x_n>0)}.$$

PROOF. Suppose such a function exists, and extend it to be zero in the domain $(t > 0)$; the smoothness of the extension $b^0(x, t)$ remains as before, thanks to condition (16). Therefore

$$\| b^0(x, t) \|_{B_p[h^l x \omega_x(h), h^l t \omega_t(h)] \, (t>0)} \leq \mathrm{const} \| b(x, t) \|_{B_p[h^l x \omega_x(h), h^l t \omega_t(h)] \, (t>0, x_n>0)}.$$

But because

$$m_x + (\Psi \varphi_x)(h) = 2(l_t + (\Psi \omega_t)(h) - 1/p) \geq \varepsilon > 0,$$

we conclude from the theorem on traces and extensions that

$$\| b^0(x, 0) \|_{B_p[h^m x \varphi_x(h)](R^n)} \leq \mathrm{const} \| b^0(x, t) \|_{B_p[h^l x \omega_x(h), h^l t \omega_t(h)] \, (t>0)}$$

$$\leq \mathrm{const} \| a(x) \|_{B_p[h^m x \varphi_x(h)] \, (x_n>0)},$$

which cannot be true, according to Corollary 1 of Theorem 2.

§3. A priori estimates for equations of parabolic type in a C-scale

Let $R^{n+1} = \{(x, t)\} = \{(x', x_n, t)\}$ be a Euclidean space. We consider the following potentials, in terms of which solutions of problems for the heat equation in half-spaces are expressed:

$$P[f] = \int_{R^{n+1}} \Gamma(x - y, t - \tau) f(y, \tau) \, dy \, d\tau,$$

$$K[u] = \int_{(t=0)} \Gamma(x - y, t) u(y) \, dy,$$

$$V[\varphi] = \int_{(x_n=0)} \Gamma(x - y', t - \tau) \varphi(y', \tau) \, dy' \, d\tau, \qquad W[\varphi] = -2 \partial_{x_n} V[\varphi],$$

where

$$\Gamma(x, t) = \begin{cases} \exp(-x^2/4t)(4\pi t)^{-n/2} & \text{for } t > 0; \\ 0 & \text{for } t < 0. \end{cases}$$

Estimates for these potentials in the usual Hölder norms are well known ([2], §11). An interpolation theorem with functional parameter ([4], Theorem 7), applied to inequalities (3.4)–(3.6) of [2], leads to the estimates

$$\|P[f]\|_{C[h^{2(m+1)}\omega^2(h),h^{m+1}\omega(h)](R^{n+1})} \leqslant \text{const}\|f\|_{C[h^{2m}\omega^2(h),h^m\omega(h)](R^{n+1})}, \quad (17)$$

$$\|K[u]\|_{C[h^{2m}\omega^2(h),h^m\omega(h)](t\geqslant 0)} \leqslant \text{const}\|u\|_{C[h^{2m}\omega^2(h)](R^n)}, \quad (18)$$

$$\|W[\varphi]\|_{C[h^{2m}\omega^2(h),h^m\omega(h)](x_n\geqslant 0)} \leqslant \text{const}\|\varphi\|_{C[h^{2m}\omega^2(h),h^m\omega(h)](x_n=0)}, \quad (19)$$

which hold for any m and $\omega(h)$, since, according to Definitions 2 and 3,

$$m + \lim_{\overline{h\to 0}} (\Psi\omega)(h) > 0.$$

We now consider, in the half-space $R^{n+1} \cap \{t \geqslant 0\}$, the Cauchy problem for the heat equation:

$$\partial_t u - \Delta u = f; \qquad u\,|_{t=0} = u_0(x). \qquad (\text{I})$$

THEOREM 3. *The solution of the Cauchy problem satisfies the a priori estimate*

$$\|u\|_{C[h^{2(m+1)}\omega^2(h),h^{m+1}\omega(h)](t\geqslant 0)}$$
$$\leqslant \text{const}\big(\|f\|_{C[h^{2m}\omega^2(h),h^m\omega(h)](t\geqslant 0)} + \|u_0\|_{C[h^{2(m+1)}\omega^2(h)](t=0)}\big).$$

PROOF. For $f(x, t)$ there exists an extension $F(x, t)$ to R^{n+1} with norm estimate ([4], Theorem 7)

$$\|F\|_{C[h^{2m}\omega^2(h),h^m\omega(h)](R^{n+1})} \leqslant \text{const}\|f\|_{C[h^{2m}\omega^2(h),h^m\omega(h)](t\geqslant 0)}.$$

It now remains to apply estimates (17)–(19) for potentials, in terms of which we express the solution of the Cauchy problem:

$$\|u\|_{C[h^{2(m+1)}\omega^2(h),h^{m+1}\omega(h)](t\geqslant 0)}$$
$$\leqslant \text{const}\big(\|P[F]\|_{C[h^{2(m+1)}\omega^2(h),h^{m+1}\omega(h)](R^{n+1})} + \|u_0\|_{C[h^{2(m+1)}\omega^2(h)](t=0)}\big)$$
$$\leqslant \text{const}\big(\|f\|_{C[h^{2m}\omega^2(h),h^m\omega(h)](t\geqslant 0)} + \|u_0\|_{C[h^{2(m+1)}\omega^2(h)](t=0)}\big). \qquad (20)$$

Consider in the domain $\{x_n \geqslant 0\}$ the boundary value problem

$$\partial_t v - \Delta v = f, \qquad v\,|_{x_n=0} = \varphi(x', t); \qquad (\text{II})$$

$$\partial_t w - \Delta w = f, \qquad \partial_{x_n} w\,|_{x_n=0} = \psi(x', t). \qquad (\text{III})$$

The arguments used in obtaining the estimate for the Cauchy problem yield the following theorem.

THEOREM 4. *The solutions of problems* (II) *and* (III) *satisfy the inequalities*

$$\|v\|_{C[h^{2(m+1)}\omega^2(h),h^{m+1}\omega(h)](x_n\geqslant 0)}$$
$$\leqslant \text{const}\big(\|f\|_{C[h^{2m}\omega^2(h),h^m\omega(h)](x_n\geqslant 0)} + \|\varphi\|_{C[h^{2(m+1)}\omega^2(h),h^{m+1}\omega(h)](x_n=0)}\big), \quad (21)$$

$$\|w\|_{C[h^{2(m+1)}\omega^2(h),h^{m+1}\omega(h)](x_n\geqslant 0)}$$
$$\leqslant \text{const}\big(\|f\|_{C[h^{2m}\omega^2(h),h^m\omega(h)](x_n\geqslant 0)} + \|\psi\|_{C[h^{2m+1}\omega^2(h),h^{m+1/2}\omega(h)](x_n=0)}\big). \quad (22)$$

In the domain $(t \geqslant 0) \cap (x_n \geqslant 0)$ let the mixed problems

$$\partial_t v - \Delta v = f, \qquad v\big|_{t=0} = v_0(x), \qquad v\big|_{x_n=0} = \varphi(x', t); \qquad \text{(IV)}$$

$$\partial_t w - \Delta w = f, \qquad w\big|_{t=0} = w_0(x), \qquad \partial_{x_n} w\big|_{s_n=0} = \psi(x', t), \qquad \text{(V)}$$

be posed. We extend the function f into the domain $(t \geqslant 0)$, and the functions v_0 and w_0 into R^n. Let v_1 and w_1 be solutions of the Cauchy problems with initial conditions

$$v_1\big|_{t=0} = v_0; \qquad w_1\big|_{t=0} = w_0.$$

Then $v = v_1 + v_2$ and $w = w_1 + w_2$, where $v_2(x, t)$ and $w_2(x, t)$ are solutions of the problems

$$\left.\begin{array}{l} \partial_t v_2 - \Delta v_2 = 0, \\[4pt] v_2\big|_{t=0} = 0, \\[4pt] v_2\big|_{x_n=0} = \varphi - v_1\big|_{x_n=0} = \Phi(x', t); \end{array}\right\}$$

$$\left.\begin{array}{l} \partial_t w_2 - \Delta w_2 = 0, \\[4pt] w_2\big|_{t=0} = 0, \\[4pt] \partial_{x_n} w_2\big|_{x_n=0} = \psi - \partial_{x_n}\omega_1\big|_{x_n=0} = \Psi(x', t) \end{array}\right\}.$$

It is well known that the solutions of these problems are given by the polynomials

$$v_2 = W[\Phi^0], \qquad w_2 = -2V[\Psi^0],$$

where

$$\Phi^0(x', t) = \begin{cases} \Phi(x', t) & \text{for } t \geqslant 0, \\ 0 & \text{for } t < 0; \end{cases}$$

$$\Psi^0(x', t) = \begin{cases} \Psi(x', t) & \text{for } t \geqslant 0, \\ 0 & \text{for } t < 0. \end{cases}$$

Thus, to guarantee the smoothness of the extensions $\Phi^0(x', t)$ and $\Psi^0(x', t)$, we should require the satisfaction of the following compatibility conditions: for problem (IV),

$$\partial_t^{(k)}\varphi\big|_{t=0} = \partial_t^{(k)}v\big|_{t=0,\, x_n=0} \qquad (23)$$

and for problem (V),

$$\partial_t^{(k)}\psi\big|_{t=0} = \partial_{x_n}\partial_t^{(k)}w\big|_{t=0,\, x_1=0}, \qquad (24)$$

where

$$\partial_t^{(k)}v\big|_{t=0} = \Delta\partial_t^{(k-1)}v\big|_{t=0} + \partial_t^{(k-1)}f\big|_{t=0}, \qquad v\big|_{t=0} = v_0(x);$$

$$\partial_{x_n}\partial_t^{(k)}w\big|_{t=0} = \Delta\partial_t^{(k-1)}w\big|_{t=0} + \partial_t^{(k-1)}f\big|_{t=0}; \qquad w\big|_{t=0} = w_0(x),$$

for $0 \leqslant k \leqslant k_{\max}$. The order of the compatibility, $k_{\max}$, in problem (IV) is determined according to Theorem 1, using $h^{m+1}\omega(h)$. It is equal to $m+1$ when $\int_0^\eta \omega(s)s^{-1}\,ds$ exists, and to m otherwise. In problem (V), the order of compatibility is determined in the same way, using $h^{m+1/2}\omega(h)$. The number m and the function $\omega_1(h)$ are also given by Theorem 1. Thus

$$\|\Phi^0\|_{C[h^{2(m_1+1)}\omega_1^2(h),\,h^{m_1+1}\omega_1(h)]\,(x_n=0)} \leqslant \mathrm{const}\|\Phi\|_{C[h^{2(m+1)}\omega^2(h),\,h^{m+1}\omega(h)]\,(t\geqslant 0,\,x_n=0)},$$

$$\|\Psi^0\|_{C[h^{2m_1+1}\omega_1^2(h),\,h^{m_1+1/2}\omega_1(h)]\,(x_n=0)} \leqslant \mathrm{const}\|\Psi\|_{C[h^{2m+1}\omega^2(h),\,h^{m+1/2}\omega(h)]\,(t\geqslant 0,\,x_n=0)}.$$

Arguing as in the estimate of the solutions of (II) or (III), we prove the following theorem.

THEOREM 5. *If, in problem* (IV), *compatibility conditions* (23) *of order* $k_{\max}$ *are satisfied, the order* $k_{\max}$ *being determined by Theorem* 1 *using* $h^{m+1}\omega(h)$, *or if, in problem* (V), *the compatibility conditions* (24) *of order* $k_{\max}$ *determined by* $h^{m+1/2}\omega(h)$ *are satisfied, then the solutions of these problems satisfy the inequalities*

$$\|v(x,t)\|_{C[h^{2(m_1+1)}\omega_1^2(h),\,h^{m_1+1}\omega_1(h)]\,(t\geqslant 0,\,x_n\geqslant 0)}$$

$$\leqslant \mathrm{const}\Big(\|f(x,t)\|_{C[h^{2m}\omega^2(h),\,h^m\omega(h)]\,(t\geqslant 0,\,x_n\geqslant 0)} + \|v_0(x)\|_{C[h^{2(m+1)}\omega^2(h)]\,(t=0,\,x_n\geqslant 0)}$$

$$+\,\|\varphi(x',t)\|_{C[h^{2(m+1)}\omega^2(h),\,h^{m+1}\omega(h)]\,(t\geqslant t0,\,x_n=0)}\Big), \qquad (25)$$

$$\|\omega(x,t)\|_{C[h^{2(m+1)}\omega_1^2(h),\,h^{m+1}\omega_1(h)]\,(t\geqslant 0,\,x_n\geqslant 0)}$$

$$\leqslant \mathrm{const}\Big(\|f(x,t)\|_{C[h^{2m}\omega^2(h),\,h^m\omega(h)]\,(t\geqslant 0,\,x_n\geqslant 0)} + \|w_0\|_{C[h^{2(m+1)}\omega^2(h)]\,(t=0,\,x_n\geqslant 0)}$$

$$+\,\|\psi(x',t)\|_{C[h^{2m+1}\omega^2(h),\,h^{m+1/2}\omega(h)]\,(t\geqslant 0,\,x_n=0)}\Big). \qquad (26)$$

The a priori estimates so obtained lead to existence and uniqueness theorems for solutions of (I)–(V) in a C-scale with functional parameter.

EXAMPLE 4. In problem (IV), suppose $\omega(h) \sim \mathrm{const}$. If

$$f(x,t) \in C[h^{2m},\,h^m] \quad (t\geqslant 0,\,x_n\geqslant 0), \qquad v_0(x) \in C[h^{2m+1}] \quad (t=0,\,x_n\geqslant 0),$$

$$\varphi(x',t) \in C[h^{2(m+1)},\,h^{m+1}] \qquad (t\geqslant 0,\,x_n=0),$$

then the order of compatibility is equal to m, and (25) holds for a solution $v(x,t)$ of class $C[h^{2(m+1)}|\ln h|^2,\,h^{m+1}|\ln h|]\,(t\geqslant 0,\,x_n\geqslant 0)$.

In problem (V), let $\omega(h) \sim h^{1/2}$. If

$$f(x,t) \in C[h^{2m+1},\,h^{m+1/2}] \quad (t\geqslant 0,\,x_n\geqslant 0),$$

$$w_0(x) \in C[h^{2m+3}] \quad (t=0,\,x_n\geqslant 0),$$

$$\psi(x',t) \in C[h^{2(m+1)},\,h^{m+1}] \qquad (t\geqslant 0,\,x_n=0),$$

then the order $k_{\max} = m$, and (26) holds for the solution

$$w(x,t) \in C[h^{2m+3}|\ln h|^2,\,h^{m+3/2}|\ln h|] \qquad (t\geqslant 0,\,x_n\geqslant 0).$$

§4. A priori estimates for equations of parabolic type
in a B-scale

Let the number α lie in the interval $(0, 1]$, and let $\omega(h) \sim h^\alpha$. Then the space

$$B_p[h^{l+\alpha}](Q) = B_p[h^r](Q) \qquad (r > 0)$$

is the well-known Besov space estimates of potentials in a B-scale with numerical parameter ([2], §12) go over with the aid of an interpolation theorem ([4], Theorem 7), to the spaces $B_p[h^m\omega(h)](Q)$. Namely:

$$\| P[f] \|_{B_p[h^{2(l+1)}\omega^2(h),h^{l+1}\omega(h)](R^{n+1})} \leqslant \mathrm{const}\| f \|_{B_p[h^{2l}\omega^2(h),h^{l}\omega(h)](R^{n+1})}, \tag{27}$$

$$\| K[u] \|_{B_p[h^{2l+2/p}\omega^2(h),h^{l+1/p}\omega(h)](t>0)} \leqslant \mathrm{const}\| u \|_{B_p[h^{2l}\omega^2(h)](t=0)}, \tag{28}$$

$$\| V[\varphi] \|_{B_p[h^{2l+1/p}\omega^2(h),h^{l+1/2+1/2p}\omega(h)](x_n>0)} \leqslant \mathrm{const}\| \varphi \|_{B_p[h^{2l}\omega^2(h),h^{l}\omega(h)](x_n=0)}, \tag{29}$$

$$\| W[\varphi] \|_{B_p[h^{2l+1/p}\omega^2(h),h^{l+1/2p}\omega(h)](x_n>0)} \leqslant \mathrm{const}\| \varphi \|_{B_p[h^{2l}\omega^2(h),h^{l}\omega(h)](x_n=0)}. \tag{30}$$

From here on, the argument is similar to that in §3.

THEOREM 6. *The solution of* (I) *satisfies the a priori estimate*

$$\| u \|_{B_p[h^{2l+2}\omega^2(h),h^{l+1}\omega(h)](t>0)}$$

$$\leqslant \mathrm{const}\Big(\| f \|_{B_p[h^{2l}\omega^2(h),h^{l}\omega(h)](t>0)} + \| u_0 \|_{B_p[h^{2(l+1-1/p)}\omega^2(h)](t=0)}\Big). \tag{31}$$

THEOREM 7. *Solutions of problems* (II) *and* (III) *satisfy the inequalities*

$$\| v(x, t) \|_{B_p[h^{2(l+1)}\omega^2(h),h^{l+1}\omega(h)](x_n>0)}$$

$$\leqslant \mathrm{const}\Big(\| f(x, t) \|_{B_p[h^{2l}\omega^2(h),h^{l}\omega(h)](x_n>0)}$$

$$+ \| \varphi(x', t) \|_{B_p[h^{2(l+1-1/2p)}\omega^2(h),h^{l+1-1/2p}\omega(h)](x_n=0)}\Big), \tag{32}$$

$$\| w(x, t) \|_{B_p[h^{2(l+1)}\omega^2(h),h^{l+1}\omega(h)](x_n>0)}$$

$$\leqslant \mathrm{const}\Big(\| f \|_{B_p[h^{2l}\omega^2(h),h^{l}\omega(h)](x_n>0)}$$

$$+ \| \psi(x', t) \|_{B_p[h^{2\cdot+1-1/p}\omega^2(h),h^{l+1/2-1/2p}\omega(h)](x_n=0)}\Big). \tag{33}$$

THEOREM 8. *Let $k_{\max}$ and $\omega_1(h)$ be defined in the following manner: For problem* (IV), *the quantities l, p, and $\omega(h)$ are used in the equality $h^{l+1-1/2p}\omega(h) = h^k\Omega(h)$ to deermine a nonnegative integer k and function $\Omega(h)$. In turn, this k and $\Omega(k)$ are used in Theorem 2 to determine an order of compatibility $k_{\max}$ and function $\Omega_1(h)$, after which $h^k\Omega_1(h)$ is represented in the form $h^{l+1-1/2p}\omega_1(h)$. For problem* (V), k *and $\Omega(h)$ are determined in the same way from $h^{l+1/2-1/2p}\omega(h)$; the quantities $k_{\max}$ and $\Omega_1(h)$ are then determined from Theorem 2, and finally the function $\omega_1(h)$ is obtained from the equality $h^k\Omega_1(h) = h^{l+1/2-1/2p}\omega_1(h)$.*

In problem (IV), *assume the conditions of compatibility* (23) *are fulfilled, and, in problem* (V), *conditions* (24) *of order* $k_{\max}$. *Then the solutions of these problems satisfy the inequalities*

$$\| v(x, t) \|_{B_p[h^{2(l+1)}\omega_1^2(h),\, h^{l+1}\omega_1(h)]\,(t>0,\, x_n>0)}$$

$$\leq \operatorname{const}\Big(\| f(x, t) \|_{B_p[h^{2l}\omega^2(h),\, h^l\omega(h)]\,(t>0,\, x_n>0)}$$

$$+ \| v_0(x) \|_{B_p[h^{2(l+1-1/p)}\omega^2(h)]\,(t=0,\, x_n>0)}$$

$$+ \| \varphi(x', t) \|_{B_p[h^{2(l+1-1/2p)}\omega^2(h),\, h^{l+1-1/2p}\omega(h)]\,(t>0,\, x_n=0)} \Big), \quad (34)$$

$$\| w(x, t) \|_{B_p[h^{2(l+1)}\omega_1^2(h),\, h^{l+1}\omega_1(h)]\,(t>0,\, x_n>0)}$$

$$\leq \operatorname{const}\Big(\| f(x, t) \|_{B_p[h^{2l}\omega^2(h),\, h^l\omega(h)]\,(t>0,\, x_n>0)}$$

$$+ \| w_0(x) \|_{B_p[h^{2(l+1-1/p)}\omega^2(h)]\,(t=0,\, x_n>0)}$$

$$+ \| \psi(x', t) \|_{B_p[h^{2l+1-1/p}\omega^2(h),\, h^{l+1/2-1/2p}\omega(h)]\,(t>0,\, x_n=0)} \Big). \quad (35)$$

REMARK 7. If the interval

$$\left[l + 1 - 1/2p + \varliminf_{h\to 0} (\Psi\omega)(h),\; l + 1 - 1/2p + \varlimsup_{h\to 0} (\Psi\omega)(h) \right]$$

contains points of the form $k + 1/p$ (problem (IV)) or the form $k - 1/2 + 1/p$ (problem (V)), where k is na integer, then the B-class in which the solution lies is described by the function $\omega_1(h)$, which is not equivalent to $\omega(h)$. Corollary 2 of Theorem 2 shows that this deterioration is inevitable.

As consequences of the estimates so obtained, we have existence and uniqueness theorems for solutions of (I)–(V) in a B-scale with functional parameter.

§5. An investigation of general boundary value problems
in cylindrical domains of finite height,
in B-classes and C-classes of functions

Let Q_0 be a rectangular parallelipiped ($0 \leq x_i \leq r_i$, $i = 1,\ldots,n$) in R^n, and let $1 < p \leq \infty$. The averaged spaces $S_{f,p}(Q_0)$ were defined ([5], §2) with the aid of factorization. An equivalent approach to them will be described below.

We set

$$\| a \|_{\mathrm{n}\tilde{S}_{f,p}(Q_0)} \sum_{i=1}^{n} \| a \|_{\mathrm{n}\tilde{S}_{f,p}(D(\Lambda_i^m),\, E)(Q_0)}$$

$$= \sum_{i=1}^{n} \left(\int_{Q_0} dx \int_0^{(r_i-x_i)/m} \left(\frac{|(I - G_i(h))^m a|}{f(h^m)} \right)^p \frac{dh}{h} \right)^{1/p};$$

$$\| a \|_{\tilde{S}_{f,p}(Q_0)} = \| a \|_E + \| a \|_{\mathrm{n}\tilde{S}_{f,p}(Q_0)}.$$

LEMMA 2. *The norms* $\|a\|_{\tilde{S}_{f,p}(Q_0)}$ *and* $\|a\|_{S_{f,p}(Q_0)}$ *are equivalent.*

PROOF. Let $a \in \tilde{S}_{f,p}(Q_0)$. Then the function $a(x)$ may be extended into a domain Q satisfying Definition 5 so that it will be zero for large $|x|$, and

$$\|\tilde{a}\|_{S_{f,p}(Q)} \leqslant \mathrm{const}\|a\|_{\tilde{S}_{f,p}(Q_0)},$$

where $\tilde{a}$ is the extension of a. The extension may be accomplished by the method of Hestenes and Whitney, followed by multiplication by a cutoff function. It is clear that $\|a\|_{\tilde{S}_{f,p}(Q_0)} \leqslant \|\tilde{a}\|_{S_{f,p}(Q)}$ for any exension $\tilde{a}$ of a. But, by definition,

$$\|a\|_{S_{f,p}(Q_0)} = \inf\|\tilde{a}\|_{S_{f,p}(Q)},$$

where the lower bound is taken with respect to all extensions of a. Consequently,

$$\|a\|_{\tilde{S}_{f,p}(Q_0)} \leqslant \|a\|_{S_{f,p}(Q_0)}$$

and

$$\|a\|_{S_{f,p}(Q_0)} \leqslant \|\tilde{a}\|_{S_{f,p}(Q)} \leqslant \mathrm{const}\|a\|_{\tilde{S}_{f,p}(Q_0)}.$$

Let a cylindrical domain $\Omega \times [0, T]$ be given in R^{n+1}, where $T > 0$ is finite and Ω is a connected (not necessarily bounded) domain in R^n whose boundary S satisfies Ljapunov conditions ([2], §13). Then there exist a partition of $\Omega \times [0, T]$ into overlapping subsets Q_k and smooth invertible mappings Q_h onto the standard domains Q_0 or Q of Definition 5.

The definition of C-spaces and B-spaces in $\Omega \times [0, T]$ and $S \times [0, T] = \Gamma$, and their norms, is carried out completely analogously to [2], §13. Lemma 2 allows one to interpolate known estimates for the traces on the boundary and for the extensions from the boundary into the domain.

In the cylinder $\Omega \times [0, T]$, consider a boundary value problem for a system of equations, parabolic in the Douglis-Nirenberg sense (see [2], Problem 4.17 in §14):

$$Lu = f, \qquad Au|_{t=0} = \varphi, \qquad Bu|_{\Gamma} = \Phi \tag{36}$$

under suitable restrictions on the operators L, A, and B. Suppose also that the necessary compatibility conditions are satisfied, their order being determined both by the smoothness of the right sides of (36) with the aid of Theorem 1 or Theorem 2, and by the operators L, A, and B. By the method of [2], §§15–18, this problem reduces to Cauchy's problem, a priori estimates for which are obtained with the aid of an interpolation theorem, and to the problem with zero initial data. In the latter problem the necessity of extending the function by zero in the region of negative t arises; then one may again apply the interpolation theorem. The a priori estimates so obtained in a C-scale or B-scale are completely analogous to the estimates of §§3 and 4.

The author expresses profound gratitude to T. D. Ventcel' for posing the problem and for her interest in the work.

Received 15/SEPT/75

Bibliography

1. V. A. Solonnikov, *A priori estimates for second-order parabolic equation*, Trudy Mat. Inst. Steklov. **70** (1964), 133–212; English transl. in Amer. Math. Soc. Transl. (2) **65** (1967).

2. ______, *On boundary value problems for linear parabolic systems of differential equations of general form*, Trudy Mat. Inst. Steklov. **83** (1965); English transl., Proc. Steklov Inst. Math. **83** (1965).

3. Einar Hille and Ralph S. Phillips, *Functional analysis and semi-groups*, rev. ed., Amer. Math. Soc., Providence, R.I., 1957.

4. T. F. Kalugina, *Interpolation of Banach spaces with a functional parameter*, Vestnik Moskov. Univ. Ser. I Mat. Meh. **1975**, no. 6, 68–77; English transl. in Moscow Univ. Math. Bull. **30** (1975).

5. ______, *A priori estimates of the norms of the solutions of a regular elliptic problem in Banach spaces*, Vestnik Moskov. Univ. Ser. I Mat. Meh. **1976**, no. 3, 14–24; English transl. in Moscow Univ. Math. Bull. **31** (1976).

6. J.-L. Lions and J. Peetre, *Sur une classe d'espaces d'interpolation*, Inst. Hautes Études Sci. Publ. Math. No. 19 (1964), 5–68.

7. V. P. Il'in, *On some properties of classes of differentiable functions defined in a domain*, Trudy Mat. Inst. Steklov. **84** (1965), 93–143; English transl. in Proc. Steklov Inst. Math. **84** (1965).

8. O. V. Besov, *Estimates for the moduli of continuity of abstract functions given in a region*, Trudy Mat. Inst. Steklov. **131** (1974), 16–24; English transl. in Proc. Steklov Inst. Math. **131** (1974).

9. M. L. Gol'dman, *Imbedding of generalized Hölder classes*, Mat. Zametki **12** (1972), 325–336; English transl. in Math. Notes **12** (1972).

10. S. M. Nikol'skiĭ, *Approximation of functions of several variables and embedding theorems*, "Nauka", Moscow, 1969; English transl., Springer-Verlag, 1974.

Translated by P. C. FIFE

Amer. Math. Soc. Transl.
(2) Vol. **118**, 1982

Almost Periodic Fourier Integral Operators
and Some of their Applications*

V. JU. KISELEV

Contents

Introduction

Let A be a uniformly elliptic operator with almost periodic coefficients. In [5] a real-valued function $N(t)$, $t \in \mathbf{R}$, is introduced that characterizes the distribution of the spectrum of A (the density distribution function of the states of A).

An asymptotic formula is given in [5] for the function $N(t)$, along with a consequent estimate of the gap lengths in the operator spectrum. Some of the main results in the present article give a more precise asymptotic expression for $N(t)$ with a best possible estimate of the remainder and a corresponding estimate of the gap lengths. To derive the asymptotics we use the method of Hörmander [2], which in the classical situation (an operator on a compact manifold without boundary) gives a best possible asymptotic expression for the number of eigenvalues of A less than t. Hörmander's method can be used because an approximation of the operator e^{-itA} is constructed here in the form of an almost periodic Fourier integral operator.

The paper is made up of three parts.

In the first part we introduce the spaces of symbols $a(x, y, \theta)$ and phase functions $\varphi(x, y, \theta)$ that are almost periodic with respect to a translation

$$f(x, y, \theta) \mapsto f(x + \lambda, y + \lambda, \theta).$$

1980 *Mathematics Subject Classification.* Primary 35F20, 42A75, 47G05; Secondary 35A30, 35F15, 35J30, 35S05, 41A60, 45D05, 47A10, 47C15, 47F05.

*Translation of Trudy Sem. Petrovsk. **3** (1978), 81–97.

A Fourier integral operator acting in spaces of almost periodic functions is then constructed for a given symbol and phase function.

The second part contains the construction of the operator $U(t) = e^{-itA}$, where A is a uniformly elliptic pseudodifferential operator of first order, with almost periodic symbol. The approximation of $U(t)$ turns out to be a Fourier integral operator of the type described above.

In the third part the approximation of $U(t)$ constructed earlier is used to derive an asymptotic formula for $N(t)$, as well as an estimate of the gap lengths in the spectrum of the original operator.

The author expresses deep gratitude to M. A. Šubin for posing the problem and for constant attention to this work, and he thanks V. Ja. Ivriĭ for useful advice.

I. Almost periodic Fourier integral operators

0. *Basic notation.* Let the set of nonnegative integers be denoted everywhere by $\mathbf{Z}_+$, and the sphere $\{\xi \in \mathbf{R}^n, |\xi| = 1\}$ by $\mathbf{S}^{n-1}$.

The space of functions $f \in C^k(\mathbf{R}^n)$ whose derivatives through order k are uniform almost periodic functions is denoted by $\mathrm{CAP}^k(\mathbf{R}^n)$.

The space of functions $f \in C^k(M)$ that are uniformly bounded along with all their derivatives through order k in the domain $M \subset \mathbf{R}^n$ is denoted by $C_b^k(M)$.

The symbol $\mathfrak{A}$ denotes an admissible algebra of functions (the term was introduced in [4]), i.e., $\mathfrak{A}$ is a subalgebra of $C_b(\mathbf{R}^n)$ satisfying the following conditions:

1) $\mathfrak{A}$ is a Banach subalgebra of $C_b(\mathbf{R}^n)$;

2) the norm in A is invariant under translation in $\mathbf{R}^n$, i.e., $f(\cdot + \lambda) \in \mathfrak{A}$ and $\|f(\cdot + \lambda)\|_{\mathfrak{A}} = \|f(\cdot)\|_{\mathfrak{A}}$ for $f \in \mathfrak{A}$ and $\lambda \in \mathbf{R}^n$;

3) $e_\xi(x) \equiv e^{i\xi x} \in \mathfrak{A}$ for $\xi \in \mathbf{R}^n$, and $\|e_\xi\|_{\mathfrak{A}} \leqslant c$ (c does not depend on ξ);

4) the set $\{e_\xi, \xi \in \mathbf{R}^n\}$ generates the algebra $\mathfrak{A}$, i.e., the linear combinations of e_ξ are norm-dense in $\mathfrak{A}$.

The Bohr compactification of $\mathbf{R}^n$ is denoted by $\mathbf{R}_B^n$, and $B^2(\mathbf{R}^n) \equiv L^2(\mathbf{R}_B^n)$ is the nonseparable Hilbert space of Besicovitch almost periodic functions.

If E is a Banach space, then $\mathcal{L}(E)$ denotes the algebra of bounded operators in E.

1. Let Ω be a domain in $\mathbf{R}^n$, $\Omega \ni 0$. We set

$$\tilde{\Omega} = \{(x, y) \in \mathbf{R}^{2n}, x - y \in \Omega\}.$$

DEFINITION 1.1. Let $\rho > 0$ and $\delta < 1$. The class

$$\{a(x, y, \theta) \in C^\infty_{x,y,\theta}, (x, y) \in \tilde{\Omega}, \theta \in \mathbf{R}^N, \text{ the function } \partial_y^\gamma \partial_x^\beta \partial_\theta^\alpha$$
$\times a(x + \cdot, y + \cdot, \theta)$ is a continuous function of the variables (x, y, θ) with values in $\mathfrak{A}$ for any multi-indices $\alpha, \beta, \gamma,$ and the estimates $\|\partial_y^\gamma \partial_x^\beta \partial_0^\alpha a(x + \cdot, y + \cdot, \theta)\|_{\mathfrak{A}} \leqslant C_{\alpha\beta\gamma}(1 + |\theta|)^{m - \rho|\alpha| + \delta(|\beta| + |\gamma|)}$, $(x, y) \in \tilde{\Omega}$, $\theta \in \mathbf{R}^N$ hold uniformly in $(x, y)\}$

of functions of the variables (x, y, θ) is denoted by $\mathfrak{A} \, FS_{\rho,\delta}^{\prime m}(\Omega, \mathbf{R}^N)$.

DEFINITION 1.2. Let $\mathfrak{A}\,\mathrm{FS}^m_{\rho,\delta}(\Omega, \mathbf{R}^N)$ be the class of functions $a \in \mathrm{FS}'^m_{\rho,\delta}(\mathbf{R}^n, \mathbf{R}^N)$ for which supp $a \subset \tilde{\Omega} \times \mathbf{R}^N$.

In the case $\mathfrak{A} = \mathrm{CAP}(\mathbf{R}^n)$ we simply write AP; for example $\mathrm{APFS}^m_{\rho,\delta}$ denotes $\mathrm{CAP}(\mathbf{R}^n)\mathrm{FS}^m_{\rho,\delta}$, and so on. In the case $\rho = 1$, $\delta = 0$ these indices will be omitted. We also omit the parentheses in notation of the type $\mathfrak{A}\,\mathrm{FS}^m_{\rho,\delta}(\Omega, \mathbf{R}^N)$ and $\mathfrak{A}\,\mathrm{FS}'^m_{\rho,\delta}(\Omega, \mathbf{R}^N)$ if it is clear or irrelevant which Ω and N are meant.

The next proposition is completely analogous to Proposition 3.3 in [4].

PROPOSITION 1.1. *Suppose that* $a_j \in \mathfrak{A}\,\mathrm{FS}^{m_j}_{\rho,\delta}$, $j \in \mathbf{Z}_+$ *and* $m_j \to -\infty$ *as* $j \to \infty$. *Let* $m'_k = \max_{j \geqslant k} m_j$. *Then there exists a symbol* $a \in \mathfrak{A}\,\mathrm{FS}^{m'_0}_{\rho,\delta}$ *such that for any* k

$$a - \sum_{j<k} a_j \in \mathfrak{A}\,\mathrm{FS}^{m'_k}_{\rho,\delta}.$$

The last relation will be written briefly as $a \sim \Sigma a_j$.

2. Below, the operators

$$(Iu)(x) = \iint e^{i\varphi(x,y,\theta)} a(x, y, \theta) u(y)\, dy d\theta, \tag{2.1}$$

where $u \in C^l_b(\mathbf{R}^n)$ for sufficiently large integers l, will be defined. Suppose that Ω is a bounded domain, $a(x, y, \theta) \in \mathfrak{A}\,\mathrm{FS}^m_{\rho,\delta}(\Omega, \mathbf{R}^N)$, and $\varphi(x, y, \theta)$ is a real-valued function on $\tilde{\Omega} \times \mathbf{R}^N$ such that φ is positively homogeneous of order 1 in θ and $\varphi \in C^\infty(\tilde{\Omega} \times (\mathbf{R}^N \setminus 0))$. Suppose that for $\theta \neq 0$ the function $\varphi(x + \cdot, y + \cdot, \theta)$ is a continuous function of (x, y, θ) with values in $\mathfrak{A}$, and that the norm $\|\varphi(x + \cdot, y + \cdot, \theta)\|_{\mathfrak{A}}$ is uniformly bounded with respect to (x, y, θ) for $\theta \in \mathbf{S}^{N-1}$. Assume also that

$$\left|\mathrm{grad}_{(y,\theta)}\varphi(x, y, \theta)\right| \geqslant \varepsilon > 0, \qquad (x, y) \in \tilde{\Omega}, \theta \in \mathbf{S}^{N-1}. \tag{2.2}$$

LEMMA 2.1. *Under the assumptions made about the functions* φ *and* a *there exists a differential operator* L *with respect to the variables* (y, θ) *of the form*

$$L = \sum_{k=1}^{N} a_k(x, y, \theta)\frac{\partial}{\partial\theta_k} + \sum_{k=N+1}^{N+n} a_k(x, y, \theta)\frac{\partial}{\partial y_{k-N}} + a_{n+N+1}(x, y, \theta),$$

such that

$${}^t L e^{i\varphi} = e^{i\varphi} \tag{2.3}$$

(*here* ${}^t L$ *is the adjoint of* L), *and* $a_k(x, y, \theta) \in \mathfrak{A}\,\mathrm{FS}'^j(\Omega, \mathbf{R}^N)$, *where* $j = 0$ *for* $k = 1,\ldots,N$, *and* $j = -1$ *for* $k = N + 1,\ldots,N + n + 1$.

PROOF. From [4] we have the following assertion:

If $f \in \mathfrak{A}$ *and* $|f(\lambda)| \geqslant \varepsilon > 0$ *for* $\lambda \in \mathbf{R}^n$, *then* $1/f(\cdot) \in \mathfrak{A}$.

By using this assertion the lemma can be proved just as Lemma (1.2.1) in [3].

3. Let $u \in C^\infty_b(\mathbf{R}^n)$. For a and φ as in §2 we define the operator I by

$$(Iu)(x) = \iint e^{i\varphi(x,y,\theta)} L^M\left(x, y, \theta, \frac{\partial}{\partial y}, \frac{\partial}{\partial\theta}\right)(a(x, y, \theta)u(y))\, dy d\theta. \tag{3.1}$$

Since $\rho > 0$ and $\delta < 1$, the number $\kappa = \min\{\rho, 1 - \delta\}$ is positive. The integral on the right-hand side of (3.1) converges absolutely for

$$M\kappa > N + m, \tag{3.2}$$

since the operator L^M maps $\mathfrak{A}\,FS^m_{\rho,\delta}$ continuously into $\mathfrak{A}\,FS^{m-M\kappa}_{\rho,\delta}$. The convergence of the integral with respect to y follows from Definition 1.2.

To define Iu for an arbitrary symbol $a(x, y, \xi) \in \mathfrak{A}\,FS^m_{\rho,\delta}$ it is thus necessary to choose an M satisfying (3.2), and to use (3.1). For a symbol $a(x, y, \theta)$ compactly supported with respect to θ the definitions of Iu by (2.1) and (3.1) give one and the same result because of (2.3), which enables us to obtain (3.1) from (2.1) by integrating by parts. We remark that it was sufficient to assume that $u \in C^M_b(\mathbf{R}^n)$.

The class of operators constructed from a given phase function $\varphi \in \mathfrak{A}\,FS'^1_{1,0}$ and all possible symbols $a \in \mathfrak{A}\,FS^m_{\rho,\delta}$ is denoted by $\mathfrak{A}\,FL^m_{\rho,\delta}(\varphi)$. The set of operators with all possible a and φ is denoted by $\mathfrak{A}\,FL^m_{\rho,\delta}$.

4. We have the formula

$$(Iu)(x + \lambda) = \iint e^{i\varphi(x+\lambda, y+\lambda, \theta)} L^M\left(x + \lambda, y + \lambda, \theta, \frac{\partial}{\partial y}, \frac{\partial}{\partial \theta}\right)$$
$$\times \left(a(x + \lambda, y + \lambda, \theta)u(y + \lambda)\right) dy d\theta,$$

and this gives us the next result in an obvious way.

PROPOSITION 4.1. *Let* $I \in \mathfrak{A}\,FL^m_{\rho,\delta}$. *The operator* I *defined by* (3.1) *acts continuously from* $\mathrm{CAP}^k(\mathbf{R}^n)$ *to* $\mathrm{CAP}^j(\mathbf{R}^n)$ *and from* $C^k_b(\mathbf{R}^n)$ *to* $C^j_b(\mathbf{R}^n)$ *for* $k\kappa > N + m + j$, *where* $\kappa = \min\{\rho, 1 - \delta\}$.

II. Construction of the operator $U(t)$

5. Following [4], we say that $a(x, \xi) \in \mathrm{APS}^m$ if $a(x, \xi) \in C^\infty(\mathbf{R}^{2n})$ and for any multi-indices α and β the function $\partial^\alpha_\xi \partial^\beta_x a(\cdot, \xi)$ is a continuous function on $\mathbf{R}^n_\xi$ with values in $\mathrm{CAP}(\mathbf{R}^n)$, and

$$|\partial^\alpha_\xi \partial^\beta_x a(x, \xi)| \leqslant C_{\alpha\beta}(1 + |\xi|)^{m-|\alpha|}, \qquad x \in \mathbf{R}^n, \xi \in \mathbf{R}^n.$$

Let us consider the pseudodifferential operator $A \in \mathrm{APL}^m$ corresponding to the symbol $a(x, \xi)$:

$$(Au)(x) = (a(x, D_x)u)(x) = \int e^{i\langle x-y, \xi\rangle} a(x, \xi)u(y) \, dy d\xi.$$

(See [4] for the definition.)

Suppose now that $a(x, \xi) \in \mathrm{APS}^1$, and that $a(x, \xi)$ expands in an asymptotic series $a \sim \Sigma^\infty_{-1} a_{-j}$, where the functions $a_{-j}(x, \xi) \in \mathrm{APS}^{-j}$ are positively homogeneous in ξ of order $-j$ for $|\xi| \geqslant r > 0$. It is also assumed that $a_1(x, \xi)$ is a real-valued function and that

$$c_1 \leqslant a_1(x, \xi) \leqslant c_2, \qquad x \in \mathbf{R}^n, \xi \in \mathbf{S}^{n-1},$$

where $c_1, c_2 > 0$ (uniform ellipticity).

Suppose that the operator $a(x, D_x) = A$ is selfadjoint in $L^2(\mathbf{R}^n)$, and let $\{E_s, s \in \mathbf{R}\}$ be its spectral resolution. We can now define the operator

$$U(t) = e^{-itA} = \int e^{-its}\, dE_s,$$

which satisfies the conditions

$$(D_t + A)U(t) = 0; \qquad U(0) = \mathrm{Id}.$$

We first find an approximation $Q(t)$ of $U(t)$ in the form of a Fourier integral operator

$$(Q(t)u)(t, x) = \int e^{i\varphi(t,x,y,\xi)} q(t, x, y, \xi) u(y)\, dy d\xi.$$

Following [2], we find φ in the form

$$\varphi(t, x, y, \xi) = \psi(x, y, \xi) - ta_1(y, \xi). \tag{5.1}$$

The function ψ is determined from the equation

$$a_1(x, \psi_x(x, y, \xi)) = a_1(y, \xi) \tag{5.2}$$

with the boundary conditions

$$\psi|_{\langle x-y,\xi\rangle=0} = 0, \qquad \psi_x|_{x=y} = \xi. \tag{5.3}$$

The function q is found in the form of an asymptotic sum $q \sim \Sigma_{j=0}^{+\infty} q_{-j}(t, x, y, \xi)$, where q_{-j} is a homogeneous function of order $-j$ in ξ, and $q_{-j}(t, \cdot, \cdot, \cdot) \in$ $\mathrm{APFS}^{-j}(\Omega, \mathbf{R}^n)$ for $\xi \neq 0$. Here Ω is a bounded domain in $\mathbf{R}^n$, and $\Omega \ni 0$. The estimates of q_{-j} in the class APFS^{-j} do not depend on $t \in [-t_0, t_0]$. Furthermore, $D_t q$ has the same properties as q. Equations are obtained for the q_{-j} by setting equal to zero the homogeneous terms of the expansion

$$e^{-i\varphi}(D_t + A)(qe^{i\varphi}) \sim (\varphi_t(t, x, y, \xi) + a_1(x, \varphi_x(t, x, y, \xi)))q_0 + \cdots, \tag{5.4}$$

where the terms having order of homogeneity $l \leqslant 0$ are denoted by dots. The expansion is obtained by the classical theorem on the action of a pseudodifferential operator on an exponential. It can be used because the classes $\mathrm{APFS}^m_{\rho,\delta}$ are contained in the Hörmander classes $S^m_{\rho,\delta}$ [3].

Since the function ψ is being sought in the class of homogeneous functions of the first order in ξ, it suffices to find it for $\xi \equiv \mathbf{S}^{n-1}$.

REMARK 5.1. It will be shown later that $\psi \in C_b^\infty(\Omega \times M)$, $M \subset\subset \mathbf{R}^n$; it then follows from (5.1) that (2.2) holds.

6. Let us consider the Cauchy problem

$$a(x, \psi_x(x, y, N)) = a(y, N) \equiv c; \tag{6.1}$$

$$\psi|_S = 0; \qquad \psi_x|_{x=y} = N, \tag{6.2}$$

where $S = \{x: \langle x - y, N\rangle = 0\}$. Here $y \in \mathbf{R}^n$ and $N \in \mathbf{S}^{n-1}$ are parameters which are assumed to be constants in solving the problem (6.1), (6.2). It is assumed that $a(x, \xi)$ is homogeneous of the first order in ξ and has the properties

$$|\partial_\xi^\beta \partial_x^\alpha a(x, \xi)| \leq C_{\alpha\beta}, \qquad x \in \mathbf{R}^n, \xi \in \mathbf{S}^{n-1}, \tag{6.3}$$

$$a(x, \xi) \geq \varepsilon > 0, \qquad x \in \mathbf{R}^n, \xi \in \mathbf{S}^{n-1}. \tag{6.4}$$

The function $a(x, \xi)$ can be associated with the Hamiltonian system

$$\dot{x} = a_\xi(x, \xi); \qquad \dot{\xi} = -a_x(x, \xi). \tag{6.5}$$

A solution $(x(t), \xi(t))$ of (6.5) (a bicharacteristic of $a(x, \xi)$) exists on the whole t-axis, as follows from (6.3) and (6.4).

Let $\Gamma_\psi = \{(x, \psi_x)\} \subset \mathbf{R}^n \times \mathbf{R}^n$. If ψ is a solution of (6.1), then, as is well known, Γ_ψ is invariant under a translation along the bicharacteristics of $a(x, \xi)$, and $a(x(t), \xi(t)) = \text{const}$ if $(x(t), \xi(t))$ is a bicharacteristic.

PROPOSITION 6.1. *Let* $(x(t), \xi(t))$ *be a bicharacteristic of the function* $a(x, \xi)$, $\psi(x)$ *a solution of* (6.1), (6.2), *and* $\xi(0) = \psi_x(x(0))$. *Then* $\psi(x(t)) = tc \ (\equiv ta(y, N))$.

PROOF. Using Euler's theorem on the derivative of a homogeneous function, we get that

$$\dot{\psi} = \psi_x \dot{x} = \psi_x a_\xi(x(t), \xi(t)) = \xi(t) a_\xi(x(t), \xi(t)) = a(x(t), \xi(t)) = c,$$

from which, considering (6.2), we get that $\psi(x(t)) = tc$.

Thus, to solve the problem (6.1), (6.2) it is necessary to find the bicharacteristics $(x(t), \xi(t))$ of $a(x, \xi)$ that correspond to the condition $a(x(t), \xi(t)) = c$ and are such that $x(0) \in S$ and $\xi(0) = N$, and then to set $\psi(x(t)) = tc$. More precisely, let g denote the mapping

$$g \colon S \times (-t_0, t_0) \to \mathbf{R}^n, \qquad g \colon (x^0, t) \mapsto x(x^0, t),$$

where $x(x^0, t)$ is the first component of the bicharacteristic of $a(x, \xi)$ with the initial condition $x(x^0, 0) = x^0 \in S$, $\xi(x^0, 0) = N$.

LEMMA 6.1. *For small* t_0 *the mapping* g *is a* C^∞*-diffeomorphism of the domain* $S \times (-t_0, t_0)$ *onto some neighborhood* U *of the hyperplane* S *in* $\mathbf{R}^n$.

The breadth of the neighborhood U in the direction of the normal N to S is not less than some positive quantity, since it follows from (6.4) and Euler's theorem on homogeneous functions that

$$\langle a_\xi(x, \xi), \xi \rangle = a(x, \xi) \geq \varepsilon > 0, \qquad \xi \in \mathbf{S}^{n-1}, x \in \mathbf{R}^n. \tag{6.6}$$

The inverse g^{-1} of g assigns to a point $x \in U$ some $x^0 \in S$ and $t \in (-t_0, t_0)$. The resulting function $t(x)$ enables us to write the solution $\psi(x)$ in the form

$$\psi(x) = ct(x),$$

according to Proposition 6.1.

Using (6.6), we can show that

$$\det g_* \geq \varepsilon_0 > 0,$$

from which it follows, in turn, that the inverse g^{-1} is given by a tuple of functions in $C_b^\infty(U)$. In particular, we have

PROPOSITION 6.2. $t(\cdot) \in C_b^\infty(U)$.

REMARK 6.1. If the constant c in (6.1) (i.e., in the equation $a(x, \psi_x) = c$) depends on the parameters, with a class C_b^∞ dependence, then the solution ψ will depend on these parameters, and it follows from Proposition 6.2 that $\psi \in C_b^\infty$ with respect to all the variables.

PROPOSITION 6.3. *For any $\varepsilon > 0$ there exists a $\delta > 0$ such that if a_1 and a_2 satisfy the conditions (6.3) and (6.4) and are homogeneous of the first order in ξ,*

$$|\partial_x^\alpha \partial_\xi^\beta(a_1(x, \xi) - a_2(x, \xi))| \leq \delta; \qquad \xi \in \mathbf{S}^{n-1};$$
$$|\alpha| + |\beta| \leq 2; \qquad |c_1 - c_2| \leq \delta,$$

and φ_1 and φ_2 are solutions of the problems

$$a_j(x, \varphi_{jx}) = c_j; \qquad \varphi_j|_S = 0, \qquad \varphi_{jx}|_{x=y} = N,$$

$j = 1, 2$, then $|\varphi_1(x) - \varphi_2(x)| \leq \varepsilon$ for any $x \in U$, where U is some neighborhood of S with constant breadth.

A proof follows from the construction of the solution and Propositions 6.1 and 6.2.

7. Let us return to our study of the function $\psi(x, y, \xi)$ that is the solution of the problem (5.2), (5.3).

PROPOSITION 7.1. $\psi(x + \cdot, y + \cdot, \xi) \in \mathrm{CAP}(\mathbf{R}^n)$.

PROOF. Let $a_{1,\lambda}(x, \xi) = a_1(x + \lambda, \xi)$ and $\psi^\lambda(x, y, \xi) = \psi(x + \lambda, y + \lambda, \xi)$. Take $\varepsilon > 0$, and let δ be obtained from ε by Proposition 6.3. Suppose that $\lambda \in \mathbf{R}^n$ is a common δ-almost period of the functions $x \mapsto \partial_\xi^\beta \partial_x^\alpha a_1(x, \xi)$, $|\alpha| + |\beta| \leq 2$. Since ψ^λ is a solution of the problem

$$a_{1,\lambda}(x, \psi_x^\lambda) = a_{1,\lambda}(y, \xi), \qquad \psi^\lambda|_{\langle x-y, \xi\rangle = 0} = 0, \qquad \psi_x^\lambda|_{x=y} = \xi,$$

it follows from Proposition 6.3 that

$$|\psi(x, y, \xi) - \psi^\lambda(x, y, \xi)| \leq \varepsilon; \qquad (x, y) \in \tilde{\Omega}, \xi \in \mathbf{S}^{n-1},$$

from which we get

$$|\psi^\mu(x, y, \xi) - \psi^{\lambda+\mu}(x, y, \xi)| \leq \varepsilon; \qquad \mu \in \mathbf{R}^n.$$

Thus, λ is an ε-almost period of the function $\mu \to \psi^\mu$. Since for any $\delta > 0$ there exists a relatively dense subset of $\mathbf{R}^n$ composed of vectors λ that are simultaneously δ-almost periods for $\partial_\xi^\beta \partial_x^\alpha a_1(x, \xi)$, $|\alpha| + |\beta| \leq 2$, it has thereby been proved that $\psi(x + \cdot, y + \cdot, \xi) \in \mathrm{CAP}(\mathbf{R}^n)$.

We have the following simple assertion:

PROPOSITION 7.2. *If $f(x, y, \xi) \in C_b^\infty(\tilde{\Omega} \times M)$, where $M \subset \mathbf{R}_\xi^n$ is a bounded set, and $f(x + \cdot, y + \cdot, \xi) \in \mathrm{CAP}(\mathbf{R}^n)$, then $\partial_x^\alpha \partial_y^\beta \partial_\xi^\gamma f(x + \cdot, y + \cdot, \xi)$ is a continuous function of x, y and ξ with values in $\mathrm{CAP}(\mathbf{R}^n)$ for any α, β and γ.*

In view of (5.5), Proposition 7.2 yields the following result.

PROPOSITION 7.3. *The phase function φ is such that*

$$\partial_t^\alpha \partial_x^\beta \partial_y^\gamma \partial_\xi^\delta \varphi(t, x + \cdot, y + \cdot, \xi)$$

is a continuous function of (t, x, y, ξ) with values in $\mathrm{CAP}(\mathbf{R}^n)$ for any α, β, γ and δ.

8. The following proposition holds, whose proof is similar to that of the corresponding proposition in [2].

PROPOSITION 8.1. *Suppose that $\varphi \in \mathrm{APFS}'^1$ is a phase function homogeneous in ξ and such that $\mathrm{grad}_\xi \varphi = 0$ only for $x = y$, and*

$$\varphi(x, y, \xi) = \langle x - y, \xi \rangle + \alpha(x, y, \xi),$$

where

$$|\alpha(x, y, \xi)| \leqslant C |x - y|^2 (1 + |\xi|), \qquad c > 0.$$

Then $\mathrm{APFL}^m(\varphi) = \mathrm{APFL}^m(\varphi_0)$, where $\varphi_0(x, y, \xi) = \langle x - y, \xi \rangle$.

REMARK 8.1. If the operator $A \in \mathrm{APFL}^m(\varphi_0)$ can be written in the form of an operator with a symbol that is expandable in an asymptotic series of functions homogeneous in ξ of order $m - j, j \in \mathbf{Z}_+$, and with the phase function φ_0, then when it is expressed with the phase function φ, the symbol can be chosen to be expandable in the same asymptotic series.

The remark follows from a modification of the same proof.

9. The function ψ chosen earlier satisfies the conditions of Proposition 8.1. Suppose that $\zeta(x) \in C_0^\infty(\Omega)$ and $\zeta \equiv (2\pi)^{-n}$ in a domain $\Omega' \subset\subset \Omega$. Then

$$\int e^{i\langle x-y, \xi \rangle} \zeta(x - y) f(y)\, dy d\xi = f(x).$$

The function $\zeta(x - y)$ can be regarded as a symbol in $\mathrm{APFS}^0(\Omega, \mathbf{R}^n)$. By Proposition 8.1, the identity operator

$$f \mapsto \int e^{i\langle x-y, \xi \rangle} \zeta(x - y) f(y)\, dy d\xi$$

can be represented as a Fourier integral operator with phase function ψ

$$f(x) = \int e^{i\psi(x,y,\xi)} I(x, y, \xi) f(y)\, dy d\xi,$$

where $I(x, y, \xi) \in \mathrm{APFS}^0(\Omega, \mathbf{R}^n)$.

Let $I \sim I_0 + I_{-1} + I_{-2} + \cdots$ be an expansion of I into functions homogeneous in ξ; $I_j \in \mathrm{APFS}^j(\Omega, \mathbf{R}^n)$.

10. PROPOSITION 10.1. *Suppose that*

$$\left(\sum_{j=1}^n a_j(x) \frac{\partial}{\partial x_j} + b(x) \right) u(x) + c(x) = 0 \tag{10.1}$$

is a first-order equation with the boundary condition

$$u|_{x_n=0} = f(x'), \tag{10.2}$$

where $x = (x', x_n)$, $x' \in \Omega \subset \mathbf{R}^{n-1}$ and $x_n \in (-A, A)$, $A > 0$. It is assumed that $a_n(x) \equiv 1$, a_j $(j = 1, 2, \ldots, n - 1)$, b, $c \in C_b^\infty(\Omega \times (-A, A))$, $f \in C_b^\infty(\Omega)$ and that the functions $a_j(x)$ are real-valued.

Then the solution $u(x)$ of (10.1), (10.2) exists, is unique, and belongs to $C_b^\infty(\Omega' \times (-A, A))$, where $\Omega' \subset \Omega$, $\Omega' = \Omega'(A)$, is a region whose size is determined by the constants in the coefficient estimates, and Ω' differs as little as desired from Ω for small A.

PROOF. The standard method for solving first-order linear equations enables us to write out an explicit expression for the solution $u(x)$ in the form of a formula containing the function $x(x^0, t)$, where the latter is the characteristic of (10.1) with initial condition $x(x^0, 0) = (x^0, 0)$. Since the $a_j(x)$ and their derivatives are bounded, all the derivatives of $x(x^0, t)$ with respect to the initial data x^0 are bounded. Using this, we get that $u \in C_b^\infty$.

REMARK 10.1. If all the coefficients of (10.1) (except a_n) and the boundary data depend on parameters, and the dependence is of class C_b^∞, then the solution u will depend on these parameters, and $u \in C_b^\infty$ with respect to all variables.

PROPOSITION 10.2. *Let there be given two Cauchy problems:*

$$\left(\sum_{i=1}^n a_i^j(x) \frac{\partial}{\partial x_i} + b^j(x) \right) u^j(x) + c^j(x) = 0; \qquad u^j|_{x_n=0} = f^j(x'),$$

where $a_n^j(x) \equiv 1$, a_i^j, b^j, $c^j \in C_b^\infty(\Omega \times (-A, A))$ and $f^j \in C_b^\infty(\Omega)$ $(i = 1, 2, \ldots, n - 1; j = 1, 2)$.

Then for any $\varepsilon > 0$ there exists a $\delta > 0$ such that if, for any $x \in \Omega \times (-A, A)$ and any $i = 1, \ldots, n - 1$,

$$|a_i^1(x) - a_i^2(x)| < \delta, \qquad |b^1(x) - b^2(x)| < \delta,$$

$$|c^1(x) - c^2(x)| < \delta, \qquad |f^1(x') - f^2(x')| < \delta,$$

then $|u^1(x) - u^2(x)| < \varepsilon$ for $x \in \Omega'$.

The proof follows from a theorem stating that the solution of an ordinary differential equation depends continuously on the right-hand side and the initial conditions.

11. The symbol $q(t, x, y, \xi)$ for the Fourier integral operator (see §5) is chosen in the form of an asymptotic series with terms homogeneous in ξ:

$$q \sim q_0' + q_{-1}' + q_{-2}' + \cdots,$$

where $q_j'(t, \cdot, \cdot, \cdot) \in \mathrm{APFS}^j$ for $\xi \neq 0$, and $q_j'(t, x, y, \xi) = \zeta(x - y) q_j(t, x, y, \xi)$. The cutoff function ζ and the functions q_j will be chosen below.

Beginning with $j = 0$, we set equal to zero successively the homogeneous (in ξ) terms in the expansion (5.4) in homogeneous functions.

For $j = 0$ we get the equation

$$D_t q_0 + \sum_{|\alpha|=1} a_1^{(\alpha)}(x, \varphi_x)(D_x^\alpha q_0) + \left(\sum_{|\alpha|=2} a_1^{(\alpha)}(x, \varphi_x)D_x^\alpha(i\varphi)\frac{1}{\alpha!} \right) q_0 = 0$$

with the initial condition

$$q_0\big|_{t=0} = I_0(x, y, \xi).$$

(Here $a^{(\alpha)} \equiv \partial_\xi^\alpha a$.) For $-j > 0$ the result is

$$D_t q_j + \sum_{|\alpha|=1} a_1^{(\alpha)}(x, \varphi_x)(D_x^\alpha q_j) + \left(\sum_{|\alpha|=2} a_1^{(\alpha)}(x, \varphi_x)D_x^\alpha(i\varphi)\frac{1}{\alpha!} \right) q_j$$

$$+ R_j(t, x, y, \xi) = 0;$$

$$q_j\big|_{t=0} = I_j, \tag{11.1}$$

where $R_j(t, x, y, \xi)$ is a known function obtained by means of algebraic operations and differentiation operations from the function s a, φ and $q_0, \dots, q_{j+1}$, which are determined in the preceding steps.

LEMMA 11.1. *All the coefficients of* (11.1) *have the form* $F(t, x, y, \xi)$, *where* $\partial_t^\alpha \partial_x^\beta \partial_y^\gamma \partial_\xi^\delta F(t, x + \cdot, y + \cdot, \xi)$ *is a continuous function of* (t, x, y, ξ) *with values in* $\mathrm{CAP}(\mathbf{R}^n)$ *for any* α, β, γ *and* δ.

The proof of the lemma for R_j follows from Proposition 11.1 below, applied to $q_0, q_{-1}, \dots, q_{j+1}$.

PROPOSITION 11.1. *The function* $\partial_t^\alpha \partial_x^\beta \partial_y^\gamma \partial_\xi^\delta q_j(t, x + \cdot, y + \cdot, \xi)$ *is continuous with respect to* (t, x, y, ξ) *with values in* $\mathrm{CAP}(\mathbf{R}^n)$ *for any* α, β, γ *and* δ.

The proof is carried out similarly to the proofs of Propositions 7.1 and 7.3 by using Propositions 10.1 and 10.2, Lemma 11.1, and Proposition 7.2.

Suppose that $\zeta \in C_0^\infty(\Omega)$, $\zeta \equiv 1$ in domain $\Omega_0 \subset\subset \Omega$. Let

$$q_j'(t, x, y, \xi) = \zeta(x - y)q_j(t, x, y, \xi).$$

The function $q_j'(t, \cdot, \cdot, \cdot)$ belongs to $\mathrm{APFS}^j(\Omega, \mathbf{R}^n)$.

12. Thus, we have constructed a Fourier integral operator $Q(t)$ for which

$$\begin{cases} (D_t + A)Q(t) = K(t); \\ Q(0) = \mathrm{Id} + k, \end{cases} \tag{12.1}$$

where $K(t)$ is an operator with kernel $\overline{K}(t, x, y) \in C^\infty((-t_0, t_0) \times \mathbf{R}^{2n})$ such that $\partial_t^\alpha \partial_x^\beta \partial_y^\gamma \overline{K}(t, x + \cdot, y + \cdot)$ is a continuous function of (t, x, y) with values in $\mathrm{CAP}(\mathbf{R}^n)$ for any α, β and $\dot{\gamma}$, and k is an operator with kernel $\overline{k}(x, y) \in C^\infty(\mathbf{R}^{2n})$, $\overline{k}(x, y) \equiv 0$ for $x - y \in \mathbf{R}^{2n} \setminus \tilde{\Omega}_0$, where $\Omega_0 \subset\subset \Omega$, and $\partial_x^\alpha \partial_y^\beta \overline{k}(x + \cdot, y + \cdot)$ is a continuous function of (x, y) with values in $\mathrm{CAP}(\mathbf{R}^n)$ for any α and β.

Let $Q_1(t) = Q(t) - k$. For $Q_1(t)$ we get from (12.1) that

$$(D_t + A)Q_1(t) = K(t) - Ak \equiv K_1(t), \qquad Q_1(0) = \mathrm{Id}, \tag{12.2}$$

where $K_1(t)$ has the same properties as $K(t)$.

Variation of the constant in (12.2) yields a Volterra equation for the determination of $U(t)$:

$$U(t) = Q_1(t) - \int_0^t U(s)K_1(t-s)\,ds. \tag{12.3}$$

The kernel $\overline{Q}_1(t, x, y)$ of the operator $Q_1(t)$ can be assumed to be a compactly supported generalized function in the variable y that depends smoothly on (t, x) as well as on the parameter [3]. It follows from (12.2) that for the kernels we have

$$(D_t + A)\overline{Q}_1(t, x, y) = \overline{K}_1(t, x, y) \in C^\infty, \tag{12.4}$$

where the operator A acts with respect to the variable x. Then

$$(D_t + A)\overline{Q}_1(t, x, y) = \zeta(x - y)(D_t + A)\overline{Q}_1(t, x, y)$$
$$+ (1 - \zeta(x - y))(D_t + A)\overline{Q}_1(t, x, y),$$

where the function $\zeta \in C_0^\infty(\Omega)$ is such that $\zeta(x - y) \equiv 1$ in a neighborhood of supp $\overline{Q}_1(t, x, y)$. The first term is equal to 0 for large values of $|x - y|$. The second term admits the estimate

$$\left| \partial_t^\alpha \partial_x^\beta \partial_y^\gamma (1 - \zeta(x - y))(D_t + A)\overline{Q}_1(t, x, y) \right| \leqslant C_{\alpha\beta\gamma N}(1 + |x - y|)^{-N},$$

$N \in \mathbf{Z}_+$, which is obtained by integration by parts. Taking (12.4) into account, we obtain

PROPOSITION 12.1. *The function* $\overline{K}_1(t, x, y)$ *admits the estimates*

$$\left| \partial_x^\alpha \partial_y^\beta \overline{K}_1(t, x, y) \right| \leqslant C_{\alpha\beta N}(1 + |x - y|)^{-N} \tag{12.5}$$

for any $\alpha, \beta \in \mathbf{Z}_+^n$, $N \in \mathbf{Z}_+$, $t \in (-t_0, t_0)$ *and* $x, y \in \mathbf{R}^n$.

A solution of (12.3) will be sought in the form $U(t) = Q_1(t) - \int_0^t Q_1(t - \tau)R(\tau)\,d\tau$, from which we obtain a Volterra equation for $R(t)$:

$$R(t) = -K_1(t) + \int_0^t K_1(t - \tau)R(\tau)\,d\tau. \tag{12.6}$$

13. PROPOSITION 13.1. *The operator* $T(t) = U(t) - Q_1(t)$ *has an infinitely differentiable kernel* $\overline{T}(t, x, y)$ *satisfying the estimates*

$$\left| \partial_x^\alpha \partial_y^\beta \overline{T}(t, x, y) \right| \leqslant T_{\alpha\beta N}(1 + |x - y|)^{-N}, \qquad N \in \mathbf{Z}_+, \tag{13.1}$$

and the functions $\partial_x^\alpha \partial_y^\beta \overline{T}(t, x + \cdot, y + \cdot)$ *are continuous functions of* (t, x, y) *with values in* CAP$(\mathbf{R}^n)$.

A proof is contained in §19.

The operator $T(t)$ is continuous in $L^2(\mathbf{R}^n)$, as follows from its form.

14. The operator e^{-itA} gives a solution in $L^2(\mathbf{R}^n)$ of the Cauchy problem

$$\begin{cases} (D_t + A)u(t, x) = 0; \\ u(0, x) = f(x). \end{cases} \tag{14.1}$$

Let $'A$ be the operator adjoint to A in the sense of the standard duality in $L^2(\mathbf{R}^n)$. The solution of the Cauchy problem

$$\begin{cases} \left(D_t - {}'A\right)v(t, x) = 0; \\ v(s, x) = g(x) \end{cases}$$

(s is fixed) is given by the operator $e^{i(t-s)'A}$. We now use the Holmgren principle ([1], Chapter 2, §2.1). That theorem implies that the solution of the problem (14.1) is unique, and this implies that the operator $U(t)$ constructed above is equal to e^{-itA}.

Summarizing the foregoing, we obtain a theorem.

THEOREM 14.1. *Let the operator $A \in \mathrm{APL}^1$ be as in §5. Then the operator $U(t) = e^{-itA}$ has the form $U(t) = Q_1(t) + T(t)$, where the Fourier integral operator $Q_1(t)$ has a symbol $q(t, x, y, \xi)$ such that the functions*

$$\partial_x^\alpha \partial_y^\beta \partial_\xi^\gamma q(t, x + \cdot, y + \cdot, \xi)$$

are continuous functions of (t, x, y, ξ) with values in $\mathrm{CAP}(\mathbf{R}^n)$. The phase functions $\varphi(t, x, y, \xi)$ of the operator $Q_1(t)$ has an analogous property. The properties of the operator $T(t)$ are described in Proposition 13.1.

III. Asymptotics of $N(t)$

15. A certain algebra of operators in the space $L^2(\mathbf{R}_B^n \times \mathbf{R}^n) \equiv B^2(\mathbf{R}^n) \otimes L^2(\mathbf{R}^n)$ that is a II_∞-factor is defined in [8] (see also [6], §1). This algebra, which we denote by $\mathcal{C}_B$, is generated as a W^*-algebra by the sets of operators $\{e_\lambda \otimes e_\lambda\}$ and $\{\mathrm{Id} \otimes T_\lambda\}$, $\lambda \in \mathbf{R}^n$, where e_λ is the operator of multiplication by $e^{i\lambda\cdot}$, and T_λ is the operator of translation by λ.

Let APL be the set of operators $A \in \mathcal{L}(L^2(\mathbf{R}^n))$ such that the vector-valued function

$$A_\lambda = T_{-\lambda} A T_\lambda \tag{15.1}$$

of the variable $\lambda \in \mathbf{R}^n$ with values in $\mathcal{L}(L^2(\mathbf{R}^n))$ is a uniform almost periodic function. This APL is a C^*-subalgebra of $\mathcal{L}(L^2(\mathbf{R}^n))$. To an operator $A \in \mathrm{APL}$ we assign an operator $A^\# \in \mathcal{C}_B$ according to the formula

$$A^\# u(\lambda, x) = \left[A_\lambda u(\lambda, \cdot)\right](x),$$

where A_λ is defined in (15.1). The mapping $A \mapsto A^\#$ is a $*$-representation of the algebra APL in $\mathcal{C}_B$.

A trace Sp is defined on the factor $\mathcal{C}_B$.

Let $P \in \mathrm{APL}^m$. As in [6] (§8), define the operator $P^\#$ as the closure of the operator defined on the set of functions $u \in \mathrm{Trig}(\mathbf{R}^n) \otimes L^2(\mathbf{R}^n)$ by the formula

$$(P^\# u)(\lambda, x) = \int e^{i\langle x-y, \xi\rangle} p(x + \lambda, \xi) u(\lambda, y)\, dy\, d\xi,$$

where $p(x, \xi)$ is the symbol of the operator P. Suppose next that P is a selfadjoint elliptic operator. According to §8 in [6], $P^\# \eta \mathcal{C}_B$, i.e. the spectral projections of $P^\#$ are in $\mathcal{C}_B$.

DEFINITION 15.1. Suppose that $P^{\#} = \int s\,dE_s$. Let

$$N(s) \equiv N_P(s) = \operatorname{Sp} E_s.$$

The trace Sp is first defined on the positive operators in $\mathcal{C}_B$. Let us consider the set D of positive operators $C \in \mathcal{C}_B$ such that $\operatorname{Sp} C < \infty$. The functional Sp can be extended by linearity to the (complex) linear span of the set D, denoted by $S_1(\mathcal{C}_B)$, which is a two-sided ideal in $\mathcal{C}_B$. This $S_1(\mathcal{C}_B)$ is called the ideal of finite-trace operators.

16. Suppose that the function $\overline{K}(x, y)$ belongs to $C^{\infty}(\mathbf{R}^{2n})$, and that

$$|\partial_x^{\alpha}\partial_y^{\beta}\overline{K}(x, y)| \leqslant C_{\alpha\beta N}(1 + |x - y|)^{-N}, \qquad x, y \in \mathbf{R}^n, \qquad (16.1)$$

with

$$\partial_x^{\alpha}\partial_y^{\beta}\overline{K}(X + \,\cdot\,, y + \,\cdot\,) \text{ a continuous function of } (x, y) \text{ with values in } \mathrm{CAP}(\mathbf{R}^n).$$

$$(16.2)$$

Then the operator K acting by the formula

$$(Kf)(x) = \int \overline{K}(x, y)f(y)\, dy,$$

is bounded in $L^2(\mathbf{R}^n)$, and the operator $K^{\#}$ (with kernel $\overline{K}(x + \lambda, y + \lambda)$) belongs to $\mathcal{C}_B$.

PROPOSITION 16.1. *Suppose that* $K_1 = K^{\#}$ *is an integral operator in* $B^2(\mathbf{R}^n) \otimes L^2(\mathbf{R}^n)$, *and the kernel* $\overline{K}(x, y)$ *of the operator* K *satisfies* (16.1) *and* (16.2). *Then* $K_1 \in S_1(\mathcal{C}_B)$ *and*

$$\operatorname{Sp} K_1 = M_{\lambda}\{\overline{K}(\lambda, \lambda)\}, \qquad (16.3)$$

where $M_{\lambda}\{f(\lambda)\}$ *denotes the mean value of the almost periodic function* $f(\lambda)$.

PROOF. It was shown in [8] that for operators of the form $\tilde{M}_{\psi} = F^{-1}M_{\psi}F$ (where $F_{x \to \xi}$ is the Fourier transformation, M_{ψ} is the operator of multiplication by the function $\psi \in L^{\infty}$, and $\psi(\xi) > 0$) the condition $\int |\psi(\xi)|\, d\xi < \infty$ is a criterion for $\mathrm{Id} \otimes \tilde{M}_{\psi} \in S_1(\mathcal{C}_B)$. In particular, the operator $(1 - \Delta_x)^{-N}$ (acting with respect to the second variable) belongs to $S_1(\mathcal{C}_B)$ when $N > n/2$. Let $\mathcal{K}$ be the space of functions satisfying (16.1) and (16.2), with the topology defined by the seminorms equal to the best possible constants in (16.1). We write K in the form $K = (1 - \Delta)^{-N}L$; $L = (1 - \Delta)^N K$. The kernel $\overline{L}(x, y)$ of the operator L belongs to $\mathcal{K}$; consequently, $L^{\#} \in \mathcal{C}_B$. Since $K_1 = (1 - \Delta_x)^{-N}L^{\#}$ and $S_1(\mathcal{C}_B)$ is an ideal in $\mathcal{C}_B$, it follows that $K_1 \in S_1(\mathcal{C}_B)$. Let us now derive (16.3). The trace Sp defines a continuous linear functional (denoted also by Sp) on $\mathcal{K}$; this is easy to show, starting from the fact that Sp is a continuous linear functional on $S_1(\mathcal{C}_B)$ and using the fact that ([9], the inequality in Theorem 7 of Part I, §6)

$$\operatorname{Sp}(|ST|) \leqslant \|S\|\operatorname{Sp}(|T|),$$

where $\|R\|$ denotes the norm of the operator R in L_2, and $R = U|R|$ is the polar decomposition of R.

If $\overline{K}(x, y) \in \mathcal{K}$, then the mean value $M_\lambda\{\overline{K}(x + \lambda, y + \lambda)\}$ exists as a limit in the topology of $\mathcal{K}$.

By the definition of the trace, $\operatorname{Sp} S = \operatorname{Sp}(USU^{-1})$, where U is any unitary operator. If $U = (\operatorname{Id} \otimes T_\lambda)$, then we get that $\operatorname{Sp} \overline{K} = \operatorname{Sp} \overline{K}(\cdot + \lambda, \cdot + \lambda)$ for any $\lambda \in \mathbf{R}^n$, and hence

$$\operatorname{Sp} M_\lambda\{\overline{K}((x - y) + \lambda, \lambda)\} = \operatorname{Sp} M_\lambda\{\overline{K}(x + \lambda, y + \lambda)\}$$
$$= M_\lambda\{\operatorname{Sp} \overline{K}(x + \lambda, y + \lambda)\} = \operatorname{Sp} \overline{K}(x, y).$$

Let $K'(z) = M_\lambda\{\overline{K}(z + \lambda, \lambda)\}$. If $U = (e_\lambda \otimes e_\lambda)$, then for any λ

$$\operatorname{Sp'} K'(z) = \operatorname{Sp'}(K'(z)e^{i\lambda z}), \tag{16.4}$$

where $z = x - y$, and $\operatorname{Sp'}$ denotes the restriction of Sp to the set of functions $f \in \mathcal{K}$ that depend only on $(x - y)$, which set coincides with the Schwartz space $S(\mathbf{R}^n_z)$. Then $\operatorname{Sp'} \in S'(\mathbf{R}^n_z)$.

Since it follows from (16.4) that the Fourier transform $\widetilde{\operatorname{Sp}}'$ of the functional Sp' is a constant, we have that $\operatorname{Sp}' = \operatorname{const} \cdot \delta(z)$, and so

$$\operatorname{Sp} K = \operatorname{Sp'} K'(x - y) = \operatorname{const} \cdot K'(0) = \operatorname{const} \cdot M_\lambda\{\overline{K}(\lambda, \lambda)\}.$$

For the normalization of the trace we choose $\operatorname{const} \doteq 1$.

17. Suppose that $A \in \operatorname{APL}^1$ is the same as in §5, $U(t) = e^{-itA}$, and $Q_1(t)$ is constructed from A in §§5–12. Let $U(t) - Q_1(t)$ be denoted by $T(t)$. It follows from Proposition 13.1 that the kernel $\overline{T}(t, x, y)$ of the operator $T(t)$ satisfies (13.1), and also that $\partial_t^\alpha \partial_x^\beta \partial_y^\gamma T(t, x + \cdot, y + \cdot)$ is a continuous function of (t, x, y) with values in $\operatorname{CAP}(\mathbf{R}^n)$ for any α, β and γ.

Let the function $\hat{\rho} \in C_0^\infty(\mathbf{R}^1)$ be such that $\hat{\rho}(t) \equiv 1$ is a neighborhood of 0, $\operatorname{supp} \hat{\rho} \subset (-t_0, t_0)$ and $\rho(s) = F^{-1}_{t \to s}\hat{\rho}(t) > 0$. The function ρ belongs to $S(\mathbf{R}^1)$. The operator $\hat{\rho}(t)T(t)$ has kernel $\overline{T}(t, x, y)\hat{\rho}(t)$ in $C^\infty(\mathbf{R}^{2n+1})$. According to Proposition 16.1,

$$\operatorname{Sp} F^{-1}_{t \to s}\big(\hat{\rho}(t)T(t)^\#\big) = O\big((1 + |s|)^{-N}\big) \tag{17.1}$$

for any $N \in \mathbf{Z}_+$. We have

$$F^{-1}_{t \to \mu}\big(\hat{\rho}(t)U(t)^\#\big) = \int \rho(s - \mu)\, dE_\mu,$$

(the E_μ are the spectral projections of $A^\#$), and

$$\operatorname{Sp} F^{-1}_{t \to \mu}\big(\hat{\rho}(t)U(t)^\#\big) = \int \rho(s - \mu)\, dN(\mu), \tag{17.2}$$

where the function $N(t)$ is constructed from the operator A. (17.2) is easy to justify by using the fact that Sp is a normal trace.

By construction, the kernel of $Q_1(t)$ is

$$\overline{Q}_1(t, x, y) = \int q(t, x, y, \xi)e^{i(\psi(x, y, \xi) - ta_1(y, \xi))}\, d\xi.$$

Let $R(s, \lambda, \xi) = (2\pi)^{-1}\int\hat{\rho}(t)q(t, \lambda, \lambda, \xi)e^{its}\, dt$. The function R admits the estimates

$$|\partial_\lambda^\alpha \partial_\xi^\beta R(s, \lambda, \xi)| \leq C_{\alpha\beta N}(1 + |\xi|)^{-|\beta|}(1 + |s|)^{-N},$$

where $N \in \mathbf{Z}_+$ is arbitrary. These estimates are obtained from the estimates for q and the fact that $\rho \in S(\mathbf{R}^1)$. From analogous estimates it follows that the integral with respect to ξ in the expression

$$F_{t \to s}^{-1}(\hat{\rho}(t)\overline{Q}_1(t, x, y)) = \int\left[(2\pi)^{-1}\int\hat{\rho}(t)q(t, x, y, \xi)e^{it(s-a_1(y,\xi))}\, dt\right]e^{i\psi(x,y,\xi)}\, d\xi$$

converges, because $a_1(x, \xi)$ is uniformly elliptic in x. Consequently, we can apply to the operator $F_{t \to s}^{-1}(\hat{\rho}(t)Q_1(t))^{\#}$ Proposition 16.1, which gives us that

$$\operatorname{Sp} F_{t \to s}^{-1}(\hat{\rho}(t)Q_1(t))^{\#} = \int M_\lambda\{R(s - a_1(\lambda, \xi), \lambda, \xi)\}\, d\xi.$$

According to (17.1)

$$\left|\int\rho(s - \mu)\, dN(\mu) - \int M_\lambda\{R(s - a_1(\lambda, \xi), \lambda, \xi)\}\, d\xi\right| = O(|s|^{-N}).$$

If we next go through arguments completely analogous to the corresponding arguments in [2], we get

PROPOSITION 17.1.

$$\left|N(s) - M_\lambda\left\{\int_{a_1(\lambda,\xi)<s}(2\pi)^{-a}\, d\xi\right\}\right| \leq c(1 + |s|)^{n-1}. \tag{17.3}$$

The estimate (17.3) gives the asymptotic behavior of the function $N(s) \equiv N_A(s)$.

18. Let $P \in \mathrm{APL}^m$ be a uniformly elliptic differential operator that is selfadjoint in $L^2(\mathbf{R}^n)$ and has leading symbol $p_m(x, \xi) > 0$. It was proved in [6] (§6) that there is an $R > 0$ such that for the operator $Q = P + R \cdot \mathrm{Id}$ it makes sense to speak of the complex powers Q^s, $s \in \mathbf{C}$; these are pseudodifferential operators with a symbol that can be expanded in an asymptotic series of functions positively homogeneous in ξ (for large $|\xi|$) of order $ms - j, j \in \mathbf{Z}_+$. We next take $R = 0$ and consider the operator Q instead of P. Let $A = P^{1/m}$. Then $a_1(x, \xi) = (p_m(x, \xi))^{1/m}$. We define $A^{\#}$ and $P^{\#}$ as in §15. Let

$$A^{\#} = \int s\, dE_s^1, \qquad P^{\#} = \int s\, dE_s$$

be their spectral decompositions. Observe that $A^{\#} = (P^{\#})^{1/m}$ and $E_\lambda = E_{\lambda^{1/m}}^1$. Further, $N(s) = N_1(s^{1/m})$, where $N(s) = \operatorname{Sp} E$ and $N_1(\mu) = \operatorname{Sp} E_\mu^1$. Applying

(17.3) to $N_1(\mu)$, we obtain

THEOREM 18.1. *If the differential operator $P \in \mathrm{APL}^m$ is uniformly elliptic and selfadjoint, then the function $N(s) \equiv N_P(s)$ corresponding to it satisfies the inequality*

$$\left| N(s) - M_\lambda \left\{ \int_{P_m(\lambda,\xi)<s} (2\pi)^{-n} \, d\xi \right\} \right| \leqslant c(1 + |s|)^{(n-1)/m}. \qquad (18.1)$$

PROPOSITION 18.1 ([6], Proposition 8.4). *Let $P \in \mathrm{APL}^m$. Then $\sigma(\bar{P}) = \sigma(P^{\#})$, where σ denotes the spectrum of an operator, and $\bar{P}$ the closure of P.*

A *gap* of the function $N(t)$ is defined to be an interval $(a, b) \subset [0, \infty)$ such that $N(t) = \mathrm{const}$ for $t \in (a, b)$. The gaps of the function $N_P(t)$ are gaps in the spectrum of the operator $P^{\#}$.

It follows from Proposition 18.1 that the gaps of the function $N_P(t)$ are also gaps in the spectrum of the operator P, and so the function $N_P(t)$ is to a certain degree a characteristic of the spectrum of P.

The next result follows from Proposition 10.1 in [6] and (18.1).

PROPOSITION 18.2. *There is a number $M > 0$ such that if $(a, a + 1)$ is a gap of the function $N_P(t)$, $a \geqslant 1$, then*

$$l \leqslant Ma^{1-1/m}.$$

REMARK 18.1. The estimate (18.1) cannot be improved, as follows from a careful examination of operators with constant coefficients and the explicit formulas for $N(t)$ that hold for these operators ([7], §8).

We remark that an analogous estimate for the gap lengths in the spectrum was obtained by V. I. Feĭgin by different methods.

19. PROOF OF PROPOSITION 13.1. Suppose that $f_1(x, y) \in C^\infty(\mathbf{R}^{2n})$ is such that (for $j = 1$)

$$\left. \begin{aligned} |\partial_x^\alpha \partial_y^\beta f_j(x, y)| &\leqslant C_{j\alpha\beta N}(1 + |x - y|)^{-N}, \qquad N \in \mathbf{Z}_+, \\ \text{and } \partial_x^\alpha \partial_y^\beta f_j(x &+ \cdot, y + \cdot) \text{ are continuous functions} \\ \text{of } (x, &y) \text{ with values in } \mathrm{CAP}(\mathbf{R}^n). \end{aligned} \right\} \qquad (19.1)$$

LEMMA 19.1. *Let K be an integral operator of the form*

$$(Ku)(x) = \int f_2(x, y)u(y) \, dy,$$

where the kernel $f_2(x, y)$ satisfies (19.1) for $j = 2$. Then the function $j_3(x, y) = (Kf_1)(x, y)$ (where K acts with respect to the first variable) satisfies (19.1) for $j = 3$, and $C_{3\alpha\beta N} = CC_{2\alpha 0(N+n+1)}C_{10\beta N}$, where C depends only on n.

PROOF. We have

$$\left| (1 + |x - y|)^N \partial_x^\alpha \partial_y^\beta \int f_2(x, z) f_1(z, y) \, dz \right|$$

$$\leqslant C_{2\alpha0(N+n+1)} C_{10\beta N} \int \frac{(1 + |x - y|)^N \, dz}{(1 + |x - z|)^{N+n+1}(1 + |z - y|)^N}$$

$$\leqslant C_{2\alpha0(N+n+1)} C_{10\beta N} \int \frac{dz}{(1 + |x - z|)^{n+1}} = CC_{2\alpha0(N+n+1)} C_{10\beta N}.$$

The almost periodicity follows from the fact that a uniform limit of almost periodic functions is almost periodic.

LEMMA 19.2. *Let I be a Fourier integral operator of the form* (2.1), *and let* $a(x, y, \xi) \in AFS_{\rho,\delta}^m(\Omega, \mathbf{R}^N)$, *where Ω is a bounded region in $\mathbf{R}^n$ and φ a phase function satisfying the conditions of §2. Then the function $f_4 = If_1$ (where I acts with respect to the first variable) satisfies* (19.1) *for $j = 4$.*

The proof follows from the representation of I in the form (3.1) with Definition 1.2 taken into account.

It follows from (12.6) that

$$T(t) = \int_0^t Q_1(t - \tau)\left(-K_1(\tau) + \sum_1^\infty R_j(\tau) \right) d\tau,$$

where

$$(-1)^j R_j(\tau) = \int_0^\tau \int_0^{s_1} \cdots \int_0^{s_{j-1}} K_1(s_j) \cdots K_1(s_1 - \tau) \, ds_j \cdots ds_1.$$

Since the estimates (12.5) hold for the kernel K_1, Lemma 19.1 gives us that the operator $R_j(\tau)$ has a kernel of class C^∞; moreover, if $\bar{R}_j(\tau, x, y)$ is this kernel, then

$$\left| \partial_x^\alpha \partial_y^\beta \bar{R}_j(\tau, x, y) \right| \leqslant \frac{1}{j!} t_0^j C^{j-1} C_{00(N+n+1)}^{j-2} C_{0\beta N} C_{\alpha0(N+n+1)}(1 + |x - y|)^{-N},$$

and the usual condition for almost periodicity is satisfied. Consequently, the series

$$\bar{R}(\tau, x, y) = \sum_1^\infty \bar{R}_j(\tau, x, y)$$

converges, and the sum satisfies the estimates

$$\left| \partial_x^\alpha \partial_y^\beta \bar{R}(\tau, x, y) \right| \leqslant R_{\alpha\beta N}(1 + |x - y|)^{-N}$$

with the usual condition for almost periodicity satisfied.

If we next apply Lemma 19.2 to $I = Q_1(t - \tau)$ and $f_1 = \bar{K}_1(\tau, x, y) + \bar{R}(\tau, x, y)$, then we get the required statement about the kernel of the operator $T(t)$.

Received 15/SEPT/75

V. JU. KISELEV

Bibliography

1. I. M. Gel'fand and G. E. Šilov, *Generalized functions*. Vol. III: *Some questions in the theory of differential equations*, Fizmatgiz, Moscow, 1958; English transl., Academic Press, 1967.

2. Lars Hörmander, *The spectral function of an elliptic operator*, Acta Math. **121** (1968), 193–218.

3. ______ , *Fourier integral operators*. I, Acta Math. **127** (1971), 79–183.

4. M. A. Šubin, *Differential and pseudodifferential operators in spaces of almost periodic functions*, Math. Sb. **95(137)** (1974), 560–587; English transl. in Math. USSR Sb. **24** (1974).

5. ______ , *Elliptic almost periodic operators and von Neumann algebras*, Funkcional. Anal. i Priložen. **9** (1975), no. 1, 89–90; English transl. in Functional Anal. Appl. **9** (1975).

6. ______ , *Pseudodifferential almost-periodic operators and von Neumann algebras*, Trudy Moskov. Mat. Obšč. **35** (1976), 103–164; English transl. in Trans. Moscow Math. Soc. **1979**, no. 1(35).

7. ______ , *The density of states of selfadjoint elliptic operators with almost periodic coefficients*, Trudy Sem. Petrovsk. **3** (1978), 243–275, English transl. in this volume.

8. L. A. Coburn, R. D. Moyer and I. M. Singer, *C*-algebras of almost periodic pseudo-differential operators*, Acta Math. **130** (1973), 279–307.

9. Jacques Dixmier, *Les algèbres d'opérateurs dans l'espace Hilbertien (Algèbres de von Neumann)*, Gauthier-Villars, Paris, 1957.

Translated by H. H. McFADEN

Amer. Math. Soc. Transl.
(2) Vol. **118**, 1982

Fundamental Solutions of Quasihyperbolic Equations and Polynomials of Several Variables*

A. A. LOKŠIN

Contents

Introduction

In his well-known paper [1], I. G. Petrovskiĭ gave necessary and sufficient conditions for the existence of stable lacunae of hyperbolic equations. These conditions state that certain cycles on the normal surface are homologous to zero. In many cases the verification of these conditions is itself a difficult problem. Borovikov [2] established a number of sufficient conditions for the absence of lacunae.

Atiyah, Bott and Gårding [3] generalized the concept of a Petrovskiĭ lacuna and established a number of properties of such lacunaes.

In this paper we shall establish necessry and sufficient conditions for the absence of weak lacunae in "annuli" lying at the base of the characteristic cone of a homogeneous, quasihyperbolic equation with constant real coefficients in the space of $t, x_1, \ldots, x_n$, where n is odd (the definition of an "annulus" is given in §1).

1980 *Mathematics Subject Classification*. Primary 35L99; Secondary 35L25.
*Translation of Trudy Sem. Petrovsk. **3** (1978), 99–116. MR **58** #1675.

A brief exposition of the work was published in [14].

We recall the definition of quasihyperbolicity introduced in [4]. A homogeneous equation of order l

$$L\left(\frac{\partial}{\partial t}, \frac{\partial}{\partial x}\right)K \equiv P_{l-m}\left(\frac{\partial}{\partial x}\right)\frac{\partial^m K}{\partial t^m} + \cdots + P_l\left(\frac{\partial}{\partial x}\right)K = 0 \tag{1}$$

is called *quasihyperbolic* if the corresponding characteristic equation

$$L(\lambda, \xi) = P_{l-m}(\xi)\lambda^m + \cdots + P_l(\xi) = 0 \tag{2}$$

(where $P_k(\xi)$ are homogeneous polynomials of degree k) has m real and distinct roots $\lambda_j(\xi)$ on the sphere $|\xi| = 1$. It is also assumed that the polynomials $P_{l-m}(\xi)$ and $P_l(\xi)$ do not vanish on this sphere (which implies that l and m are even), and hence the normal surface is bounded. We henceforth assume that $P_{l-m}(\xi) \geq 0$.

In the case $l = m$ our equation is obviously hyperbolic.

We consider the fundamental solution of the Cauchy problem for equation (1), i.e., the solution satisfying the initial conditions

$$\left(\partial^i/\partial t^i\right)K(0, x) = 0; \qquad i = 0, 1, \ldots, m - 2,$$

$$\left(\partial^{m-1}/\partial t^{m-1}\right)K(0, x) = \delta(x). \tag{3}$$

It is shown in [5] that formulas of Herglotz-Petrovskiĭ type are valid for $K(t, x)$; namely,

$$K(t, x) = c_{mn}\int_{L(1,\xi)=0}\frac{(-1)^N f_{mn}(x\xi + t)P_{l-m}(\xi)}{|\operatorname{grad} L(1, \xi)|}\,dS, \tag{4}$$

where the integration goes over all $m/2$ ovals O_N of the surface $L(1, \xi) = 0$ (the origin is a point of multiplicity $l - m$ for this surface); N is the number of the oval counting in order away from the origin. For n odd we have $f_{mn}(s) = s^{m-n-1}\operatorname{sign} s$ if $m - 1 \geq n$, and $f_{mn}(s) = \delta^{(n-m)}(s)$ if $m - 1 < n$; dS is the surface area element of the surface $L(1, \xi) = 0$, and $c_{mn} = \text{const}$.

REMARK 1. It is shown in [6] that the fundamental solution (4) is distinguished in the class of all fundamental solutions by the condition that it tends to zero as $|x| \to \infty$.

We now present the definition of a weak lacuna [7].

A region G is called a *weak lacuna of order* p for equation (1) if the fundamental solution $K(t, x)$ of the Cauchy problem is a polynomial in t of degree $p - 1$ in this region.

We shall not study weak lacunae in arbitrary regions cut out by the characteristic cone (with vertex at the origin) on the place $t = 1$, but rather only in connected components of the sets G_k. The sets G_k admit the following description [2]. We consider the ovals O_i in the section of the normal cone by the plane $\lambda = 1$, and denote their dual characteristic ovals by C_i. We set $Q_k = \bigcup_{i \geq k+1} C_i$. We denote by A_k the set of those points of the space $x_1, \ldots, x_n$ for which the plane $x\xi + 1 = 0$ does not intersect $\bigcup_{i \leq k} O_i$, while the component of the complement to Q_k which contains infinity we denote by B_k. We set $G_k = A_k \cap B_k$.

§1. Theorems on the decomposition of operators in the case of the existence of a weak lacuna in an "annulus"

We set $S_1 = \bigcup_{i \geq k+1} O_i$ and $S_2 = \bigcup_{j \leq k} O_j$.

THEOREM 1. *Suppose that at least one of the connected components of the set G_k is a weak lacuna of the operator L. Then $L = L_1 L_2$, where L_1 is a hyperbolic operator whose normal surface is S_1, and L_2 is a quasihyperbolic operator whose normal surface consists of S_2 and possibly the origin (the origin belongs to the normal surface of L_2 if $l > m$).*

REMARK 1. The preceding theorem is a generalization of Theorem 3 of [2] pertaining to the case of strong lacunae of strictly hyperbolic equations.

REMARK 2. Suppose that the hypotheses of Theorem 1 are satisfied. The assertion of the theorem then implies that the region G_k is contained between two closed, nonintersecting, convex surfaces and is therefore connected. Indeed, the outer characteristic oval C' of the hyperbolic operator L_1 is convex ([8], Chapter VI, §3.3). Let C'' be the image of the convex hull of the outer oval of the normal surface for the operator L_2. Then C'' is also convex and encompasses C'. The region G_k is obviously contained between C' and C''; we call it in this case an "annulus". Thus, if the set G_k is not an "annulus", then none of its connected components is a weak lacuna.

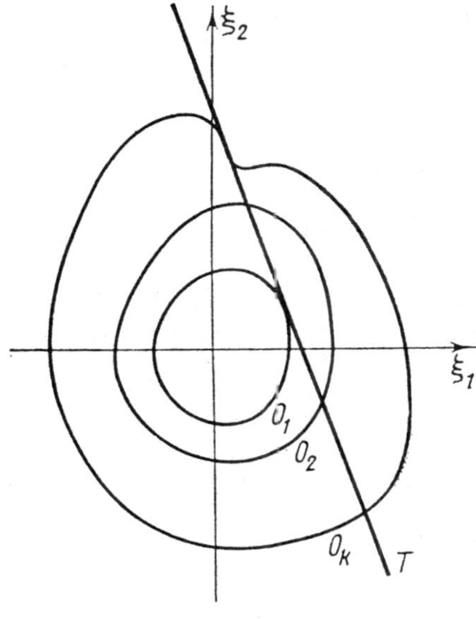

FIGURE 1

We remark that when (1) is hyperbolic the set G_1 is always connected. Indeed, it suffices to show that the characteristic ovals C_1 and C_i $(i = 2,\dots,m/2)$ do not intersect. This is equivalent to the assertion that the ovals O_1 and O_i $(i = 2,\dots,m/2)$ of the normal surface have no common tangents. We suppose that O_1

and O_{i_0} have a common tangent T. Then the number of points of intersection (counting multiplicities) of the line T with the normal surface is greater than the degree of the characteristic equation (Figure 1). Hence G_1 is connected.

PROOF. Using (4), we can represent the fundamental solution in the form $K(t, x) = K_1(t, x) + K_2(t, x)$, where K_1 is an integral over S_1 and K_2 is an integral over S_2. It is evident from (4) that in each component of G_k the difference $K = K_1 = K_2$ is a polynomial of degree at most $m - n - 1$.

Suppose that the order of the weak lacuna in question is equal to $p_0 \geqslant 0$. We choose any $p \geqslant \max\{p_0, m - n - 1\}$. We note that the functions $\partial_t^p K_1(0, x)$ must be concentrated at the origin. Indeed, $K_1(t, x)$ has singularities only on the set $\bigcup_{j \geqslant k+1} C_j$ of inner characteristic ovals ([2], p. 249). This means that the function $\partial_t^p K_1(t, x)$, which is equal to zero in our connected component of G_k, can be extended analytically to the exterior of the characteristic cone of equation (1), whence it follows that $\partial_t^p K_1(0, x)$ is concentrated at the origin.

It is not hard to see that the function

$$\partial_t^p K_1(t, x) \equiv \frac{\partial^p}{\partial t^p} c \int_{S_1} \frac{(-1)^j f_{mn}(x\xi + t) P_{l-m}(\xi)}{|\operatorname{grad} L(1, \xi)|} \, dS$$

is a solution of (1), i.e.,

$$L(\partial/\partial t, \partial/\partial x)\partial_t^p K_1(t, x) = 0.$$

We take the Fourier transform of this equality and then set $t = 0$. We obtain

$$P_{l-m}(\sigma)A_{p+m}(\sigma) + \cdots + P_l(\sigma)A_p(\sigma) = 0, \tag{5}$$

where

$$A_p = (i)^p \partial_t^p \tilde{K}_1(0, \sigma). \tag{6}$$

We know that the functions $\partial_t^p K_1(0, x)$ must be concentrated at the origin for $p \geqslant \max\{p_0, m - n - 1\}$; hence, the functions $A_p(\sigma)$ must be polynomials for these p. This fact and (5) imply that all the $A_p(\sigma)$ $(p = 0, 1, \ldots)$ must belong to the field of rational functions of σ. In [6] the explicit formulas[1]

$$A_p = {\sum_j}' \frac{P_{l-m}(\sigma)\lambda_j^p}{L'_\lambda(\lambda_j, \sigma)},$$

$$L'_\lambda(\lambda_j, \sigma) = (\partial/\partial\lambda)L(\lambda, \sigma)|_{\lambda=\lambda_j} \tag{7}$$

are obtained, where the prime on the sum means that the summation goes over those roots $\lambda_j(\sigma)$ of the characteristic equation (2) which correspond to outer ovals of the normal surface (i.e., to the ovals $O_{k+1}, \ldots, O_{m/2}$). We have

$$L'_\lambda(\lambda_j, \sigma) = \sum_{k=1}^m k P_{l-k}(\sigma)\lambda_j^{k-1},$$

---

[1] The author uses this opportunity to point out that in [6], p. 146, dS is the surface element of $Q(1, \xi) = 0$ and $d\omega_\eta$ is the surface element of the sphere $|\eta| = 1$; on p. 152 $v_j(\sigma)$ $(j = 1, \ldots, m)$ denote the roots of the equation $Q(\lambda, \sigma) = 0$ for $|\sigma| = 1$.

and hence

$$\frac{1}{P_{l-m}(\sigma)} \sum_{k=1}^{m} k P_{l-k}(\sigma) A_{p+k-1}(\sigma) = {\sum_{j}}' \lambda_j^p \tag{8}$$

is a rational function, whence we see that all the elementary symmetric functions of the roots λ_j corresponding to the outer ovals of the normal surface belong to the field of rational functions of σ ([11], §39). Hence, $L(\lambda, \sigma)$ decomposes in the ring of polynomials in λ over the field of rational functions of σ. However, it is known from algebra [11] that in this case $L(\lambda, \sigma)$ decomposes in the ring of polynomials of all the variables $\lambda, \sigma_1, \ldots, \sigma_n$, i.e., $L(\lambda, \sigma) = L_1(\lambda, \sigma) L_2(\lambda, \sigma)$, where $L_i(\lambda, \sigma)$ $(i = 1, 2)$ are quasihyperbolic operators, and the characteristic roots of $L_1(\lambda, \sigma)$ correspond to the outer ovals of the normal surface, while those of $L_2(\lambda, \sigma)$ correspond to the inner ovals.

We shall now show that $L_1(\lambda, \sigma)$ is hyperbolic (it may always be assumed that $L_1(\lambda, \sigma)$ is primitive as a polynomial in λ, since a common factor of its coefficients depending on σ can be assigned to the polynomial $L_2(\lambda, \sigma)$). We have

$$A_{p+i} = \sum_{j=1}^{m_1} \lambda_j^i \frac{\lambda_j^p P_{l-m}(\sigma)}{L_1'(\lambda_j, \sigma) L_2(\lambda_j, \sigma)} = Q_i(\sigma); \qquad i = 0, 1, \ldots, m_1 - 1, \tag{9}$$

where $\lambda_1, \ldots, \lambda_{m_1}$ are the roots of the polynomial $L_1(\lambda, \sigma)$ indexed in increasing order, $Q_i(\sigma)$ are polynomials, and m_1 is the degree of $L_1(\lambda, \sigma)$ in λ. Solving the system (9), we obtain

$$\frac{P_{l-m}(\sigma)\lambda_j^p}{L_1'(\lambda_j, \sigma) L_2(\lambda_j, \sigma)} = \frac{\begin{vmatrix} 1 & \cdots & Q_0(\sigma) & \cdots & 1 \\ \vdots & & \vdots & & \vdots \\ \vdots & & \vdots & & \vdots \\ \lambda_1^{m_1-1} & \cdots & Q_{m_1-1} & \cdots & \lambda_{m_1}^{m_1-1} \end{vmatrix}}{\begin{vmatrix} 1 & \cdots & 1 & \cdots & 1 \\ \vdots & & & & \vdots \\ \lambda_1^{m_1-1} & \cdots & & \cdots & \lambda_{m_1}^{m_1-1} \end{vmatrix}}. \tag{10}$$

However, $L_1'(\lambda_j, \sigma) = P_0(\sigma) \prod_{j \neq i}(\lambda_j - \lambda_i)$, where $P_0(\sigma)$ is the leading coefficient of $L_1(\lambda, \sigma)$. Therefore,

$$\frac{1}{L_2(\lambda_j, \sigma)} \lambda_j^p = \frac{P_0(\sigma)}{P_{l-m}(\sigma)} R(\lambda_1, \ldots, \lambda_{j-1}, \lambda_{j+1} \cdots \lambda_{m_1}; \sigma),$$

where the degree of the polynomial R in λ_i does not exceed m_1, i.e., a fixed number not depending on p. Hence, the $\lambda_j(\sigma)$ are entire algebraic functions ([12], Chapter IX, §1). The theorem is proved.

THEOREM 2. *The exterior of the characteristic cone of equation* (1) (*i.e., the region between the plane $t = 0$ and the sheet of the characteristic surface closest to it*) *is a weak lacuna if and only if*

$$L = P_{l-m}\left(\frac{\partial}{\partial x}\right) L_1\left(\frac{\partial}{\partial t}, \frac{\partial}{\partial x}\right),$$

where $P_{l-m}(\partial/\partial x)$, *as we know, is elliptic, and* $L_1(\partial/\partial t, \partial/\partial x)$ *is a strictly hyperbolic operator.*

REMARK 1. The theorem implies that the weak lacuna of equation (1) in question is a lacuna in the sense of Petrovskiĭ ([1], p. 290). Indeed, the exterior of the characteristic cone of the operator $L(\partial/\partial t, \partial/\partial x)$ is simultaneously the exterior of the characteristic cone of the hyperbolic operator $L_1(\partial/\partial t, \partial/\partial x)$, while the fundamental solutions of the form (4) for two operators differing by the factor $P_{l-m}(\partial/\partial x)$ obviously coincide.

REMARK 2. In contrast to the other theorems of this paper, Theorem 2 is valid in even-dimensional as well as in odd-dimensional space.

PROOF. Suppose that the order of the weak lacuna in question is p_0. We choose $p \geqslant p_0$. Then the functions

$$A_{p+i} = \sum_{j=1}^{m} \frac{\lambda_j^{p+i} P_{l-m}(\sigma)}{L_\lambda'(\lambda_j, \sigma)}$$

must be polynomials for all $i \geqslant 0$. We obtain the system

$$\sum_{j=1}^{m} \lambda_j^i \frac{\lambda_j^p P_{l-m}(\sigma)}{L_\lambda'(\lambda_j, \sigma)} = Q_j(\sigma); \qquad i = 0, 1, \ldots, m - 1,$$

where the Q_i are polynomials in σ. By Cramer's formulas

$$\frac{P_{l-m}(\sigma)}{L_\lambda'(\lambda_j, \sigma)} \lambda_j^p = \frac{\begin{vmatrix} 1 & \cdots & Q_0 & \cdots & 1 \\ \cdots & \cdots & \cdots & \cdots & \cdots \\ \lambda_1^{m-1} & \cdots & Q_{m-1} & \cdots & \lambda_m^{m-1} \end{vmatrix}}{\begin{vmatrix} 1 & \cdots & \cdots & 1 \\ \cdots & \cdots & \cdots & \cdots \\ \lambda_1^{m-1} & \cdots & \cdots & \lambda_m^{m-1} \end{vmatrix}},$$

whence $\lambda_j^p = Q(\lambda_1, \ldots, \lambda_{j-1}, \lambda_{j+1}, \ldots, \lambda_n; \sigma)$, and the degree of the polynomial Q in the variables λ_i does not exceed m, i.e., a fixed number not depending on p. As in the preceding theorem, from this we see that all the $\lambda_i(\sigma)$ are entire algebraic functions. Hence, $P_{l-m}(\sigma)$ must be a common factor of all the $P_j(\sigma)$. The theorem is proved.

§2. Auxiliary lemmas on polynomials. Necessary and sufficient conditions for the existence of a weak lacuna in an "annulus"

Our next problem is to establish necessary and sufficient conditions for the existence of weak lacunae in "annuli".

Thus, suppose that there is a weak lacuna in the "annulus" G_k. Then by Theorem 1 $L = L_1 L_2$, where $L_1 = \partial^{m_1}/\partial t^{m_1} + \cdots + Q_{m_1}(\partial/\partial x)$ is a homogeneous hyperbolic operator of order m_1.

It is not hard to see that

$$L_1\left(\frac{\partial}{\partial t}, \frac{\partial}{\partial x}\right)\left\{ c \int_{L(1,\xi)=0} \frac{(-1)^N f_{mn}(x\xi + t) P_{l-m}(\xi)}{|\operatorname{grad} L(1,\xi)|}\, dS\right\} = 0,$$

whence

$$\partial_t^{p-m_1} L_1(\partial/\partial t, \partial/\partial x) K_1(t, x) = 0. \tag{11}$$

We set $(-1)^p \partial_t^p \tilde{K}_1(0, \sigma) = A_p(\sigma)$. The Fourier transform of (11) then gives for $t = 0$

$$A_p(\sigma) + Q_1(\sigma) A_{p-1}(\sigma) + \cdots + Q_{m_1}(\sigma) A_{p-m_1}(\sigma) = 0. \tag{12}$$

The initial functions for the relation (12) considered as a difference equation in p are the symmetric rational functions of the roots λ_j (this is evident from (9)), and hence

$$A_0, \dots, A_{m_1 - 1} \tag{13}$$

are rational functions of σ.

In order that in the "annulus" G_k there be a weak lacuna it is necessary and sufficient (because of the analyticity of $K_1(t, x)$ off the characteristic cone of L_1) that, starting with some index, $\partial_t^i K_1(0, x)$ be concentrated at the origin. This is equivalent to the $A_i(\sigma)$ being polynomials starting with this index.

Thus, the question of the existence of a weak lacuna in an "annulus" reduces to the following: suppose that there is given the recurrent sequence (12) with polynomial coefficients, and the initial functions (13) are rational. It is required to find conditions under which our sequence gives only polynomials beginning with some index.

LEMMA 1. *Let P and Q be polynomials in $\sigma_1, \dots, \sigma_n$ with complex coefficients, and suppose that PQ^{-1} is not a polynomial. Then it is possible to select a line in the space of $\sigma_1, \dots, \sigma_n$ such that the restriction of PQ^{-1} to this line is also not a polynomial.*

PROOF. We suppose that the quotient PQ^{-1} is reduced. We shall first prove that there exists a point $\sigma^0 = (\sigma_1^0, \dots, \sigma_n^0)$ such that $Q(\sigma^0) = 0$ and $P(\sigma^0) \neq 0$. Indeed, if this were not so, then by Hilbert's Nullstellensatz ([12], Chapter X, §2) our quotient would be reducible, contrary to hypothesis. We now draw a line through σ^0 such that the restriction of Q to this line is not identically zero. This line is obviously the desired line, since at σ^0 the ratio PQ^{-1} has a pole, and hence it cannot be a polynomial.

LEMMA 2. *If a sequence of solutions of problem* (12), (13) *for* $p \geqslant p_0$ *consists only of polynomials, then* $p_0 \leqslant n_0 = m_1 M$, *where* m_1 *is the length of the recurrence relation* (12), *and* M *is the maximal degree of the denominators of the initial functions* (13).

PROOF. We suppose that the lemma has been proved in the one-dimensional case. It is easy to see that this same n_0 also works in the multidimensional case (for the same m_1 and M). Indeed, otherwise we apply Lemma 1 and obtain a contradiction.

Thus, it suffices to solve the problem in the one-dimensional case. We expand the initial functions into partial fractions in the complex domain (some of the coefficients of the expansion may be equal to zero):

$$A_i = R_i + \frac{a_i^1}{\sigma - \alpha} + \cdots + \frac{a_i^l}{(\sigma - \alpha)^l} + \frac{b_i^1}{\sigma - \beta} + \cdots + \frac{b_i^l}{(\sigma - \beta)^l} + \cdots . \tag{14}$$

Here the R_i are polynomials, and $l \leqslant M$; α and β are roots of the denominators of all the A_i, $i = 0, 1, \ldots, m_1 - 1$.

We choose an arbitrary root, say α, from the expansion of A_i by (14). It is evident from (12) and (14) that the coefficients with which $(\sigma - \alpha)^{-l}$ enters the expansion of $A_p, \ldots, A_{p-m_1}$ are related by the formula

$$a_p^l = Q_1(\alpha)a_{p-1}^l + \cdots + Q_{m_1}(\alpha)a_{p-m_1}^l. \tag{15}$$

If we wish that for $p \geqslant p_0$ the sequence $\{A_p\}$ consist only of polynomials, then we must have $a_p^l = 0$ for $p \geqslant p_0$. Suppose that $a_p^l = 0$ for $p \geqslant p_0$. Suppose further that $Q_{m_1-s}(\alpha) \neq 0$, while $Q_{m_1-s+1}(\alpha), \ldots, Q_{m_1}(\alpha)$ are all zero. We choose $p > p_0 + m_1 - s - 1$ and solve (15) for the term $a_{p+m_1-s-1}^l$. All terms on the right side of the equality thus obtained are then obviously zero. Therefore, $a_{p+m_1-s-1}^l = 0$. Replacing the index p by $p - 1, p - 2, \ldots$ successively, we find that

$$a_i^l = 0; \qquad i \geqslant m_1. \tag{16}$$

On expanding Q_j by Taylor's formula in powers of $\sigma - \alpha$ and substituting this into (12), we find, by noting (14), that

$$a_p^{l-1} = \left\{ Q_1(\alpha)a_{p-1}^{l-1} + \cdots + Q_{m_1}(\alpha)a_{p-m_1}^{l-1} \right\}$$
$$+ \frac{1}{1!} \left\{ Q_1'(\alpha)a_{p-1}^l + \cdots + Q_{m_1}'(\alpha)a_{p-m_1}^l \right\}. \tag{17}$$

The second brace vanishes by (16) for $p \geqslant 2m_1$. From this it is evident that we must have $a_p^{l-1} = 0$ for $p \geqslant 2m_1$. On continuing this process using the Taylor expansion of the polynomials Q_j at the point α, we obtain in similar fashion $a_p^1 = 0$ for $p \geqslant lm$. Setting $n_0 = Mm$, we obtain the required result.

LEMMA 3. *In order that a sequence $\{A_p(\sigma)\}$ of solutions of problem (12), (13) consist only of polynomials starting from some index, it is necessary and sufficient that the functions $A_{n_0}(\sigma)$, $A_{n_0+1}(\sigma),\ldots,A_{n_0+m_1-1}(\sigma)$ be polynomials.*

PROOF. Sufficiency is obvious, since the recurrence relation (12) possesses polynomial coefficients. Necessity follows immediately from Lemma 2.

THEOREM 3. *For the existence of a weak lacuna in the "annulus" G_k it is necessary and sufficient that the following condition hold:*

a) $L = L_1 L_2$, *where the L_i $(i = 1, 2)$ are the operators described in Theorem 1.*

b) *The rational functions*

$$A_p = \sum_{i=1}^{m_1} \frac{\lambda_i^p}{L_1'(\lambda_i, \sigma) L_2'(\lambda_i, \sigma)}, \qquad p = n_0,\ldots,n_0 + m_1 - 1,$$

are polynomials in σ (here m_1 is the order of the operator L_1 and λ_i are its characteristic roots).

Moreover, the order of the weak lacuna in question is less than or equal to $n_0 < m^3$ (n_0 is defined in Lemma 2).

The necessity of conditions a) and b) follows from Theorem 1 and Lemma 3. Sufficiency follows from Lemma 3.

§3. The explicit form of the operators in some special cases

Suppose that in the "annulus" $G_{m/2-1}$ of the homogeneous, hyperbolic operator L there is a weak lacuna. Then, by Theorem 1, $L = L_1 L_2$, where L_1 is a homogeneous hyperbolic operator of second order. By dilation of the axes x_i $(i = 1,\ldots,n)$ and rotation about the t-axis L_1 can always be brought to the form

$$L_1 = \left(\frac{\partial}{\partial t} + \sum_{i=1}^{n} \varepsilon_i \frac{\partial}{\partial x_i} \right)^2 - \sum_{i=1}^{n} \frac{\partial^2}{\partial x_i^2}, \tag{18}$$

and its characteristic roots are

$$\lambda_{1,2}(\sigma) = - \sum_{i=1}^{n} \varepsilon_i \sigma_i \pm \rho, \tag{19}$$

where $\rho = \sqrt{\sum_{1 \leqslant i \leqslant n} \sigma_i^2}$.

THEOREM 4. *Let $L = L_1 L_2$, where L_1 has the form (18), while the normal surface of L_2 is contained strictly within the normal surface of L_1. Then in order that the "annulus" $G_{m/2-1}$ be a weak lacuna it is necessary and sufficient that*

$$L_2(\lambda_i(\sigma), \sigma) = c\lambda_i^{m-2}(\sigma), \tag{20}$$

where $\lambda_i(\sigma)$ is either of the two characteristic roots of the operator L_1, and $c = $ const.

Moreover, the order of the weak lacuna in question is $\leqslant m - 2$ (m is the order of the operator L).

Proof. In [6] the following explicit formulas were obtained:

$$c\partial_t^k \tilde{K}_1(0, \sigma) = \frac{\lambda_1^k}{2\rho L_2(\lambda_1, \sigma)} + \frac{\lambda_2^k}{-2\rho L_2(\lambda_2, \sigma)}. \tag{21}$$

Because of Theorem 3, for the existence of a weak lacuna it is necessary and sufficient that the following two equalities hold:

$$\frac{1}{\rho}\left\{ \frac{\left(-\sum_{i=1}^n \varepsilon_i \sigma_i + \rho\right)^k}{L_2(-\sum \varepsilon_i \sigma_i + \rho, \sigma)} - \frac{\left(-\sum_{i=1}^n \varepsilon_i \sigma_i - \rho\right)^k}{L_2(-\sum_{i=1}^n \varepsilon_i \sigma_i - \rho, \sigma)} \right\} = P_1(\sigma), \tag{22}$$

$$\frac{1}{\rho}\left\{ \frac{\left(-\sum_{i=1}^n \varepsilon_i \sigma_i + \rho\right)^{k+1}}{L_2(-\sum_{i=1}^n \varepsilon_i \sigma_i + \rho; \sigma)} - \frac{\left(-\sum_{i=1}^n \varepsilon_i \sigma_i - \rho\right)^{k+1}}{L_2(-\sum_{i=1}^n \varepsilon_i \sigma_i - \rho, \sigma)} \right\} = P_2(\sigma), \tag{23}$$

where the $P_i(\sigma)$ are polynomials, $i = 1, 2$, and $k \geqslant n_0$.

We multiply (22) by $(-\sum_1^n \varepsilon_i \sigma_i - \rho)$ and subtract it from (23). We obtain

$$\frac{\left(-\sum_{i=1}^n \varepsilon_i \sigma_i + \rho\right)^k}{L_2(-\sum_{i=1}^n \varepsilon_i \sigma_i + \rho, \sigma)} = P(\rho, \sigma), \tag{24}$$

where $P(\rho, \sigma)$ is a polynomial in ρ and σ. We let $\sigma_i/\rho = \omega_i$, and we then set $\rho = 1$. From (24) we obtain

$$\frac{\left(-\sum_{i=1}^n \varepsilon_i \omega_i + 1\right)^k}{L_2(-\sum_{i=1}^n \varepsilon_i \omega_i + 1, \omega)} = P(1, \omega). \tag{25}$$

We successively make the two changes of variables

$$\omega_n' = \omega_n + 1; \qquad \omega_i' = \omega_i \quad (i = 1, \dots, n - 1).$$

$$y_i = \frac{\omega_i'}{\sum_{i=1}^n (\omega_i')^2}. \tag{26}$$

The sphere $\omega^2 = 1$ hereby goes over into the plane $y_n = \frac{1}{2}$, while (25) can be rewritten as follows:

$$\frac{\left(-\varepsilon_n(y_n/|y|^2 - 1) - \sum_{i=1}^n \varepsilon_i y_i/|y|^2 + 1\right)^k}{L_2(-\varepsilon_n(y_n/|y|^2 - 1) - \sum_{i=1}^n \varepsilon_i y_i/|y|^2 + 1; y/|y|^2)} = P(1, y/|y|^2). \tag{27}$$

We set

$$f(y) = -\varepsilon_n(y_n - |y|^2) - \sum_{i=1}^n \varepsilon_i y_i + |y|^2. \tag{28}$$

Then (27) can be rewritten as follows:

$$(f(y))^k (|y|^2)^{m-2-k+p} / L_2(f(y), y) = P_0(y), \tag{29}$$

where $P_0(y) = P(1, y/|y|^2)|y|^{2p}$ is a polynomial.

It is not hard to see that the polynomial (28) with $y_n = \frac{1}{2}$ is irreducible for $n \geqslant 3$ over the field of real numbers.

Therefore, from (29) and the theorem on the uniqueness of the decomposition into prime factors in the ring of polynomials of several variables it follows that

$$L_2(f(y), y) = C_1(f(y))^{q_1}(|y|^2)^{q_2}, \tag{30}$$

where

$$q_1 + q_2 = m - 2, \qquad q_i \geqslant 0, \qquad C_1 = \text{const.}$$

Returning to the variables σ_i, we find from (30) on noting (28)

$$L_2\left(-\sum_{i=1}^{n} \varepsilon_i\sigma_i + \rho, \sigma\right) = \left(-\sum_{i=1}^{n} \varepsilon_i\sigma_i + \rho\right)^{q_1} \rho^{m-2-q_1}C_1. \tag{31}$$

Analogous computations for the second root (19) give

$$L_2\left(-\sum_{i=1}^{n} \varepsilon_i\sigma_i - \rho, \sigma\right) = \left(-\sum_{i=1}^{n} \varepsilon_i\sigma_i - \rho\right)^{p_1} \rho^{m-2-p_1}C_2. \tag{32}$$

We now observe that

$$L_2(\lambda_1(-\sigma), -\sigma) = L_2(-\lambda_2(\sigma), -\sigma) = L_2(\lambda_2(\sigma), \sigma). \tag{33}$$

Substituting (33) into (32) and (31), we obtain

$$\left(\sum_{i=1}^{n} \varepsilon_i\sigma_i + \rho\right)^{q_1} \rho^{m-2-q_1}C_1 = (-1)^{p_1}\left(\sum_{i=1}^{n} \varepsilon_i\sigma_i + \rho\right)^{p_1} \rho^{m-2-p_1}C_2, \tag{34}$$

whence

$$q_1 = p_1; \qquad C_1 = (-1)^{q_1}C_2. \tag{35}$$

Using (35), we now substitute (31) and (32) into the formula for $\partial_t^k \tilde{K}_1(0, \sigma)$:

$$\partial_t^k \tilde{K}_1(0, \sigma) = \frac{1}{2C\rho^{m-1-q_1}}\left\{(-\sum \varepsilon_i\sigma_i + \rho)^{k-q_1} + (-1)^{k-1}(\sum \varepsilon_i\sigma_i + \rho)^{k-q_1}\right\}. \tag{36}$$

We expand the terms in braces in the numerator of (36). Depending on the parity of q_1, either the coefficient of ρ^0 or of ρ^1 is nonzero; we therefore have two cases:

$$\partial_t^k \tilde{K}_1(0, \sigma) = \frac{A_1(\sum \varepsilon_i\sigma_i)^{k-q_1} + A_2(\sum \varepsilon_i\sigma_i)^{k-q_1-2}\rho^2 + \cdots}{\rho^{m-q_1-1}}, \tag{37}$$

$$\partial_t^k \tilde{K}_1(0, \sigma) = \frac{B_1(\sum \varepsilon_i\sigma_i)^{k-q_1-1}\rho + B_2(\sum \varepsilon_i\sigma_i)^{k-q_1-3}\rho^3 + \cdots}{\rho^{m-q_1-1}}, \tag{38}$$

where the A_i and B_i are nonzero constants and $0 \leqslant q_1 \leqslant m - 2$.

Suppose that all $\varepsilon_i = 0$. Taking (35) into account, (31) and (32) then give the condition

$$L_2(\pm\rho, \sigma) = C(\pm\rho)^{m-2}. \tag{39}$$

Suppose now that some $\varepsilon_i \neq 0$. We first consider the case (37). We set $j = m - q_1 - 2 \geq 0$. We have

$$\rho^j \partial_t^k \tilde{K}_1(0, \sigma) = \frac{A_1(\Sigma \, \varepsilon_i \sigma_i)^{k-q_1}}{\rho} + R(\rho, \sigma), \tag{40}$$

where $R(\rho, \sigma)$ is a polynomial in the variables ρ and σ. From the proof of Lemma 1 it follows that the polynomial ρ^2 has a complex zero σ^0 which is not a zero of $\Sigma \, \varepsilon_i \sigma_i$. Hence, the function $\partial_t^k \tilde{K}_1(0, \sigma)$ cannot be a polynomial.

In the case (38) we have

$$\rho^j \partial_t^k \tilde{K}_1(0, \sigma) = B_1(\Sigma \, \varepsilon_i \sigma_i)^{k-q_1-1} + B_2(\Sigma \, \varepsilon_i \sigma_i)^{k-q_1-3} \rho^2 + \cdots . \tag{41}$$

If $j > 0$, we divide both sides of the last equality by ρ. We obtain an equality analogous to (40), from which it follows that $\partial_t^k \tilde{K}(0, \sigma)$ cannot be a polynomial. We have thus found that in order that the functions $\partial_t^k \tilde{K}_1(0, \sigma)$ for all sufficiently large k be polynomials it is necessary that $q_1 = m - 2$; the relations (31) and (32) hereby assume the form (20).

On the other hand, if $q_1 = m - 2$, then in the numerator of (36) there remain only terms containing odd powers of ρ, and (36) is a polynomial for $k \geq m - 2$. The theorem is proved.

REMARK 1. It may always be assumed that Theorem 4 is not formulated in the special system of coordinates where $L_1(\partial/\partial t, \partial/\partial x)$ has the form (18) but rather in arbitrary coordinates. It follows from (20) that the polynomial $L_2(\lambda, \sigma) - c\lambda^{m-2}$ has the same roots $\lambda_1(\sigma)$ and $\lambda_2(\sigma)$ as the irreducible polynomial of second degree $L_1(\lambda, \sigma)$. Therefore, (20) can be given the form

$$L_2(\lambda, \sigma) = L_1(\lambda, \sigma)Q(\lambda, \sigma) + c\lambda^{m-2}, \tag{42}$$

where $Q(\lambda, \sigma)$ is a polynomial.

REMARK 2. It is not hard to see that Theorem 4 is also true for the "annulus" G_1 in the case where the normal surface of the operator $L_2(\partial/\partial t, \partial/\partial x)$ encompasses the normal surface of $L_1(\partial/\partial t, \partial/\partial x)$.

THEOREM 5. *Let $L(\partial/\partial t, \partial/\partial x)$ be a homogeneous hyperbolic operator (with a bounded surface). In order that each of the regions in the intersection of its characteristic cone with the plane $t = 1$ be a weak lacuna it is necessary and sufficient that L have the form*

$$L\left(\frac{\partial}{\partial t}, \frac{\partial}{\partial x}\right) = \prod_{i=1}^{m/2} \left\{ (1 - c_i)L_1\left(\frac{\partial}{\partial t}, \frac{\partial}{\partial x}\right) + c_i \frac{\partial^2}{\partial t^2} \right\}, \tag{43}$$

where $L_1(\partial/\partial t, \partial/\partial x)$ is a hyperbolic operator of second order, and $c_i = $ const.

PROOF. We observe that it suffices to prove that $L(\partial/\partial t, \partial/\partial x)$ has the form (43) if and only if all the regions G_k are weak lacunae. Indeed, if all the regions G_k are weak lacunae, then by Theorem 1

$$L\left(\frac{\partial}{\partial t}, \frac{\partial}{\partial x}\right) = \prod_{i=1}^{m/2} L_i\left(\frac{\partial}{\partial t}, \frac{\partial}{\partial x}\right), \tag{44}$$

where the $L_i(\partial/\partial t, \partial/\partial x)$ are hyperbolic operators of second order. Hence, the characteristic cone of $L(\partial/\partial t, \partial/\partial x)$ consists of convex ovals contained in one another. This means that the plane $t = 1$ is decomposed by the characteristic cone into "annuli", an exterior, and a core. The exterior of the characteristic cone is always a lacuna for a hyperbolic operator. The core is a weak lacuna or a lacuna, since the dimension of the space is odd [2]. Therefore, the function $\partial_t^p K(t, x)$ is concentrated on the characteristic cone if we take $p > \max\{p_i\}$, where the p_i are the orders of the weak lemma in the core and "annuli"; this gives the assertion of the theorem.

We thus suppose that all the regions G_k are weak lacunae. In the coordinate system we chose in the preceding theorem we have $L = L_1 \prod_{i \geqslant 2} L_i$, where L_1 has the form (18). The condition that in the "annulus" G_1 there is a weak lacuna can be written (see (42)) as

$$L_2 \cdots L_{m/2}\left(-\sum_{j=1}^{n} \varepsilon_j \sigma_j + \rho; \sigma\right) = c\left(-\sum_{j=1}^{n} \varepsilon_j \sigma_j + \rho\right)^{m-2}, \tag{45}$$

where m is the order of L and $c = $ const. Let

$$L_i(\lambda, \sigma) = \lambda^2 + P_1^i(\sigma)\lambda + P_2^i(\sigma). \tag{46}$$

The expression (45) can then be rewritten

$$\prod_{i=2}^{m/2}\left\{\left(-\sum_{j=1}^{n} \varepsilon_j \sigma_j + \rho\right)^2 + p_1^i(\sigma)\left(-\sum_{j=1}^{n} \varepsilon_j \sigma_j + \rho\right) + P_2^i(\sigma)\right\}$$

$$= c\left(-\sum_{j=1}^{n} \varepsilon_j \sigma_j + \rho\right)^{m-2}. \tag{47}$$

In (47) we set $\rho = 1$. Then

$$\prod_{i=2}^{m/2}\left\{\left(-\sum_{j=1}^{n} \varepsilon_j \omega_j + 1\right)^2 + p_1^i(\omega)\left(-\sum_{j=1}^{n} \varepsilon_j \omega_j + 1\right) + P_2^i(\omega)\right\}$$

$$= c\left(-\sum_{j=1}^{n} \varepsilon_j \omega_j + 1\right)^{m-2}. \tag{48}$$

We now make the change of variables (26) and then use the theorem on the uniqueness of the decomposition of polynomials into irreducible factors; hence

$$L_i\left(-\sum_{j=1}^{n} \varepsilon_j \sigma_j + \rho, \sigma\right) = c_i\left(-\sum_{j=1}^{n} \varepsilon_j \sigma_j + \rho\right)^2, \tag{49}$$

where $c_i = $ const, $i = 2, \ldots, m/2$. We now find the form of L_i from (49). Using Remark 1 to Theorem 4 and equating coefficients of ρ, we obtain

$$L_i(\partial/\partial t, \partial/\partial x) = (1 - c_i)L_1(\partial/\partial t, \partial/\partial x) + c_i \partial^2/\partial t^2. \tag{50}$$

Hence, $L(\partial/\partial t, \partial/\partial x)$ has the form (43).

Conversely, suppose that L has the form (43). We shall show that in this case all the "annuli" G_k are weak lacunae. We consider $G_{m/2-2}$. By the formulas obtained in [6], we have in the "annulus" in question

$$c\partial_t^p \tilde{K}_1(0, \sigma)$$

$$= \left\{ \frac{\lambda_1^p}{L_1'(\lambda_1, \sigma)L_2(\lambda_1, \sigma) \cdots L_{m/2}(\lambda_1, \sigma)} + \frac{\lambda_2^p}{L_1'(\lambda_2, \sigma)L_2(\lambda_2, \sigma) \cdots L_{m/2}(\lambda_2, \sigma)} \right\}$$

$$+ \left\{ \frac{\mu_1^p}{L_1(\mu_1, \sigma)L_2'(\mu_1, \sigma) \cdots L_{m/2}(\mu_1, \sigma)} + \frac{\mu_2^p}{L_1(\mu_2, \sigma)L_2'(\mu_2, \sigma) \cdots L_{m/2}(\mu_2, \sigma)} \right\},$$

$$\tag{51}$$

where λ_1 and λ_2 are the characteristic roots of $L_1(\partial/\partial t, \partial/\partial x)$, and μ_1 and μ_2 are roots of $L_2(\partial/\partial t, \partial/\partial x)$. From (43) we have

$$\prod_{i=2}^{m/2} L_i(\lambda_k(\sigma), \sigma) = \lambda_k^{m-2}(\sigma) \prod_{i=2}^{m/2} c_i, \qquad \prod_{\substack{i=1 \\ i\neq 2}}^{m/2} L_i(\mu_k(\sigma), \sigma) = \mu_k^{m-2}(\sigma) \prod_{\substack{i=1 \\ i\neq 2}}^{m/2} c_i,$$

$$\tag{52}$$

where $k = 1, 2$. It is clear from (52) that each of the braces in (51) is a polynomial in the variables σ for $p \geqslant m - 2$.

The case of an arbitrary "annulus" $G_{m/2-i}$ is considered similarly; here the formula corresponding to (51) contains i braces. The theorem is proved.

§4. Another form of Theorem 3 in the case of hyperbolic equations

THEOREM 6. *Let $L = L_1 L_2$ be a hyperbolic operator, and let G_k be an "annulus" separating the characteristic surfaces of L_1 and L_2. In order that the region G_k be a weak lacuna of L it is necessary and sufficient that there exist polynomials $R_i(\lambda, \sigma)$ such that*

$$L_1(\lambda, \sigma)R_1(\lambda, \sigma) + L_2(\lambda, \sigma)R_2(\lambda, \sigma) = \lambda^{n_0} \tag{53}$$

where n_0 is the critical number defined in Lemma 2.

PROOF. By Theorem 3, for the existence of a weak lacuna in the "annulus" G_k it is necessary and sufficient that $A_p(\sigma)$, $p = n_0, \ldots, n_0 + m_1 - 1$, be polynomials in σ (m_1 is the order of L_1).

Using the explicit formulas (9) for $A_p(\sigma)$, in which it is necessary to set $P_{l-m}(\sigma) \equiv 1$ because of hyperbolicity, we obtain (10).

Since

$$L_1'(\lambda_i, \sigma) = \prod_{\substack{j \\ j\neq i}} (\lambda_i - \lambda_j)$$

$$\frac{\lambda_i^{n_0}}{L_2(\lambda_i, \sigma)} = \frac{L_1'(\lambda_i, \sigma)(-1)^{i-1}}{\Pi_{j>k}(\lambda_j - \lambda_k)} \begin{vmatrix} 1 & \cdots & Q_0 \\ \cdots & \cdots & \cdots \\ \lambda_1^{m_1-1} & \cdots & Q_{m_1-1} \end{vmatrix}, \tag{54}$$

on cancelling by $\prod_{j>k}(\lambda_j - \lambda_k)$, we obtain

$$\lambda_i^{n_0}/L_2(\lambda_i, \sigma) = T(\lambda_1,\ldots,\lambda_{i-1}, \lambda_{i+1} \cdots \lambda_{m_1}; \sigma), \tag{55}$$

where T is a polynomial which is symmetric in $\lambda_1,\ldots,\lambda_{i-1}, \lambda_{i+1},\ldots,\lambda_{m_1}$.

We shall show that T can be represented in the form $R_2(\lambda_i, \sigma)$, where R_2 is a polynomial in the variables λ_i and $\sigma_1,\ldots,\sigma_n$. By the fundamental theorem of the theory of symmetric functions ([11], §38) it suffices to demonstrate this in the case where T is an elementary symmetric polynomial. Let $T = \lambda_i,\ldots,\lambda_{m_1-1}$, $i = m_1$. We rewrite T in the form

$$T = \left(\lambda_1 \cdots \lambda_{m_1-1} + \lambda_2 \cdots \lambda_{m_1} + \lambda_1\lambda_3 \cdots \lambda_{m_1} + \cdots \right)$$
$$- \left(\lambda_2 \cdots \lambda_{m_1} + \cdots + \lambda_1\lambda_3 \cdots \lambda_{m_1}\right)$$
$$= P_1(\sigma) - \lambda_{m_1}T_1(\lambda_1,\ldots,\lambda_{m_1-1}),$$

where $P_1(\sigma)$ is a polynomial and T_1 is an elementary symmetric polynomial of degree $m_1 - 2$. The polynomial T_1, in turn, can be represented in the form $P_2(\sigma) - \lambda_{m_1}T_2(\lambda_1,\ldots,\lambda_{m_1-1})$, where $P_2(\sigma)$ is a polynomial and T_2 is an elementary symmetric polynomial of degree $m_1 - 3$. Continuing this process, we find that $T = R_2(\lambda_{m_1}, \sigma)$. It is obvious from the proof that this same result holds for the T_i $(i = 1,\ldots,m_1 - 1)$, i.e., for all the elementary symmetric polynomials.

Thus,

$$\lambda_i^{n_0}/L_2(\lambda_i, \sigma) = R_2(\lambda_i, \sigma), \tag{56}$$

whence $L_2(\lambda, \sigma)R_2(\lambda, \sigma) - \lambda^{n_0}$ is divisible by the polynomial $L_1(\lambda, \sigma)$, which gives (53).

Conversely, suppose (53) holds. We shall show that, starting with some index p, the functions

$$A_p = \sum_{i=1}^{m_1} \frac{\lambda_i^F}{L_1'(\lambda_i, \sigma)L_2(\lambda_i, \sigma)} \tag{57}$$

are polynomials in σ. We substitute the root λ_i into (53), and find that

$$L_2(\lambda_i, \sigma)R_2(\lambda_i, \sigma) = \lambda_i^{n_0}. \tag{58}$$

Let $R_2(\lambda, \sigma) = \sum_j a_j(\sigma)\lambda^j$. It is then possible to rewrite (57) as follows:

$$A_p = \sum_j a_j(\sigma) \sum_i \frac{\lambda_i^{p+j-n_0}}{L_1'(\lambda_i, \sigma)}. \tag{59}$$

This is a polynomial if $p \geq n_0$, since the sum

$$\sum_i \frac{\lambda_i^{p+j-n_0}}{L_1'(\lambda_i, \sigma)}$$

is the residue at infinity of the expression $\lambda^{p+j-n_0}/L_1(\lambda, \sigma)$, and this residue is obviously a polynomial in σ. The theorem is proved.

REMARK 1. Let

$$L_1(\partial/\partial t, \partial/\partial x) = H_1(\partial/\partial t, \partial/\partial x)H_2(\partial/\partial t, \partial/\partial x).$$

It follows from Theorem 6 that a perturbation of the operator $L = L_2 H_1 H_2$ under which the factor H_1 goes over into a homogeneous factor

$$H_1^* = H_1(\partial/\partial t, \partial/\partial x) + \varepsilon(\partial/\partial t, \partial/\partial x) L_2(\partial/\partial t, \partial/\partial x),$$

while the rest of the factors remain unchanged preserves a relation of the form (53), and hence also a weak lacuna. (It is assumed that the coefficients of the operator $\varepsilon(\partial/\partial t, \partial/\partial x)$ are sufficiently small.)

REMARK 2. Condition (53) of Theorem 6 is equivalent to the condition that the polynomials $L_1(1, \sigma)$ and $L_2(1, \sigma)$ have no common roots in the complex domain ([13], §75).

REMARK 3. Let $L = L_1 L_2 H_1 H_2$, and suppose that the normal surface of the operator $L_1 L_2$ encompasses the normal surface of $H_1 H_2$. We denote by $G = G(L)$ the "annulus" separating the characteristic surfaces of $L_1 L_2$ and $H_1 H_2$. Let $G_{ij} = G_{ij}(L_i H_j)$ be an "annulus" separating the characteristic surfaces of L_i and H_j. It follows from Remark 2 to Theorem 6 that G is a weak lacuna of L if and only the G_{ij} are weak lacunae of the operators $L_i H_j$ ($i = 1, 2; j = 1, 2$).

THEOREM 7. *Let* $L = L_1 L_2 H$ *be a homogeneous hyperbolic operator with a bounded normal surface, and suppose that* L_1 *and* L_2 *are operators of second order which are not connected by the relation* (50). *Then none of the* G_k *is a weak lacuna.*

PROOF. We set $H = H_1 Q$, where H_1 is the product of all those irreducible factors of H having a characteristic surface which is not entirely contained in the "annulus" G^0 between the characteristic surfaces of L_1 and L_2.

We now consider an arbitrary "annulus" G_j of L. If L does not decompose into two operators having G_j between their characteristic surfaces, then by Theorem 1 there is no weak lacuna in G_j. If L does decompose into the operators indicated in Theorem 1, then there are obviously only two cases:

$$\text{a) } G_j \subseteq G^0; \qquad \text{b) } G_j \cap G^0 = \varnothing.$$

Indeed, if a) is not satisfied, then G_j cannot intersect G^0, since the images of the convex hulls of the ovals Q_i are contained in one another.

In case a) it follows from Theorem 6 and Remark 3 to Theorem 6 that the region G_j cannot be a weak lacuna of L.

In case b), also because of Remark 3 to Theorem 6, it suffices to consider in place of H the operator consisting of all its factors with characteristic surface lying entirely on the other side of G_j than G^0. This can always be accomplished, since Theorem 1 entitles us to assume with no loss of generality their characteristic surfaces. We denote the set of remaining factors by H_2 (obviously, H_2 is a divisor of H_1).

We consider the operators $H_2 L_2$ and $H_2 L_1$. If G_j is a weak lacuna of L, then for each operator $H_2 L_i$ ($i = 1, 2$) the "annulus" G^i separating the characteristic surfaces of H_2 and L_i is a weak lacuna. Hence, by Remark 1 to Theorem 4,

$$H_2(\lambda, \sigma) = L_1 Q_1 + c_1 \lambda^k, \tag{60}$$

$$H_2(\lambda, \sigma) = L_2 Q_2 + c_2 \lambda^k, \tag{61}$$

where k is the degree of the polynomial $H_2(\lambda, \sigma)$, Q_1 and Q_2 are polynomials, and $c_1, c_2 = \text{const}$.

We assume with no loss generality that L_1 has the special form indicated in Theorem 4. Let $\lambda_1 = -\Sigma \varepsilon_i \sigma_i + \rho$ be a root of the characteristic equation of L_1. We substitute $\lambda_1(\sigma)$ into (60) and (61). Subtracting (61) from (60), we obtain

$$(c_1 - c_2)\lambda_1^k = L_2(\lambda_1, \sigma)Q_2(\lambda_1, \sigma). \tag{62}$$

We first consider the case where $c_1 \neq c_2$. We set $\rho = 1$, and we make the change of variables (26). Here the sphere $\omega^2 = 1$ goes over into the plane $y_n = \frac{1}{2}$. Using the uniqueness of the decomposition into prime factors in the ring of polynomials of several variables, we find that

$$L_2(\lambda_1(\sigma), \sigma) = (c_1 - c_2)\lambda_1^2(\sigma), \tag{63}$$

which contradicts the hypothesis of the theorem (see Remark 1 to Theorem 4).

Suppose now that $c_1 = c_2$. In this case (60) and (61) give

$$L_1 Q_1 = L_2 Q_2. \tag{64}$$

Therefore, $Q_2(\lambda, \sigma)$ is divisible by $L_1(\lambda, \sigma)$. Hence, H_2 has the form

$$H_2(\lambda, \sigma) = L_1 L_2 P - c\lambda^k, \tag{65}$$

where $P(\lambda, \sigma)$ is a polynomial and $c = \text{const}$.

By construction, H_2 must satisfy the following two conditions:

a) H_2 is hyperbolic.

b) The normal surface of H_2 lies entirely on one side of the region contained between the normal surfaces of L_1 and L_2. (This region is dual to the "annulus" G^0.)

We shall show that if conditions a) and b) are satisfied, then (65) is impossible. Indeed, we consider the graph of the function $\varphi(\sigma_1) = L_1 L_2 P(1, \sigma_1, 0, \ldots, 0)$. Condition b) affords two possibilities (Figures 2 and 3). It is evident from the figures that the polynomial $\varphi(\sigma_1)$ with degree at most k must have at least $k + 1$ extrema, which is a contradiction. The theorem is proved.

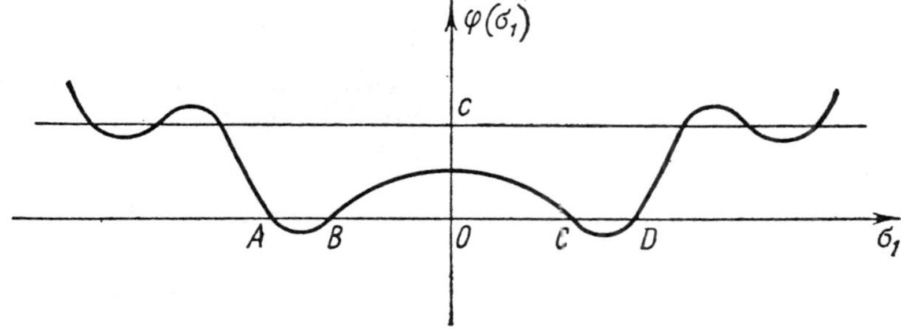

FIGURE 2. A and D are roots of $L_1(1, \sigma_1, 0, \ldots, 0)$;
B and C are roots of $L_2(1, \sigma, 0, \ldots, 0)$

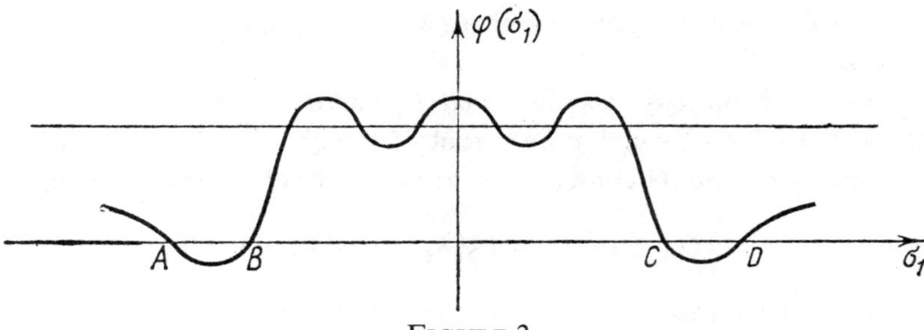

FIGURE 3

COROLLARY. *Suppose that a homogeneous hyperbolic operator L with a bounded normal surface is the product of operators of second order. If at least one "annulus" G_k contains a weak lacuna, then L has the form (43), and there are weak lacunae in all the "annuli".*

REMARK. In the case where n is even there are no weak lacunae in the regions G_k. This follows immediately from the results of Borovikov [2], since his proofs given for lacunae in the sense of Petrovskiĭ and hyperbolic equations are also valid for weak lacunae and quasihyperbolic equations.

In conclusion the author thanks S. A. Gal'pern for posing the problem and for his attention to its solution.

Received 1/FEB/75

BIBLIOGRAPHY

1. I. G. Petrovskiĭ, *On the diffusion of waves and lacunae for hyperbolic equations*, Mat. Sb. **17(59)** (1945), 289–370.

2. V. A. Borovikov, *Some sufficient conditions for the absence of lacunae*, Mat. Sb. **55(97)** (1961), 237–254. (Russian)

3. M. F. Atiyah, R. Bott and L. Gårding, *Lacunas for hyperbolic differential operators with constant coefficients*. I, Acta Math. **124** (1970), 109–189.

4. S. A. Gal'pern, *The Cauchy problem for general systems of linear partial differential equations*, Uspehi Mat. Nauk **18** (1963), no. 2(11), 239–249. (Russian)

5. ______, *Fundamental solutions and lacunae of quasihyperbolic equations*, Uspehi Mat. Nauk **29** (1974), no. 2(176), 154–165; English transl. in Russian Math. Surveys **29** (1974).

6. A. A. Lokšin, *Quasihyperbolic equations*, Trudy Sem. Petrovsk. **1** (1975), 145–153. (Russian)

7. S. A. Gal'pern and V. E. Kondrašov, *The Cauchy problem for differential operators decomposing into wave factors*, Trudy Moskov. Mat. Obšč. **16** (1967), 109–136; English transl. in Trans. Moscow Math. Soc. **16** (1967).

8. Richard Courant, *Methods of mathematical physics*. Vol. II: *Partial differential equations*, Interscience, 1962.

9. A. B. Mitel'man, *The absence of weak lacunae between ovals of the characteristic surface of a hyperbolic operator*, Uspehi Mat. Nauk **27** (1972), no. 6(168), 239–240.

10. ______, *Condition for the absence of weak lacunae in a certain hyperbolic equation*, Vestnik Moskov. Univ. Ser. I Mat. Meh. **1966**, no. 6, 28–33. (Russian)

11. A. G. Kuroš, *A course of higher algebra*, 4th rev. ed., GITTL, Moscow, 1955; English transl. of 10th ed., "Mir", Moscow, 1975.

12. Serge Lang, *Algebra*, Addison-Wesley, 1965.

13. B. L. van der Waerden, *Moderne algebra*. Vol. II, Springer-Verlag, 1931; English transl., Ungar, New York, 1950.

14. A. A. Lokšin, *Conditions for the existence of weak lacunae*, Uspehi Mat. Nauk **30** (1975), no. 3(183), 165–166. (Russian)

Translated by J. R. SCHULENBERGER

Amer. Math. Soc. Transl.
(2) Vol. **118**, 1982

On the Behavior of Solutions
of Inhomogeneous Elliptic Systems
in Unbounded Domains*

O. A. OLEĬNIK AND N. O. MAKSIMOVA

In this paper, estimates are obtained for solutions of inhomogeneous elliptic systems with general inhomogeneous boundary conditions in unbounded domains. Moreover, estimates are obtained for solutions of inhomogeneous elliptic systems depending on a parameter when the parameter tends to infinity. Similar methods, based on the analyticity of solutions of certain auxiliary elliptic systems, were used earlier in [1]–[3]. This paper contains detailed proofs and generalizations of some of the results given in [4].

Let ω be a domain in $R^n_x = (x_1,\ldots,x_n)$ with boundary $\partial\omega$, let

$$\Omega = \omega \times \{|x_0| < a\}, \qquad \Omega \subset R^{n+1}_{x_0,x} = (x_0, x) \equiv (x_0, x_1,\ldots,x_n),$$

let $\partial\Omega$ be the boundary of Ω, and let $a = \mathrm{const} > 0$. In Ω we consider a system of differential equations of the form

$$\sum_{j=1}^{N} \sum_{\alpha_0+|\alpha|\leqslant m_{lj}} a_{lj}^{\alpha_0\alpha}(x)D_{x_0}^{\alpha_0}D_x^{\alpha}v_j = F_l(x_0, x), \qquad l = 1,\ldots,N, \tag{1}$$

where

$$D_{x_j} = -i\frac{\partial}{\partial x_j}, \qquad j = 0, 1,\ldots,n, \alpha = (\alpha_1,\ldots,\alpha_n),$$

$$|\alpha| = \alpha_1 + \cdots + \alpha_n, \qquad D_x^{\alpha} = D_{x_1}^{\alpha_1} \cdots D_{x_n}^{\alpha_n}.$$

We shall assume that in Ω the system (1) is uniformly elliptic. This means that there exist two collections of integers $s_1,\ldots,s_N$ and $t_1,\ldots,t_N$ such that for any l,

---

1980 *Mathematics Subject Classification*. Primary 35J55; Secondary 35B40.
* Translation of Trudy Sem. Petrovsk. **3** (1978), 117–137. MR **80** a 35044.

$j = 1,\ldots,N$, the inequality $m_{lj} \leqslant s_l + t_j$ holds, and

$$\sum_{j=1}^{N} (s_j + t_j) = 2m,$$

where $m > 0$ is an integer, $m_{lj} = 0$ if $s_l + t_j \leqslant 0$, and moreover for any vector $(\xi_0, \xi_1,\ldots,\xi_n) \equiv (\xi_0, \xi) \in R_{\xi_0,\xi}^{n+1}$ and any point $(x_0, x) \in \Omega$ the inequality

$$\lambda\left(|\xi_0|^2 + |\xi|^2\right)^m \leqslant |P(x, \xi_0, \xi)| \leqslant \lambda^{-1}\left(|\xi_0|^2 + |\xi|^2\right)^m$$

holds, where

$$P(x, \xi_0, \xi) \equiv \det\|p_{lj}(x, \xi_0, \xi)\|,$$

$$p_{lj}(x, \xi_0, \xi) = \sum_{\alpha_0 + |\alpha| = s_l + t_j} a_{lj}^{\alpha_0\alpha}(x)\xi_0^{\alpha_0}\xi^{\alpha}, \qquad \xi^{\alpha} = \xi_1^{\alpha_1} \cdots \xi_n^{\alpha_n}.$$

The positive number λ is called the *constant of ellipticity* of the system (1). It may be assumed that $s_j \leqslant 0$ and $t_j \geqslant 0$. We denote $\max\{t_j\}$ by t'.

On the subset $\Gamma = \gamma \times \{|x_0| < a\}$ of $\partial\Omega$, where $\gamma \subset \partial\omega$, let boundary conditions of the form

$$\sum_{j=1}^{N} \sum_{\alpha_0 + |\alpha| \leqslant l_{hj}} b_{hj}^{\alpha_0\alpha}(x)D_{x_0}^{\alpha_0}D_x^{\alpha}v_j = \Phi_h(x_0, x), \qquad h = 1,\ldots,N, \qquad (2)$$

be given, and assume there exists a collection of integers $r_1,\ldots,r_N$ such that for any $j = 1,\ldots,N$ and $h = 1,\ldots,m$ we have $l_{hj} \leqslant r_n + t_j$, with $l_{hj} = 0$ when $r_n + t_j = 0$. We shall assume that γ is an open subset of $\partial\omega$. In particular, γ may be the empty set, or may coincide with the entire boundary $\partial\omega$.

We consider the matrix $\|B_{hj}\|$, where

$$B_{hj}(x, \xi_0, \xi) \equiv \sum_{\alpha_0 + |\alpha| = r_h + t_j} b_{hj}^{\alpha_0\alpha}(x)\xi_0^{\alpha_0}\xi^{\alpha}.$$

We denote by $\|L^{kj}(x, \xi_0, \xi)\|$ the matrix of principal minors of the matrix $\|p_{lj}(x, \xi_0, \xi)\|$. Let $T(x_0, x)$ be the collection of vectors (ξ_0, ξ) orthogonal to the normal vector $(0, \nu)$ to Γ at the point (x_0, x) and such that $|\xi_0|^2 + |\xi|^2 = 1$. Let

$$M^+(x, \xi_0, \xi, \tau) \equiv \prod_{j=1}^{m} \left(\tau - \tau_j^+(x, \xi_0, \xi)\right),$$

where $\tau_j^+(x, \xi_0, \xi)$ are the roots with positive imaginary part of the polynomial $P(x, \xi_0, \xi + \tau\nu)$ in τ, $(x_0, x) \in \Gamma$, and $(\xi_0, \xi) \in T(x_0, x)$. It is known that these roots exist for $n \geqslant 2$ (see, for example, [5]). In the case of two independent variables (x_0, x_1) the presence of m such roots τ_j^+ is an additional requirement on the system (1). Let

$$\sum_{j=1}^{N} B_{hj}(x, \xi_0, \xi + \tau\nu)L^{jl}(x, \xi_0, \xi + \tau\nu) \equiv \sum_{s=0}^{m-1} q_h^{ls}(x, \xi_0, \xi)\tau^s$$

$$(\bmod\, M(x, \xi_0, \xi, \tau)).$$

We consider the matrix $Q = \|q_h^{ls}\|$ with m rows ($h = 1,\ldots,m$) and mN columns ($s = 0, 1,\ldots, m - 1$; $l = 1,\ldots, N$). We shall assume that at each point $(x_0, x) \in \Gamma$ the rank of the matrix Q is equal to m (the Šapiro-Lopatinskiĭ condition). This means that if M_j ($j = 1,\ldots,s'$) are the minors of order m of the matrix Q, then

$$\Delta(x, \xi_0, \xi) = \max_j \left|M^j(x, \xi_0, \xi)\right| \neq 0$$

for any vector $(\xi_0, \xi) \in T(x_0, x)$. We set

$$\Delta_\Gamma = \inf_{(x_0, x) \in \Gamma, (\xi_0, \xi) \in T(x_0, x)} \Delta(x, \xi_0, \xi).$$

The system (1) and boundary conditions (2) will play an auxiliary role in the sequel. They will be used to study general elliptic systems in a domain ω in the space R_x^n.

In the ball $S_\rho \subset R_{x_0,x}^{n+1}$ of radius ρ, or in a half-ball Σ_ρ of radius ρ, $\rho \leqslant 1$, we define seminorms

$$[f]_\sigma^{p,s} = \sum_{\alpha_0 + |\alpha| = s} \sup_\sigma \left(r^{p+s}(x_0, x)\left|D_{x_0}^{\alpha_0}D_x^\alpha f(x_0, x)\right|\right),$$

where $r(x_0, x)$ denotes the distance from (x_0, x) to the spherical part of the boundary, and the supremum is taken over $\sigma = S_\rho$ or $\sigma = \Sigma_\rho$ respectively. Let

$$|f|_\sigma^{p,l} = \sum_{s=0}^l [f]_\sigma^{p,s}.$$

The function $f(x)$ belongs to $C^l(G)$ if f has continuous derivatives in G up to order l inclusive and

$$\|f\|_G^l = \sup_G \sum_{|\alpha| \leqslant l} |D^\alpha f| < \infty.$$

LEMMA 1. *Let* $v = (v_1,\ldots,v_N)$ *be a solution of* (1) *in the half-ball* Σ_ρ, $\rho \leqslant 1$, *satisfying conditions* (2) *on the plane portion* c_ρ *of the boundary of* Σ_ρ, *with* $v_j \in C^{l_0+t_j+1}(\Sigma_\rho), j = 1,\ldots,N, l_0 = \max(0, r_1,\ldots,r_m)$. *Assume that for some* $q \geqslant 0$

$$\left\|a_{lj}^{\alpha_0\alpha}\right\|_{\Sigma_\rho}^{l_0+q-s_l+1} + \left\|b_{hj}^{\alpha_0'\alpha'}\right\|_{\sigma_\rho}^{l_0+q-r_h+1} \leqslant M,$$

$$M = \text{const} > 0\, l, \qquad j = 1,\ldots,N, \qquad h = 1,\ldots,m,$$

$$\alpha_0 + |\alpha| \leqslant m_{lj}, \qquad \alpha_0' + |\alpha'| \leqslant l_{hj},$$

$$\|F_l\|_{\Sigma_\rho}^{l_0+q-s_l+1} < \infty, \qquad \|\Phi_h\|_{\sigma_\rho}^{l_0+q-r_h+1} < \infty.$$

Then $v_j \in C^{l_0+q+t_j}(\Sigma_\rho)$, *and*

$$|v_j|_{\Sigma_\rho}^{t'-t_j, l_0+q+t_j} \leqslant C_1 \left[\sum_{l=1}^N \left(|v_l|_{\Sigma_\rho}^{t'-t_l,0} + \|F_l\|_{\Sigma_\rho}^{l_0+q-s_l+1}\right) + \sum_{h=1}^m \|\Phi_h\|_{\sigma_\rho}^{l_0+q-r_h+1}\right],$$

where the constant C_1 *depends only on* $n, N, m, q, \lambda, \Delta_{\sigma_\rho}$, *and* M, *but not on* ρ.

Lemma 1 follows from Theorem 9.2 of [6].

LEMMA 2. *Let* $v = (v_1, \ldots, v_N)$ *be a solution of* (1) *in the ball* $S\rho$, $\rho \leqslant 1$, *with* $v_j \in C^{t_j+1}(S_\rho)$. *Suppose for some* $q \geqslant 0$, *that* $\|a_{lj}^{\alpha_0\alpha}\|_{S_\rho}^{q-s_l+1} \leqslant M$, $\|F_l\|_{S_\rho}^{q-s_l+1} < \infty$, $l, j = 1, \ldots, N$, $\alpha_0 + |\alpha| \leqslant m_{lj}$. *Then* $v_j \in C^{q+t_j}(S_\rho)$, *and*

$$|v_j|_{S_\rho}^{t'-t_j, q+t_j} \leqslant C_2 \left[\sum_{l=1}^N \left(|v_l|_{S_\rho}^{t'-t_l, 0} + \|F_l\|_{S_\rho}^{q-s_l+1} \right) \right],$$

where the constant C_2 *depends on* n, N, m, q, λ *and* M, *but not on* ρ.

This lemma is a consequence of Theorem 4 of [7].

Let the boundary γ belong to class $C^{l_0+q+t'+1}$, $q \geqslant 0$. This means that for any point $P \in \gamma$ there is a j ($j = 1, \ldots, n$) such that in some neighborhood G of P the boundary γ of ω may be represented in the form

$$x_j = \psi_j(x_1, \ldots, x_{j-1}, x_{j+1}, \ldots, x_n),$$

where $\psi_j \in C^{l_0+q+t'+1}(G)$.

We assume that for any point $P \in \gamma$ there exists a neighborhood ω_P such that ω_P contains a ball of radius d_1 with center at the point P, the set $\omega_P \cap \bar\omega$ is homeomorphic under some transformation F_P to a closed half-ball of radius $\rho(P) \leqslant 1$, and the set $\omega_P \cap \partial\omega$ is homeomorphic under the transformation F_P to the plane portion of the boundary of the half-ball. The coordinate transformation F_P and its inverse F_P^{-1} are given in local coordinates by functions whose norms in $C^{l_0+q+t'+1}$ are bounded by a constant K_1. If the constants d_1 and K_1 do not depend on $P \in \gamma$, we shall denote them by $d_1(\gamma)$ and $K_1(\gamma)$. It may be assumed that $d_1(\gamma) \leqslant 1$ and $K_1(\gamma) \geqslant 1$.

We denote by $r(E_1, E_2)$ the distance between the sets E_1 and E_2 ($E_1, E_2 \subset R_x^n$).

LEMMA 3. *Let* (1) *be a uniformly elliptic system defined in the domain* $\Omega = \omega \times \{|x_0| < a\}$, $a = \text{const} > 2$, *with boundary conditions* (2) *on* $\Gamma = \gamma \times \{|x_0| < a\}$, *and let* $\Delta_\Gamma > 0$. *Assume that* γ *belongs to the class* $C^{l_0+q+t'+1}$ *with constants* $d_1(\gamma)$ *and* $K_1(\gamma)$, $q = \text{const} > 0$, *and that*

$$\left\|a_{lj}^{\alpha_0\alpha}\right\|_\gamma^{l_0+q-s_l+1} + \left\|b_{hj}^{\alpha_0'\alpha'}\right\|_\gamma^{l_0+q-r_h+1} \leqslant M, \qquad l, j = 1, \ldots, N,$$

$$h = 1, \ldots, m, \qquad \alpha_0 + |\alpha| \leqslant m_{lj}, \qquad \alpha_0' + |\alpha'| \leqslant l_{hj},$$

$$\|F_l\|_\Omega^{l_0+q-s_l+1} < \infty, \quad l = 1, \ldots, N, \qquad \|\Phi_h\|_\Gamma^{l_0+q-r_h+1} < \infty, \quad h = 1, \ldots, m.$$

Then for any solution $v = (v_1, \ldots, v_N)$ *of the problem* (1), (2) *such that* $v_j \in C^{l_0+t_j+1}(\Omega \cup \Gamma)$ *the following estimate holds for* $\alpha_0 + |\alpha| \leqslant l_0 + q + t_j$:

$$d^{t'-t_j+\alpha_0+|\alpha|}\left\|D_{x_0}^{\alpha_0} D_x^\alpha v_j\right\|_{\Omega_1}^0$$

$$\leqslant C\left[\sum_{l=1}^N \left(d^{t'-t_l}\|v_l\|_\Omega^0 + \|F_l\|_\Omega^{l_0+q-s_l+1} \right) + \sum_{h=1}^m \|\Phi_h\|_\Gamma^{l_0+q-r_h+1} \right], \qquad (3)$$

where $j = 1, \ldots, N$, $\Omega_1 = \omega_1 \times \{|x_0| < a - d\}$, $\omega_1 \subset \omega$, $\bar{\omega}_1 \cap \partial\omega = \gamma_1 \subset \gamma$, $d = \min[1, r(\omega_1, \partial\omega \setminus \gamma)]$ if $\partial\omega \setminus \gamma$ is nonempty, and $d = 1$ if $\gamma = \partial\omega$. The constant C depends on n, N, m, l_0, q, M, Δ_Γ, $d_1(\Gamma)$, $K_1(\Gamma)$ and λ, but not on a or d.

PROOF. For any point $P \in \Gamma_2 = \Gamma \cap \bar{\Omega}_2$, where $\Omega_2 = \{x_0, x; \ (x_0, x) \in \Omega; \ r(x, \omega_1) < d/2, |x_0| < a - d/2\}$, by virtue of the conditions of the theorem there exist a neighborhood Ω_P that contains a ball of radius $d_1(\gamma) \leqslant 1$ with center at P, and for which there is a homeomorphism F_P: $y_0 = x_0$, $y = F_P(x)$ mapping $\Omega_P \cap \Omega$ onto the half-ball $V_{P'}$ of radius $\rho(P) \leqslant 1$ with center at the point $P' = F_P(P)$ and $\Omega_P \cap \partial\Omega$ onto the plane part of the boundary of $V_{P'}$. The functions yielding the transformation F_P and its inverse F_P^{-1} have derivatives through order $l_0 + q + t' + 1$ bounded by a constant $K_1(\gamma) \geqslant 1$. Let $V_{P'}^1$ and $V_{P'}^2$ be half-balls in the space $R_{y_0, y}^{n+1} = (y_0, y)$ with center at P' and radii $dd_1/2K_1 n$ and $dd_1/4K_1 n$ respectively. Let $W_P^1 = F^{-1}(V_{P'}^1)$ and $W_P^2 = F_P^{-1}(V_{P'}^2)$. It is easy to see that by our choice of radii of the half-balls $V_{P'}^1$ and $V_{P'}^2$, the set $W_P^1 \cap (\partial\Omega \setminus \Gamma)$ is empty and the distance from any point $P_1 \in \Omega_1 \setminus G_1$, where $G_1 = \bigcup_{P \in \Gamma_2} W_P^2 \cap \Omega_1$, to Γ is not less than $d_1 d/4(K_1 n)^2$. Consequently, for any point $P_1 \in \Omega_1 \setminus G_1$ the ball S_{P_1} of radius $dd_1/4(K_1 n)^2$ and center at P_1 does not intersect $\partial\Omega$. Applying Lemma 2 to S_{P_1}, we obtain that for $\alpha_0 + |\alpha| \leqslant q + t_j + l_0$

$$d^{t' - t_j + \alpha_0 + |\alpha|}\left|D_{x_0}^{\alpha_0} D_x^\alpha v_j(P_1)\right| \leqslant C_2\left(\sum_{l=1}^{N} d^{t' - t_l}\|v_l\|_{S_{P_1}}^0 + \sum_{l=1}^{N} \|F_l\|_{S_{P_1}}^{l_0 + q + t_j + 1}\right). \quad (4)$$

Applying Lemma 1 and transforming from variables (y_0, y) to the variables (x_0, x), we obtain for any point $P_1 \in G_1$ and for $\alpha_0 + |\alpha| \leqslant l_0 + q + t_j$

$$d^{t' - t_j + \alpha_0 + |\alpha|}\left|D_{x_0}^{\alpha_0} D_x^\alpha v_j(P_1)\right|$$

$$\leqslant C_1\left[\sum_{l=1}^{N}\left(d^{t' - t_l}\|v_l\|_{W_P^1}^0 + \|F_l\|_{W_P^1}^{l_0 + q - s_l + 1}\right) + \sum_{h=1}^{m} \|\Phi_h\|_\Gamma^{l_0 + q - r_h + 1}\right], \quad (5)$$

where $j = 1, \ldots, N$ and $P_1 \in W_P^2$. The constants C_1 and C_2 depend on n, N, m, l_0, q, M, Δ_Γ, λ, $d_1(\gamma)$ and $K_1(\gamma)$. Taking (4) and (5) into account, we obtain that for $\alpha_0 + |\alpha| \leqslant l_0 + q + t_j$

$$d^{t' - t_j + \alpha_0 + |\alpha|}\left\|D_{x_0}^{\alpha_0} D_x^\alpha v_j\right\|_{\Omega_1}^0$$

$$\leqslant C\left[\sum_{l=1}^{N} d^{t' - t_l}\|v_l\|_\Omega^0 + \sum_{l=1}^{N} \|F_l\|_\Omega^{l_0 + q - s_l + 1} + \sum_{h=1}^{m} \|\Phi_h\|_\Gamma^{l_0 + q - r_h + 1}\right],$$

$$j = 1, \ldots, N.$$

The constant C depends on the same quantities as C_1 and C_2. Lemma 3 is proved.

We now consider the analytic continuation of solutions of the system (1) for complex values $x_0 + iy_0$. For this purpose, we introduce the notation

$$Q_\delta(\sigma) = \{x_0 + iy_0, x; \ (x_0, x) \in \sigma, |y_0| < \delta\}.$$

Theorem 1. *Let system (1) be defined in the domain $\Omega = \omega \times \{|x_0| < a\}$, $a > 2$, and boundary conditions (2) on $\Gamma = \gamma \times \{|x_0| < a\}$. Assume the conditions of Lemma 3 are fulfilled. Let $\Omega_1 = \omega_1 \times \{|x_0| < a - 2d\}$ and $\Gamma_1 = \gamma_1 \times \{|x_0| < a - 2d\}$, where $d = \min(1, r(\omega_1, \partial\omega \setminus \gamma))$ if $\partial\omega \setminus \gamma$ is nonempty, and $d = 1$ and $\omega_1 = \omega$ if $\partial\omega = \gamma$. Assume that F_l $(l = 1, \dots, N)$ together with all their derivatives through the order $p_l = l_0 + q - s_l + 1$ and Φ_h $(h = 1, \dots, m)$ together with all their derivatives through order $q_h = l_0 + q - r_h + 1$ are continued analytically in $x_0 + iy_0$ into the domains $Q_{\delta_1}(\Omega)$ and $Q_{\delta_1}(\Gamma)$ respectively.*

Then any solution $v = (v_1, \dots, v_N)$ of the problem (1), (2) in Ω such that $v_j \in C^{l_0 + t_j + 1}(\overline{\Omega})$ for $j = 1, \dots, N$ may be continued analytically with respect to $x_0 + iy_0$ into the domain $Q_\delta(\Omega_1)$ together with all derivatives of the v_j through order $|\alpha| \le l_0 + q + t_j - 1$. Furthermore, for $|\alpha| \le l_0 + q + t_j - 1$

$$\left\| D_x^\alpha v_j \right\|_{Q_\delta(\Omega)}^0 \le C_0 \left[\sum_{l=1}^N \left(\|v_l\|_\Omega^0 + \|F_l\|_{Q_{\delta_1}(\Omega)}^{p_l} \right) + \sum_{h=1}^m \|\Phi_h\|_{Q_{\delta_1}(\gamma)}^{q_h} \right], \tag{6}$$

where the constants δ and C_0 depend on n, N, m, l_0, q, M, Δ_Γ, λ, d, $d_1(\gamma)$, $K_1(\gamma)$ and δ_1.

Proof. For any positive integers N_1 and $k \ge 0$, $k \le N_1$, we set

$$\Omega_{k,N_1} = \left\{ x_0, x; (x_0, x) \in \Omega, r(x, \omega_1) < (1 - k/N_1)d, |x_0| < a - kd/N_1 - d \right\}$$

and $\Gamma_{k,N_1} = \overline{\Omega}_{k,N_1} \cap \Gamma$. Clearly,

$$r\left(\Omega_{k,N_1}, \partial\Omega_{k-1,N_1} \setminus \Gamma_{k-1,N_1}\right) \ge d/N_1, \quad \Omega_{N_1,N_1} = \Omega_1 \quad \text{and} \quad \Omega_{k,N_1} \subset \Omega_{k-1,N_1}.$$

We apply Lemma 3 to the domain Ω_{k,N_1} and the domain Ω_{k-1,N_1}, containing it. From (3) with $\alpha_0 = 1$ and $|\alpha| = 0$ we obtain

$$\sum_{l=1}^N \left(\frac{d}{N_1} \right)^{t'-t_l} \left\| D_{x_0} v_j \right\|_{\Omega_{k,N_1}}^0$$

$$\le CN \left(\frac{d}{N_1} \right)^{-1} \left[\sum_{l=1}^N \left(\frac{d}{N_1} \right)^{t'-t_l} \|v_l\|_{\Omega_{k-1,N_1}}^0 + \sum_{l=1}^N \|F_l\|_\Omega^{p_l} + \sum_{h=1}^m \|\Phi_h\|_\Gamma^{q_h} \right] \tag{7}$$

for any solution $v = (v_1, \dots, v_N)$ of (1) and (2), provided that $v_j \in C^{l_0 + t_j + 1}(\overline{\Omega})$. Let $\kappa(t) \in C_0^\infty(R_t^1)$, $\kappa(t) > 0$ for $|t| < 1$, $\kappa(t) \equiv 0$ for $|t| \ge 1$, and

$$\int_{-\infty}^{+\infty} \kappa(t)\, dt = 1.$$

We set

$$v_j^\varepsilon(x_0, x) = \int_{-\infty}^{+\infty} v_j(t, x)\kappa_\varepsilon(x_0 - t)\, dt,$$

where $\kappa_\varepsilon(t) = \varepsilon^{-1}\kappa(t/\varepsilon)$. It is immediate that, for $\varepsilon < d$ and any s, $(D_{x_0}^s v_1^\varepsilon, \dots, D_{x_0}^s v_N^\varepsilon)$ is also a solution of the system (1) in Ω_{0,N_1} with $D_{x_0}^s F_l^\varepsilon$ on the right side of the system and boundary conditions (2) with functions $D_{x_0}^s \Phi_h^\varepsilon$ on the right side of (2). From (7) we obtain that for all $\varepsilon < d$ and $k \leqslant N_1$

$$\sum_{j=1}^{N} \left(\frac{d}{N_1}\right)^{t'-t_j} \left\|D_0^k v_j^\varepsilon\right\|_{\Omega_{k,N_1}}^0$$

$$\leqslant CN\left(\frac{d}{N_1}\right)^{-1}\left[\sum_{l=1}^{N} \left(\frac{d}{N_1}\right)^{t'-t_l} \left\|D_{x_0}^{k-1} v_l^\varepsilon\right\|_{\Omega_{k-1,N_1}}^0 \right.$$

$$\left. + \sum_{l=1}^{N} \left\|D_{x_0}^{k-1} F_l^\varepsilon\right\|_\Omega^{p_l} + \sum_{h=1}^{m} \left\|D_{x_0}^{k-1} \Phi_h^\varepsilon\right\|_\Gamma^{q_h}\right]. \qquad (8)$$

Suppose that for any $k \leqslant k_0 < N_1$ an estimate of the type

$$\sum_{j=1}^{N} \left(\frac{d}{N_1}\right)^{t'-t_j} \left\|D_{x_0}^k v_j^\varepsilon\right\|_{\Omega_{k,N_1}}^0$$

$$\leqslant \left[\left(CNN_1 d^{-1}\right)^k \sum_{l=1}^{N} \left(\frac{d}{N_1}\right)^{t'-t_l} \left\|v_l^\varepsilon\right\|_\Omega^0 \right.$$

$$\left. + \sum_{s=0}^{k-1} \left(CNN_1 d^{-1}\right)^{k-s} \left(\sum_{l=1}^{N} \left\|D_{x_0}^s F_l^\varepsilon\right\|_\Omega^{p_l} + \sum_{h=1}^{m} \left\|D_{x_0}^s \Phi_h^\varepsilon\right\|_\Gamma^{q_h}\right)\right] \qquad (9)$$

holds. From (8) it follows that under this assumption (9) holds also for $k = k_0 + 1$ and $\varepsilon < d$. Since, by (7), (9) holds for $k = 1$, by induction (9) is also valid for any $k \leqslant N_1$. From (3) it follows that for any N_1 and $|\alpha| \leqslant l_0 + q + t_j - 1$ we have

$$\left\|D_x^\alpha D_{x_0}^{N_1} v_j^\varepsilon\right\|_{\Omega_1}^0$$

$$\leqslant C\left(N_1 d^{-1}\right)^{t'+|\alpha|-t_j+1}$$

$$\times \left[\sum_{l=1}^{N} \left(\left\|D_{x_0}^{N_1-1} F_l^\varepsilon\right\|_\Omega^{p_l} + \left(\frac{d}{N_1}\right)^{t'-t_l} \left\|D_{x_0}^{N_1-1} v_l^\varepsilon\right\|_{\Omega_{N_1-1,N_1}}^0\right) + \sum_{h=1}^{m} \left\|D_{x_0}^{N_1-1} \Phi_h^\varepsilon\right\|_\Gamma^{q_h}\right].$$

$$(10)$$

And from (9) and (10) it follows that for any N_1 and $|\alpha| \leqslant l_0 + q + t_j - 1$ we have

$$\left\|D_x^\alpha D_{x_0}^{N_1} v_j^\varepsilon\right\|_{\Omega_1}^0 \leqslant \left(CNN_1 d^{-1}\right)^{N_1+t'-t_j+|\alpha|+1}$$

$$\times \left[\sum_{j=1}^{N} \left\|v_j^\varepsilon\right\|_\Omega^0 + \sum_{s=0}^{N_1-1} \left(CNN_1 d^{-1}\right)^{-s} \left(\sum_{l=1}^{N} \left\|D_{x_0}^s F_l^\varepsilon\right\|_\Omega^{p_l} + \sum_{h=1}^{m} \left\|D_{x_0}^s \Phi_h^\varepsilon\right\|_\Gamma^{q_h}\right)\right].$$

$$(11)$$

Since v_j^ε, $j = 1, \ldots, N$, converge in $C^0(\Omega_1)$ to the functions v as $\varepsilon \to 0$, and since

$$\left\| D_{x_0}^{s-1} F_l^\varepsilon \right\|_\Omega^{p_l} \leqslant \left\| D_{x_0}^{s-1} F_l \right\|_\Omega^{p_l}, \qquad \left\| D_{x_0}^{s-1} \Phi_h^\varepsilon \right\|_\Gamma^{q_h} \leqslant \left\| D_{x_0}^{s-1} \Phi_h \right\|_\Gamma^{q_h}, \qquad \left\| v_j^\varepsilon \right\|_\Omega^0 \leqslant \left\| v_j \right\|_\Omega^0,$$

it follows that (11) holds also for the functions v_j. By Cauchy's theorem for a function $f(z)$ analytic in G we have

$$\frac{d^p f(z)}{dz^p} = \frac{p!}{2\pi i} \int_{\partial G} \frac{f(\xi)}{(\xi - z)^{p+1}} \, d\xi, \qquad p = 1, 2, \ldots.$$

Since, according to the conditions of the theorem, all functions $F_l(x_0, x)$ with derivatives to order p_l may be continued analytically in $x_0 + iy_0$ into $Q_{\delta_1}(\Omega)$, by Cauchy's theorem we obtain that for $|\alpha| \leqslant p_l$, for any p, and for $(x_0, x) \in \Omega$

$$\left| D_{x_0}^p D_x^\alpha F_l(x_0, x) \right| \leqslant (p!/\delta_1^p) \left\| D_x^\alpha F_l(x_0 + iy_0, x) \right\|_{Q_{\delta_1}(\Omega)}^0. \tag{12}$$

Exactly the same way, we obtain that for $|\alpha| \leqslant q_h$ and $(x_0, x) \in \Gamma$

$$\left| D_{x_0}^p D_x^\alpha \Phi_h(x_0, x) \right| \leqslant (p!/\delta_1^p) \left\| D_x^\alpha \Phi_h(x_0 + iy_0, x) \right\|_{Q_{\delta_1}(\Gamma)}^0. \tag{13}$$

By Stirling's formula,

$$A_1(n/e)^n < n! < A_2(n/e)^{n+1}, \tag{14}$$

where A_1 and A_2 are positive constants independent of n. From (11), bearing in mind the inequality $N_1^{-s} s! \leqslant 1$ for $N_1 \geqslant s$ and (12)–(14), for $|\alpha| \leqslant l_0 + q + t_j - 1$ we obtain

$$\left\| D_x^\alpha D_{x_0}^{N_1} v_j \right\|_{\Omega_1}^0 \leqslant K_1 (CNd^{-1})^{N_1 + t' - t_j + |\alpha| + 1} N_1^{t' - t_j + |\alpha| + 1} N_1! e^{N_1}$$
$$\times \left[\sum_{j=1}^N \| v_j \|_\Omega^0 + \sum_{s=0}^{N_1-1} (CN\delta_1 d^{-1})^{-s} \left(\sum_{l=1}^N \| F_l \|_{Q_{\delta_1}(\Omega)}^{p_l} + \sum_{h=1}^m \| \Phi_h \|_{Q_{\delta_1}(\Gamma)}^{q_h} \right) \right].$$

From this it follows that

$$\left\| D_x^\alpha D_{x_0}^{N_1} v_j \right\|_{\Omega_1}^0 \leqslant K_2 N_1! (CNd^{-1}e)^{N_1} (CNd^{-1}N_1)^{t' - t_j + |\alpha| + 1}$$
$$\times \sum_{s=0}^{N_1-1} (CN\delta_1 d^{-1})^{-s} \left[\sum_{j=1}^N \left(\| v_j \|_\Omega^0 + \| F_j \|_{Q_{\delta_1}(\Omega)}^{p_j} \right) + \sum_{h=1}^m \| \Phi_h \|_{Q_{\delta_1}(\Gamma)}^{q_h} \right].$$

Here K_1 and K_2 are constants independent of N_1. If $CN\delta_1 d^{-1} > 1$, then

$$\left\| D_x^\alpha D_{x_0}^{N_1} v_j \right\|_{\Omega_1}^0 \leqslant K_3 N_1! B^{N_1} \left[\sum_{l=1}^N \left(\| v_l \|_\Omega^0 + \| F_l \|_{Q_{\delta_1}(\Omega)}^{p_l} \right) + \sum_{h=1}^m \| \Phi_h \|_{Q_{\delta_1}(\Gamma)}^{q_h} \right],$$

where $B = CNd^{-1} \exp(t' - t_j + |\alpha| + 2)$ and the constant K_3 is independent of N_1.

If $CN\delta_1 d^{-1} \leqslant 1$, then

$$\left\| D_x^\alpha D_{x_0}^{N_1} v_j \right\|_{\Omega_1}^0 \leqslant K_4 N_1! \, (CNd^{-1}e)^{N_1} (CNd^{-1}N_1)^{t'-t_j+|\alpha|+1}$$

$$\times (CN\delta_1 d^{-1})^{-N_1} \sum_{s=0}^{N_1-1} (CN\delta_1 d^{-1})^{-s+N_1}$$

$$\times \left[\sum_{l=1}^N \left(\|v_l\|_\Omega^0 + \|F_l\|_{Q_{\delta_1}(\Omega)}^{p_l} \right) + \sum_{h=1}^m \|\Phi_h\|_{Q_{\delta_1}(\Gamma)}^{q_h} \right]$$

$$\leqslant K_5 N_1! B_1^{N_1} \left[\sum_{l=1}^N \left(\|v_l\|_\Omega^0 + \|F_l\|_{Q_{\delta_1}(\Omega)}^{p_l} \right) + \sum_{h=1}^m \|\Phi_h\|_{Q_{\delta_1}(\Gamma)}^{q_h} \right],$$

where $B_1 = \delta_1^{-1} \exp\{t' - t_j + |\alpha| + 3\}$ and K_4 and K_5 are constants independent of N_1.

Let $B_0 = \max_j \{B, B_1\}$. Then

$$\left\| D_x^\alpha D_{x_0}^{N_1} v_j \right\|_{\Omega_1}^0 \leqslant K N_1! B_0^{N_1} \left[\sum_{l=1}^N \left(\|v_l\|_\Omega^0 + \|F_l\|_{Q_{\delta_1}(\Omega)}^{p_l} \right) + \sum_{h=1}^m \|\Phi_h\|_{Q_{\delta_1}(\Gamma)}^{q_h} \right],$$

$$K = \max\{K_3, K_5\}, \qquad N_1 = 1, 2, \ldots. \tag{15}$$

For any point $(x_0', x) \in \Omega_1$ and for $|\alpha| \leqslant l_0 + q + t_j - 1$ we have

$$D_x^\alpha v_j(x_0, x) = \sum_{N_1=0}^{k-1} \frac{1}{N_1!} D_{x_0}^{N_1} D_x^\alpha v_j(x_0', x)(x_0 - x_0')^{N_1}$$

$$+ (1/k!) D_{x_0}^k D_x^\alpha v_j(\theta, x)(x_0 - x_0')^k,$$

where $|\theta - x_0'| \leqslant |x - x_0'|$. We show that the remainder in the Taylor series expansion tends to zero as $k \to \infty$. Let $|x_0 - x_0'| < \varepsilon$, and let ε be so small that $(x_0, x) \in \Omega_1$ for $|x_0 - x_0'| < \varepsilon$. For the remainder, according to (15), the estimate

$$\left| \frac{1}{k!} D_{x_0}^k D_x^\alpha v_j(\theta, x)(x_0 - x_0')^k \right| \leqslant K_6 B_0^k \varepsilon^k$$

holds, with K_6 independent of k, since $(\theta, x) \in \Omega_1$. Let $\varepsilon < 1/B_0$. Then the remainder tends to zero as $k \to \infty$. This means that the functions $D_x^\alpha v_j$ for $|\alpha| \leqslant l_0 + q + t_j - 1$ are analytic functions for x_0 for each point $(x_0, x) \in \Omega_1$. We now estimate the radius of convergence of the Taylor series for $D_x^\alpha v_j$ and the moduli of the functions $D_x^\alpha v_j$ inside the circle of convergence for $|\alpha| \leqslant l_0 + q + t_j - 1$. For any point $(x_0', x) \in \Omega_1$ the Taylor series

$$\sum_{N_1=0}^\infty \frac{1}{N_1!} D_{x_0}^{N_1} D_x^\alpha v_j(x_0', x)(x_0 - x_0')^{N_1}$$

according to (15) is majorized by the series

$$\sum_{N_1=0}^\infty K B_0^{N_1} |x_0 - x_0'|^{N_1} \left[\sum_{l=1}^N \left(\|v_l\|_\Omega^0 + \|F_l\|_{Q_{\delta_1}(\Omega)}^{p_l} \right) + \sum_{h=1}^m \|\Phi_h\|_{Q_{\delta_1}(\Gamma)}^{q_h} \right],$$

which converges for $|x_0 - x_0'| < 1/B_0$. Thus we may set $\delta = 1/2B_0$. For $|z_0 - x_0'| \leqslant 1/2B_0$ we have

$$\left| D_x^\alpha v_j(z_0, x) \right| \leqslant 2K \left[\sum_{l=1}^{N} \left(\|v_l\|_\Omega^0 + \|F_l\|_{Q_{\delta_1}(\Omega)}^{p_l} \right) + \sum_{h=1}^{m} \|\Phi_h\|_{Q_{\delta_1}(\Gamma)}^{q_h} \right], \qquad z_0 = x_0 + iy_0.$$

The theorem is proved.

REMARK 1. From the expressions for B and B_1 it is easy to see that if $\delta_1 < d/CN$, then $\delta = a_1\delta_1$, where a_1 is a constant independent of δ_1. If, however, $\delta_1 > d/CN$, then δ does not depend on δ_1.

REMARK 2. We consider now the special case when

$$F_l(x_0, x) = f_l(x)e^{i\mu x_0}, \qquad \Phi_h(x_0, x) = \varphi_h(x)e^{i\mu x_0}, \quad \text{where } \mu = \mu_1 + i\mu_2.$$
$$(16)$$

Then the estimate (11) for the functions v_j takes the form

$$\left\| D_x^\alpha D_{x_0}^{N_1} v_j \right\|_{\Omega_1}^0 \leqslant \left(CNN_1 d^{-1} \right)^{N_1 + t' - t_j + |\alpha| + 1}$$
$$\times \left[\sum_{j=1}^{N} \|v_j\|_\Omega^0 + \sum_{s=0}^{N_1-1} \left(CNN_1 d^{-1} \right)^{-s} \left(\sum_{l=1}^{N} \left\| D_{x_0}^s F_l \right\|_\Omega^{p_l} + \sum_{h=1}^{m} \left\| D_{x_0}^s \Phi_h \right\|_\Gamma^{q_h} \right) \right].$$
$$(17)$$

Clearly

$$\left\| D_{x_0}^s F_l \right\|_\Omega^{p_l} \leqslant C_1(1 + |\mu|)^\kappa |\mu|^s e^{a|\mu_2|} \|f_l\|_\omega^{p_l},$$

$$\left\| D_{x_0}^s \Phi_h \right\|_\Gamma^{q_h} \leqslant C_1(1 + |\mu|)^\kappa |\mu|^s e^{a|\mu_2|} \|\varphi_h\|_\gamma^{q_h},$$

where $\kappa = \max_{l,h}\{p_l, q_h\}$ and C_1 is a constant depending only on κ.

From (17) we obtain for $|\alpha| \leqslant l_0 + q + t_j - 1$

$$\left\| D_x^\alpha D_{x_0}^{N_1} v_j \right\|_{\Omega_1}^0 \leqslant \left(CNN_1 d^{-1} \right)^{N_1 + t' - t_j + |\alpha| + 1}$$
$$\times \left[\sum_{j=1}^{N} \|v_j\|_\Omega^0 + C_1 e^{a|\mu_2|}(1 + |\mu|)^\kappa \sum_{s=0}^{N_1-1} \left(CNN_1 d^{-1} \right)^{-s} |\mu|^s \right.$$
$$\left. \times \left(\sum_{l=1}^{N} \|f_l\|_\omega^{p_l} + \sum_{h=1}^{m} \|\varphi_h\|_\gamma^{q_h} \right) \right]. (18)$$

Since $N_1^s \geqslant s!$ for $s \leqslant N_1$, it follows that

$$\left\| D_x^\alpha D_{x_0}^{N_1} v_j \right\|_{\Omega_1}^0 \leqslant K' N_1! B^{N_1}$$

$$\times \left[\sum_{j=1}^{N} \|v_j\|_\Omega^0 + C_1 e^{a|\mu_2|}(1 + |\mu|)^\kappa \sum_{s=0}^{N_1-1} \frac{1}{s!} \left(\frac{d|\mu|}{CN} \right)^s \left(\sum_{l=1}^{N} \|f_l\|_\omega^{p_l} + \sum_{h=1}^{m} \|\varphi_h\|_\gamma^{q_h} \right) \right]$$

$$\leqslant K' N_1' B^{N_1}$$

$$\times \left[\sum_{j=1}^{N} \|v_j\|_\Omega^0 + C_1(1 + |\mu|)^\kappa \exp\left\{ a|\mu_2| + \frac{d}{CN}|\mu| \right\} \left(\sum_{l=1}^{N} \|f_l\|_\omega^{p_l} + \sum_{h=1}^{m} \|\varphi_h\|_\gamma^{q_h} \right) \right],$$

where $B = CNd^{-1}\exp(t' - t_j + |\alpha| + 2)$ and the constant K' does not depend on N_1. Thus, if (16) is fulfilled, we may set $\delta = 1/2B$. Therefore, in case (16), inequality (6) takes the form

$$\left\| D_x^\alpha v_j \right\|_{Q_\delta(\Omega_1)}^0$$

$$\leqslant C_0 \left[\sum_{l=1}^{N} \|v_l\|_{\Omega}^0 + (1 + |\mu|)^\kappa \exp\left\{ a|\mu_2| + \frac{d}{CN}|\mu| \right\} \left(\sum_{l=1}^{N} \|f_l\|_{\omega}^{p_l} + \sum_{h=1}^{m} \|\varphi_h\|_{\gamma}^{q_h} \right) \right],$$

$$(19)$$

where the constants δ, C, and C_0 depend only on n, N, m, l_0, q, m, λ, Δ_Γ, d, $d_1(\gamma)$ and $K_1(\gamma)$.

The next theorems will be obtained as a consequence of Theorem 1 on the analytic continuation of the solutions of (1).

We denote $\omega_R = \omega \cap \{|x| < R\}$ and $\partial \omega_R = \partial \omega \cap \{|x| < R\}$.

THEOREM 2. *In a domain $\omega \subset R_x^n$ with boundary $\partial\omega$, let the system of equations*

$$\sum_{j=1}^{N} \sum_{\alpha_0 + |\alpha| \leqslant m_{lj}} a_{lj}^{\alpha_0 \alpha}(x)\mu^{\alpha_0} D_x^\alpha u_j'' = f_l(x), \qquad l = 1,\ldots,N, \qquad (20)$$

be given, with boundary conditions on $\partial\omega$ of the form

$$\sum_{j=1}^{N} \sum_{\alpha_0 + |\alpha| \leqslant l_{hj}} b_{hj}^{\alpha_0 \alpha}(x)\mu^{\alpha_0} D_x^\alpha u_j = \varphi_h(x), \qquad h = 1,\ldots,m, \qquad (21)$$

where the parameter $\mu = \mu_1 + i\mu_2$; μ_1 and μ_2 are real numbers.

Assume the system (1) corresponding to (20) in the domain $\Omega = \omega \times R_{x_0}^1$, with boundary conditions (2) corresponding to (21) on $\Gamma = \partial\omega \times R_{x_0}^1$, is uniformly elliptic in Ω, and $\Delta_\Gamma > 0$. Assume that the boundary of ω belongs to class $C^{l_0 + t_j + 1}$, with

$$\left\| a_{lj}^{\alpha_0 \alpha} \right\|_{\omega}^{l_0 - s_l + 1} + \left\| b_{hj}^{\alpha_0 \alpha'} \right\|_{\partial\omega}^{l_0 - r_h + 1} \leqslant M, \qquad M = \text{const} > 0,$$

$$\alpha_0 + |\alpha| \leqslant m_{lj}, \qquad \alpha_0' + |\alpha'| \leqslant l_{hj}, \qquad l, j = 1,\ldots,N, \qquad h = 1,\ldots,m;$$

$$d_1(\partial\omega) > 0, \qquad K_1(\partial\omega) < \infty, \qquad q = 0$$

and that for any $R > 0$

$$\sum_{l=1}^{N} \|f_l\|_{\omega_R}^{p_l} + \sum_{h=1}^{m} \|\varphi_h\|_{\partial\omega_R}^{q_h} \leqslant M_1 \exp\{\delta_2 R\}, \qquad (22)$$

where $\delta_2 = \text{const} > 0$. Let $u = (u_1,\ldots,u_n)$ be a solution of (20) and (21), with $u_j \in C^{l_0 + t_j + 1}(\omega')$ in any finite domain $\omega' \subset \omega$. If, for any $R > 0$,

$$\sum_{l=1}^{N} \|u_l\|_{\omega_R}^0 \leqslant M_2 \exp\{\delta_3 R\}, \qquad \delta_3 = \text{const} > 0, \qquad (23)$$

and $\mu = \mu_1 + i\mu_2$ satisfies the condition

$$\delta|\mu_1| > \ln C_0 + |\mu_2| + \delta_2, \qquad \delta|\mu_1| > \ln C_0 + |\mu_2| + \delta_3, \qquad (24)$$

270   O. A. OLEĬNIK AND N. O. MAKSIMOVA

where the constants δ and C_0 are determined by the problem (1), (2) *and depend only on* $n, N, m, l_0, M, \Delta_\Gamma, \lambda, d_1(\partial\omega)$ *and* $K_1(\partial\omega)$, *then for any* $R > 0$

$$\sum_{l=1}^{N} \|u_l\|_{\omega_R}^0 \leqslant M_3 \exp\{\delta_2 R\}.$$

Here M_1, M_2 *and* M_3 *are constants independent of* R.

PROOF. Clearly $v(x_0, x) = \exp\{i\mu x_0\}u(x)$ is a solution of (1) in Ω for $F_l(x_0, x) = f_l(x)e^{i\mu x_0}$ with boundary conditions (2) on Γ and $\Phi_h(x_0, x) = \varphi_h(x)e^{i\mu x_0}$, $l = 1,\ldots,N$; $h = 1,\ldots,m$, if $u(x)$ is a solution of (20) in ω with boundary conditions (21) on $\gamma = \partial\omega$. We set $\Omega_R = \omega_R \times \{|x_0| < R\}$ and $\Gamma_R = \partial\omega_R \times \{|x_0| < R\}$. By virtue of Theorem 1,

$$\sum_{l=1}^{N} \|v_l\|_{Q_\delta(\Omega_{R_0})}^0 \leqslant C_0\left[\sum_{l=1}^{N}\left(\|v_l\|_{\Omega_{r_0+1}}^0 + \|f_l(x)e^{i\mu x_0}\|_{Q_{\delta_1}(\Omega_{R_0+1})}^{p_l}\right)\right.$$

$$\left. + \sum_{h=1}^{m} \|\varphi_h(x)e^{i\mu x_0}\|_{Q_{\delta_1}(\Gamma_{R_0+1})}^{q_h}\right],$$

where δ_1 may be taken to be any positive number. From this it follows that

$$\sum_{l=1}^{N} \|u_l\|_{\omega_{R_0}}^0 \leqslant C_0 \exp\{-R_0|\mu_2| - \delta|\mu_1|\}$$

$$\times\left[\exp\{(R_0 + 1)|\mu_2|\}\sum_{l=1}^{N}\|u_l\|_{\omega_{R_0+1}}^0 + C_1\exp\{(R_0 + 1)|\mu_2| + \delta_1|\mu_1|\}\right.$$

$$\left.\times\left(\sum_{l=1}^{N}\|f_l\|_{\omega_{R_0+1}}^{p_l} + \sum_{h=1}^{m}\|\varphi_h\|_{\partial\omega_{R_0+1}}^{q_h}\right)(1 + |\mu|)^\kappa\right],$$

where $\kappa = \max_{l,h}\{p_l, q_h\}$ and C_1 is a constant depending only on p_l and q_h. Furthermore,

$$\sum_{l=1}^{N} \|u_l\|_{\omega_{R_0}} \leqslant \exp\{|\mu_2| + \ln C_0 - \delta|\mu_1|\}\sum_{l=1}^{N}\|u_l\|_{\omega_{R_0+1}}^0$$

$$+ \exp\{\ln C_0 C_1 + |\mu_2| - \delta|\mu_1| + \delta_1|\mu_1|\}(1 + |\mu|)^\kappa$$

$$\times\left(\sum_{l=1}^{N}\|f_l\|_{\omega_{R_0+1}}^{p_l} + \sum_{h=1}^{m}\|\varphi_h\|_{\partial\omega_{R_0+1}}^{q_h}\right). \tag{25}$$

Let $H = |\mu_2| + \ln C_0 - \delta|\mu_1|$ and

$$H_1 = \exp\{\ln C_0 C_1 + |\mu_2| - \delta|\mu_1| + \delta_1|\mu_1|\}(1 + |\mu|)^\kappa.$$

From (25) we obtain that

$$\sum_{l=1}^{N} \|u_l\|_{\omega_{R_0}}^0 \leqslant e^H \sum_{l=1}^{N}\|u_l\|_{\omega_{R_0+1}}^0 + H_1\left(\sum_{l=1}^{N}\|f_l\|_{\omega_{R_0+1}}^{p_l} + \sum_{h=1}^{m}\|\varphi_h\|_{\omega_{R_0+1}}^{q_h}\right).$$

Applying this inequality successively to the domains $\omega_{R_0+1}, \omega_{R_0+2}, \ldots, \omega_{R_0+s}$, we obtain

$$\sum_{l=1}^{N} \|u_l\|_{\omega_{R_0}}^0 \leq e^{Hs} \sum_{i=1}^{N} \|u_l\|_{\omega_{R_0+s}}^0 + H_1 \sum_{j=1}^{s} e^{H(j-1)} \left(\sum_{l=1}^{N} \|f_l\|_{\omega_{R_0+j}}^{p_l} + \sum_{h=1}^{m} \|\varphi_h\|_{\partial\omega_{R_0+j}}^{q_h} \right).$$

(26)

Taking the assumptions of Theorem (22) and (23) into account, we deduce from (26) that

$$\sum_{l=1}^{N} \|u_l\|_{\omega_{R_0}}^0 \leq C_2 \exp\{Hs + \delta_3(R_0 + s)\} + C_3 H_1 \sum_{j=1}^{s} e^{H(j-1)} e^{\delta_2(R_0+j)}$$

$$= C_2 e^{\delta_3 R_0} e^{(H+\delta_3)s} + C_3 H_1 e^{\delta_2 R_0 - H} \sum_{j=1}^{s} e^{(H+\delta_2)j}.$$

Let $s \to \infty$. Then, by virtue of condition (24) on μ,

$$\sum_{l=1}^{N} \|u_l\|_{\omega_{R_0}}^0 \leq C_4 \exp\{\delta_2 R_0\},$$

since $H + \delta_3 < 0$, $H + \delta_2 < 0$ and $\exp\{(H + \delta_3)s\} \to 0$ as $s \to \infty$, $\sum_{j=1}^{n} e^{(H+\delta_2)j} < C_5$, and the constants C_4 and C_5 do not depend on R_0. The theorem is proved.

Let ω_1 be a bounded subdomain of ω. We introduce the function $p(x)$, equal to the distance from the point $x \in \omega_1$ to $\partial\omega \setminus \gamma$. We denote by h_ρ^x the intersection of ω with the ball of radius ρ with center at x, and set

$$H_\rho^x = h_\rho^x \times \{|x_0| < \rho\}, \qquad G_\rho^x = \Gamma \cap H_\rho^x, \qquad g_\rho^x = \gamma \cap h_\rho^x.$$

THEOREM 3. *Let the system (20) be given in a domain $\omega \subset R_x^n$ with boundary $\partial\omega$. Let boundary conditions on γ of the form (21) be given, γ being an open set on $\partial\omega$. Let the system (1), corresponding to (20) in $\Omega = \omega \times R_{x_0}^1$, with corresponding boundary conditions (2) on $\Gamma = \gamma \times R_{x_0}^1$, be uniformly elliptic in Ω, and let $\Delta_\Gamma > 0$. Assume that the boundary of ω belongs to the class $C^{l_0+q+l'+1}$, with*

$$\left\|a_{lj}^{\alpha_0\alpha}\right\|_\omega^{l_0+q-s_l+1} + \left\|b_{hj}^{\alpha_0'\alpha'}\right\|_\gamma^{l_0+q-r_h+1} < M, \qquad M = \text{const},$$

$$l, j = 1, \ldots, N, \qquad h = 1, \ldots, m, \qquad \alpha_0 + |\alpha| \leq m_{lj},$$

$$\alpha_0' + |\alpha'| \leq l_{hj}, \qquad d_1(\gamma) > 0, \qquad K_1(\gamma) < \infty,$$

$$\left.\begin{array}{ll} \displaystyle\sum_{|\alpha| \leq p_l} \sum_{l=1}^{N} \left|D_x^\alpha f_l^{(x)}\right| \leq C_1 \exp\{\delta_2|x|\}, & p_l = l_0 + q - s_l + 1, \\[4mm] \displaystyle\sum_{|\alpha| \leq q_h} \sum_{h=1}^{m} \left|D_x^\alpha \varphi_l^{(x)}\right| \leq C_1 \exp\{\delta_2|x|\}, & q_h = l_0 + q - r_h + 1, \end{array}\right\}, \qquad (27)$$

where $C_1 = \mathrm{const}$, $\delta_2 = \mathrm{const} > 0$ and $q \geqslant 0$. Let $u = (u_1, \ldots, u_N)$ be a solution of (20) and (21), $u_j \in C^{l_0 + q + t_j + 1}(\omega')$ in any finite domain ω' such that $\bar{\omega}' \subset \omega \cup \gamma$. If, in ω,

$$\sum_{l=1}^{N} |u_l(x)| \leqslant C_2 \exp\{\delta_3|x|\}, \qquad \delta_3 = \mathrm{const} > 0, \tag{28}$$

and $\mu = \mu_1 + i\mu_2$ satisfies the condition

$$H + \delta_2 < 0, \tag{29}$$

where $H = \ln C_0 + |\mu_2| - \delta|\mu_1|$ and the constants δ and C_0 are determined by the problem (1), (2) and depend only on n, N, m, λ, l_0, Δ_Γ, M, $d_1(\partial\omega)$ and $K_1(\partial\omega)$, then for any $x \in \omega_1$

$$\sum_{l=1}^{N} |u_l(x)| \leqslant C_3\big(\exp\{(H + \delta_3)p(x) + \delta_3|x|\} + \exp\{\delta_2|x|\}\big). \tag{30}$$

If condition (28) holds and

$$H + \delta_2 > 0, \tag{31}$$

then for any $x \in \omega_1$

$$\sum_{l=1}^{N} |u_l(x)| \leqslant C_4\big[\exp\{(H + \delta_3)p(x) + \delta_3|x|\} + \exp\{(H + \delta_2)p(x) + \delta_2|x|\}\big],$$

$$C_2, C_3, C_4 = \mathrm{const}. \tag{32}$$

PROOF. Clearly $v(x_0, x) = \exp\{i\mu x_0\}u(x)$ is a solution of (1) in Ω with boundary conditions (2) on Γ with

$$F_l(x_0, x) = f_l(x)e^{i\mu x_0},$$

$$\Phi_h(x_0, x) = \varphi_h(x)e^{i\mu x_0}, \qquad l = 1, \ldots, N, h = 1, \ldots, m,$$

when $u(x)$ is a solution of (20) in ω with boundary conditions (21) on γ. Let $x \in \omega_1$. Then, according to Theorem 1, for $|\alpha| \leqslant l_0 + q + t_j - 1$ and some $\delta_1 > 0$ we have

$$\left\| D_x^\alpha v_j \right\|_{Q_\delta(H_s^x)}^0 \leqslant C_0 \left[\sum_{l=1}^{N} \left(\|v_l\|_{H_{s+1}^x}^0 + \|F_l\|_{Q_{\delta_1}(H_{s+1}^x)}^{p_e} \right) + \sum_{h=1}^{m} \|\Phi_h\|_{Q_{\delta_1}(G_{s+1}^x)}^{q_h} \right].$$

From this it follows that

$$\left\| D_x^\alpha u_j \right\|_{h_s^x}^0 \leqslant \exp\{\ln C_0 + |\mu_2| - \delta|\mu_1|\} \sum_{l=1}^{N} \|u_l\|_{h_{s+1}^x}^0$$

$$+ \exp\{\ln C'C_0 + |\mu_2| - \delta|\mu_1| + \delta_1|\mu_1|\}(1 + |\mu|)^\kappa$$

$$\times \left(\sum_{l=1}^{N} \|f_l\|_{h_{s+1}^x}^{p_l} + \sum_{h=1}^{m} \|\varphi_h\|_{g_{s+1}^x}^{q_h} \right), \tag{33}$$

where C' is a constant depending only on p_l and q_h, and where $\kappa = \max_{l,h}\{p_l, q_h\}$. We set

$$H_1 = \exp\left\{\ln C'C_0 + |\mu_2| - \delta|\mu_1| + \delta_1|\mu_1|\right\}(1 + |\mu|)^\kappa.$$

Then from (33) it follows that for $|\alpha| \leqslant l_0 + q + t_j - 1$

$$\left\|D_x^\alpha u_j\right\|_{h_s^x}^0 \leqslant e^H \sum_{j=1}^N \|u_j\|_{h_{s+1}^x}^0 + H_1\left[\sum_{l=1}^N \|f_l\|_{h_{s+1}^x}^{p_l} + \sum_{h=1}^m \|\varphi_h\|_{g_{s+1}^x}^{q_h}\right].$$

Applying this estimate successively for $s = 1,\ldots,[p(x)] - 1$, for $|\alpha| \leqslant l_0 + q + t_j - 1$ we obtain

$$\left\|D_x^\alpha u_j\right\|_{h_1^x}^0 \leqslant \exp\{H([p(x)] - 1)\} \sum_{l=1}^N \|u_l\|_{h_{[p(x)]}^x}^0$$

$$+ H_1\left[\sum_{j=1}^{[p(x)]-1} e^{H(j-1)}\left(\sum_{l=1}^N \|f_l\|_{h_{j+1}^x}^{p_l} + \sum_{h=1}^m \|\varphi_k\|_{g_{j+1}^x}^{q_h}\right)\right].$$

Taking (27) and (28) into account, we obtain

$$\left\|D_x^\alpha u_j\right\|_{h_1^x}^0 \leqslant C_5\left[\exp\{H([p(x)] - 1) + \delta_3[p(x)] + \delta_3|x|\}\right.$$

$$\left. + H_1 \sum_{j=1}^{[p(x)]-1} \exp\{H(j - 1) + \delta_2|x| + \delta_2 j\}\right]$$

$$\leqslant C_6\left[\exp\{(H + \delta_3)p(x) + \delta_3|x|\} + \exp\{\delta_2|x|\} \sum_{j=1}^{[p(x)]-1} \exp\{(H + \delta_2)j\}\right].$$

$$\tag{34}$$

If condition (29) is fulfilled, then from (34) we deduce that for $|\alpha| \leqslant q + l_0 + t_j - 1$

$$|D_x^\alpha u_j(x)| \leqslant C_7\left[\exp\{(H + \delta_3)p(x) + \delta_3|x|\} + \exp\{\delta_2|x|\}\right].$$

If condition (31) is fulfilled, then from (34) we find that for $|\alpha| \leqslant t_j + q + l_0 - 1$

$$|D_x^\alpha u_j(x)| \leqslant C_8\left[\exp\{(H + \delta_3)p(x) + \delta_3|x|\} + \exp\{(H + \delta_2)p(x) + \delta_2|x|\}\right].$$

From Theorem 3 we obtain the assertion on the behavior of the inhomogeneous system of the form (20) with boundary conditions (21) in a neighborhood of infinity. Let $\gamma = \partial\omega \cap \{|x| > R\}$. In this case we may set $p(x) = |x| - R$. Inequality (30) takes the form

$$\sum_{l=1}^N |u_l(x)| \leqslant C_9\left[\exp\{(H + \delta_3)|x| + \delta_3|x|\} + \exp\{\delta_2|x|\}\right],$$

and (32) is transformed into the estimate

$$\sum_{l=1}^{N} |u_l(x)| \leqslant C_{10}\big[\exp\{(H + 2\delta_3)|x|\} + \exp\{(H + 2\delta_2)|x|\}\big],$$

$$C_j = \text{const}, \qquad j = 5,\ldots,10.$$

Now we establish a theorem on the behavior of solutions of elliptic systems in unbounded domains under certain conditions on the behavior of the coefficients of the system as $|x| \to \infty$.

We denote by A^ρ the image of the set $A \subset R_x^n$ under a coordinate transformation of the type

$$y = \rho^{-1}x, \qquad \rho = \text{const} > 0.$$

We set

$$\sigma_\rho = \{x;\ \rho/4 < |x| < 4\rho\}, \qquad \sigma_\rho = \{x;\ \rho/2 < |x| < 2\rho\}.$$

We consider a function $\varphi(s)$, positive for $s > 0$ and satisfying the following condition: for any $\rho > 1$ and for any $s > 0$ such that $\rho/4 < s < 4\rho$ we have

$$h_1\varphi(\rho) \leqslant \varphi(s) \leqslant h_2\varphi(\rho), \tag{35}$$

where h_1 and h_2 are constants which do not depend on s or ρ. We consider elliptic systems with growing or decaying coefficients as $|x| \to \infty$, the behavior being determined by the function $\varphi(s)$.

THEOREM 4. *Let* $\omega = \{x;\ |x| > R\}$, $R = \text{const} > 1$, *and* $\Omega = \omega \times R_{x_0}^1$. *Assume that for any* $(\xi_0, \xi) \in R_{\xi_0,\xi}^{n+1}$

$$\det \left\| \sum_{\alpha_0 + |\alpha| = s_l + t_j} a_{lj}^{\alpha_0\alpha}(x)\varphi^{-\alpha_0}(|x|)\xi_0^{\alpha_0}\xi^{\alpha} \right\| \geqslant \lambda_1\big(|\xi_0|^2 + |\xi|^2\big)^m, \tag{36}$$

where the constant $\lambda_1 > 0$, *and for any* $x \in \omega$ *and* $|\beta| \leqslant q - s_l + 1$, $q = \text{const} \geqslant 0$,

$$\big|D_x^\beta a_{lj}^{\alpha_0\alpha}(x)\big| \leqslant M_1|x|^{-|\beta|+\alpha_0+|\alpha|-s_l-t_j}\big(\varphi(|x|)\big)^{\alpha_0}, \tag{37}$$

$$l,j = 1,\ldots,N, \qquad \alpha_0 + |\alpha| \leqslant s_l + t_j, \qquad M = \text{const}.$$

Let $u = (u_1,\ldots,u_N)$ *be a solution of* (20) *with* $\mu = 1$ *in* ω *and* $u_j(x) \in C^{t_j+1}(\omega')$ *for any bounded domain* $\omega' \subset \omega$, $j = 1,\ldots,N$. *Suppose, in* ω, *that*

$$\sum_{|\alpha| \leqslant -s_l+q+1} |D_x^\alpha f_l| \leqslant C_1 \exp\{\delta_2|x|\varphi(|x|)\}, \qquad C_1 = \text{const}, \delta_2 = \text{const}, \tag{38}$$

$$|u_j(x)| \leqslant C_2 \exp\{\delta_3|x|\varphi(|x|)\}, \qquad C_2 = \text{const} > 0, \delta_3 = \text{const}. \tag{39}$$

Then, in ω, *for* $|\alpha| \leqslant q + t_j - 1$ *and for* $j = 1,\ldots,N$,

$$|D_x^\alpha u_j(x)| \leqslant C_3\Big[|x|^{t_j-|\alpha|} \exp\{(4h_2\delta_3 - \delta)|x|\varphi(|x|)\}$$

$$+ |x|^{\beta_j-|\alpha|}\big(1 + |x|\varphi(|x|)\big)^{1+\tau'+q} \exp\{(\delta_1 - \delta + 4h_2\delta_2)|x|\varphi(|x|)\}\Big],$$

when $\delta_2, \delta_3 > 0$;

$$\left| D_x^\alpha u_j(x) \right|$$

$$\leqslant C_3 \Big[|x|^{\tau_j - |\alpha|} \exp\{ (4h_2\delta_3 - \delta) |x| \varphi(|x|) \}$$

$$+ |x|^{\beta_j - |\alpha|} \left(1 + |x| \varphi(|x|) \right)^{1 + \tau' + q} \exp\{ (\delta_1 - \delta - \tfrac{1}{4} h_1 |\delta_2|) |x| \varphi(|x|) \} \Big],$$

when $\delta_3 > 0$ and $\delta_2 < 0$; and

$$\left| D_x^\alpha u_j(x) \right|$$

$$\leqslant C_3 \Big[|x|^{\tau_j - |\alpha|} \exp\{ (-\delta - \tfrac{1}{4} h_1 |\delta_3|) |x| \varphi(|x|) \}$$

$$+ |x|^{\beta_j - |\alpha|} \left(1 + |x| \varphi(|x|) \right)^{1 + \tau' + q} \exp\{ (\delta_1 - \delta - \tfrac{1}{4} h_1 |\delta_2|) |x| \varphi(|x|) \} \Big],$$

when $\delta_2, \delta_3 < 0$. Here

$$\tau_j = \max_k \{ t_j - t_k \}, \qquad \beta_j = \max_l \{ t_j + s_l \}, \qquad \tau' = \max_l \{ -s_l \},$$

δ_1 is an arbitrary constant, $C_3 = \mathrm{const} > 0$, and the constant $\delta > 0$ depends only on n, N, m, q, λ, M, h_1, h_2 and δ_1.

PROOF. Let $\hat{x} \in \omega$, $|\hat{x}| > R + 1$ and $\rho = |\hat{x}|$. It is easy to see that $v(y_0, y) = (v_1(y_0, y), \ldots, v_N(y_0, y))$ satisfies the system

$$\sum_{j=1}^{N} \sum_{\alpha_0 + |\alpha| \leqslant m_{lj}} a_{lj}^{\alpha_0 \alpha}(\rho y) \rho^{s_l + t_j - \alpha_0 - |\alpha|} \varphi^{-\alpha_0}(\rho) D_{y_0}^{\alpha_0} D_y^\alpha v_j = f_l \cdot \rho^{s_l} \exp\{ i\rho \varphi(\rho) y_0 \},$$

$$l = 1, \ldots, N, \tag{40}$$

in the domain

$$\Omega_\rho = (\sigma_\rho)^\rho \times \{ |y_0| < 2 \} = \{ y; \tfrac{1}{4} < y < 4 \} \times \{ |y_0| < 2 \},$$

provided that

$$v_j(y_0, y) = \rho^{-t_j} \exp\{ i\rho \varphi(\rho) y_0 \} u_j(\rho y)$$

and also provided that $u(x) = (u_1(x), \ldots, u_N(x))$ is a solution of (20) in σ_ρ with $\mu = 1$. Let $\Omega_\rho' = (\sigma_\rho')^\rho \times \{ |y_0| < 1 \}$. According to (35) and (36), the system (40) is uniformly elliptic in Ω_ρ and the constant of ellipticity does not depend on ρ. Moreover, from (37) it follows that

$$\left\| a_{lj}^{\alpha_0 \alpha}(\rho y) \rho^{s_l + t_j - \alpha_0 - |\alpha|} \varphi^{-\alpha_0}(\rho) \right\|_{(\sigma_\rho)^\rho}^{q - s_l + 1} \leqslant M,$$

where $l, j = 1, \ldots, N$, $\alpha_0 + |\alpha| \leqslant s_l + t_j$, and the constant M does not depend on ρ. Therefore, according to Theorem 1,

$$\left\| D_y^\alpha v_j \right\|_{Q_\delta(\Omega_\rho')}^0 \leqslant C_0 \left[\sum_{k=1}^{N} \| v_k \|_{\Omega_\rho}^0 - \sum_{l=1}^{N} \| F_l \|_{Q_{\delta_1}(\Omega_\rho)}^{q - s_l + 1} \right] \tag{41}$$

for $|\alpha| \leqslant q + t_j - 1$, $j = 1, \ldots, N$, where δ_1 may be an arbitrary number, $C_0 = \text{const}$, and δ and C_0 do not depend on ρ, though they depend on δ_1. Here also

$$Q_\delta(\Omega'_\rho) = \{ y_0 + iy_0^*, y; (y_0, y) \in \Omega'_\rho, |y_0^*| < \delta \},$$

$$Q_{\delta_1}(\Omega_\rho) = \{ y_0 + iy_0^*, y; (y_0, y) \in \Omega_\rho, |y_0^*| < \delta_1 \}.$$

From (41) it follows that for $|\alpha| \leqslant q + t_j - 1$

$$\left\| D_x^\alpha u_j \right\|_{\sigma'_\rho}^0 \leqslant C_1' \rho^{t_j - |\alpha|} \exp\{ -\rho\varphi(\rho)\delta \}$$

$$\times \left[\sum_{k=1}^N \|u_k\|_{\sigma_\rho}^0 \rho^{-t_k} + \sum_{l=1}^N \|f_l\|_{\sigma_\rho}^{q - s_l + 1} \rho^{s_l}(1 + \rho\varphi(\rho))^{q - s_l + 1} \exp\{ \delta_1\rho\varphi(\rho) \} \right].$$

$$(42)$$

Let $\delta_2, \delta_3 > 0$. Then, using (38) and (39), we obtain from (42) that

$$\left\| D_x^\alpha u_j \right\|_{\sigma'_\rho}^0 \leqslant C_2' \rho^{t_j - |\alpha|} \exp\{ -\rho\varphi(\rho)\delta \}$$

$$\times \left[\sum_{k=1}^N \rho^{-t_k} \exp\{ 4\delta_3 h_2 \rho\varphi(\rho) \} \right.$$

$$\left. + \sum_{l=1}^N \rho^{s_l}(1 + \rho\varphi(\rho))^{q - s_l + 1} \exp\{ \delta_1\rho\varphi(\rho) + 4h_2\delta_2\rho\varphi(\rho) \} \right], \quad (43)$$

where the constants C_1' and C_2' do not depend on ρ. If $\delta_2 < 0$ and $\delta_3 > 0$, then from (42) it follows that

$$\left\| D_x^\alpha u_j \right\|_{\sigma'_\rho}^0 \leqslant C_2' \rho^{t_j - |\alpha|}$$

$$\times \left[\sum_{k=1}^N \rho^{-t_k} \exp\{ (4\delta_3 h_2 - \delta)\rho\varphi(\rho) \} \right.$$

$$\left. + \sum_{l=1}^N \rho^{s_l}(1 + \rho\varphi(\rho))^{q - s_l + 1} \exp\left\{ \left(\delta_1 - \frac{1}{4}h_1|\delta_2| - \delta \right)\rho\varphi(\rho) \right\} \right].$$

$$(44)$$

If $\delta_2, \delta_3 < 0$, in exactly the same way we obtain

$$\left\| D_x^\alpha u_j \right\|_{\sigma'_\rho}^0 \leqslant C_2' \rho^{t_j - |\alpha|}$$

$$\times \left[\sum_{k=1}^N \rho^{-t_k} \exp\left\{ \left(-\delta - \frac{1}{4}|\delta_3|h_1 \right)\rho\varphi(\rho) \right\} \right.$$

$$\left. + \sum_{l=1}^N \rho^{s_l}(1 + \rho\varphi(\rho))^{1 - s_l + q} \exp\left\{ \left(\delta_1 - \delta - \frac{1}{4}h_1|\delta_2| \right)\rho\varphi(\rho) \right\} \right].$$

$$(45)$$

Since $\rho = |\hat{x}|$, (43)–(45) imply the assertion of the theorem.

Analogous theorems hold for (20) with boundary conditions (21) for $\mu = 1$ in ω, when $\partial\omega \cap \{|x| > R\} = \gamma$. One may use Remark 2 following Theorem 1 to refine estimates (43)–(45).

THEOREM 5. *Let the assumptions of Theorem 4 hold. Then, in ω, for $|\alpha| \leqslant t_j + q - 1, j = 1, \ldots, N$,*

$$|D_x^\alpha u_j(x)| \leqslant C_3 \Big[|x|^{\tau_j - |\alpha|} \exp\{(4h_2\delta_3 - \delta)|x|\varphi(|x|)\}$$
$$+ |x|^{\beta_j - |\alpha|} \big(1 + |x|\varphi(|x|)\big)^\kappa \exp\{(b - \delta + 4h_2\delta_2)|x|\varphi(|x|)\}\Big],$$

when $\delta_2, \delta_3 > 0$;

$$|D_x^\alpha u_j(x)| \leqslant C_3 \Big[|x|^{\tau_j - |\alpha|} \exp\{(4h_2\delta_3 - \delta)|x|\varphi(|x|)\}$$
$$+ |x|^{\beta_j - |\alpha|} \big(1 + |x|\varphi(|x|)\big)^\kappa \exp\{(b - \delta - \tfrac{1}{4}h_1|\delta_2|)|x|\varphi(|x|)\}\Big],$$

when $\delta_2 < 0$ and $\delta_3 > 0$; and

$$|D_x^\alpha u_j(x)| \leqslant C_3 \Big[|x|^{\tau_j - |\alpha|} \exp\{(-\tfrac{1}{4}h_1|\delta_3| - \delta)|x|\varphi(|x|)\}$$
$$+ |x|^{\beta_j - |\alpha|} \big(1 + |x|\varphi(|x|)\big)^\kappa \exp\Big\{\Big(b - \delta - \frac{1}{4}h_1|\delta_2|\Big)|x|\varphi(|x|)\Big\}\Big]$$

when $\delta_2, \delta_3 < 0$. Here, as in Theorem 4,

$$\tau_j = \max_k \{t_j - t_k\}, \qquad \beta_j = \max_l \{t_j + s_l\}, \qquad \kappa = \max_{l,h} \{p_l, q_h\},$$

C_3 *is a constant, and the constants b and δ depend on n, N, m, q, λ_1 and M_1.*

PROOF. Since the right side of (40) takes the form (16), where $\mu = i\rho\varphi(\rho)$, the solution $v(y_0, y)$ of (40) considered in the proof of Theorem 4 is subject to estimate (19). According to this estimate,

$$\|D_y^\alpha v_j\|^0_{Q_\delta(\Omega_\rho')} \leqslant C_0 \Bigg[\sum_{l=1}^N \|v_l\|^0_{\Omega_\rho} + (1 + \rho\varphi(\rho))^\kappa \exp\{b\rho\varphi(\rho)\} \sum_{l=1}^N \rho^{s_l}\|f_l\|^{p_l}_{\sigma_\rho} \Bigg],$$
$$b = d/CN. \tag{46}$$

From (38), (39) and (46) it follows that, for some constants C_1' and C_2' independent of ρ,

$$\|D_x^\alpha u_j\|^0_{\sigma_\rho'} \leqslant \rho^{t_j - |\alpha|} \exp\{-\delta\rho\varphi(\rho)\}$$

$$\times \Bigg[C_1' \exp\{\delta_3 4h_2 \rho\varphi(\rho)\} \sum_{l=1}^N \rho^{-t_l}$$

$$+ C_2'(1 + \rho\varphi(\rho))^\kappa \exp\{(b + 4h_2\delta_2)\rho\varphi(\rho)\} \sum_{l=1}^N \rho^{s_l} \Bigg],$$

when $\delta_2, \delta_3 > 0$,

$$\left\| D_x^\alpha u_j \right\|_{\sigma_\rho'}^0 \leqslant \rho^{t_j - |\alpha|} \exp\{-\delta\rho\varphi(\rho)\}$$

$$\times \left[C_1' \exp\left\{ -\frac{1}{4} h_1 |\delta_3| \rho\varphi(\rho) \right\} \sum_{l=1}^N \rho^{-t_l} \right.$$

$$\left. + C_2'(1 + \rho\varphi(\rho))^\kappa \exp\left\{ \left(b - \frac{1}{4} h_1 |\delta_2| \right) \rho\varphi(\rho) \right\} \sum_{l=1}^N \rho^{s_l} \right],$$

when $\delta_2, \delta_3 < 0$, and

$$\left\| D_x^\alpha u_j \right\|_{\sigma_\rho'}^0 \leqslant \rho^{t_j - |\alpha|} \exp\{-\delta\rho\varphi(\rho)\}$$

$$\times \left[C_1' \exp\{\delta_3 4 h_2 \rho\varphi(\rho)\} \sum_{l=1}^N \rho^{-t_l} \right.$$

$$\left. + C_2'(1 + \rho\varphi(\rho))^\kappa \exp\left\{ \left(b - \frac{1}{4} h_1 |\delta_2| \right) \rho\varphi(\rho) \right\} \sum_{l=1}^N \rho^{s_l} \right],$$

when $\delta_3 > 0$ and $\delta_2 < 0$. These inequalities imply the assertion of the theorem.

As a consequence of Theorem 1, we also obtain a theorem on the behavior of solutions of (20) with boundary conditions (21) on γ as $|\mu_1| \to \infty$.

THEOREM 6. *Assume that the domain ω, the set γ on $\partial\omega$, the domain ω_1, and the system (1) in Ω with boundary conditions (2) on Γ satisfy the conditions of Theorem 1. Then, for any solution $u(x) = (u_1(x), \ldots, u_N(x))$ of (20) in ω with boundary conditions (21) on γ such that $u_j(x) \in C^{l_0 + t_j + 1}(\omega')$, $j = 1, \ldots, N$, if $\overline{\omega}' \subset \omega \cup \gamma$, then for $|\alpha| \leqslant l_0 + q + t_j - 1$*

$$\left\| D_x^\alpha u_j \right\|_{\omega_1}^0 \leqslant \exp\{\ln C_0 - \delta|\mu_1| + 2d|\mu_2|\} \sum_{l=1}^N \|u_l\|_\omega^0$$

$$+ (1 + |\mu|)^\kappa \exp\{\ln C_0 C' - \delta|\mu_1| + 2d|\mu_2| + \delta_1|\mu_1|\}$$

$$\times \left(\sum_{l=1}^N \|f_l\|_\omega^{p_l} + \sum_{h=1}^m \|\varphi_h\|_\gamma^{q_h} \right), \tag{47}$$

where the constant C' depends only on p_l and q_h, $l = 1, \ldots, N$, $h = 1, \ldots, m$, δ_1 is an arbitrary positive number, $\kappa = \max_{l,h}\{p_l, q_h\}$, and the constants δ and C_0 depend on n, N, m, l_0, q, λ, Δ_Γ, M, $d_1(\gamma)$, $K_1(\gamma)$, d and δ_1.

PROOF. Estimate (47) follows immediately from Theorem 1. In fact, if $u(x) = (u_1(x), \ldots, u_N(x))$ is a solution of (20) with boundary conditions (21) on γ, then

$$v(x_0, x) = (v_1(x_0, x), \ldots, v_N(x_0, x)),$$

where

$$v_j(x_0, x) = \exp\{i\mu x_0\} u_j(x), \qquad j = 1, \ldots, N,$$

is a solution of (1) in Ω with boundary conditions (2) on Γ. From Theorem 1 we have that for $|\alpha| \leq l_0 + q + t_j - 1, j = 1,\ldots,N$,

$$\left\| D_x^\alpha v_j \right\|_{Q_\delta(\Omega_1)}^0 \leq C_0 \left[\sum_{l=1}^N \left\| v_l \right\|_\Omega^0 + \sum_{l=1}^N \left\| f_l e^{i\mu x_0} \right\|_{Q_{\delta_1}(\Omega)}^{p_l} + \sum_{h=1}^m \left\| \varphi_h e^{i\mu x_0} \right\|_{Q_{\delta_1}(\Gamma)}^{q_h} \right].$$

And from this we obtain (47) for $u(x)$. The theorem is proved.

Estimates of the type (47) with $f_l \equiv 0$ and $\varphi_h \equiv 0$ were used in [8] in studying generalized analyticity of solutions of elliptic and parabolic equations.

THEOREM 7. *Let the assumptions of Theorem 6 be fulfilled. Then, for any solution* $u(x) = (u_1(x),\ldots,u_N(x))$ *of* (20) *in* ω *with boundary conditions* (21) *on* γ *such that* $u_j(x) \in C^{l_0+t_j+1}(\omega')$ *for* $\overline{\omega}' \subset \omega \cup \gamma$, *the estimate*

$$\left\| D^\alpha u_j \right\|_{\omega_1}^0 \leq \exp\left\{ 2d|\mu_2| - \delta|\mu_1| \right\}$$

$$\times \left[C_0 \sum_{l=1}^N \left\| u_l \right\|_\omega^0 + C_0'(1+|\mu|)^\kappa \exp\{b|\mu|\} \left(\sum_{l=1}^N \left\| f_l \right\|_\omega^{p_l} + \sum_{h=1}^m \left\| \varphi_h \right\|_\gamma^{q_h} \right) \right],$$

$$(48)$$

holds for $|\alpha| \leq l_0 + q + t_j - 1, j = 1,\ldots,N$, *where* b, C_0, C_0' *and* δ *are positive constants depending only on* n, N, m, l_0, q, λ, Δ_Γ, M, $d_1(\gamma)$, $K_1(\gamma)$ *and* d; *the constant* $\kappa = \max_{l,h}\{p_l, q_h\}$, $p_l = l_0 + q - s_l + 1$, $q_h = l_0 + q - r_n + 1$, $d = \min(1, r(\omega_1, \partial\omega \setminus \gamma))$ *if* $\partial\omega \setminus \gamma$ *is nonempty, and* $d = 1$ *if* $\partial\omega = \gamma$.

PROOF. For a solution $v(x_0, x) = (v_1(x_0, x),\ldots,v_N(x_0, x))$ (where $v_j(x_0, x) = \exp\{i\mu x_0\}u_j(x))$ of system (1) in Ω with conditions (2) on Γ and $F_l = f_l \exp\{i\mu x_0\}$, $\Phi_h = \varphi_h \exp\{i\mu x_0\}$, $l = 1,\ldots,N$, $h = 1,\ldots,m$, the estimate (19) obtained in Remark 2 to Theorem 1 holds. According to this estimate, for $|\alpha| \leq l_0 + q + t_j - 1$ we have

$$\left\| D_x^\alpha v_j \right\|_{Q_\delta(\Omega_1)}^0$$

$$\leq C_0 \left[\sum_{l=1}^N \left\| v_l \right\|_\Omega^0 + \exp\{a|\mu_2| + b|\mu|\}(1+|\mu|)^\kappa \left(\sum_{l=1}^N \left\| f_l \right\|_\omega^{p_l} + \sum_{h=1}^m \left\| \varphi_h \right\|_\gamma^{q_h} \right) \right].$$

For a solution $u(x) = (u_1(x) + \cdots + u_N(x))$ of (20) and (21) it follows from this that

$$\left\| D_x^\alpha u_j \right\|_{\omega_1}^0 \leq \exp\left\{ -\delta|\mu_1| + 2d|\mu_2| \right\}$$

$$\times \left[C_0 \sum_{l=1}^N \left\| u_l \right\|_\omega^0 + C_0'(1+|\mu|)^\kappa \exp\{b|\mu|\} \left(\sum_{l=1}^N \left\| f_l \right\|_\omega^{p_l} + \sum_{h=1}^m \left\| \varphi_h \right\|_\gamma^{q_h} \right) \right].$$

$$(49)$$

From (49) we obtain the required inequality (48).

Received 15/SEPT/75

BIBLIOGRAPHY

1. O. A. Oleĭnik and E. V. Radkevič, *The behavior at infinity of the solutions of certain systems of partial differential equations*, Uspehi Mat. Nauk **28** (1973), no. 5 (173), 249–250. (Russian)

2. O. A. Oleĭnik, *Certain properties of the solutions of second order equations with nonnegative characteristic form*, Vestnik Moskov. Univ. Ser. I Mat. Meh. **1974**, no. 1, 125–134; English transl. in Moscow Univ. Math. Bull. **29** (1974).

3. O. A. Oleĭnik and E. V. Radkevič, *Analyticity and theorems of Liouville and Phragmén-Lindelöf type for general parabolic systems of differential equations*, Funkcional. Anal. i Priložen. **8** (1974), no. 4, 59–70; English transl. in Functional Anal. Appl. **8** (1974).

4. O. A. Oleĭnik, *The analyticity of solutions of partial differential equations and its applications*, Trends in Applications of Pure Mathematics to Mechanics (Conf., Univ. Lecce, Lecce, 1975), Pitman, London, 1976, pp. 281–298.

5. S. Agmon, A. Douglis and L. Nirenberg, *Estimates near the boundary for solutions of elliptic partial differential equations satisfying general boundary conditions.* I, Comm. Pure Appl. Math. **12** (1959), 623–727.

6. ______, *Estimates near the boundary for solutions of elliptic partial differential equations satisfying general boundary conditions.* II, Comm. Pure Appl. Math. **17** (1964), 35–92.

7. Avner Douglis and Louis Nirenberg, *Interior estimates for elliptic systems of partial differential equations*, Comm. Pure Appl. Math. **8** (1955), 503–538.

8. E. M. Landis and O. A. Oleĭnik, *Generalized analyticity and some related properties of solutions of elliptic and parabolic equations*, Uspehi Mat. Nauk **29** (1974), no. 2 (176), 190–206; English transl. in Russian Math. Surveys **29** (1974).

Translated by P. C. FIFE

Amer. Math. Soc. Transl.
(2) Vol. **118**, 1982

First Integrals and Integrability
of Systems of Quasilinear Equations*

A. V. FURSIKOV

Contents

We consider the system of quasilinear equations

$$\frac{\partial u(t, x)}{\partial t} + \sum_{j=1}^{n} f_j(u(t, x)) \frac{\partial u(t, x)}{\partial x_j} = 0, \qquad t > 0, x = (x_1, \ldots, x_n) \in R^n,$$

$$(0.1)$$

where $u(t, x) = (u_1(t, x), \ldots, u_p(t, x))$ is a vector-valued function, and $f_j(u)$ ($j = 1, \ldots, n$) are given $p \times p$ matrices. It is assumed that the solution $u(t, x)$ of the system (0.1) for $t = 0$ satisfies the initial condition

$$u(t, x)|_{t=0} = u_0(x) \in C_0^1(R^n) \qquad (0.2)$$

($C_0^k(R^n)$ is the space of compactly supported functions which are k times continuously differentiable).

Suppose that $t \in [0, T]$, where T is some positive number, there exists a continuously differentiable solution $u(t, x)$ of the Cauchy problem (0.1), (0.2). We ask under what conditions on the system (0.1) is it possible to completely integrate it, i.e., to express the solution $u(t, x)$ ($0 < t < T$) by means of formulas in terms of the initial condition $u_0(x)$. It turns out that only a rather restricted class of systems of the form (0.1) can be completely integrated. Therefore, in addition to conditions for complete integrability of the system (0.1), we shall be

---

1980 *Mathematics Subject Classification.* Primary 35F25; Secondary 35R99.
*Translation of Trudy Sem. Petrovsk. **3** (1978), 197–222. MR **80**a 35082.

interested also in the possibility of obtaining less detailed information on the solution $u(t, x)$ for $0 < T \leqslant T$. More precisely, we shall be interested in the question of conditions on the system (0.1) under which it is possible to construct expressions of the form $g(u(t, x))$ for certain functions $g(v)$, $v \in R^p$, and also expressions of the form

$$\int_\Gamma g(u(t, x)) \, dx', \qquad \int_{R^n} \langle \omega, x \rangle^k g(u(t, x)) \, dx, \qquad (0.3)$$

where in the first of the integrals (0.3) the integration goes over some hyperplane $\Gamma \subset R^n$, and in the second integral $\omega \in R^n$ is a fixed vector, k is a nonnegative integer, and $\langle \cdot, \cdot \rangle$ is the inner product in R^n.

It is proved that for smooth solutions $u(t, x)$ of the Cauchy problem (0.1), (0.2) we have

$$\int e^{i\langle x, \xi \rangle} g(u(t, x)) \, dx = \int e^{i\langle x, \xi \rangle} \int_0^{u_0(x)} \langle g_v(v), e^{it\langle \xi, f(v) \rangle} \, dv \rangle \, dx, \qquad (0.4)$$

$$\int \langle \omega, x \rangle^k g(u(t, x)) \, dx$$

$$= \sum_{j=0}^{k} \frac{k! \, t^j}{j! \, (k-j)!} \int \langle \omega, x \rangle^{k-j} \int_0^{u_0(x)} \langle g_v(v), \langle \omega, f(v) \rangle^j \, dv \rangle \, dx, \qquad (0.5)$$

where

$$g_v(v) = (\partial g / \partial v_1, \ldots, \partial g / \partial v_p), \qquad dv = (dv_1, \ldots, dv_p),$$

$$\xi = (\xi_1, \ldots, \xi_n), \qquad \langle \xi, f(v) \rangle = \sum_{j=1}^{n} \xi_j f_j(v),$$

and the inner integrals on the right sides of (0.4) and (0.5) are taken over some contour in R^p joining the points 0 and $u_0(x)$. Since the contours in the inner integrals in (0.4) and (0.5) are not indicated, the right sides of (0.4) and (0.5) are meaningful if the inner integrals do not depend on the path of integration joining 0 and $u_0(x)$. We note that just this condition of independence of the path of integration is essentially the condition under which (0.4) and (0.5) are valid, i.e., this condition is a condition for the integrability of the system (0.1).

We note that we assume beforehand the existence of a smooth solution $u(t, x)$ of the Cauchy problem (0.1), (0.2), and, aside from the condition that certain integrals be independent of the path of integration, we impose no conditions on the system (0.1). In order to guarantee the existence of a solution of the Cauchy problem (0.1), (0.2) for a sufficiently smooth initial condition, it suffices to impose a condition of hyperbolicity in the sense of Petrovskiĭ [1] on (0.1).

The formulas (0.4) and (0.5) are obtained in the following fashion. In §1 these formulas are derived in the case of a single one-dimensional equation. It is not difficult to formally generalize these formulas to the case of systems. However, in order to prove the validity of the resulting formulas, it is necessary to use first integrals of quasilinear systems. In §2 the definition and simple but important properties of first integrals are presented, and in §3 certain first integrals $\Phi(t, u)$ are constructed by means of which the validity of (0.4) and (0.5) is demonstrated.

It is interesting to observe that these first integrals $\Phi(t, u)$ are defined for all $t \in R^1$ and $u(x) \in C_0^1(R^n)$. We note that earlier in [2] and [3] first integrals $\Psi(t, u)$ of parabolic equations and systems were constructed in the form of series in u and were defined for all $t \in R^1$ only in some neighborhood of zero of the corresponding function space where they are also analytic in u. The first integrals $\Phi(t, u)$ constructed in §3 are entire analytic functionals of u if the matrices $f_j(v)$ of the system (0.1) are entire analytic functions of their arguments.

In §4 examples are presented of systems for which the integrability conditions are satisfied and (0.4) and (0.5) are valid.

The main results of this work were announced in [4].

§1. Derivation of the formulas in the scalar case

In this section we consider the equation

$$\partial u(t, x)/\partial t + f(u(t, x))\partial u(t, x)/\partial x = 0, \qquad x \in R^1, t \geqslant 0, \qquad (1.1)$$

where $f(v)$ is an infinitely smooth, real-valued function of the variable $v \in R^1$. We consider solutions $u(t, x)$ of this equation which satisfy the initial condition

$$u(t, x)\big|_{t=0} = u_0(x) \in C_0^1(R^1). \qquad (1.2)$$

These solutions are considered for those values of t for which $u(t, x) \in C_0^1(R_x^1)$. In this section a formula is derived for the Fourier transform $\hat{u}(t, \xi)$ of the solution $u(t, x)$ of problem (1.1), (1.2); a number of other formulas are also derived. In following sections these formulas are generalized to certain classes of systems of quasilinear equations.

We first construct a solution $u(t, x)$ of problem (1.1), (1.2) by means of the well-known method of characteristics.

Let $u(t, x)$ be a solution of problem (1.1), (1.2), and let $x(t, x_0)$ be a solution of the Cauchy problem

$$\partial x(t, x_0)/\partial t = f(u(t, x(t, x_0))), \qquad x(t, x_0)\big|_{t=0} = x_0. \qquad (1.3)$$

Then by (1.1)

$$(d/dt)u(t, x(t, x_0)) = 0,$$

and hence for any (sufficiently small) $t > 0$

$$u(t, x(t, x_0)) = u_0(x_0). \qquad (1.4)$$

Taking (1.4) into account, it is easy to see that the solution $x(t, x_0)$ of (1.3) is given by

$$x(t, x_0) = x_0 + tf(u_0(x_0)). \qquad (1.5)$$

The formulas (1.4) and (1.5) uniquely determine the solution $u(t, x)$ of problem (1.1), (1.2) for those t for which the function $x(t, x_0)$ defined by (1.5) has an inverse function $x_0(t, x)$.

Let $g(x)$, $H(x) \in C^1(R^1)$, $h(x) \in C^0(R^1)$ with $g(0) = 0$, $dH(x)/dx = h(x)$, and suppose that $u(t, x)$ is a smooth solution of problem (1.1), (1.2). Then

$$\int h(x)g(u(t, x))\, dx = -\int H(x)\frac{\partial g(u(t, x))}{\partial x}\, dx. \tag{1.6}$$

Making the change of variable $x = x(t, x_0)$ on the right side of this equality, where $x(t, x_0)$ is defined by (1.5), and noting that

$$\frac{\partial u(t, x)}{\partial x}\frac{\partial x(t, x_0)}{\partial x_0} = \frac{\partial u(t, x(t, x_0))}{\partial x_0} = \frac{\partial u_0(x_0)}{\partial x_0},$$

we obtain

$$\int h(x)g(u(t, x))\, dx = -\int H(x_0 + tf(u_0(x_0)))\frac{\partial g(u_0(x_0))}{\partial x_0}\, dx_0. \tag{1.7}$$

The left side of (1.7), which depends on the solution of problem (1.1), (1.2), is thus expressed in terms of the initial condition $u_0(x)$. We shall henceforth consider the cases $h(x) = e^{ix\xi}$ ($\xi \in R^1$) and $h(x) = x^m$ (m an integer).

LEMMA 1.1. *Let $u(x) \in C_0^1(R^1)$ and $\alpha(x) \in C^0(R^1)$. Then*

$$\int_{-\infty}^{x}\alpha(u(x))\frac{\partial u(x)}{\partial x}\, dx = \int_{0}^{u(x)}\alpha(v)\, dv. \tag{1.8}$$

PROOF. For negative x sufficiently large in modulus, both sides of (1.8) are equal to zero. Since the derivatives with respect to x of both sides of (1.8) coincide, it is valid for all $x \in R^1$.

In (1.7) we set $h(x) = e^{ix\xi}$, $\xi \in R^1$. Integrating by parts on the right side of (1.7) and using Lemma 1.1, we then obtain the desired formula

$$\int e^{ix\xi}g(u(t, x))\, dx = \int e^{ix\xi}\int_{-\infty}^{x} e^{it\xi f(u_0(y))}\frac{\partial g(u_0(y))}{\partial y}\, dy\, dx$$

$$= \int e^{ix\xi}\int_{0}^{u_0(x)} e^{it\xi f(v)}g_v(v)\, dv\, dx, \tag{1.9}$$

where $g_v(v) = dg(v)/dv$. We note that (1.9) has been proved only for those t for which the function (1.5) has an inverse. In any case those $t \in (0, t_+)$, where

$$t_+ = \sup\left\{\tau: \tau \geq 0, 1 + \rho\frac{\partial f(u_0(x))}{\partial x} > 0, \forall x \in R^1\right\}, \tag{1.10}$$

satisfy this condition. Setting $g(v) = v$ in (1.9), we obtain an expression for the Fourier transform

$$\hat{u}(t, \xi) = \int e^{ix\xi}u(t, x)\, dx$$

of the solution $u(t, x)$ of problem (1.1), (1.2) in terms of the initial condition $u_0(x)$.

THEOREM 1.1. *Let $u(t, x)$ be a solution of problem (1.1), (1.2) with $t \in (0, t_+)$, where t_+ is defined by (1.10). Then its Fourier transform $\hat{u}(t, \xi)$ is given by*

$$\hat{u}(t, \xi) = \int e^{ix\xi} \int_0^{u_0(x)} e^{it\xi f(v)} \, dv \, dx. \tag{1.11}$$

REMARK 1.1. In general, for $t > t_+$ (1.11) does not define the Fourier transform of the solution $u(t, x)$ of problem (1.1), (1.2). To show this, we consider a solution $u(t, x)$ of the equation

$$\partial u/\partial t + u\partial u/\partial x = 0, \tag{1.12}$$

satisfying the initial condition

$$u(t, x)|_{t=0} = u_0(x) = \begin{cases} 0 & \text{for } |x| > 1, \\ x + 1 & \text{for } -1 < x < 0, \\ 1 - x & \text{for } 0 < x < 1. \end{cases} \tag{1.13}$$

(Of course, Theorem 1.1 is also valid for initial conditions of the type (1.13) with a piecewise continuous derivative.)

Computing $\hat{u}(t, \xi)$ by means (1.11), we find that

$$\hat{u}(t, \xi) = \left(\frac{e^{i\xi t} - e^{-i\xi}}{1 + t} + \frac{e^{i\xi} - e^{i\xi t}}{1 - t} - e^{i\xi} + e^{-i\xi} \right) \Big/ (i\xi)^2 t. \tag{1.14}$$

In the present case $t_+ = 1$, and so for $0 \leqslant t \leqslant 1$ the function (1.14) defines the Fourier transform of the solution of problems (1.12), (1.13). If we apply the inverse Fourier transform to the function (1.14) for $t > 1$, we find that

$$u(t, x) = \begin{cases} 0 & \text{for } x < -1 \text{ and } x > t, \\ (1 + x)/(1 + t) & \text{for } -1 < x < 1, \\ 2(t - x)/(t^2 - 1) & \text{for } 1 < x < t. \end{cases} \tag{1.15}$$

It is easy to see that for $1 \leqslant x \leqslant t$ the function $u(t, x)$ does not satisfy (1.12).

REMARK 1.2. It is also easy to obtain an explicit formula expressing the solution $u(t, x)$ of problem (1.1), (1.2) in terms of the initial condition $u_0(x)$. Integrating the right side of (1.11) by parts, we obtain

$$\hat{u}(t, \xi) = -\frac{1}{i\xi} \int e^{i\xi(y + tf(u_0(y)))} \frac{\partial u_0(y)}{\partial y} \, dy$$

$$= \int e^{i\xi(y + tf(u_0(y)))} \left(1 + t\frac{\partial f(u_0(y))}{\partial y} \right) u_0(y) \, dy. \tag{1.16}$$

Applying the inverse Fourier transform to both sides of (1.16), we obtain

$$u(t, x) = \int \delta(y + tf(u_0(y)) - x)\left(1 + t\frac{\partial f(u_0(y))}{\partial y} \right) u_0(y) \, dy, \tag{1.17}$$

where $\delta(z)$ is the Dirac delta function (the integral on the right side of (1.17) is understood in the sense of the theory of generalized functions). From (1.17) it is easy to deduce that

$$u(t, x) = \int u_0(y)\, d\theta(y + tf(u_0(y)) - x), \qquad (1.18)$$

where $\theta(z)$ is the Heaviside function equal to zero for $z < 0$ and to one for $z > 0$.

The formulas (1.17) and (1.18) are valid for $0 < t < t_+$, where t_+ is defined by (1.10). However, (1.18) can be modified so that it also delivers the solution $u(t, x)$ of (1.1) for $t \geq t_+$. Of course, for such t, $u(t, x)$ will be a multivalued function of x. Thus, the solution $u(t, x)$ of problem (1.12), (1.13) for $t > t_+ = 1$ is given by

$$u(t, x) = \begin{cases} 0 & \text{for } x \leq -1, \\ (x + 1)/(t + 1) & \text{for } -1 < x \leq t, \\ (x - 1)/(t - 1) & \text{for } t \geq x > 1, \\ 0 & \text{for } x \geq 1. \end{cases} \qquad (1.19)$$

We note that the function (1.15) for $1 < x < t$ is equal to

$$(x + 1)/(t + 1) - (x - 1)/(t - 1) + 0,$$

i.e., to the alternating sum of the values of the multivalued function (1.19).

We now consider the case where the function $h(x)$ of (1.7) is equal to x^m, where m is a nonnegative integer. Using Newton's binomial expansion, integrating by parts on the right side of (1.7), and using Lemma 1.1, in this case we obtain

$$\int x^m g(u(t, x))\, dx = \sum_{j=0}^{m} \frac{m!\, t^j}{j!\,(m - j)!} \int x^{m-j} \int_0^{u_0(x)} f^j(v) g_v(v)\, dv\, dx, \qquad (1.20)$$

where, as previously, $g_v(v) = dg(v)/dv$. The analogue of (1.20) for the case of systems of quasilinear equations will be considered below.

§2. Preliminary facts regarding first integrals of systems of quasilinear equations

1. We consider a system of quasilinear equations of the form

$$\frac{\partial u(t, x)}{\partial t} + \sum_{j=1}^{n} f_j(u(t, x)) \frac{\partial u(t, x)}{\partial x_j} = 0, \quad t > 0, x = (x_1, \ldots, x_n) \in R^n, \qquad (2.1)$$

where $u(t, x) = (u_1(t, x), \ldots, u_p(t, x))$ is a vector-valued function, and $f_j(v) = \| f_{j,kl}(v) \|$ $(j = 1, \ldots, n;\ k, l = 1, \ldots, p)$ is a $p \times p$ matrix depending continuously on $v \in R^p$. It is assumed that the solution $u(t, x)$ of (2.1) at $t = 0$ satisfies the condition

$$u(t, x)|_{t=0} = u_0(x) = (u_{0,1}(x), \ldots, u_{0,p}(x)) \in (C_0^1(R^n))^p. \qquad (2.2)$$

We here consider only solutions $u(t, x)$ defined on the set $(t, x) \in [0, T] \times R^n$ (T is some positive number) which are continuously differentiable with respect to the variables (t, x) and are equal to zero for $|x| > R$, where R is a sufficiently large number not depending on $t \in [0, T]$ but depending on $u(t, x)$. We denote the space of these functions by $DC_0^1([0, T] \times R^n)$.

Integrability conditions for the system (2.1) will be obtained below, i.e., conditions under which (0.4) and (0.5) are valid. First integrals of (2.1) will be used.

Now we present those facts regarding first integrals which we shall need in the sequel.

DEFINITION 2.1. A *first integral* of the system (2.1) is a functional $\Phi(t, u) = \Phi(t, u_1, \ldots, u_p)$ defined on $R^1 \times (C_0^0(R^n))^p$ such that for any solution $u(\tau, x) \in DC_0^1([0, T] \times R^n)$ and any $t \in R^1$ the quantity $\Phi(t + \tau, u(\tau, \cdot))$ does not depend on $\tau \in [0, T]$.

DEFINITION 2.2. A functional $\Phi(t, u) = \Phi(t, u_1, \ldots, u_p)$ defined on $R^1 \times (C_0^0(R^n))^p$ is *differentiable with respect to all variables at the point* (t, u) if for any increment

$$(\Delta t, \delta u) = (\Delta t, \delta u_1, \ldots, \delta u_p) \in R^1 \times (C_0^0(R^n))^p$$

we have

$$\Phi(t + \Delta t, u + \delta u) - \Phi(t, u)$$
$$= \frac{\partial \Phi(t, u)}{\partial t} \Delta t + \int \left\langle \frac{\delta \Phi(t, u)}{\delta u(x)} \cdot \delta u(x) \right\rangle dx + Q(t, u, \Delta t, \delta u), \qquad (2.3)$$

where

$$\frac{\delta \Phi(t, u)}{\delta u(x)} = \left(\frac{\delta \Phi(t, u)}{\delta u_1(x)}, \ldots, \frac{\delta \Phi(t, u)}{\delta u_p(x)} \right),$$

$$\left\langle \frac{\delta \Phi(t, u)}{\delta u(x)}, \delta u(x) \right\rangle = \sum_{j=1}^{p} \frac{\delta \Phi(t, u)}{\delta u_j(x)} \delta u_j(x),$$

and $\delta \Phi(t, u)/\delta u_j(x)$ $(j = 1, \ldots, p)$ is an element of the space dual to $C_0^0(R^n)$ which depends continuously on $t \in R^1$ (the integral on the right side of (2.3) is understood in the sense of the theory of generalized functions). It is here assumed that for any sequence

$$(\Delta t_k, \delta u_{(k)}(x)) = (\Delta t_k, \delta u_{1(k)}(x), \ldots, \delta u_{p,(k)}(x)),$$

which tends to zero in the sense of the topology of $R^1 \times (C_0^0(R^n))^p$ as $k \to \infty$ the quantity

$$\frac{Q(t, u, \Delta t_k, \delta u_{(k)}(x))}{|\Delta t| + \Sigma_{j=1}^{p} \sup_x |\delta u_{j,(k)}(x)|}$$

tends to zero as $k \to \infty$.

We remark that the generalized function $\delta \Phi(t, u)/\delta u_j(x)$ $(j = 1, \ldots, p)$ is called the *variational derivative* of the functional $\Phi(t, u)$ with respect to the variable $u_j(x)$.

Suppose that a first integral $\Phi(t, u)$ of (2.1) for any $(t, u) \in R^1 \times (C_0^0(R^n))^p$ is differentiable with respect to all variables in the sense of Definition 2.2. Then for any solution $u(\tau, x) \in DC_0^1([0, T] \times R^n)$ of (2.1) we have

$$0 = \frac{d}{d\tau} \Phi(t + \tau, u(\tau, \cdot))$$

$$= \frac{\partial \Phi(t + \tau, u(\tau, \cdot))}{\partial \tau} + \int \left\langle \frac{\delta \Phi(t + \tau, u(\tau, \cdot))}{\delta u(x)}, \frac{\partial u(\tau, x)}{\partial \tau} \right\rangle dx. \quad (2.4)$$

Expressing the derivative $\partial u(\tau, x)/\partial \tau$ in (2.4) in terms of (2.1) and setting in (2.4) $\tau = 0$ and $u(0, x) = u(x)$, we obtain

$$\frac{\partial \Phi(t, u)}{\partial t} - \int \left\langle \frac{\delta \Phi(t, u)}{\delta u(x)}, \sum_{j=1}^{n} f_j(u(x)) \frac{\partial u(x)}{\partial x_j} \right\rangle dx = 0. \quad (2.5)$$

Thus, any first integral $\Phi(t, u)$ differentiable in all variables satisfies (2.5). The converse assertion is also true.

PROPOSITION 2.1. *Suppose that a functional $\Phi(t, u)$ for each $(t, u) \in R^1 \times (C_0^0(R^n))^p$ is differentiable in all variables and, moreover, satisfies (2.5). Then $\Phi(t, u)$ is a first integral of the system (2.1).*

PROOF. Let $u(\tau, x) \in DC_0^1([0, T] \times R^n)$ be a solution of (2.1). We must show that $\Phi(t + \tau, u(\tau, \cdot))$ does not depend on $\tau \in [0, T]$. Differentiating this expression with respect to τ, we obtain the right side of (2.4). Inserting $\partial u(x, \tau)/\partial \tau$ from (2.1) and using the fact that $\Phi(t, u)$ satisfies (2.5), we find that $d\Phi(t + \tau, u(\tau, \cdot))/d\tau = 0$.

A relation which follows directly from Definition 2.1 will be useful in the sequel.

PROPOSITION 2.2. *Suppose that a functional $\Phi(t, u)$ defined on $R^1 \times (C_0^0(R^n))^p$ is a first integral of the system (2.1) and that $u(t, x) \in DC_0^1([0, T] \times R^n)$ is a solution of (2.1). Then for $t \in [0, T]$*

$$\Phi(-t, u(0, \cdot)) = \Phi(0, u(t, \cdot)). \quad (2.6)$$

2. Before proceeding to a derivation of the integrability conditions and the proof of (0.4) and (0.5), we prove a lemma which is needed below.

LEMMA 2.1. *Suppose that the differential form*

$$\langle h(t, v), dv \rangle = \sum_{j=1}^{p} h_j(t, v)\, dv_j,$$

where

$$h(t, v) = \left(h_1(t, v), \ldots, h_p(t, v) \right) \in \left(C^1([0, T] \times R^n) \right)^p,$$

for each $t \in [0, T]$, is the total differential of some function, and that $h(t, v)$ is twice continuously differentiable with respect to t. Suppose that the functional $\Psi(t, v)$ defined on $[0, T] \times (C_0^0(R^n))^p$ is given by

$$\Psi(t, u) = \int \chi(x) \int_0^{u(x)} \langle h(t, v), dv \rangle \, dx, \qquad (2.7)$$

where $\chi(x) \in C^0(R^n)$, $u(x) = (u_1(x), \ldots, u_p(x)) \in (C_0^0(R^n))^p$, and the inner integral in (2.7) is taken over any path in R^p joining the points 0 and $u(x)$ (since it does not depend on the path of integration). Then the functional $\Psi(t, u)$ is differentiable with respect to all variables $(t, u_1(x), \ldots, u_p(x))$, and the variational derivative $\delta\Psi(t, u)/\delta u_j(x)$ of Ψ is given by

$$\delta\Psi(t, u)/\delta u_j(x) = \chi(x)h_j(t, u(x)). \qquad (2.8)$$

PROOF. Let

$$(\Delta t, \delta u(x)) = (\Delta t, \delta u_1(x), \ldots, \delta u_p(x)) \in R^1 \times (C_0^0(R^n))^p.$$

We write the increment of $\Psi(t, u)$ in the form

$$\Psi(t + \Delta t, u + \delta u) - \Psi(t, u) = \Psi(t + \Delta t, u + \delta u) - \Psi(t, u + \delta u)$$

$$+ \sum_{j=1}^{p} \left(\Psi(t, u_1, \ldots, u_{j-1}, u_j + \delta u_j, \ldots, u_p + \delta u_p) \right.$$

$$\left. - \Psi(t, u_1, \ldots, u_{j-1}, u_j, u_{j+1} + \delta u_{j+1}, \ldots, u_p + \delta u_p) \right). \quad (2.9)$$

It follows from (2.9) that to prove the differentiability of $\Psi(t, u)$ it suffices to show that

$$\Psi(t + \Delta t, u) - \Psi(t, u) = \frac{\partial \Psi(t, u)}{\partial t} \Delta t + Q_0(t, \Delta t, u), \qquad (2.10)$$

$$\Psi(t, u_1, \ldots, u_{j-1}, u_j + \delta u_j, u_{j+1}, \ldots, u_p) - \Psi(t, u_1, \ldots, u_{j-1}, u_j, u_{j+1}, \ldots, u_p)$$

$$= \int \frac{\delta\Psi(t, u)}{\delta u_j(x)} \delta u_j(x) \, dx + Q_j(t, u, \delta u_j) \qquad (j = 1, \ldots, p),$$

$$(2.11)$$

while $|Q_0(t, \Delta t, u)|/|\Delta t|$ (respectively, $Q_j(t, u, \delta u_j)/\sup_x |\delta u_j(x)|$) tends to zero uniformly relative to any set

$$U = \left\{ u(x) \in (C_0^0(R^n))^p : \operatorname{supp} u(x) \subset K, \ \sup_{x \in K} |u(x)| < c \right\}, \qquad (2.12)$$

where K is some compact set in R^n and $C > 0$ is a constant, if $\Delta t \to 0$ (respectively, δu_j tends to zero in the topology of $C_0^0(R^n)$).

We first prove the validity of the decomposition (2.11) and compute the variational derivative $\delta\Psi(t, u)/\delta u_j(x)$.

Let $\delta_j(x) = (0,\ldots,\delta u_j(x),\ldots,0)$ be the p-dimensional vector having all coordinates except the jth equal to zero, while the jth coordinate is equal to $\delta u_j(x) \in C_0^0(R^n)$. Then

$$\Psi(u + \delta_j) - \Psi(u) = \int \chi(x) \int_{u(x)}^{u(x)+\delta_j(x)} \langle h(t, v), dv \rangle \, dx. \qquad (2.13)$$

It may be assumed that in the inner integral in (2.13) the integration goes over a segment joining the points $u(x)$ and $u(x) + \delta_j(x)$.

Since

$$\int_{u(x)}^{u(x)+\delta_j(x)} \langle h(t, v), dv \rangle$$

$$= h_j(t, u(x))\delta u_j(x) + \int_{u(x)}^{u(x)+\delta_j(x)} \int_0^1 \frac{d}{ds} h_j(t, u(x) + (v - u(x))s) \, ds \, dv_j,$$

it follows that

$$\Psi(u + \delta_j) - \Psi(u) = \int \chi(x) h_j(t, u(x))\delta u_j(x) \, dx + Q_j(t, u, \delta u_j),$$

$$(2.14)$$

where

$$Q_j(t, u, \delta u_j) = \int \chi(x) \int_{u(x)}^{u(x)+\delta_j(x)} \int_0^1 \frac{d}{ds} h_j(t, u(x) + (v - u(x))s) \, ds \, dv_j \, dx.$$

$$(2.15)$$

Thus, to prove (2.8) it suffices to estimate the remainder Q_j defined by (2.15). More precisely, we must show that if a sequence of functions $\delta u_{j,(n)}(x) \in C_0^0(R^n)$ and

$$\|\delta u_{j,(n)}(x)\| = \sup_{x \in R^m} |\delta u_{j,(n)}(x)| \to 0, \quad \text{as } n \to \infty,$$

where all the functions $\delta u_{j,(n)}(x)$ have the same compact support, then

$$|Q_j(t, u, \delta u_{j,(n)})| / \|\delta u_{j,(n)}(x)\| \to 0 \quad \text{as } n \to \infty \qquad (2.16)$$

uniformly relative to any set U of the form (2.12).

Since $u(x) \in U$ and $h_j(t, v) \in C^1([0, T] \times R^p)$, it follows that for any $|v| < C_1$ and $0 \leqslant s \leqslant 1$,

$$|(d/ds)h_j(t, u(x)) + (v - u(x)s)| \leqslant C_2 |v - u(x)|,$$

where C_1 and C_2 are constants, and

$$|v - u(x)|^2 = \sum_{j=1}^p |v_j - u_j(x)|^2.$$

Therefore, from (2.15) it follows that

$$|Q_j(t, u, \delta u_j)| \leqslant C \int |\chi(x)| \left| \int_{u(x)}^{u(x)+\delta_j(x)} |v - u(x)| dv_j \right| dx. \qquad (2.17)$$

Making the change of variable $v - u(x) = w$ in the inner integral in (2.17), we find by (2.17) that

$$\left| Q_j(t, u, \delta u_j) \right| \leqslant C \int |\chi(x)| \int_0^{|\delta u_j(x)|} w_j \, dw_j \, dx \leqslant \left| C \int |\chi(x)| \delta u_j(x) \right|^2 dx. \qquad (2.18)$$

If all the functions $\delta u_{j,(n)}(x)$ have the same compact support, (2.16) follows from (2.18).

To estimate the remainder $Q_0(t, \Delta t, u)$ in (2.10), we note that

$$Q_0(t, \Delta t, u) = \int \chi(x) \int_0^{u(x)} \left\langle \int_0^1 (1 - s) \frac{d^2}{ds^2} h(t + \Delta ts, v) \, ds, \, dv \right\rangle dx.$$

Using this formula and arguing as in the estimate of Q_j, it is easy to obtain the required estimate for Q_0 as well.

§3. First integrals and integrability conditions for systems of quasilinear equations

1. We proceed to the formulation of integrability conditions for the system (2.1). We first introduce some notation. If A is a matrix, then A^* henceforth denotes the transpose of A; if $h = (h_1, \ldots, h_p)$ is a vector, then $(h)_l$ denotes the lth coordinate of the vector h; the n-dimensional unit sphere is

$$S^n = \left\{ \omega = (\omega_1, \ldots, \omega_n) \in R^n : \sum_{j=1}^{n} \omega_j^2 = 1 \right\}.$$

We denote by $f(v)$ the vector having as coordinates the matrices $f_j(v)$ that occur in (2.1): $f(v) = (f_1(v), \ldots, f_n(v))$; if $\xi = (\xi_1, \ldots, \xi_n) \in R^n$, then

$$\langle \xi, f(v) \rangle = \sum_{j=1}^{n} \xi_j f_j(v).$$

Below, $dv = (dv_1, \ldots, dv_p)$. If $A(v) = \| A_{ij}(v) \|$ is a matrix, then $A(v)dv$ denotes a vector whose coordinates are differential forms, where the ith coordinate is equal to $\sum_1^p A_{ij}(v)dv_j$. If $g(v)$ is a scalar function of the variable $v = (v_1, \ldots, v_p) \in R^p$, then $g_v(v) = (\partial g/\partial v_1, \ldots, \partial g/\partial v_p)$ and

$$\langle g_v(v), A(v) \, dv \rangle = \sum_{i,j=1}^{p} A_{ij}(v) \frac{\partial g}{\partial v_i} dv_j.$$

DEFINITION 3.1. We say that for the system (2.1), the function $g(v) \in C^1(R^p)$, and the vector $\omega \in S^n$ *integrability conditions to order N are satisfied* if the differential forms

$$\left\langle g_v(v), \langle \omega, f(v) \rangle^k f_j(v) \, dv \right\rangle, \qquad j = 1, \ldots, n, \qquad (3.1)$$

for $k = 0, 1, \ldots, N$ and $v \in R^p$ are total differentials of some functions (i.e., the forms (3.1) are exact).

2. To prove (0.4) and (0.5) we need to construct certain first integrals of (2.1). We assume that integrability conditions to order $N = \infty$ are satisfied for $g(v) \in C^1(R^p)$ and $\omega \in S^n$. Multiplying the jth form of (3.1) by ω_j and summing on j, it is then easy to see that the form

$$\left\langle g_v(v), \left\langle \omega, f(v) \right\rangle^k dv \right\rangle \tag{3.1'}$$

is exact for any positive integer k. Therefore, for any $\alpha \in R^1$ the differential form $\left\langle g_v(v), e^{i\alpha\langle \omega, f(v)\rangle} dv \right\rangle$ is exact. Indeed, by expanding this form in a series in α, it is easy to see that each term of the series obtained has the form (3.1') and is hence exact. Since the integral of an exact form along any closed contour is equal to zero, for any $a, b \in R^p$ the integral

$$\int_a^b \left\langle g_v(v), e^{i\alpha\langle \omega f(v)\rangle} dv \right\rangle$$

is well defined.

We consider the functional defined on $R^1 \times (C_0^0(R^n))^p$ and depending on the parameter $\xi \in R^n$

$$\Phi(t, u, \xi) = \int e^{i\langle x, \xi\rangle} \int_0^{u(x)} \left\langle g_v(v), e^{-it\langle \xi, f(v)\rangle} dv \right\rangle dx, \tag{3.2}$$

where $g(0) = 0$ and the vector $\xi \in R^n$ is such that $\xi/|\xi| = \omega$. We note that the right side of (3.2) is analogous to the right side of the second of the equalities in (1.9). We shall show that the functional in (3.2) is a first integral of the system (2.1).

THEOREM 3.1. *Suppose that integrability conditions to order $N = \infty$ are satisfied for the system* (2.1), *the vector $\omega \in S^n$, and the function $g(v) \in C^2(R^p)$ such that $g(0) = 0$. Then the functional defined by* (3.2) *is a first integral of* (2.1).

PROOF. By Proposition 2.1 it suffices to show that the functional (3.2) is differentiable in all variables (t, u) and satisfies (2.5). The differentiability of (3.2) follows directly from Lemma 2.1. We shall show that it satisfies (2.5). By Lemma 2.1

$$\delta\Phi(t, u)/\delta u_j(x) = e^{i\langle x, \xi\rangle}\left(\left(e^{-it\langle \xi, f(u(x))\rangle}\right)*g_v(u(x))\right)_j, \tag{3.3}$$

where we have used the notation introduced at the beginning of this section. From (3.3) it follows that

$$\int \left\langle \frac{\delta\Phi(t, u)}{\delta u(x)}, \sum_{j=1}^n f_j(u)\frac{\partial u}{\partial x_j} \right\rangle dx$$

$$= \sum_{j=1}^n \int e^{i\langle x, \xi\rangle} \left\langle g_v(u(x)), e^{-it\langle \xi, f(u(x))\rangle} f_j(u(x))\frac{\partial u}{\partial x_j} \right\rangle dx. \tag{3.4}$$

By the hypotheses of the theorem the differential form $\langle g_v(v), e^{-it\langle \xi, f(v)\rangle} f_j(v)\, dv\rangle$ $(j = 1,\ldots,n)$ is a total differential of some function $H_j(v)$. This function is uniquely determined up to an additive constant. We choose this constant so that $H_j(0) = 0$. Then

$$H_j(u) = \int_0^u \left\langle g_v(v), e^{-it\langle \xi, f(v)\rangle} f_j(v)\, dv \right\rangle, \tag{3.5}$$

and the integral in (3.5) is taken over any path in R^p joining the points 0 and u, since it does not depend on the choice of path. Noting that

$$(\partial/\partial x_j)H_j(u) = \left\langle g_v(u(x)), e^{-it\langle \xi, f(u(x))\rangle} f_j(u(x))\partial u(x)/\partial x_j \right\rangle,$$

from (3.4) and (3.5) we find that

$$\int \left\langle \frac{\delta\Phi(t, u)}{\delta u(x)}, \sum_{j=1}^n f_j(u(x))\frac{\partial u(x)}{\partial x_j} \right\rangle dx$$

$$= \sum_{j=1}^n \int e^{i\langle x,\xi\rangle}\frac{\partial}{\partial x_j} H_j(u(x))\, dx = -i\int e^{i\langle x,\xi\rangle} \sum_{j=1}^n \xi_j H_j(u(x))\, dx$$

$$= -i\int e^{i\langle x,\xi\rangle} \int_0^{u(x)} \left\langle g_v(v), e^{-it\langle \xi, f(v)\rangle} \langle \xi, f(v)\rangle\, dv \right\rangle dx. \tag{3.6}$$

Differentiating (3.2) with respect to t, we obtain

$$\frac{\partial\Phi(t, u)}{\partial t} = -i\int e^{i\langle x,\xi\rangle} \int_0^{u(x)} \left\langle g_v(v), e^{-it\langle \xi, f(v)\rangle} \langle \xi, f(v)\rangle\, dv \right\rangle dx. \tag{3.7}$$

It follows from (3.6) and (3.7) that the functional (3.2) satisfies (2.5).

We construct some further first integrals of the system (2.1) by using (1.15).

THEOREM 3.2. *Suppose that integrability conditions to some finite order N are satisfied for the system (2.1), the vector $\omega \in S^n$, and the function $g(v) \in C^2(R^p)$ such that $g(0) = 0$. Then the functionals*

$$\Phi_k(t, u) = \sum_{j=0}^k \frac{k!\,(-t)^j}{j!\,(k-j)!} \int \langle x, w\rangle^{k-j} \int_0^{u(x)} \left\langle g_v(v), \langle \omega, f(v)\rangle^j\, dv \right\rangle dx \tag{3.8}$$

with $k = 0, 1,\ldots,N$ are first integrals of the system (2.1).

PROOF. Since the forms (3.1′) with $k = 0, 1,\ldots,N - 1$ are exact, the inner integral on the right side of (3.8) is well defined. By Proposition 2.1 and Lemma 2.1 it suffices to show that the functional (3.8) satisfies (2.5). By Lemma 2.1

$$\frac{\delta\Phi_k(t, u)}{\delta u_l(x)} = \sum_{j=1}^k \frac{k!\,(-t)^j}{j!\,(k-j)!} \langle \omega, x\rangle^{k-j}\left(\left(\langle \omega, f(u(x))\rangle^j\right)^* g_v(v)\right)_l. \tag{3.9}$$

294   A. V. FURSIKOV

Using (3.9) and arguing as in the derivation of (3.6), we obtain

$$
\int \left\langle \frac{\delta \Phi(t, u)}{\delta u(x)}, \sum_{l=1}^{n} f_l(u(x)) \frac{\partial u(x)}{\partial x_l} \right\rangle dx
$$

$$
= \sum_{j=1}^{k} \frac{k!\,(-t)^j}{j!\,(k-j)!} \int \langle x, \omega \rangle^{k-j} \sum_{l=1}^{n} \frac{\partial}{\partial x_l} \int_0^{u(x)} \left\langle g_v(v), \langle \omega, f(v) \rangle^j f_l(v)\, dv \right\rangle dx
$$

$$
= -\sum_{j=1}^{k} \frac{k!\,(-t)^j}{j!\,(k-j-1)!} \sum_{l=1}^{n} \omega_l \int \langle \omega, x \rangle^{k-j-1}
$$

$$
\times \int_0^{u(x)} \left\langle g_v(v), \langle \omega, f(v) \rangle^j f_l(v)\, dv \right\rangle dx
$$

$$
= -\sum_{j=0}^{k-1} \frac{k!\,(-t)^j}{j!\,(k-j-1)!} \int \langle \omega, x \rangle^{k-j-1} \int_0^{u(x)} \left\langle g_v(v), \langle \omega, f(v) \rangle^{j+1}\, dv \right\rangle dx.
$$

From this it follows that

$$
\int \left\langle \frac{\delta \Phi(t, u)}{\delta u(x)}, \sum_{l=1}^{n} f_l(u(x)) \frac{\partial u(x)}{\partial x_l} \right\rangle dx
$$

$$
= -\sum_{j=1}^{k} \frac{k!\,(-t)^{j-1}}{(j-1)!\,(k-j)!} \int \langle \omega, x \rangle^{k-j} \int_0^{u(x)} \left\langle g_v(v), \langle \omega, f(v) \rangle^j\, dv \right\rangle dx.
$$

$$
\tag{3.10}
$$

If we differentiate (3.8) with respect to t, we get an expression which coincides with the right side of (3.10).

REMARK 3.1. It may seem that the integrability condition to the appropriate order N for the system (2.1), the vector $\omega \in S^n$, and the function $g(v)$ was introduced in order to lend meaning to the expressions (3.2) and (3.8). These expressions may be given meaning even when the integrability conditions are not satisfied. Thus, integrating by parts in the right side of (3.2) with respect to x_l, we obtain

$$
-i\xi_l \Phi(t, u, \xi) = \int e^{i\langle x, \xi \rangle} \left\langle g_v(u(x)), e^{-it\langle \xi, f(u(x)) \rangle} \frac{\partial u(x)}{\partial x_l} \right\rangle dx. \tag{3.11}
$$

The right side of (3.11) is meaningful for any f, g and ξ. However, substituting the functional (3.11) into (2.5), we can easily see that it satisfies this equation only if the integrability conditions to order $N = \infty$ are satisfied. (The variational derivatives of (3.11) are easily computed if we note that

$$
\frac{\delta}{\delta u_l(x)} h(u(y)) = h_l(u(x))\delta(x-y);
$$

$$
\frac{\delta}{\delta u_l(x)} \frac{\partial h(u(y))}{\partial y_k} = h_l(u(x)) \frac{\partial}{\partial y_k} \delta(x-y),
$$

where $h(v) = h(v_1,\ldots,v_p)$ is a smooth function of p variables, $h_1(v) = \partial h(v)/\partial v_l$, $h(u(y))$ is a functional which assigns to the function $u(z) = (u_1(z),\ldots,u_p(z))$ the value of the function $h(u(z))$ at the point $z = y$, and $\delta(z)$ is the Dirac delta function.)

REMARK 3.2. If in Theorems 3.1 and 3.2 it is additionally assumed that the function $g(v)$ and the matrices $f_j(v)$ ($j = 1,\ldots,n$) are entire functions of the p complex variables $(v_1,\ldots,v_p) \in \mathbf{C}^p$, then (3.2) and (3.8) are entire functionals relative to $u(x) \in (C_0^0(R^n))^p$. (Here $C_0^0(R^n)$ is a space of complex-valued functions.) This implies, for example, that the functional (3.2) for any fixed $t \in R^1$ and $\xi \in R^n$ ($\xi/|\xi| = \omega$) can be expanded in a convergent series in a neighborhood of any function $u(x) \in (C_0^0(R^n))^p$:

$$\Phi(t, u + \Delta u, \xi) - \Phi(t, u, \xi)$$

$$= \sum_{r=1}^{\infty} \int \Psi_r(x_1,\ldots,x_r, u)[\Delta u(x_1),\ldots,\Delta u(x_r)]\,dx_1,\ldots,dx_r, \qquad (3.12)$$

where $\Psi_r(x_1,\ldots,x_r, u)[v_1,\ldots,v_r]$, a functional in u and r, is linear form in the $v_j \in \mathbf{C}^p$ ($j = 1,\ldots,r$), while for fixed $u \in (C_0^0(R^n))^p$ and $v_j \in \mathbf{C}^p$ ($j = 1,\ldots,r$) Ψ_r is a generalized function belonging to the space dual to $C_0^0(R^{n'})$.

Indeed, because of the above conditions, the form

$$\left\langle g_v(v),\, e^{-it\langle \xi, f(v)\rangle}\, dv \right\rangle$$

is the total differential of some entire function $H(v)$, and hence

$$\Phi(t, u + \Delta u, \xi) - \Phi(t, u, \xi) = \int e^{i\langle x,\xi\rangle} \int_{u(x)}^{u(x)+\Delta u(x)} \left\langle g_v(v),\, e^{-it\langle \xi, f(v)\rangle}\, dv \right\rangle dx$$

$$= \int e^{i\langle x,\xi\rangle}\big(H(u(x) + \Delta u(x)) - H(u(x))\big)\,dx.$$

$$(3.13)$$

Expanding $H(u + \Delta u)$ in a series in Δu, we find from (3.13) that

$$\Phi(t, u + \Delta u, \xi) - \Phi(t, u, \xi) = \sum_{r=1}^{\infty} \sum_{|\alpha|=r} \int e^{i\langle x,\xi\rangle} \frac{1}{\alpha!} D_u^\alpha H(u(x))\Delta u^\alpha(x)\,dx,$$

$$(3.14)$$

where $\alpha = (\alpha_1,\ldots,\alpha_p)$ (the α_j are nonnegative integers), $|\alpha| = \alpha_1 + \cdots + \alpha_p$, $\alpha! = \alpha_1! \cdots \alpha_p!$, and $D_v^\alpha H(v) = \partial^{|\alpha|} H(v)/\partial v_1^{\alpha_1} \cdots \partial v_p^{\alpha_p}$. To obtain (3.12) from (3.14) it now suffices to note that for any function $\chi(x) \in C^0(R^n)$ and $\Delta u_1(x),\ldots,\Delta u_r(x) \in C_0^0(R^n)$

$$\int \chi(x)\Delta u_1(x) \cdots \Delta u_r(x)\,dx$$

$$= \int \chi(x_1)\delta(x_2 - x_1)\delta(x_3 - x_2) \cdots \delta(x_r - x_{r-1})\Delta u_1(x_1)$$

$$\cdots \Delta u_r(x_r)\,dx_1 \cdots dx_r, \qquad (3.15)$$

where $\delta(z)$ is the Dirac delta function, and the integral on the right side of (3.15) is understood in the sense of the theory of generalized functions. The convergence of the series for any $\Delta u \in (C_0^0(R^n))^p$ is obvious.

We note that in the case of parabolic equations and systems in [2] and [3] first integrals $\Phi(t, u)$ are constructed which for all $t \in R^1$ are defined and analytic in u in a sufficiently small neighborhood of zero in the corresponding function space. In [5] examples of first integrals $\Phi(t, u)$ were constructed which can be expanded in series in u that diverge for sufficiently large u. By Remark 3.2 the functional Φ defined by (3.2) (and also the functional (3.8)) is an example of a first integral which can be expanded in a series in u that converges for all $u \in (C_0^0(R^n))^p$.

3. We now are in a position to derive formulas expressing certain functionals of the solutions $u(t, x)$ of (2.1) at some positive time t in terms of functionals of u at the time $t = 0$.

THEOREM 3.3. *Suppose that the hypotheses of Theorem 3.2 are satisfied and* $u(t, x) \in DC_0^1([0, T] \times R^n)$ *is a solution of the system* (2.1) *which at* $t = 0$ *is equal to the given function*

$$u(t, x)\big|_{t=0} = u_0(x). \tag{3.16}$$

Then for any $t \in [0, T]$

$$\int \langle \omega, x \rangle^k g(u(t, x)) \, dx$$

$$= \sum_{j=0}^{k} \frac{k! \, t^j}{j! \, (k-j)!} \int \langle \omega, x \rangle^{k-j} \int_0^{u_0(x)} \langle g_v(v), \langle \omega, f(v) \rangle^j \, dv \rangle \, dx, \tag{3.17}$$

where $k = 0, 1, \ldots, N$.

PROOF. By Theorem 3.2 the functional $\Phi_k(t, u)$ defined by (3.8) is a first integral of (2.1). Moreover, at $t = 0$

$$\Phi_k(0, u) = \int \langle \omega, x \rangle^k \int_0^{u(x)} \langle g_v(v), dv \rangle \, dx = \int \langle \omega, x \rangle^k g(u(x)) \, dx.$$

Therefore, (3.17) follows immediately from Proposition 2.2.

The next result is proved in an altogether similar way.

THEOREM 3.4. *Suppose that the hypotheses of Theorem 3.1 are satisfied and* $u(t, x) \in DC_0^1([0, T] \times R^n)$ *is a solution of the Cauchy problem* (2.1), (3.16). *Then for any* $t \in [0, T]$

$$\int e^{i\langle x, \xi \rangle} g(u(t, x)) \, dx = \int e^{i\langle x, \xi \rangle} \int_0^{u_0(x)} \langle g_v(v), e^{it\langle \xi, f(v) \rangle} \, dv \rangle \, dx. \tag{3.18}$$

REMARK 3.3. The hypothesis in Theorems 3.3 and 3.4 that the solution $u(t, x)$ of the Cauchy problem (2.10), (3.16) belongs to $DC_0^1([0, T] \times R^n)$ can be relaxed somewhat. Namely, it suffices to suppose that $u(t, x)$ can be represented in the form

$$u(t, x) = u_\infty + u_1(x),$$

where $u_1(t, x) \in DC_0^1([0, T] \times R^n)$ and $u_\infty \in R^p$ is a constant vector. Of course now the condition $g(0) = 0$ must be replaced by the condition $g(u_\infty) = 0$.

REMARK 3.4. In all the foregoing arguments it was assumed that the matrices $f_j(v)$ and the function $g(v)$ are defined for all $v \in R^p$, and it was required that the integrability conditions also be satisfied for all $v \in R^p$. Instead of this it is possible to assume that $f_j(v)$ and $g(v)$ are defined for $v \in \Omega$, where $\Omega \subset R^p$ is some domain, and to require that the integrability conditions be satisfied in this domain. Of course, in this case in Theorems 3.3 and 3.4 it is necessary to require that the solution $u(t, x)$ of problem (2.1), (3.16) for all (t, x) in its domain take values in Ω.

REMARK 3.5. In Theorems 3.3 and 3.4 it was assumed that the Cauchy problem (2.1), (3.16) has a solution $u(t, x) \in DC_0^1([0, T] \times R^n)$. Let us suppose that the system (2.1) satisfies the condition of hyperbolicity in the sense of Petrovskiĭ. More precisely, we assume that for all $\omega \in S^n$ and $v \in R^p$ the matrix $\langle \omega, f(v) \rangle$ has real, distinct eigenvalues $\lambda_1(\omega, v), \ldots, \lambda_p(\omega, v)$. It then follows from the results of [1] that if the initial condition $u_0(x)$ belongs to $C_0^k(R^n)$, where k is a sufficiently large number, then problem (2.1), (3.16) has a unique solution $u(t, x) \in DC_0^1([0, T] \times R^n)$, where $T > 0$ is a sufficiently small number. In Theorems 3.3 and 3.4 it is then possible to replace the assumption of the existence of a solution of problem (2.1), (3.16) by the conditions formulated in this remark.

We now present a simple corollary of Theorem 3.4.

THEOREM 3.5.   *Suppose that for the system* (2.1) *there exist functions* $g^1(v), \ldots, g^p(v) \in C^2(R^p)$ *satisfying the following conditions:*

1) *The mapping* $G(v) = (g^1(v), \ldots, g^p(v))$ *is a diffeomorphism of* R^p *onto itself, and* $G(0) = 0$.

2) *For all* $k = 1, \ldots, p$ *the system* (2.1), *the function* g^k, *and any vector* $\omega \in S^n$ *satisfy the integrability condition to order* $N = \infty$.

Let $u(t, x) \in DC_0^1([0, T] \times R^n)$ *be a solution of the Cauchy problem* (2.1), (3.16). *Then for each* $t \in [0, T]$ *the solution* $u(t, x)$ *can be expressed in terms of the initial condition* $u_0(x)$ *by means of the operations of integrating and computing functions.*

PROOF. By Theorem 3.4, using these operations we can compute the Fourier transforms of the functions $g^k(u(t, x))$ and hence these functions themselves. Applying now the mapping G^{-1} to the functions $g^1(u(t, x)), \ldots, g^p(u(t, x))$, we obtain the function $u(t, x)$.

It is interesting to observe that by using (3.18) we can define a time T such that for $0 < t < T$ the solution $u(t, x)$ of problem (2.1), (3.16) is a smooth function. Here, in addition to the conditions of Theorem 3.5, it is necessary to require that the system (2.1) be hyperbolic in the sense of Petrovskiĭ. For simplicity we moreover assume that all the functions considered are infinitely differentiable.

Let $\lambda_k(\omega, v)$ $(k = 1, \ldots, p)$ be the eigenvalues of the matrix $\langle \omega, f(v) \rangle$. Let

$$h_k(t, \omega, y) = \langle \omega, y \rangle + t\lambda_k(\omega, u_0(y)), \tag{3.19}$$

where $\omega \in S^n$ and $u_0(y) \in C_0^\infty(R^n)$. We define the numbers t_k as the suprema of those $t > 0$ such that $\mathrm{grad}_y\, h_k(\tau, \omega, y) \neq 0$ for all $0 \leqslant \tau \leqslant t$, $\omega \in S^n$, $y \in R^n$:

$$t_k = \sup\{t\colon \mathrm{grad}_y\, h_k(\tau, \omega, y) \neq 0, \quad \forall 0 \leqslant \tau \leqslant t, y \in R^n, \omega \in S^n\}. \quad (3.20)$$

Let

$$t_+ = \min\{t_k, k = 1, \ldots, p\}. \tag{3.21}$$

We shall show that it is possible to take t_+ for T. Denoting the left side of (3.18) by $\hat{g}^k(t, \xi)$ and integrating by parts on x_1 on the right side of this equality, we find that

$$-i\xi_l \hat{g}^k(t, \xi) = \int e^{i\langle x, \xi \rangle} \left\langle g_v^k(u_0(x)), e^{it\langle \xi, f(u_0(x)) \rangle} \frac{\partial u_0(x)}{\partial x} \right\rangle dx. \tag{3.22}$$

LEMMA 3.1. *Suppose that for all* $v \in R^p$ *and* $\omega \in S^n$ *the eigenvalues of the matrix* $\langle \omega, f(v) \rangle$ *are real and distinct. Suppose that the functions* $g^k(v) \in C^\infty(R^p)$*, the matrices* $f_j(v)$ *are infinitely differentiable with respect to* v*, and* $u_0(x) \in (C_0^\infty(R^n))^p$*. Then for any* $t \in (0, t_+)$*, where* t_+ *is defined by* (3.20) *and* (3.21)*, the function* $\hat{g}^k(t, \xi)$ *defined by* (3.22) *is the Fourier transform of an infinitely differentiable function.*

PROOF. The matrix $\langle \omega, f(v) \rangle$ can be brought to diagonal form by means of an invertible transformation $T(\omega, v)$ which is infinitely differentiable with respect to all variables and for which the inverse transformation $T^{-1}(\omega, v) = S(\omega, v)$ is also infinitely differentiable with respect to all variables. Here

$$\langle \omega, f(v) \rangle = T(\omega, v) \Lambda(\omega, v) T^{-1}(\omega, v), \tag{3.23}$$

where $\Lambda(\omega, v)$ is a diagonal matrix having as diagonal elements $\lambda_k(\omega, v)$ the eigenvalues of the matrix $\langle \omega, f(v) \rangle$. Denoting the elements of the matrices $T(\omega, v)$ and $S(\omega, v)$ by $T_{ij}(\omega, v)$ and $S_{ij}(\omega, v)$ $(i, j = 1, \ldots, p)$ and using (3.23), we rewrite (3.22) as follows:

$$-i\xi_l \hat{g}^k(t, \xi)$$
$$= \sum_{r, j, m = 1}^{p} \int g_r^k(u_0(x)) T_{rl}(\omega, u_0(x)) e^{i|\xi| h_l(t, \omega, x)} S_{lm}(\omega, u_0(x)) \frac{\partial u_{0,m}(x)}{\partial x^l} dx,$$

where $g_r^k(v) = \partial g^k(v)/\partial v_r$, the functions $h_l(t, \omega, x)$ are defined by (3.19), and $u_{0,m}(x)$ is the mth coordinate of the vector $u_0(x)$. If $t \in (0, t_+)$, then by (3.20) and (3.21) for each $l = 1, \ldots, p$ we have $\mathrm{grad}_x\, h_l(t, \omega, x) \neq 0$ for any $\omega \in S^n$ and $x \in R^n$. Therefore, as is well known [6], the right side of (3.24) decays as $|\xi| \to \infty$ faster than any negative power of $|\xi|$. This means that $\hat{g}^k(t, \xi)$ is the Fourier transform of an infinitely differentiable function.

REMARK 3.6. It is also easy to show that if the hypotheses of Lemma 3.1 are satisfied, then for any t the function $\hat{g}^k(t, \xi)$ defined by (3.22) is the Fourier transform of a compactly supported function. Indeed, it is clear that $\hat{g}^k(t, \xi)$ is an entire function of the variable $\xi \in \mathbf{C}^n$. Moreover, using the hyperbolicity of the matrix $\langle \xi, f(v) \rangle$, we can show that

$$|-i\xi \hat{g}_k(t, \xi)| \leqslant C e^{M|\operatorname{Im} \xi|}, \tag{3.24}$$

where the constants $C > 0$ and $M > 0$ do not depend on $\xi \in \mathbf{C}^n$. The assertion of Remark 3.6 now follows from (3.23) by the Paley-Wiener theorem.

From Lemma 3.1 and Remark 3.6 it follows that the functions

$$g^k(t, x) = \frac{1}{(2\pi)^n} \int e^{-i\langle x, \xi \rangle} \hat{g}^k(t, \xi) \, d\xi$$

for any $t \in (0, t_+)$ belong to $C_0^\infty(R^n)$. Hence, for these t the function $u(t, x) = G^{-1}(g^1(t, x), \dots, g^p(t, x))$ also belongs to $(C_0^\infty(R^n))^p$. Using this fact, we can prove by the methods of [1] the existence of a smooth solution of the Cauchy problem (2.1), (3.16) for $t \in (0, t_+)$, where t_+ is defined by (3.20) and (3.21), if it is assumed that the hypotheses of Theorem 3.5 and Lemma 3.1 are satisfied.

We observe that the uniqueness of the solution to the Cauchy problem (2.1), (3.16) follows directly from Theorem 3.5. Indeed, by (3.18) the Fourier transform of $g^k(u(t, x))$ is uniquely determined by the initial condition $u_0(x)$. Therefore, $g^k(u(t, x))$ and hence also the solution $u(t, x)$ are uniquely determined.

§4. Examples of systems for which the integrability conditions are satisfied

1. We shall present examples of systems for which the integrability conditions are satisfied. Suppose that (2.1) is not a system but a scalar equation, i.e., $f_j(v)$ are not matrices depending on $v \in R^p$ but rather scalar functions depending on $v \in R^1$. Then for any $g(v)$, k, and $\omega \in S^n$ the forms (3.1) have the form $h(v)\,dv$, where $h(v)$ is some function of $v \in R^1$ and, of course, they are total differentials of some function. Therefore, in the case of a scalar equation the integrability conditions to any order N are satisfied for any $g(v)$ and $\omega \in S^n$.

Suppose that the system (2.1) is linear, i.e., the matrices f_j do not depend on $v = (v_1, \dots, v_p)$. Let $g(v) = v_k$. Then for any $\omega \in S^n$ and k the forms (3.1) have the form $c_1 dv_1 + \cdots + c_p dv_p$, where $c_1, \dots, c_p$ are constants, and they are therefore total differentials. The Fourier transform of the coordinates $u_j(t, x)$ of the solution $u(t, x)$ of the linear system can then be expressed in terms of the initial condition $u(0, x)$ by means of simple formulas (as is well known).

2. We now consider a few examples of one-dimensional systems, i.e., systems of the form

$$\partial u(t, x)/\partial t + f(u(t, x))\partial u(t, x)/\partial x = 0, \qquad x \in R^1, \tag{4.1}$$

where $f(u)$ is a $p \times p$ matrix and $u(t, x)$ is a p-dimensional vector.

Since the integrability condition is or is not satisfied simultaneously for the vectors ω and $-\omega$, in the one-dimensional case the vector ω does not enter the integrability conditions.

In the case of integrability to order $N = 0$ the integrability condition coincides with the well-known condition that the system (4.1) have conservation laws ([7], p. 36).

Suppose that (4.1) consists of two equations ($p = 2$). For the function $g(v) = g(v_1, v_2)$ the integrability condition to order $N = 0$ is then satisfied if the differential form

$$\left(f_{11}(v)\frac{\partial g}{\partial v_1} + f_{21}(v)\frac{\partial g}{\partial v_2} \right) dv_1 + \left(f_{22}(v)\frac{\partial g}{\partial v_1} + f_{22}(v)\frac{\partial g}{\partial v_2} \right) dv_2 \qquad (4.2)$$

is exact, where $f_{i,j}(v)$ are the elements of the matrix $f(v)$. The form (4.2) is exact if $g(v)$ satisfies the equation

$$\frac{\partial}{\partial v_2}\left(f_{11}\frac{\partial g}{\partial v_1} + f_{21}\frac{\partial g}{\partial v_2} \right) = \frac{\partial}{\partial v_1}\left(f_{12}\frac{\partial g}{\partial v_1} + f_{22}\frac{\partial g}{\partial v_2} \right). \qquad (4.3)$$

In general, (4.3) has an infinite set of solutions, and therefore the system (4.1) consisting of two equations has an infinite set of conservation laws.

In the case $p \geqslant 3$, in general, there exist no functions $g(v) \not\equiv$ const for which the integrability condition is satisfied even to order $N = 0$; however there are such functions for a number of important systems (for example, for the system of equations of gas dynamics).

3. For the system

$$\frac{\partial u_1}{\partial t} + u_2\frac{\partial u_1}{\partial x} = 0, \qquad \frac{\partial u_2}{\partial t} + u_1\frac{\partial u_2}{\partial x} = 0,$$

we shall construct a function $g(v) = g(v_1, v_2)$ which satisfies the integrability condition to order $N = 1$. For this it is necessary to find a function satisfying the equations

$$(v_2 - v_1)\frac{\partial^2 g}{\partial v_1 \partial v_2} + \frac{\partial g}{\partial v_1} - \frac{\partial g}{\partial v_2} = 0,$$

$$(v_2^2 - v_1^2)\frac{\partial^2 g}{\partial v_1 \partial v_2} + 2v_2\frac{\partial g}{\partial v_1} - 2v_1\frac{\partial g}{\partial v_2} = 0. \qquad (4.4)$$

Multiplying the first of equations (4.4) by $v_2 + v_1$ and subtracting the result from the second equation, we find that $g(v)$ must satisfy the equation $\partial g/\partial v_1 + \partial g/\partial v_2 = 0$, and hence $g(v_1, v_2) = g(v_1 - v_2)$. Substituting this function into (4.4), we find that it must satisfy the equation $\alpha g''(\alpha) + 2g'(\alpha) = 0$, and hence $g(\alpha) = c_1 + c_2\alpha^{-1}$, where c_1 and c_2 are arbitrary constants. Since $g(\alpha)$ has a singularity at $\alpha = 0$, by Remarks 3.3 and 3.4 we can evaluate the integral $\int x^k g(u(t, x))\, dx$ ($k = 0, 1$) on the class of functions $\{u(x)\} = \{u_1(x), u_2(x)\}$ which for sufficiently large $|x|$ become a fixed vector $u_\infty \in R^2 \setminus \{0\}$ and such that $u_1(x) \neq u_2(x)$ for each x. We must choose c_1 and c_2 in $g(\alpha)$ so that $g(u_1(\infty) - u_2(\infty)) = 0$.

4. We shall present examples of one-dimensional systems (4.1) for which there exist functions $g(v)$ satisfying the integrability condition for finite N and also for $N = \infty$. These examples actually provide a description of integrable systems.

Suppose that for the system (4.1) and the function $g(v) = g_1(v)$ the integrability conditions are satisfied to some order $N > 0$. It is easy to see that this implies that the vector fields

$$f^*(v)\nabla g_1(v), (f^2(v))^*\nabla g_1(v), \ldots, (f^{N+1}(v))^*\nabla g_1(v), \tag{4.5}$$

where $\nabla g_1(v) = (\partial g_1/\partial v_1, \ldots, \partial g_1/\partial v_p)$, are gradients of some functions $g_2(v), \ldots, g_{N+2}(v)$:

$$f^*(v)\nabla g_1(v) = \nabla g_2(v),$$
$$(f^2(v))^*\nabla g_1(v) = \nabla g_3(v), \ldots, (f^{N+1}(v))^*\nabla g_1(v) = \nabla g_{N+2}(v). \tag{4.6}$$

The equalities (4.6) can be rewritten as follows:

$$\nabla g_2(v) = f^*(v)\nabla g_1(v),$$
$$\nabla g_3(v) = f^*(v)\nabla g_2(v), \ldots, \nabla g_{N+2}(v) = f^*(v)\nabla g_{N+1}(v), \tag{4.7}$$

where $f^*(v)\nabla g_{N+2}(v)$ is no longer a gradient vector field.

Let $N + 2 \leqslant p$, where p is the number of equations in (4.1). We suppose that the vectors $\nabla g_1(v), \ldots, \nabla g_{N+2}(v)$ are linearly independent for any $v \in R^p$. To the vector fields $\nabla g_1(v), \ldots, \nabla g_{N+2}(v)$ we add vector fields $\nabla g_{N+3}(v), \ldots, \nabla g_p(v)$ so that $g(v) = (g_1(v), \ldots, g_p(v))$ is a smooth, one-to-one mapping of R^p onto R^p. In (4.1) we go over from the functions $u(t, x)$ to the functions $g(t, x) = g(u(t, x))$.

It is then easy to see that $g(t, x)$ satisfies the system of equations

$$\frac{\partial g(t, x)}{\partial t} + A(g(t, x))\frac{\partial g(t, x)}{\partial x} = 0.$$

Here $A(g) = (Dg/Du)f(u)(Du/Dg)$, where Dg/Du is the Jacobian of the mapping $g(u)$, and $(Du/Dg) = (Dg/Du)^{-1}$ is the Jacobian of the mapping $u(g) = g^{-1}(g)$. We consider the matrix

$$A^*(g) = (Du/Dg)^*f^*(u)(Dg/Du)^*.$$

By (4.7) the first $N + 1$ columns of the matrix $f^*(u)(Dg/Du)^*$ are equal respectively to $\nabla g_2(v), \ldots, \nabla g_{N+2}(v)$. Therefore, noting that $(Du/Dg)^* = ((Dg/Du)^*)^{-1}$, we can easily see that the first $N + 1$ rows of the matrix $A(g)$ have the following structure:

$$\begin{pmatrix} 0, & 1, & 0, & . & . & . & 0 \\ 0, & 0, & 1, & . & . & . & 0 \\ \cdots & \cdots & \cdots & \cdots & \cdots & \cdots & \cdots \\ 0, & \cdots & 0, & 1, & 0, & \cdots & 0 \end{pmatrix},$$

i.e., in the kth row ($k \leqslant N + 1$) there are zeros everywhere except at the $(k + 1)$st place, and at the $(k + 1)$st place there is a one. We have thus proved the following result.

PROPOSITION 4.1. *Suppose that for the system* (4.1) *and the function* $g_1(v)$ *the integrability condition is satisfied to order* N *with* $N + 2 \leqslant p$, *where* p *is the number of equations in* (4.1). *Suppose for any* $v \in R^p$ *the vectors* (4.5) *are linearly independent. Then there exists a change* $g(u)$ *of dependent variables* $u(t, x)$, $g(t, x) = g(u(t, x))$, *transforming* (4.1) *into a system of equations in which the first* $N + 1$ *equations have the form*

$$\frac{\partial g_1(t, x)}{\partial t} + \frac{\partial g_2(t, x)}{\partial x} = 0, \dots, \frac{\partial g_{N+1}(t, x)}{\partial t} + \frac{\partial g_{N+2}(t, x)}{\partial x} = 0.$$

We note that the arguments presented at the beginning of this subsection make it possible also to construct examples of systems for which there exists a function $g(v)$ satisfying the integrability conditions to order $N = \infty$. Indeed, suppose that (4.1) and the function $g_1(v)$ are such that (4.6) holds with $N + 2 \leqslant p$. We assume that one of the following two conditions is satisfied:

a) $f^*(v) \nabla g_{N+2}(v) = \sum_{j=1}^{N+2} c_j \nabla g_j(v)$, c_j constants,

b) $f^*(v) \nabla g_{N+2}(v) = h(g_{N+2}(v)) \nabla g_{N+2}(v)$, $h(g) \in C^0(R^1)$.

In this case it is easy to see that each of the functions $g_j(v)$ $(j = 1, \dots, N + 2)$ satisfies the integrability condition to order $N = \infty$.

5. We now consider the system of equations of gas dynamics [8]

$$\partial \rho(t, x) / \partial t + \operatorname{div}(\rho u(t, x)) = 0, \tag{4.8}$$

$$\partial u(t, x) / \partial t + \langle u, \nabla \rangle u + \rho^{-1} \operatorname{div} P = 0, \tag{4.9}$$

where $\rho(t, x)$ is the density of the gas, $u(t, x) = (u_1(t, x), \dots, u_n(t, x))$ is the velocity vector of the gas at the point $x = (x_1, \dots, x_n)$ (the number n in (4.8) and (4.9) is usually equal to 1, 2, or 3), $\langle u, \nabla \rangle u$ is the n-dimensional vector with jth coordinate $(\langle u, \nabla \rangle u)_j$ given by

$$(\langle u, \nabla \rangle u)_j = \sum_{i=1}^{n} u_i \frac{\partial u_j}{\partial x_i}; \qquad \operatorname{div}(\rho u) = \sum_{i=1}^{n} \frac{\partial \rho u_i}{\partial x_i},$$

$P = \| P_{ij}(t, x) \|$ $(i, j = 1, \dots, n)$ is the tensor of internal stresses, and $\operatorname{div} P = \sum_i \partial P_{ij}(t, x) / \partial x_i$. Below we shall sometimes find it convenient to respect P as a collection of vectors $(P_1, \dots, P_n)$, where $P_i = (P_{i1}, \dots, P_{in})$.

The system (4.8), (4.9) is underdetermined, since it consists of $n + 1$ equations and contains $n^2 + n + 1$ unknown functions. In order that it be determined, additional conditions are usually imposed on the tensor P. We shall assume that P is a given function of ρ and u. We shall show that for any given function $P(\rho, u)$ the system (4.8), (4.9) satisfies an integrability condition with $g(\rho, u) = \rho - \rho_\infty$ and $N = 1$, where we assume that, for sufficiently large $|x|$, $\rho(t, x) = \rho_\infty$ and $u(t, x) = 0$. It is easy to write the system (4.8), (4.9) in the form (2.1) with $f_i = u_i I + F_i$ $(i = 1, \dots, n)$, where I is the identity matrix, and $F_i = \| F_{i,jl} \|$ $(j, l = 0, 1, \dots, n)$ is an $(n + 1) \times (n + 1)$ matrix having ρ in the zeroth row at the ith place and zeros at the remaining sites of the zeroth row; the element of the matrix $F_{i,jl}$ at the jth $(j = 1, \dots, n)$ row and lth $(l = 0, 1, \dots, n)$ column is equal to $\rho^{-1} \partial P_{ij} / \partial u_l$, where $\partial / \partial u_0 = \partial / \partial \rho$.

If $g(\rho, u) = \rho - \rho_\infty$, then $\nabla g(\rho, u) = (1, 0, \ldots, 0)$, and it is easy to see that

$$f_i^* \nabla(\rho - \rho_\infty) = \nabla(\rho u_i), \qquad (4.10)$$

and for any vector $\omega \in S^n$

$$\langle \omega, f^* \rangle \nabla(\rho - \rho_\infty) = \nabla(\rho \langle \omega, u \rangle). \qquad (4.11)$$

Here for any $i = 1, \ldots, n$

$$f_i^* \nabla(\rho \langle \omega, u \rangle) = \nabla(\rho u_i \langle \omega, u \rangle + P_i \omega), \qquad (4.12)$$

where $P_i \omega = \Sigma_1^n P_{ij} \omega_j$. Thus, it has been shown that the system (4.8), (4.9) satisfies the integrability condition with $g(\rho, u) = \rho - \rho_\infty$, $N = 1$, and any $\omega \in S^n$; hence the following result holds:

PROPOSITION 4.2. *If*

$$\big(\rho(t, x) - \rho_\infty, u_1(t, x), \ldots, u_n(t, x)\big) \in DC_0^1([0, T] \times R^n)$$

and $(\rho(t, x), u_1(t, x), \ldots, u_n(t, x))$ is a solution of the system (4.8), (4.9) satisfying the initial condition $(\rho(0, x), u_1(x, 0), \ldots, u_n(0, x))$, for any $\omega \in S^n$ and $t \in [0, T]$

$$\int \langle \omega, x \rangle (\rho(t, x) - \rho_\infty) \, dx$$

$$= \int (\rho(0, x) - \rho_\infty) \langle \omega, x \rangle \, dx + t \int \rho(0, x) \langle \omega, u(0, x) \rangle \, dx \qquad (4.13)$$

Formula (4.13) is a consequence of (3.17).

Using Proposition 4.2 with $\rho_\infty = 0$ and $\rho(0, x) \geqslant 0$, it is easy to find the center of gravity of the gas for positive t. Indeed, taking in (4.13) as ω the vector ω_k with the kth coordinate equal to one and all other coordinates equal to zero, we obtain a formula expressing the coordinates of the center of gravity of the gas at time t in terms of $\rho(0, x)$ and $u(0, x)$:

$$\int x_k \rho(t, x) \, dx \Big/ \int \rho(t, x) \, dx, \qquad k = 1, \ldots, n.$$

(In order that the system (4.8), (4.9) be meaningful for $\rho_\infty = 0$, it is necessary to go over from the functions ρ and u to ρ and ρu.)

We shall now show that, under additional assumptions regarding the function of state $P(\rho, u)$ expressing the stress tensor in terms of the density ρ and the velocity u, it is possible to obtain more complete information regarding the solution $(\rho(t, x), u(t, x))$ of the system (4.8), (4.9).

PROPOSITION 4.3. *Suppose that the function of state $P(\rho, u)$ is such that for some $\eta \in S^n$*

$$\langle P\eta, \eta \rangle = \sum_{i,j=1}^n P_{ij} \eta_i \eta_j = -\rho \langle \eta, u \rangle^2 + A\rho \langle \eta, u \rangle + B\rho + C, \qquad (4.14)$$

where A, B, and C are constants. Then for the system (4.8), (4.9) an integrability condition to order $N = \infty$ is valid with $g(\rho, u) = \rho - \rho_\infty$ and $\omega = \eta$. Therefore, if

$$\big(\rho(t, x) - \rho_\infty, u_1(t, x), \ldots, u_n(t, x)\big) \in DC_0^1([0, T] \times R^n)$$

and $(\rho(t, x), u_1(t, x), \ldots, u_n(t, x))$ *is a solution of the system* (4.8), (4.9) *satisfying the initial condition* $(\rho(0, x), u_1(0, x), \ldots, u_n(0, x))$, *then for* $t \in [0, T]$ *and* $\xi = |\xi| \eta$ $(|\xi| \in R^1)$

$$\int e^{i\langle x, \xi\rangle}(\rho(t, x) - \rho_\infty)\, dx = \frac{\alpha_2 e^{it|\xi|\alpha_1} - \alpha_1 e^{it|\xi|\alpha_2}}{\alpha_2 - \alpha_1} \int e^{i\langle x, \xi\rangle}(\rho(0, x) - \rho_\infty)\, dx$$

$$+ \frac{e^{it|\xi|\alpha_2} - e^{it|\xi|\alpha_1}}{\alpha_2 - \alpha_1} \int e^{i\langle x, \xi\rangle}\langle \eta, u(0, x)\rangle \rho(0, x)\, dx, \qquad (4.15)$$

where α_1 *and* α_2 *are the roots of the equation* $\alpha^2 - A\alpha - B = 0$. *It is assumed that* $\alpha_1 \neq \alpha_2$.

PROOF. Multiplying both sides of (4.12) by ω_i and summing on i, we obtain

$$\langle \omega, f^*\rangle \nabla(\rho\langle \omega, u\rangle) = \nabla(\rho\langle \omega, u\rangle^2 + \langle P\omega, \omega\rangle). \qquad (4.16)$$

Setting $\omega = \eta$ in (4.16) and recalling (4.14), we obtain

$$\langle \eta, f^*\rangle \nabla(\rho\langle \eta, u\rangle) = A\nabla(\rho\langle \eta, u\rangle) + B\nabla(\rho - \rho_\infty). \qquad (4.17)$$

It follows from (4.10)–(4.12) and (4.17) that the integrability condition for $g(\rho, u) = \rho - \rho_\infty$ and $\omega = \eta$ is satisfied to order $N = \infty$.

We now prove (4.15). We first compute the vector-valued function

$$H(\tau) = e^{\tau\langle \eta, f^*\rangle}\nabla(\rho - \rho_\infty). \qquad (4.18)$$

By (4.11) and (4.17) the function $H(\tau)$ satisfies the differential equation

$$(d^2/d\tau^2)H(\tau) = A(d/d\tau)H(\tau) + BH(\tau)$$

and the initial conditions

$$H(0) = \nabla(\rho - \rho_\infty), \qquad (d/d\tau)H(\tau)|_{\tau=0} = \rho\langle \eta, u\rangle.$$

Therefore,

$$H(\tau) = \left(\frac{\alpha_2}{\alpha_2 - \alpha_1}e^{\alpha_1\tau} - \frac{\alpha_1}{\alpha_2 - \alpha_1}e^{\alpha_2\tau}\right)\nabla(\rho - \rho_\infty)$$

$$+ \frac{e^{\alpha_2\tau} - e^{\alpha_1\tau}}{\alpha_2 - \alpha_1}\nabla(\rho\langle \eta, u\rangle), \qquad (4.19)$$

where α_1 and α_2 are the roots of the equation $\alpha^2 - A\alpha - B = 0$. (Here we assume that $\alpha_1 \neq \alpha_2$; of course, for $\alpha_1 = \alpha_2$ it is also easy to derive a formula analogous to (4.19).) Equality (4.15) follows from (4.18) and (4.19).

In conclusion we consider the system of equations of gas dynamics (4.8), (4.9) in the case where $n = 1$:

$$\partial\rho/\partial t + u\partial\rho/\partial x + \rho\partial u/\partial x = 0,$$
$$\partial u/\partial t + u\partial u/\partial x + \rho^{-1}\partial P/\partial x = 0. \qquad (4.20)$$

Along with (4.20) we also consider the equation obtained from the law of conservation of energy:

$$\partial\varepsilon/\partial t + P\rho^{-1}\partial u/\partial x + u\partial\varepsilon/\partial x = 0, \qquad (4.21)$$

where ε is the internal energy of the gas. We assume that the pressure P is a function of ρ and ε: $P = P(\rho, \xi)$.

PROPOSITION 4.4. *Suppose that in the system* (4.20), (4.21) *the pressure* $P(\rho, \varepsilon) = 2\rho\varepsilon$. *An integrability condition to order* $N = 2$ *with* $g(\rho, u, \varepsilon) = \rho - \rho_\infty$ *is then valid for the system* (4.20), (4.21). *Therefore, if*

$$(\rho(t, x) - \rho_\infty, u(t, x), \varepsilon(t, x)) \in D_0^1([0, T] \times R^n)$$

and $(\rho(t, x), u(t, x), \varepsilon(t, x))$ *is a solution of the system* (4.20), (4.21) *satisfying the initial condition* $(\rho(0, x), u(0, x), \varepsilon(0, x))$, *then for* $t \in [0, T]$

$$\int x^2(\rho(t, x) - \rho_\infty)\, dx = \int x^2(\rho(0, x) - \rho_\infty)\,dx + 2t \int x\rho(0, x)u(0, x)\, dx$$

$$+ 4t^2 \int \left(\rho(0, x)\varepsilon(0, x) + \frac{u^2(0, x)}{2} \right) dx.$$

To prove this it suffices to verify the integrability condition and use (3.17).

For the system (4.20), (4.21) with $\rho_\infty = 0$ and $\rho(0, x) \geqslant 0$,[1] of course a formula defining the center of gravity holds.

$$M(t) = \int x\rho(t, x)\, dx \Big/ \int \rho(t, x)\, dx$$

$$= \left\{ t \int \rho(0, x)u(0, x)\, dx + \int x\rho(0, x)\, dx \right\} \Big/ \int \rho(0, x)\, dx.$$

Using Proposition 4.4, it is easy to compute the quantity

$$D(t) = \int (x - M(t))^2 \rho(t, x)\, dx,$$

which gives information on the degree of concentration of the gas about its center of gravity if $\rho(t, x) \geqslant 0$.

Finally, we consider the system (4.20) assuming that the pressure P is a function of ρ.

PROPOSITION 4.5. *Suppose that in the system* (4.20) *the pressure is* $P(\rho) = c\rho^3$, *where c is an arbitrary constant. Then an integrability condition holds for the system* (4.20) *to order* $N = \infty$, *with* $g(\rho, u) = \rho$ *and with* $g(\rho, u) = u$.

PROOF. The conditions for the integrability of the system (4.20) can be verified directly. To compute $(f^k)^*$ it is convenient to first reduce the matrix

$$f = \begin{pmatrix} u, & \rho \\ c\rho, & u \end{pmatrix}$$

to diagonal form.

---

[1] More precisely, for the corresponding system of equations written in the form of conservation laws.

By Proposition 4.5 it is possible to express the Fourier transform $(\hat{\rho}(t, \xi), \hat{u}(t, \xi))$ of the solution $(\rho(t, x), u(t, x))$ of the system (4.20) in terms of the initial values $(\rho(0, x), u(0, x))$. Here $0 < t < t_+$, where t_+ is defined by a formula of the form (3.21). Using the arguments presented in Remark 1.2, it is also possible to obtain formulas of the form (1.17), (1.18) expressing the solution $(\rho(t, x), u(t, x))$ itself in terms of the initial condition $(\rho(0, x), u(0, x))$.

Received 17/FEB/75

BIBLIOGRAPHY

1. I. Petrowsky [I. G. Petrovskiĭ], *Über das Cauchysche Problem für Systeme von partiellen Differentialgleichungen*, Mat. Sb. **2(44)** (1937), 815–870.

2. M. I. Višik and A. V. Fursikov, *Analytic first integrals of nonlinear parabolic equations, and their applications*, Mat. Sb. **92(134)** (1973), 347–377; English transl. in Math. USSR Sb. **21** (1973).

3. M. I. Višik and A. V. Fursikov, *Analytic first integrals of nonlinear systems of differential equations that are parabolic in the sense of I. G. Petrovskiĭ, and their applications*, Uspehi Mat. Nauk **29** (1974), no. 2(176), 123–153; English transl. in Russian Math. Surveys **29** (1974).

4. A. V. Fursikov, *Formulas for certain functionals of the smooth solutions of a class of systems of quasilinear equations*, Uspehi Mat. Nauk **31** (1976), no. 1(187), 265–266. (Russian)

5. M. I. Višik and A. V. Fursikov, *Asymptotic expansion of moment functions of solutions of nonlinear parabolic equations*, Mat. Sb. **95(137)** (1974), 588–605; English transl. in Math. USSR Sb. **24** (1974).

6. Lars Hörmander, *Fourier integral operators*. I, Acta Math. **127** (1971), 79–183.

7. B. L. Roždestvenskiĭ and N. N. Janenko, *Systems of quasilinear equations and their applications to the dynamics of gases*, "Nauka", Moscow, 1968. (Russian)

8. L. G. Loĭcjanskiĭ, *The mechanics of fluids and gases*, 4th ed., "Nauka", Moscow, 1973; English transl. of 2nd ed., Pergamon Press, 1966.

Translated by J. R. SCHULENBERGER

The Density of States
of Selfadjoint Elliptic Operators
with Almost Periodic Coefficients*

M. A. ŠUBIN

Contents

Introduction

The density of states and several other spectral invariants of selfadjoint elliptic operators with almost periodic coefficients are studied in the paper.

From a physical point of view the density of states is a function of the energy E, equal to the number of possible energy levels in the interval $(E, E + \Delta E)$ divided by ΔE and by the volume of the body considered. The volume is here considered sufficiently large and the length ΔE of the interval of energies sufficiently small. The density of states is extensively used by physicists in the quantum theory of solids [1], [2] and noncrystalline substances [3], [4]. Further, while in the theory of solids they deal with periodic structures, in the theory of noncrystalline substances it is necessary to consider various violations of periodicity. These violations may have random character, and they also may be random perturbations of the frequencies. The first leads to the necessity of studying a Schrödinger operator with a random potential, while the second implies the

---

1980 *Mathematics Subject Classification.* Primary 35J40; Secondary 35P99.
* Translation of Trudy Sem. Petrovsk. **3** (1978), 243–275. MR **58** #17587.

occurrence of an almost periodic potential which, from the viewpoint of certain physicists [5], can be utilized in the quantum theory of liquids and alloys.

From the mathematical point of view this physical definition of the density of states contains two limit processes: it is necessary to let the volume tend to $+\infty$ and ΔE to 0. They are both in need of justification; however, the first of them has more content and is the more important. We restrict our consideration only to this limit process; it already affords the possibility of considering the density of states as a Borel measure on the line, which we shall identify with its distribution function. The second limit process then reduces to taking the density (or derivative) of this measure with respect to Lebesgue measure.

We note that there are a number of mathematical works devoted to the Schrödinger operator with a random potential and containing a proof of the existence of a limit of the distribution function as the volume expands (see [6] and the references given there). In [6], Chapter III, the technique is expounded which is used in these works to prove "self-averageability", as the nonrandom limit indicated above is called in [6]. This technique consists in making a relatively elementary estimate of the distribution function, which affords the possibility of applying Helly's theorem of choice, and then proving the existence of a limit for the Laplace transform of the distribution function. The significant feature here is the second step, which uses an explicit representation of the Laplace transform in terms of the Wiener integral.

We shall use the term "averageability" to denote the existence of the limit over the volume (this is shorter than the term "self-averageability" used in [6]).

In this paper the averageability of the density of states and a number of other important quantities (in particular, $A_V(\lambda, k)$ and $S_V^{\alpha\beta}(\lambda, \mu)$ of [6]) is proved for arbitrary selfadjoint elliptic operators (of higher order) with smooth, almost periodic coefficients and general elliptic boundary conditions which satisfy a condition of ellipticity with a parameter [7] uniformly in volume. Here, the limit, of course, does not depend on the boundary conditions or on the sequence of expanding domains (for sufficiently regular expansion, for example, homothetic expansion). Just as in [6], we go over to the Laplace transform, but completely different ideas are used to study it (in principle the technique of [6] does not work for equations of higher order). Namely, by the method of the Levi parametrix [8] we construct a fundamental solution of the corresponding parabolic Cauchy problem, we establish its almost periodicity in directions parallel to the diagonal, and we then estimate the difference between this fundamental solution and the Green function for the boundary value problem by using results of Èĭdel'man and Ivasišen [9].

Further, in this paper we show that it is possible to give a definition of the density of states not containing a limit process over the volume. This is accomplished by means of the concept of the relative trace in a certain Murray-von Neumann II_∞-factor. This definition was given by the author previously in [10] and [11], in which, by the way, neither the definition of the present paper nor even the term "density of states" is mentioned. The definition in terms of the relative trace is very convenient and makes it possible to indicate directly the connection between the density of states and the spectrum of the operator in the whole space (it coincides with the set of points of increase of the distribution function) and also to prove in a relatively simple way an asymptotic formula for the distribution function for large energies (see the formulation for the general case in [10] and the proof for the Schrödinger operator with a precise remainder estimate in [12]). We note here that the role of the concept of the relative trace in a II_∞-factor in the theory of operators with almost periodic coefficients was first observed in connection with the theory of the index (see [13] and [14]).

Finally, in this paper we write out explicit formulas for the density of states in the following two cases: for operators with constant coefficients and for the one-dimensional Schrödinger operator with a periodic potential (in the latter case the density of states can be expressed in terms of the trace of the monodromy matrix). Apparently, both these formulas are known, but for the sake of completeness they are presented here with proofs, because the author was not able to find them in the mathematical literature.

I wish to thank F. A. Berezin, who conjectured the assertion on the connection of the distribution function introduced by the author in [10] and [11] and the density of states; and also L. A. Bagirov for many helpful discussions.

§1. The fundamental solution of the Cauchy problem

Let A be a differential operator of the form

$$A = \sum_{|\alpha| \leqslant m} a_\alpha(x) D^\alpha, \tag{1.1}$$

where $x \in \mathbf{R}^n$, $\alpha = (\alpha_1, \dots, \alpha_n)$, the $\alpha_j \geqslant 0$ are integers, $D^\alpha = D_1^{\alpha_1} \cdots D_n^{\alpha_n}$ and $D_j = i^{-1}\partial/\partial x_j$. We denote by $C_b^\infty(\mathbf{R}^n)$ the space of functions $f(x) \in C^\infty(\mathbf{R}^n)$ such that

$$\sup_{x \in \mathbf{R}^n} |D^\beta f(x)| < +\infty \tag{1.2}$$

for any multi-index β. We henceforth assume that $a_\alpha(x) \in C_b^\infty(\mathbf{R}^n)$ for all α with $|\alpha| \leqslant m$.

We consider the Cauchy problem

$$\partial u/\partial t = -Au, \tag{1.3}$$

$$u|_{t=0} = \varphi(x), \tag{1.4}$$

where $u = u(t, x)$, $t > 0$ and $x \in \mathbf{R}^n$. Defining the principal symbol $a_m(x, \xi)$ of the operator A by the formula

$$a_m(x, \xi) = \sum_{|\alpha|=m} a_\alpha(x)\xi^\alpha, \tag{1.5}$$

we write out the condition for uniform parabolicity of this problem: there exists an $\varepsilon > 0$ such that

$$\operatorname{Re} a_m(x, \xi) \geqslant \varepsilon |\xi|^m. \tag{1.6}$$

A fundamental solution of the Cauchy problem (1.3), (1.4) is a function $\Gamma(t, x, y) \in C^\infty(\mathbf{R}_+ \times \mathbf{R}^n \times \mathbf{R}^n)$, where $\mathbf{R}_+ = \{t: t > 0\}$, such that for any function $\varphi(x) \in C_0^\infty(\mathbf{R}^n)$ the integral

$$u(t, x) = \int_{\mathbf{R}^n} \Gamma(t, x, y)\varphi(y) \, dy \tag{1.7}$$

gives a solution of the Cauchy problem (1.3), (1.4) which is bounded on $[0, T] \times \mathbf{R}^n$ for any finite $T > 0$.

We summarize in a form we need a well-known result on the existence, uniqueness, and behavior of the fundamental solution ([8], Chapter IX).

THEOREM 1.1. *Suppose that A is an operator of the form* (1.1), $a_\alpha(x) \in C_b^\infty(\mathbf{R}^n)$, $|\alpha| \leq m$, *and the condition of uniform parabolicity* (1.6) *is satisfied. Then a fundamental solution of the Cauchy problem* (1.3), (1.4) *exists, is unique, and satisfies the estimate*

$$\left| D_x^\alpha D_y^\beta D_t^\gamma \Gamma(t, x, y) \right| \leq C' t^{-(n+|\alpha|+|\beta|+m\gamma)/m} \exp\left\{ -C\left(\frac{|x - y|^m}{t} \right)^{1/(m-1)} \right\},$$

$$\tag{1.8}$$

where $t \in (0, T]$, $x, y \in \mathbf{R}^n$, α *and* β *are multi-indices,* $\gamma \geq 0$ *is an integer, and C and C′ are positive constants depending on* α, β, γ *and T.*

The object of this section is to investigate the effect of almost periodicity of the coefficients $a_\alpha(x)$ on the fundamental solution $\Gamma(t, x, y)$. We denote by $\mathrm{CAP}(\mathbf{R}^n)$ the space of all uniform, almost periodic (a.p.) functions on $\mathbf{R}^n$ (see [15]), which is a Banach algebra with norm

$$\|f\| = \sup_{x \in \mathbf{R}^n} |f(x)|. \tag{1.9}$$

We shall also make use of the space $\mathrm{CAP}^\infty(\mathbf{R}^n)$ consisting of functions $f(x) \in C^\infty(\mathbf{R}^n)$ such that $D^\alpha f(x) \in \mathrm{CAP}(\mathbf{R}^n)$ for any multi-index α. We note here that

$$\mathrm{CAP}^\infty(\mathbf{R}^n) = \mathrm{CAP}(\mathbf{R}^n) \cap C_b^\infty(\mathbf{R}^n). \tag{1.10}$$

THEOREM 1.2. *Let A be an operator satisfying the hypotheses of Theorem* 1.1, *and suppose further that* $a_\alpha(x) \in \mathrm{CAP}(\mathbf{R}^n)$ *for all* α *with* $\lceil \alpha \rceil \leq m$. *Then the fundamental solution* $\Gamma(t, x, y)$ *of the Cauchy problem* (1.3), (1.4) *satisfies the following condition: for any multi-indices* α *and* β *and any integer* $\gamma \geq 0$

$$D_x^\alpha D_y^\beta D_t^\gamma \Gamma(t, x + z, y + z) \in \mathrm{CAP}(\mathbf{R}_z^n) \tag{1.11}$$

and, moreover, the left side of (1.11) *defines a continuous function of t, x and y on* $\mathbf{R}_+ \times \mathbf{R}^n \times \mathbf{R}^n$ *with values in* $\mathrm{CAP}(\mathbf{R}_z^n)$ (*continuity is here understood as the continuity of a vector-valued function of t, x and y with values in the space* $\mathrm{CAP}(\mathbf{R}^n)$ *equipped with the norm* (1.9)).

REMARK. We have restricted ourselves in the formulation and in the proof to the simplest version of Theorem 1.2. However, by analogous arguments it is possible to obtain a generalization of Theorem 1.2 to the case where instead of (1.3) we consider the system

$$\partial u/\partial t = -A(t, x, D_x)u, \qquad t \in [0, T], x \in \mathbf{R}^n,$$

which is uniformly parabolic in the Petrovskiĭ sense and such that all derivatives of the coefficients are continuous functions of t with values in $\mathrm{CAP}(\mathbf{R}^n_x)$. In this case the fundamental solution is a matrix-valued function $\Gamma(t, x, \tau, y)$ defined for $0 \leqslant \tau < t \leqslant T$ which satisfies estimates analogous to (1.8) and is such that $D_x^\alpha D_y^\beta D_t^\gamma D_\tau^\delta \Gamma(t, x + z, \tau, y + z)$ is a continuous function of t, x, τ and y with values in $\mathrm{CAP}(\mathbf{R}^n_z)$.

We remark that this immediately implies the fact, proved by another method in [16], that if $\varphi(x) \in \mathrm{CAP}^\infty(\mathbf{R}^n)$, then for the unique bounded solution $u(t, x)$ of the Cauchy problem with initial data $\varphi(x)$ any derivative $D_x^\alpha D_t^\gamma u(t, x)$ is a continuous function of t with values in $\mathrm{CAP}(\mathbf{R}^n)$.

It is also not difficult to consider the case of finite smoothness of the coefficients. We omit the corresponding formulation since it is involved.

PROOF OF THEOREM 1.2. 1. We shall show that it suffices to verify the inclusion

$$\Gamma(t, x + z, y + z) \in \mathrm{CAP}(\mathbf{R}^n_z), \qquad t > 0, x, y \in \mathbf{R}^n. \tag{1.12}$$

Indeed, to prove all assertions of the theorem it suffices to show that $\Gamma(t, x + z, y + z)$ is an infinitely differentiable function of t, x and y with values in $\mathrm{CAP}(\mathbf{R}^n_z)$. It follows from (1.12) that this is equivalent to the infinite differentiability of $\Gamma(t, x + z, y + z)$ as a function of t, x, y with values in the space $C_b(\mathbf{R}^n)$ of continuous, bounded functions on $\mathbf{R}^n$ with norm (1.9). This, in turn, is equivalent to the fact that all derivatives with respect to t, x, and y are limits, uniform in z, of the corresponding difference quotients. The latter follows immediately from (1.8).

2. To prove (1.12) we trace the construction of the fundamental solution by the method of the Levi parametrix in the version presented in Chapter 9 of [8].

We first freeze the coefficients of the principal part of the operator A at some point $y \in \mathbf{R}^n$, and we consider the corresponding fundamental solution $Z(t, x, y)$ given by

$$Z(t, x, y) = (2\pi)^{-n} \int_{\mathbf{R}^n} e^{-t a_m(y, \xi) + i(x-y)\cdot \xi} \, d\xi. \tag{1.13}$$

From [8], Chapter 9, §3, Lemma 3, or directly from (1.13) it is easily found that if

$$|a_m(y + z, \xi) - a_m(y, \xi)| \leqslant \varepsilon |\xi|^m, \tag{1.14}$$

then

$$|Z(t, x + z, y + z) - Z(t, x, y)| \leqslant C\varepsilon t^{-n/m} \exp\left\{ -c\left(\frac{|x - y|^m}{t} \right)^{1/(m-1)} \right\},$$

$$\tag{1.15}$$

where C and c do not depend on t, x, y, or ε, but only on a_m. This implies that

$$Z(t, x + z, y + z) \in \text{CAP}(\mathbf{R}_z^n). \tag{1.16}$$

We observe that the same estimates (1.8) are satisfied for $Z(t, x, y)$ as for $\Gamma(t, x, y)$, whence, in particular, it follows that the arguments of part 1 of this proof are applicable to $Z(t, x, y)$, and hence the fact that the assertion of Theorem 1.2 is also true for the function $\Gamma(t, x, y)$.

3. We write $\Gamma(t, x, y)$ in the form ([8], Chapter 9, §4)

$$\Gamma(t, x, y) = Z(t, x, y) + \int_0^t d\sigma \int_{\mathbf{R}^n} Z(t - \sigma, x, \eta)\Phi(\sigma, \eta, y)\, d\eta. \tag{1.17}$$

Setting

$$K(t, x, y) = -\left(\frac{\partial}{\partial t} + \sum_{|\alpha| \leqslant m} a_\alpha(x)D_x^\alpha\right)Z(t, x, y), \tag{1.18}$$

we can write for Φ the integral equation

$$\Phi(t, x, y) = K(t, x, y) + \int_0^t d\sigma \int_{\mathbf{R}^n} K(t - \sigma, x, \eta)\Phi(\sigma, \eta, y)\, d\eta. \tag{1.19}$$

The function $\Phi(t, x, y)$ is obtained by solving (1.19) by the method of successive approximations in the form of the sum of the series

$$\Phi(t, x, y) = \sum_{l=1}^{\infty} K_l(t, x, y), \tag{1.20}$$

where $K_1 = K$ and

$$K_l(t, x, y) = \int_0^t d\sigma \int_{\mathbf{R}^n} K_1(t - \sigma, x, \eta)K_{l-1}(\sigma, \eta, y)\, d\eta. \tag{1.21}$$

The series (1.20) converges uniformly in t, x and y for $t \in [t_0, T]$, where $0 < t_0 < T < +\infty$, and its terms admit the estimates ([8], Chapter 9, §6)

$$|K_l(t, x, y)| \leqslant \frac{C_l}{t^{(n+m-l)/m}} \exp\left\{-c\left(\frac{|x-y|^m}{t}\right)^{1/(m-1)}\right\}, \tag{1.22}$$

where $c > 0$ and $C_l > 0$ (more precisely, $C_l = C^l/\Gamma(1 + l\kappa/m)$, where $0 < \kappa < 1$, C depends on κ but does not depend on l, and Γ in the formula for C_l is the Γ function of Euler). The function Φ itself satisfies the estimate

$$|\Phi(t, x, y)| \leqslant \frac{c}{t^{(n+m-1)/m}} \exp\left\{-c\left(\frac{|x-y|^m}{t}\right)^{1/(m-1)}\right\}. \tag{1.23}$$

4. We shall show that the functions $K_l(t, x, y)$, $l = 1, 2, \ldots$, satisfy a condition of almost periodicity of the type (1.16); namely,

$$K_l(t, x + z, y + z) \in \text{CAP}(\mathbf{R}_z^n) \tag{1.24}$$

and, moreover, $K_l(t, x + z, y + z)$ is a continuous function of t, x and y with values in $\text{CAP}(\mathbf{R}_z^n)$.

We carry out the proof by induction on l. For $l = 1$ this follows from part 2 of this proof. Suppose the assertion has been proved for K_{l-1}. From (1.21) we find that

$$K_l(t, x + z, y + z)$$

$$= \int_0^t d\sigma \int_{\mathbf{R}^n} K_1(t - \sigma, x + z, \eta + z) K_{l-1}(\sigma, \eta + z, y + z) \, d\eta. \quad (1.25)$$

It is convenient to consider (1.25) as an integral of a continuous, vector-valued function of the variables t, σ, x, y and η with values in $\mathrm{CAP}(\mathbf{R}^n_z)$. The integration here goes over σ and η, while the remaining variables are parameters. Unfortunately, continuity is destroyed on the boundary of the region of integration (for $\sigma = 0$ and for $\sigma = t$), while the region of integration itself is infinite in η and, moreover, depends on t. The latter inconvenience is easily eliminated by the change of variable $\sigma = t\lambda$ which transforms (1.25) to the form

$$K_l(t, x + z, y + z)$$

$$= \int_0^1 d\lambda \int_{\mathbf{R}^n} t K_1(t(1 - \lambda), x + z, \eta + z) K_{l-1}(t\lambda, \eta + z, y + z) \, d\eta. \quad (1.26)$$

This integral must be considered as an improper integral of a vector-valued function with values in $\mathrm{CAP}(\mathbf{R}^n_z)$. The required assertion will be proved if we show that the integral is uniformly convergent on $(0, 1) \times \mathbf{R}^n$ under the condition that $t \in [t_0, T]$, $0 < t_0 < T < +\infty$, $|x| \le R$, $|y| \le R$, $0 < R < +\infty$, which is equivalent to the uniform convergence of (1.26) as an integral of a numerical function with parameters t, x, y and z, provided only that we mean uniformity in t, x, y and z for $z \in \mathbf{R}^n$ and the same t, x and y. The latter is easily obtained from the estimates (1.22), which depend only on $x - y$, and therefore in estimating the integrand of (1.26) we never obtain z, so that essentially everything reduces to the standard argument ([8], the proofs of Lemmas 6 and 7 of Chapter 9, §4).

5. It is now clear that

$$\Phi(t, x + z, y + z) \in \mathrm{CAP}(\mathbf{R}^n_z) \quad (1.27)$$

and, moreover, $\Phi(t, x + z, y + z)$ is a continuous function of t, x and y with values in $\mathrm{CAP}(\mathbf{R}^n_z)$. The estimate (1.23) and the analogous estimate for $Z(t, x, y)$ now enable us to obtain the required inclusion (1.12) for the function $\Gamma(t, x, y)$ by considering (1.17) in a manner analogous to the arguments of part 4.

§2. The Green function of the boundary value problem
and the fundamental solution

In this section we formulate in the context we require the result of [9] on the estimate of the difference between the Green function of a parabolic boundary value problem and the fundamental solution of the Cauchy problem, and we also prove an analogue of this result for the case of periodic coefficients and boundary conditions.

1. Let Ω be a bounded domain in $\mathbf{R}^n$ with smooth boundary S. Suppose there is given a differential operator A of the form (1.1) with coefficients $a_\alpha(x) \in C_b^\infty(\mathbf{R}^n)$ satisfying the condition of uniform parabolicity (1.6) for the Cauchy problem (1.3), (1.4). We note that in this case the number m is even. Suppose there are given differential operators $B_j = B_j(x, D_x), j = 1, \ldots, m/2$, defined for $x \in S$ and having order $\leq m - 1$, i.e., they are polynomials in D_x of degree $\leq m - 1$ with coefficients in $C^\infty(S)$. In the cylinder $Q = [0, +\infty) \times \Omega$ with lateral boundary $[0, +\infty) \times S$, which we denote by Γ, we consider the boundary value problem

$$\partial u/\partial t = -Au, \tag{2.1}$$

$$u|_{t=0} = \varphi(x), \tag{2.2}$$

$$B_j u|_\Gamma = 0, \qquad j = 1, \ldots, m/2. \tag{2.3}$$

We suppose that problem (2.1)–(2.3) is parabolic (see, for example, [7]).

The Green function of problem (2.1)–(2.3) is a function $\Gamma_\Omega(t, x, y)$ defined for $t > 0$, $x \in \Omega$ and $y \in \Omega$, such that for any function $\varphi(x) \in C_0^\infty(\Omega)$ the solution $u(t, x)$ of problem (2.1)–(2.3) is given by

$$u(t, x) = \int_\Omega \Gamma_\Omega(t, x, y)\varphi(y)\, dy. \tag{2.4}$$

We further introduce the notation

$$v_\Omega(t, x, y) = \Gamma_\Omega(t, x, y) - \Gamma(t, x, y), \tag{2.5}$$

where $\Gamma(t, x, y)$ is the fundamental solution of the Cauchy problem (1.3), (1.4).

We shall be interested in the behavior of $\Gamma_\Omega(t, x, y)$ as Ω varies; we therefore now describe what "uniformity with respect to Ω" means.

Suppose there is given some family of domains Ω and on the boundary S of each of them there are given boundary operators $B_j, j = 1, \ldots, m/2$. We say that problem (2.1)–(2.3) is *parabolic uniformly with respect to* Ω if the following conditions are satisfied:

1) Any point $x_0 \in S$ has a spherical neighborhood of radius $r_0 > 0$ such that in some Cartesian coordinate system (depending on x_0) the boundary S of the domain Ω is given by the equation $x_n = f(x_1, \ldots, x_{n-1})$, in which $|D^\kappa f| \leq C_\kappa$ for any $(n - 1)$-dimensional multi-index κ, while the constants r_0 and C_κ do not depend on the choice of the domain Ω belonging to the family in question.

2) If $b(x_1, \ldots, x_{n-1})$ is one of the coefficients of the boundary operators B_j expressed in terms of $x_1, \ldots, x_{n-1}$ in the coordinates of 1), then $|D^\kappa b| \leq C_\kappa$, where C_κ does not depend on the choice of the domain Ω or the system of coordinates of the type described.

3) The condition of parabolicity of problem (2.1)–(2.3) is satisfied at all points $x_0 \in S$ uniformly with respect to the choice of x_0 and the domain Ω of the family considered.

The last condition means that a certain canonical determinant defining the parabolicity of the problem must be bounded away from zero (see [7] and [17]–[19]), while the constant bounding it from zero must not depend on the domain Ω of the point $x_0 \in S$.

The next theorem is a special case of Theorem 1.1 of [9] which we require.

THEOREM 2.1. *Suppose there is given a family of domains Ω and boundary value problems (2.1)–(2.3), and these problems are uniformly parabolic with respect to Ω (i.e., conditions 1)–3) are satisfied). Then for the Green functions Γ_Ω of these problems and the functions v_Ω defined by (2.5) the estimates*

$$|D_x^\alpha D_t^\gamma \Gamma_\Omega(t, x, y)| \leqslant Ct^{-(n+|\alpha|+m\gamma)/m} \exp\left\{-c\left(\frac{|x-y|^m}{t}\right)^{1/(m-1)}\right\}, \quad (2.6)$$

$$|D_x^\alpha D_t^\gamma v_\Omega(t, x, y)| \leqslant Ct^{-(n+|\alpha|+m\gamma)/m} \exp\left\{-c\left[\frac{(|x-y|+d(y,S))^m}{t}\right]^{1/(m-1)}\right\}$$

$$(2.7)$$

hold, where $d(y, S)$ is the distance from the point y to the boundary S, and the constants C and c do not depend on the choice of domain Ω belonging to the family in question.

We note here that in view of the obvious inequality

$$d(x, S) \leqslant |x - y| + d(y, S) \tag{2.8}$$

in the argument of the exponential function in (2.7) in place of $|x - y| + d(y, S)$ it is possible to write $|x - y| + d(x, S) + d(y, S)$.

2. We now consider the periodic case. Let $\mathbf{Z}^n$ be the set of points of $\mathbf{R}^n$ having integer coordinates, and suppose that the coefficients $a_\alpha(x)$ of the operator A have $\mathbf{Z}^n$ as period lattice, i.e.,

$$a_\alpha(x + z) = a_\alpha(x), \quad z \in \mathbf{Z}^n. \tag{2.9}$$

Suppose now that $N_1, \ldots, N_n$ is a collection of natural numbers and $\Omega = \Omega_{N_1, \ldots, N_n}$ is a rectangular parallelepiped: $\Omega = \{x: 0 \leqslant x_j \leqslant N_j, j = 1, \ldots, n\}$. The periodicity conditions (or, as physicists say, the Born-Kármán conditions) imply the periodicity of a function in each variable x_j with period N_j. In other words, (2.1) with the boundary conditions (2.2) is considered on the product $[0, +\infty) \times T^n$, where $T^n = T^n_{N_1, \ldots, N_n}$ is the n-dimensional torus which is the quotient space of $\mathbf{R}^n$ modulo lattice $\mathbf{Z}^n_{N_1, \ldots, N_n} = \{k_1 N_1 e_1 + \cdots + k_n N_n e_n\}$, where $e_1, \ldots, e_n$ are the canonical basis vectors in $\mathbf{R}^n$ and $k_1, \ldots, k_n$ are arbitrary integers.

The Green function of problem (2.1), (2.2) on Ω with periodicity conditions on S is a function $\Gamma_\Omega(t, x, y) \in C^\infty(\mathbf{R}_+ \times T^n \times T^n)$ such that if $\varphi(x) \in C^\infty(T^n)$, then the solution $u(t, x)$ ($t > 0$, $x \in T^n$) of problem (2.1), (2.2) is given by

$$u(t, x) = \int_{T^n} \Gamma_\Omega(t, x, y)\varphi(y)\, dy, \tag{2.10}$$

where dy is the measure on T^n induced by the Lebesgue measure in $\mathbf{R}^n$.

We consider $\Gamma_\Omega(t, x, y)$ as a function on $(0, +\infty) \times \mathbf{R}^n \times \mathbf{R}^n$ which is periodic in x_j and y_j with period N_j. As before, $\Gamma(t, x, y)$ denotes the fundamental solution of the Cauchy problem.

THEOREM 2.2. *Suppose that the operator A of the form (1.1) has coefficients $a_\alpha(x) \in C^\infty(\mathbf{R}^n)$ satisfying the periodicity condition (2.9) and the condition of uniform parabolicity (1.6). Let $\Gamma_\Omega(t, x, y)$ be the Green function of problem (2.1), (2.2) with periodic boundary conditions on the boundary of $\Omega = \Omega_{N_1,\ldots,N_n}$, and let $v_\Omega(t, x, y)$ be defined by (2.5). Then (2.6) and (2.7) hold for Γ_Ω and v_Ω, and the constants in these estimates do not depend on the choice of $N_1,\ldots,N_n$.*

PROOF. Because of the condition of periodicity of the coefficients (2.9), the fundamental solution $\Gamma(t, x, y)$ of the Cauchy problem possesses the property

$$\Gamma(t, x + z, y + z) = \Gamma(t, x, y), \qquad z \in \mathbf{Z}^n, t > 0, x, y \in \mathbf{R}^n. \quad (2.11)$$

In particular, this is true for $z \in \mathbf{Z}^n_{N_1,\ldots,N_n}$. It is then easy to verify that

$$\Gamma_\Omega(t, x, y) = \sum_{z \in Z^n_{N_1,\ldots,N_n}} \Gamma(t, x + z, y). \quad (2.12)$$

The convergence of (2.12) and the estimates (2.6) and (2.7) follow easily from the estimates of Theorem 1.1 for the fundamental solution. Moreover, it follows from (2.11) that $\Gamma_\Omega(t, x, y)$ is periodic in x and y. The fact expressed by (2.10) follows easily from the fact that $\Gamma(t, x, y)$ is the fundamental solution of the Cauchy problem.

§3. Families of expanding domains and limit relations
for the Green functions

In this section we formulate a convenient interpretation of the relation $\Omega \to \infty$, where Ω always denotes a domain (an open but not necessarily connected subset), in $\mathbf{R}^n$, and we consider some family of domains. We always assume that on a given family of domains there is a partial ordering denoted by $\Omega' > \Omega$ and that this ordering must make the given family a directed set, i.e., for any Ω and Ω' there must exist a domain Ω'' such that $\Omega'' > \Omega$ and $\Omega'' > \Omega'$. An example is the family of balls or the family of rectangular parallelepipeds of the form $\Omega_{N_1,\ldots,N_n}$ as in §2, ordered by inclusion or, what will be more natural for us, by diameter. The latter means that $\Omega' > \Omega$ if $d(\Omega') > d(\Omega)$, where here and henceforth $d(\Omega)$ denotes the diameter of the domain Ω.

All possible limits will be taken with respect to some directed family of domains (this is written $\Omega \to \infty$) on which additional restrictions will be imposed; we now proceed to formulate the latter.

Let $\mu(\Omega)$ denote the volume of Ω (in the sense of Lebesgue measure in $\mathbf{R}^n$), and let $\mu_h(\Omega)$ be the volume of the part Ω_h of Ω consisting of all points at distance $< h$ from the boundary S of Ω.

The weakest restriction has the form (P): $\mu(\Omega) \to +\infty$ as $\Omega \to \infty$, and there exists a function $h = h(\Omega)$ defined on the given family of domains such that for $\Omega \to \infty$ also $h(\Omega) \to +\infty$ and

$$\mu_{h(\Omega)}(\Omega)/\mu(\Omega) \to 0. \quad (3.1)$$

A stronger condition known in statistical mechanics is Fisher's condition [20]:
(F): $\mu(\Omega) \to +\infty$ as $\Omega \to \infty$, and there exists a function $\pi(\varepsilon)$ defined on $(0, \varepsilon_0)$ for some $\varepsilon_0 > 0$ such that $\lim_{\varepsilon \to 0} \pi(\varepsilon) = 0$, and

$$\mu_{\varepsilon d(\Omega)}(\Omega)/\mu(\Omega) \leqslant \pi(\varepsilon) \tag{3.2}$$

for any domain Ω of the family in question.

PROPOSITION 3.1. (F) *implies* (P).

PROOF. We observe that the condition $\mu(\Omega) \to +\infty$ as $\Omega \to \infty$ implies that $d(\Omega) \to +\infty$ as $\Omega \to \infty$. It is easy to see that condition (F) implies that (P) is satisfied with $h(\Omega) = \sqrt{d(\Omega)}$.

REMARK. Stronger versions of Fisher's condition in which additional conditions on the function $\pi(\varepsilon)$ are imposed will also be used below. We note here that for a family obtained by similarity transformations (for example, homothetic blow-up) from the same domain with a smooth or piecewise smooth boundary Fisher's condition is satisfied with a function $\pi(\varepsilon) = C\varepsilon$ for sufficiently large $C > 0$.

PROPOSITION 3.2. *If condition* (P) *is satisfied and* $f \in$ CAP($\mathbf{R}^n$), *then*

$$M\{f\} = \lim_{\Omega \to \infty} \frac{1}{\mu(\Omega)} \int_{\Omega} f(x)\, dx \tag{3.3}$$

(*the limit on the right exists and is equal to the mean value* $M\{f\}$ *of the a.p. function* f).

PROOF. By considering the standard partition of $\mathbf{R}^n$ into cubes with edge length $h(\Omega)/3\sqrt{n}$, we find that there exists a domain Ω' (depending on Ω) contained in Ω which is the union of a finite number of such cubes and such that as $\Omega \to \infty$

$$\mu(\Omega \setminus \Omega')/\mu(\Omega) \to 0. \tag{3.4}$$

Since it is obvious that

$$\lim_{\Omega \to \infty} \frac{1}{\mu(\Omega')} \int_{\Omega'} f(x)\, dx = M\{f\},$$

(3.3) follows from the inequalities

$$\left| \frac{1}{\mu(\Omega)} \int_{\Omega} f(x)\, dx - \frac{1}{\mu(\Omega')} \int_{\Omega'} f(x)\, dx \right|$$

$$\leqslant \frac{1}{\mu(\Omega)} \left| \int_{\Omega \setminus \Omega'} f(x)\, dx \right| + \frac{\mu(\Omega \setminus \Omega')}{\mu(\Omega)} \cdot \frac{1}{\mu(\Omega')} \left| \int_{\Omega'} f(x)\, dx \right|$$

$$\leqslant 2 \sup_{x \in \mathbf{R}^n} |f(x)| \frac{\mu(\Omega \setminus \Omega')}{\mu(\Omega)}$$

(see also Theorem S.2 in the Supplement to [21]).
We proceed to the main theorems in this section.

THEOREM 3.1. *Suppose there is given an operator A of the form (1.1) with coefficients $a_\alpha(x) \in \mathrm{CAP}^\infty(\mathbf{R}^n)$, and that the parabolicity condition (1.6) is satisfied. Assume also that there is given a directed family of domains possessing the following properties*:

a) *Each domain either has a smooth boundary or is a rectangular parallelepiped of the form $\Omega_{N_1,\dots,N_n}$ described in §2 (the latter possibility is considered only when the periodicity conditions (2.9) are satisfied).*

b) *Condition (P) is satisfied.*

Suppose that on the boundary of each domain of the first type there are given boundary operators $B_j, j = 1,\dots,m/2$, on the boundaries of the parallelepipeds there are given periodicity conditions, and problem (2.1)–(2.3) is parabolic uniformly with respect to all domains with smooth boundary of this family, i.e., conditions 1), 2), and 3) of §2 are satisfied.

Denote by $\Gamma_\Omega(t, x, y)$ the corresponding Green functions, and by $\Gamma(t, x, y)$ the fundamental solution of the Cauchy problem (1.3), (1.4). Then for any differential operator $L = L(x, D_x)$ with coefficients in $\mathrm{CAP}(\mathbf{R}^n)$ and for any $t > 0$ there exists the limit

$$\lim_{\Omega \to \infty} \frac{1}{\mu(\Omega)} \int_\Omega [L(x, D_x)\Gamma_\Omega(t, x, y)]\big|_{y=x}\, dx = M_x\big\{[L(x, D_x)\Gamma(t, x, y)]\big|_{y=x}\big\}.$$

$$(3.5)$$

PROOF. Theorem 1.2 and Proposition 3.2 show that the limit relation (3.5) is true if Γ_Ω is replaced there by Γ. The theorem will thus be proved if we show that

$$\lim_{\Omega \to \infty} \frac{1}{\mu(\Omega)} \int_\Omega \big|[L(x, D_x)v_\Omega(t, x, y)]\big|_{y=x}\big|\, dx = 0 \qquad (3.6)$$

in the terminology of §2. For fixed $t > 0$ and any $h > 0$ by Theorems 2.1. and 2.2 we can write the following estimate for the expression subject to the limit sign in (3.6):

$$\frac{1}{\mu(\Omega)} \int_\Omega \cdots \leqslant \frac{1}{\mu(\Omega)} \left(\int_{\Omega_h} \cdots + \int_{\Omega \setminus \Omega_h} \cdots \right) \leqslant C\left[\frac{\mu_h(\Omega)}{\mu(\Omega)} + \exp\{-ch^{m/(m-1)}\} \right],$$

$$(3.7)$$

where C and c are quantities not depending on Ω or h (they depend on t, the operator A, and the family of domains considered). Setting now $h = h(\Omega)$ (see condition (P)), we see that the right side of (3.7) tends to 0 as $\Omega \to \infty$, which proves (3.6).

COROLLARY 3.1. *Under the hypotheses of Theorem 3.1 there exists the limit*

$$\lim_{\Omega \to \infty} \frac{1}{\mu(\Omega)} \int_\Omega \Gamma_\Omega(t, x, x)\, dx = M_x\{\Gamma(t, x, x)\}, \qquad t > 0. \qquad (3.8)$$

THEOREM 3.2. *Suppose that all the hypotheses of Theorem 3.1 are satisfied, but in place of condition* (P) *one requires convexity of all domains* Ω *and that condition* (F) *is satisfied with a function* $\pi(\varepsilon)$ *such that*

$$\lim_{\varepsilon \to 0} \pi(\varepsilon) \cdot |\ln \varepsilon|^{n(m-1)/m} = 0.$$

Then for any differential operator $L = L(x, D_x)$ *with coefficients in* $\mathrm{CAP}(\mathbf{R}^n)$, *for any function* $f \in \mathrm{CAP}(\mathbf{R}^n)$, *and for any* $t > 0$ *there exists the limit*

$$\lim_{\Omega \to \infty} \frac{1}{\mu(\Omega)} \int_{\Omega \times \Omega} f(y) L(x, D_x) \Gamma_\Omega(t, x, y)\, dx\, dy$$

$$= 2^n \int_{\mathbf{R}^n} M_z \left\{ f(z - w) L\left(z + w, \frac{1}{2}(D_z + D_w)\right) \Gamma(t, z + w, z - w) \right\}\, dw.$$

$$(3.9)$$

PROOF. Theorem 3.2 will be proved if we verify that the right side of (3.9) is equal to the expression obtained from the left side by replacing Γ_Ω by Γ, and, moreover, that

$$\lim_{\Omega \to \infty} \frac{1}{\mu(\Omega)} \int_{\Omega \times \Omega} |L(x, D_x) v_\Omega(t, x, y)|\, dx\, dy = 0. \qquad (3.10)$$

1. We first prove (3.10). By Theorems 2.1 and 2.2 the integrand can be estimated in terms of

$$C \exp\left\{ -c \left[|x - y|^q + d(x, S)^q + d(y, S)^q \right] \right\},$$

where $q = m/(m - 1)$, and the positive constants C and c do not depend on Ω. The region of integration can therefore naturally be broken into 8 parts in each of which three inequalities consisting of one from each of the following three pairs are satisfied:

1) $|x - y| \geq l, |x - y| < l$;
2) $d(x, S) \geq h, d(x, S) < h$;
3) $d(y, S) \geq h, d(y, S) < h$.

Here l and h are positive numbers depending on Ω which will be chosen later.

We denote the left column of these inequalities by the letter a and the right column by the letter b; by $\Omega_{i_1 i_2 i_3}$, where each i_j is equal to a or b, we denote the subset of $\Omega \times \Omega$ on which of the kth pair the inequality standing in the column i_k is satisfied ($k = 1, 2, 3$). Finally, we denote by $I_{i_1 i_2 i_3}$ the expression subject to the limit sign in (3.10) in which the integral over $\Omega \times \Omega$ is replaced by an integral over $\Omega_{i_1 i_2 i_3}$.

Then the following estimates hold:

$$|I_{aaa}| \leq C\mu(\Omega) e^{-c(l^q + h^q)}; \qquad |I_{aab}| + |I_{aba}| \leq C\mu_h(\Omega) e^{-c(l^q + h^q)};$$

$$|I_{abb}| \leq C\left(\mu_h^2(\Omega)/\mu(\Omega)\right) e^{-cl^q}; \qquad |I_{baa}| \leq Cl^n e^{-ch^q};$$

$$|I_{bab}| + |I_{bba}| \leq Cl^n\left(\mu_h(\Omega)/\mu(\Omega)\right) e^{-ch^q};$$

$$|I_{bbb}| \leq C \min\{l^n \mu_h(\Omega)/\mu(\Omega), \mu_h^2(\Omega)/\mu(\Omega)\}.$$

Here the positive constants C and c do not depend on Ω, h or l, so that it now suffices to demonstrate the possibility of choosing $h = h(\Omega)$ and $l = l(\Omega)$ so that the right sides of all the above estimates tend to 0 as $\Omega \to \infty$. It is easy to verify that this is satisfied if we set $h = \sqrt{d(\Omega)}$ and $l = A(\ln d(\Omega))^{1/q}$, where the constant $A > 0$ must be chosen sufficiently large. Indeed, for all the integrals except I_{bbb} everything is obvious, while to estimate I_{bbb} it suffices to verify that

$$\lim_{\Omega \to \infty} \frac{l^n \mu_h(\Omega)}{\mu(\Omega)} = 0.$$

But with our choice of $l(\Omega)$ and $h(\Omega)$

$$l^n \mu_h(\Omega)/\mu(\Omega) \leqslant \pi\left(d(\Omega)^{-1/2}\right) \cdot A^n (\ln d(\Omega))^{n/q}$$

$$= 2^{n/q} A^n \pi\left(d(\Omega)^{-1/2}\right) \left|\ln(d(\Omega))^{-1/2}\right|^{n/q} \to 0$$

as $d(\Omega) \to +\infty$ by the assumption on $\pi(\varepsilon)$ formulated in the hypothesis of the theorem. Thus, (3.10) has been proved.

2. For simplicity we now suppose that $f \equiv 1$ and $L(x, D_x)$ is the identity operator (the general case is considered in the same way). If I is the right side of (3.9), then

$$I = \lim_{\Omega \to \infty} \frac{2^n}{\mu(\Omega)} \int_{\Omega} dz \int_{\mathbf{R}^n} \Gamma(t, z + w, z - w)\, dw. \tag{3.11}$$

It must be verified that

$$I = \lim_{\Omega \to \infty} \frac{1}{\mu(\Omega)} \int_{\Omega \times \Omega} \Gamma(t, x, y)\, dx\, dy. \tag{3.12}$$

We denote by I_Ω and I'_Ω the expressions subject to the limit sign in (3.11) and (3.12). Setting $x = z + w$ and $y = z - w$, we obtain

$$I'_\Omega = \frac{2^n}{\mu(\Omega)} \int_{\Omega} dz \int_{\Omega_z} \Gamma(t, z + w, z - w)\, dw, \tag{3.13}$$

where $\Omega_z = \{w : z + w \in \Omega \text{ and } z - w \in \Omega\}$. The integral with respect to z in (3.13) is taken over Ω, since $x \in \Omega$ and $y \in \Omega$ imply that $z = (x + y)/2 \in \Omega$ by the convexity of Ω. We now estimate the integral

$$I_{z,\Omega} = \int_{\mathbf{R}^n \setminus \Omega_z} \Gamma(t, z + w, z - w)\, dw.$$

The condition $w \in \mathbf{R}^n \setminus \Omega_z$ implies that $|w| \geqslant d(z, S)$. Therefore, for fixed $t > 0$

$$|I_{z,\Omega}| \leqslant C' \int_{|w| \geqslant d(z,S)} e^{-c'|w|^q}\, dw \leqslant C e^{-cd(z,S)^q}, \tag{3.14}$$

where C', c', C and c are positive constants. Using (3.14), it is easy to prove that

$$\lim_{\Omega \to \infty} \frac{1}{\mu(\Omega)} \int_{\Omega} |I_{z,\Omega}|\, dz = 0$$

(in a manner similar to the proof of Theorem 3.1), and hence

$$\lim_{\Omega \to \infty} |I_\Omega - I_\Omega'| = \lim_{\Omega \to \infty} \left| \frac{2}{\mu(\Omega)} \int_\Omega I_{z,\Omega}\, dz \right| = 0,$$

whence (3.12) follows.

COROLLARY 3.2. *Under the hypotheses of Theorem 3.2 there exists the limit*

$$\lim_{\Omega \to \infty} \frac{1}{\mu(\Omega)} \int_{\Omega \times \Omega} g(x)f(y)\Gamma_\Omega(t, x, y)\, dx\, dy$$

$$= 2^n \int_{\mathbf{R}^n} M_z\{g(z + w)f(z - w)\Gamma(t, z + w, z - w)\}\, dw, \qquad (3.15)$$

where $f, g \in \mathrm{CAP}(\mathbf{R}^n)$ and $t > 0$.

THEOREM 3.3. *Suppose that the hypotheses of Theorem 3.2 are satisfied for two operators A', A'' and two families of boundary conditions but for the same family of domains Ω, so that we have the two families of Green functions $\Gamma_\Omega'(t, x, y)$ and $\Gamma_\Omega''(t, x, y)$ and two fundamental solutions $\Gamma'(t, x, y)$ and $\Gamma''(t, x, y)$.*

Then for any two differential operators L' and L'' with coefficients in $\mathrm{CAP}(\mathbf{R}^n)$ and for any positive t and s there exists the limit

$$\lim_{\Omega \to \infty} \frac{1}{\mu(\Omega)} \int_{\Omega \times \Omega} \left[L'(x, D_x)\Gamma_\Omega'(t, x, y) \right] \cdot \left[L''(y, D_y)\Gamma_\Omega''(s, y, x) \right] dx\, dy$$

$$= 2^n \int_{\mathbf{R}^n} M_z \left\{ \left[L'\left(z + w, \frac{1}{2}(D_z + D_w) \right) \Gamma'(t, z + w, z - w) \right] \right.$$

$$\left. \times \left[L''\left(z - w, \frac{1}{2}(D_z - D_w) \right) \Gamma''(s, z - w, z + w) \right] \right\} dw. \quad (3.16)$$

The proof is completely analogous to the proof of Theorem 3.2.

COROLLARY 3.3. *Suppose that the hypotheses of Theorem 3.2 are satisfied. Then there exists the limit*

$$\lim_{\Omega \to \infty} \frac{1}{\mu(\Omega)} \int_{\Omega \times \Omega} \frac{\partial \Gamma_\Omega(t, x, y)}{\partial x_i} \frac{\partial \Gamma_\Omega(s, y, x)}{\partial y_j}\, dx\, dy$$

$$= 2^{n-2} \int_{\mathbf{R}^n} M_z \left\{ \left[\left(\frac{\partial}{\partial z_i} + \frac{\partial}{\partial w_i} \right) \Gamma(t, z + w, z - w) \right] \right.$$

$$\left. \times \left[\left(\frac{\partial}{\partial z_j} - \frac{c}{\partial w_j} \right) \Gamma(s, z - w, z + w) \right] \right\} dz, \qquad (3.17)$$

where $t, s > 0$ and $i, j = 1, \ldots, n$.

REMARK. Many other limit relations for the Green functions can be obtained in similar fashion. We restrict our attention to the relations of Theorems 3.1–3.3, since these are precisely the ones that lead to physically important consequences.

§4. Two lemmas on elliptic boundary value problems

The object of this section is an explicit formulation of two assertions, stating that a priori estimates of an elliptic boundary value problem with or without a parameter are uniform relative to the choice of domain Ω if conditions 1) and 2) of §2 and, of course, the condition of uniform ellipticity of the problem are satisfied for this family and boundary conditions. This uniformity is well known (for example, it can be extracted from the arguments of [22] and [23]), but for completeness we formulate the results explicitly and present a sketch of the proof.

LEMMA 4.1. *Let A be a uniformly elliptic operator of the form* (1.1) *on* $\mathbf{R}^n$ *with coefficients* $a_\alpha(x) \in C_b^\infty(\mathbf{R}^n)$, *and suppose there is given a family of domains* Ω *and boundary operators* B_j, $j = 1,\ldots,m/2$, *on their boundaries S (see §2) and that conditions 1) and 2) of §2 and also the uniform (with respect to Ω and the points of S) condition of ellipticity of the boundary value problem in Ω defined by the operator A and the boundary operators B_j are satisfied (see, for example, [24] or [25], Chapter 10). Then for any integer $s \geq m$ there exists a constant $C > 0$ not depending on Ω such that the a priori estimate*

$$\|u\|_s \leq C\left(\|Au\|_{s-m} + \sum_{j=1}^{m/2} \|B_j u\|'_{s-m_j-1/2} + \|u\|_0 \right) \tag{4.1}$$

holds, where $u \in C^\infty(\overline{\Omega})$, m_j is the order of the operator B_j $(m_j \leq m - 1)$, $\|\cdot\|_s$ is the Sobolev-Slobodeckiĭ norm of order s in the domain Ω, and $\|\cdot\|'_s$ is the analogous norm on the boundary S of the domain Ω.

PROOF. If the function u is concentrated in a ball of radius 1, then the uniformity of the estimate (4.1) is deduced from the assumptions by repeating the arguments of [24] and [25]. In the general case we consider the partition of unity

$$1 \equiv \sum_l \varphi_l^2(x), \qquad x \in \overline{\Omega}, \tag{4.2}$$

consisting of functions $\varphi_l(x) \in C^\infty(\overline{\Omega})$ such that the support of each of them is contained in some ball of radius 1, $|D^\alpha \varphi_l| \leq C_\alpha$, where C_α does not depend on l and Ω, and there exists an N, also not depending on Ω, such that each ball of radius 1 intersects the supports of no more than N of the functions φ_l. We can write

$$\|u\|_s^2 \leq C_1 \sum_l \|\varphi_l u\|_s^2 \leq C_2 \sum_l \left(\|A(\varphi_l u)\|_{s-m}^2 + \sum_{j=1}^{m/2} \|B_j(\varphi_l u)\|'^2_{s-m_j-1/2} + \|\varphi_l u\|_0^2 \right)$$

$$\leq C_3 \left(\sum_l \|\varphi_l Au\|_{s-m}^2 + \sum_{j=1}^{m/2} \|\varphi_l(B_j u)\|'^2_{s-m_j-1/2} + \|u\|_{s-1}^2 + \|u\|'^2_{s-1-m_j-1/2} \right)$$

$$\leq C_4 \left(\|Au\|_{s-m}^2 + \sum_{j=1}^{m/2} \|B_j u\|'^2_{s-m_j-1/2} + \|u\|_{s-1}^2 \right),$$

where the constants C_k, $k = 1,\ldots,4$, do not depend on Ω.

The passage from $\| \cdot \|_{s-1}$ to $\| \cdot \|_0$ in this estimate can be realised, for example, by means of the standard inequality

$$\|u\|_{s-1} \leqslant \varepsilon \|u\|_s + C_\varepsilon \|u\|_0, \qquad s \geqslant 1, \varepsilon > 0,$$

in which C_ε does not depend on Ω. Lemma 4.1 is proved.

LEMMA 4.2. *Suppose that the hypotheses of Lemma 4.1 are satisfied, but in place of the uniform condition of ellipticity the stronger uniform condition of ellipticity with a parameter is required for the problem defined by the operator $A + q^m$ and the boundary operators $B_j, j = 1, \ldots, m/2$, while the parameter runs through some angle Q with vertex at the point 0 in the complex plane. Then there exists an $R > 0$, not depending on Ω, such that for $|q| \geqslant R$, $q \in Q$, the boundary value problem is uniquely solvable in the corresponding Sobolev-Slobodeckiĭ spaces, and the a priori estimate*

$$\| u \|_s \leqslant C \left(\|(A + q^m)u\|_{s-m} + \sum_{j=1}^{m/2} \|B_j u\|'_{s-m_j-1/2} \right) \tag{4.3}$$

holds, where s is an integer, $s \geqslant m$, the constant C does not depend on Ω, and $\|\cdot\|$ are norms with a parameter [7]

$$\| u \|_s^2 = \|u\|_s^2 + |q|^{2s} \|u\|_0^2,$$

while $\|\cdot\|$ is the analogous norm on the boundary.

The proof consists of an argument analogous to the proof of Lemma 4.1 and making use of the results of [7]. We omit the details.

§5. The existence of the density of states

A *distribution function* is by definition a nondecreasing, left-continuous function $N(\lambda)$ on the line $(-\infty, +\infty)$ such that $\lim_{\lambda \to -\infty} N(\lambda) = 0$ (no conditions as $\lambda \to +\infty$ are imposed for the time being).

Suppose there is given a sequence of distribution functions $N_k(\lambda)$, $k = 1, 2, \ldots$. We say that $N_k(\lambda)$ *converges weakly* to $N(\lambda)$ as $k \to +\infty$ (and write $N_k(\lambda) \overset{w}{\to} N(\lambda)$) if $\lim_{k \to +\infty} N_k(\lambda) = N(\lambda)$ at all points of continuity of the distribution function $N(\lambda)$. It is known that for this it suffices that $N_k(\lambda)$ converge on a countable dense set of points.

For us conditions for weak convergence in terms of the Laplace transform

$$\tilde{N}(t) = \int_{-\infty}^{+\infty} e^{-\lambda t} \, dN(\lambda), \qquad t > 0, \tag{5.1}$$

are important. All distribution functions $N(\lambda)$ we consider possess the property

$$N(\lambda) = 0 \quad \text{for } \lambda \leqslant a. \tag{5.2}$$

Conditions for the convergence of the integral (5.1) are characterized by the following lemma.

LEMMA 5.1. *Suppose that the distribution function $N(\lambda)$ satisfies (5.2). Then the integral (5.1) converges or any $t > 0$ if and only if for any $\varepsilon > 0$*

$$N(\lambda) \leqslant C_\varepsilon e^{\varepsilon \lambda}. \tag{5.3}$$

Moreover, in (5.3) it is possible to take $C = \tilde{N}(\varepsilon)$.

PROOF. The convergence of the integral (5.1) under the condition (5.3) is obvious, and it remains only to verify the inequality

$$N(\lambda) \leqslant \tilde{N}(\varepsilon) e^{\varepsilon \lambda}. \tag{5.4}$$

This follows from the obvious chain of inequalities

$$\tilde{N}(\varepsilon) = \int_{-\infty}^{+\infty} e^{-\varepsilon \lambda}\, dN(\lambda) \geqslant \int_{-\infty}^{\lambda} e^{-\varepsilon \mu}\, dN(\mu) \geqslant e^{-\varepsilon \lambda} \int_{-\infty}^{\lambda} dN(\mu) = N(\lambda) e^{-\varepsilon \lambda}.$$

LEMMA 5.2. *Let $N_k(\lambda)$ be a sequence of distribution functions satisfying the following conditions:*
   a) *Condition (5.2) is satisfied with a constant a not depending on k ($k = 1, 2, \ldots$).*
   b) $|\tilde{N}_k(t)| \leqslant C(t)$, $\qquad\qquad\qquad\qquad\qquad\qquad\qquad$ (5.5)
where $C(t)$ does not depend on k.
   c) $\lim_{k \to +\infty} \tilde{N}_k(t)$ *exists for each $t > 0$.*
   Then $N_k(\lambda) \xrightarrow{\text{w}} N(\lambda)$, where $N(\lambda)$ is a distribution function, and $\tilde{N}(t) = \lim_{k \to +\infty} \tilde{N}_k(t), \ t > 0$.

PROOF. Because of condition b) Lemma 5.1 implies the uniform boundedness of $N_k(\lambda)$ on each interval $(-\infty, \lambda_0)$ and also the uniform convergence of all integrals of the form (5.1) defining $\tilde{N}_k(t)$, $t > 0$. Therefore, Lemma 5.2 follows from the well-known Helly selection theorems (see, for example, [26], Chapter VI, §6).

We proceed to formulate conditions for the existence of the density of states. Let A be a differential operator of the form (1.1). We define a formal adjoint operator A^+ by the formula

$$A^+ u(x) = \sum_{|\alpha| \leqslant m} D^\alpha \big(\overline{a_\alpha(x)}\, u(x) \big). \tag{5.6}$$

We henceforth assume that A is formally selfadjoint, i.e., $A = A^+$. In particular, this implies that the principal symbol $a_m(x, \xi)$ is real, and the condition (1.6) can be rewritten in the form

$$a_m(x, \xi) \geqslant \varepsilon |\xi|^m, \qquad \varepsilon > 0. \tag{5.7}$$

Let Ω be a bounded domain in $\mathbf{R}^n$ with smooth boundary, and suppose that S boundary operators B_j, $j = 1, \ldots, m/2$, of orders $\leqslant m - 1$ on the boundary are given (see §2). We shall also consider parallelepipeds with conditions of periodicity. We denote by $H^m(\Omega)$ the Sobolev space of functions $f \in L^2(\Omega)$ for which $D^\alpha f \in L^2(\Omega)$ for $|\alpha| \leqslant m$. We shall study the operator A_Ω, defined as an unbounded operator in $L^2(\Omega)$ with domain

$$D_\Omega = \big\{ u\colon u \in H^m(\Omega), B_j u|_S = 0, j = 1, \ldots, m/2 \big\}. \tag{5.8}$$

In the case of periodic conditions we set $D_\Omega = H^m(T^n)$.

We say that the boundary conditions defined by the operators B_j are *selfadjoint* if the operator A_Ω defined above is selfadjoint (a detail discussion of this condition can be found in [27]). Periodic boundary conditions are selfadjoint if $A = A^+$.

Conditions for the lower semiboundedness of the operator A_Ω are important for us. For this there is a natural sufficient algebraic condition: the condition of ellipticity with parameter for the operator $A - \lambda I$ and the system of boundary operators B_j for $\mathrm{Re}\,\lambda \leqslant 0$ or, in the terminology of [7], ellipticity with parameter of the operator $A + q^m I$ and the system of boundary operators B_j in the angle $|\arg q| \leqslant \pi/2m$. We note that under the condition of selfadjointness of the boundary value problem defined by the operators B_j it suffices to require this ellipticity with parameter only for $\lambda \leqslant 0$ or correspondingly for $q \geqslant 0$ (this implies the corresponding condition for $\mathrm{Re}\,\lambda \leqslant 0$ or $|\arg q| \leqslant \pi/2m$). Moreover, it is important that this condition is equivalent to the condition of parabolicity of the problem (2.1)–(2.3). For periodic conditions ellipticity with parameter is ensured by the condition (5.7).

Finally, we introduce the distribution function $N_\Omega(\lambda)$ equal to the number of eigenvalues of the operator A_Ω less than λ (it is well known that A_Ω has discrete spectrum; see, for example, [27]).

THEOREM 5.1. *Suppose there is given a formally selfadjoint operator A of the form (1.1) with coefficients in $\mathrm{CAP}^\infty(\mathbf{R}^n)$ satisfying the condition (5.7). Suppose that there is given a sequence of domains Ω_k, $k = 1, 2, \ldots$, and of selfadjoint boundary conditions on their boundaries, while for the family $\{\Omega_k\}$ all the hypotheses of Theorem 3.1 are satisfied. Then as $k \to +\infty$*

$$\mu(\Omega_k)^{-1} N_{\Omega_k}(\lambda) \underset{w}{\to} N(\lambda), \tag{5.9}$$

where the distribution function $N(\lambda)$ does not depend on the choice of the admissible sequence of domains and boundary conditions; moreover,

$$\tilde{N}(t) = M_x\{\Gamma(t, x, x)\}, \qquad t > 0, \tag{5.10}$$

where $\Gamma(t, x, y)$ is the fundamental solution of the Cauchy problem (1.3), (1.4).

REMARK. The measure on the line $\mathbf{R}^1$ defined by the distribution function $N(\lambda)$ or its density with respect to Lebesgue measure is the mathematical equivalent of the physical concept of the density of states.

PROOF OF THEOREM 5.1. We shall verify the applicability of Lemma 5.2 to the sequence $N_k(\lambda) = \mu(\Omega_k)^{-1} N_{\Omega_k}(\lambda)$.

Condition a) of this lemma follows from Lemma 4.2.

We now note that

$$\tilde{N}_\Omega(t) = \sum_{l=1}^\infty \exp(-\lambda_l t), \tag{5.11}$$

326   M. A. ŠUBIN

where the summation goes over all eigenvalues of the operator A_Ω (taking account of multiplicity). Since, on the other hand,

$$\Gamma_\Omega(t, x, y) = \sum_{l=1}^{\infty} e^{-\lambda_l t} \psi_l(x) \overline{\psi_l(y)}, \tag{5.12}$$

where $\psi_l(x)$, $l = 1, 2, \ldots$, is the orthonormal set of all eigenfunctions of the operator A_Ω, and (5.11) converges uniformly for any $t > 0$, it follows that

$$\tilde{N}_\Omega(t) = \int_\Omega \Gamma_\Omega(t, x, x)\, dx. \tag{5.13}$$

Condition c) of Lemma 5.2 now follows from Theorem 3.1, while condition b) follows from the estimate (2.6) of Theorem 2.1 (which also holds for periodic conditions by Theorem 2.2). Formula (5.10) now follows from Lemma 5.2 and Theorem 3.1.

Finally, the fact that the limit in (5.9) is independent of the choice of the sequence of domains and boundary conditions can be proved by mixing two different sequences, or it can be derived from the uniqueness of the Laplace transform, since by (5.10) $\tilde{N}(t)$ does not depend on these factors.

§6. Averageability of other spectral invariants

In this section we shall obtain a number of theorems on averageability of the type of Theorem 5.1. In these theorems, however, it is not a question of weak limits of nondecreasing distribution functions (or, equivalently, of weak limits of positive Radon measures on the line) but rather of weak limits of arbitrary complex-valued functions of locally bounded variation (or, equivalently, of complex-valued Radon measures on the line) and of weak limits of Radon measures on the plane.

1. Let $F(\lambda)$ be a function of locally bounded variation on the line. We call it a *complex-valued distribution function* (c.d.f.) if it is left continuous and $\lim_{\lambda \to -\infty} F(\lambda) = 0$. As in §5, we say that a sequence of c.d.f. $F_k(\lambda)$, $k = 1, 2, \ldots$, converges weakly (to a c.d.f. $F(\lambda)$) and write

$$F_k(\lambda) \underset{w}{\to} F(\lambda),$$

if $\lim_{k \to +\infty} F_k(\lambda) = F(\lambda)$ at all points of continuity of $F(\lambda)$.

We form an ordinary distribution function $N(\lambda; F)$ by the formula

$$N(\lambda; F) = \operatorname*{Var}_{\mu \in (-\infty, \lambda)} F(\mu), \tag{6.1}$$

where Var denotes the total variation.

The Laplace transform $\tilde{F}(t)$ of the c.d.f. $F(\lambda)$ is defined by (5.1), in which F is inserted in place of N.

LEMMA 6.1. *Suppose there is given a sequence of c.d.f. $F_k(\lambda)$, $k = 1, 2, \ldots$, and suppose that the functions $N(\lambda; F_k)$ satisfy conditions* a) *and* b) *of Lemma 5.2 and, moreover,* $\lim_{k \to +\infty} \tilde{F}_k(t)$ *exists for all $t > 0$. Then $F_k(\lambda) \underset{w}{\to} F(\lambda)$, where $F(\lambda)$ is a c.d.f., and $\tilde{F}(t) = \lim_{k \to +\infty} \tilde{F}_k(t)$.*

The proof is similar to the proof of Lemma 5.2.

The next result generalizes Theorem 5.1.

THEOREM 6.1. *Suppose that the operator A, the sequence of domains Ω_k $(k = 1, 2, \ldots)$, and the boundary conditions satisfy the hypotheses of Theorem 5.1. Let $L(x, D_x)$ be a differential operator with coefficients in $\mathrm{CAP}(\mathbf{R}^n)$. Let $\{\psi_l(x)\}_{l=1,2,\ldots}$ be a complete orthonormal set of eigenfunctions of the operator A_Ω defined in §5, and let λ_l be the corresponding eigenvalues. Form the c.d.f.*

$$F_{L,\Omega}(\lambda) = \sum_{\lambda_l < \lambda} (L\psi_l, \psi_l) = \sum_{\lambda_l > \lambda} \int_\Omega [L(x, D_x)\psi_l(x)] \overline{\psi_l(x)} \, dx. \qquad (6.2)$$

Then as $k \to +\infty$

$$\mu(\Omega_k)^{-1} F_{L,\Omega_k}(\lambda) \underset{w}{\to} F_L(\lambda),$$

where the c.d.f. $F_L(\lambda)$ does not depend on the choice of the sequence Ω_k or the boundary conditions. Moreover,

$$\tilde{F}_L(t) = M_x\{[L(x, D_x)\Gamma(t, x, y)]|_{y=x}\}, \qquad t > 0. \qquad (6.3)$$

PROOF. We make use of Lemma 6.1. It is obvious that

$$N(\lambda; F_{L,\Omega}) = \sum_{\lambda_l < \lambda} |(L\psi_l, \psi_l)| \leqslant \sum_{\lambda_l < \lambda} \|L\psi_l\|, \qquad (6.4)$$

where the norm is taken in $L^2(\Omega)$. Denoting by d the order of the operator L, from Lemma 4.1 we find that if $d > m$, then

$$\|L\psi_l\| \leqslant C_1\|\psi_l\|_d \leqslant C_2(\|\lambda_l\psi_l\|_{d-m} + 1), \qquad (6.5)$$

while if $d \leqslant m$, then

$$\|L\psi_l\| \leqslant C_1\|\psi_l\|_d \leqslant C_2\|\psi_l\|_m \leqslant C_3(|\lambda_l| + 1), \qquad (6.6)$$

where the constants C_j do no depend on l or Ω (for $\Omega = \Omega_k$).

Iterating (6.5) and applying (6.6) at the final step, we obtain for any d

$$\|L\psi_l\| \leqslant C(1 + |\lambda_l|)^p, \qquad (6.7)$$

where $p = [d/m] + 1$ (the brackets denote the integral part), and C does not depend on l or Ω (for $\Omega = \Omega_k$).

We now observe that $N(\lambda; F_{L,\Omega_k}) = 0$ for $\lambda < a$, where a is the same as for the functions $N_{\Omega_k}(\lambda)$ in Theorem 5.1. In addition, from (6.4) and (6.7) it is clear that

$$|N(\lambda; F_{L,\Omega_k})| \leqslant C(1 + |\lambda|)^p N_{\Omega_k}(\lambda) \quad \text{for } \lambda \geqslant |a|. \qquad (6.8)$$

Here the constant C does not depend on k. Therefore, from this it follows that for $\mu(\Omega_k)^{-1} N(\lambda; F_{L,\Omega_k})$ conditions a) and b) of Lemma 5.2 are satisfied.

It remains to note that the series (5.12) defining $\Gamma_\Omega(t, x, y)$ preserves uniform convergence on differentiating to any order with respect to x and y, so that

$$L(x, D_x)\Gamma_\Omega(t, x, y) = \sum_{l=1}^\infty e^{-\lambda_l t} L(x, D_x)\psi_l(x) \overline{\psi_l(y)}. \qquad (6.9)$$

Setting $y = x$ and integrating with respect to x, we obtain

$$\tilde{F}_{L,\Omega}(t) = \sum_{l=1}^{\infty} e^{-\lambda_l t}(L\psi_l, \psi_l) = \int_{\Omega} \left[L(x, D_x)\Gamma_{\Omega}(t, x, y) \right]\big|_{y=x} dx,$$

and it remains to use Theorem 3.1.

THEOREM 6.2. *Let A, Ω_k, and L be the same as in Theorem 5.1, while for the sequence $\{\Omega_k\}$ the hypotheses of Theorem 3.2 are satisfied. Let f, $g \in \mathrm{CAP}(\mathbf{R}^n)$. Form the c.d.f.*

$$F_{f,g,\Omega}(\lambda) = \sum_{\lambda_l < \lambda} (f, \psi_l)(\psi_l, g).$$

Then as $k \to +\infty$

$$\mu(\Omega_k)^{-1} F_{f,g,\Omega_k}(\lambda) \underset{w}{\to} F_{f,g}(\lambda), \tag{6.10}$$

where the c.d.f. $F_{f,g}(\lambda)$ does not depend on the choice of the sequence of Ω_k and boundary conditions, and

$$\tilde{F}_{f,g}(t) = 2^n \int_{\mathbf{R}^n} M_z\{ \overline{g(z + w)} f(z - w)\Gamma(t, z + w, z - w)\}\, dw, \tag{6.11}$$

PROOF. We set

$$f_l = (f, \psi_l) = \int_{\Omega} f(x)\overline{\psi_l(x)}\, dx,$$

so that f_l are the Fourier coefficients of f with respect to the orthonormal system $\{\psi_l\}$, $l = 1, 2, \ldots$. Defining g analogously, we obtain

$$F_{f,g,\Omega}(\lambda) = \sum_{\lambda_l < \lambda} f_l \cdot \overline{g_l},$$

Therefore

$$N(\lambda; F_{f,g,\Omega}) = \sum_{\lambda_l < \lambda} |f_l| \cdot |g_l|$$

$$\leqslant \frac{1}{2} \sum_{\lambda_l < \lambda} |f_l|^2 + \frac{1}{2} \sum_{\lambda_l < \lambda} |g_l|^2 \leqslant \frac{1}{2}\left(\|f\|^2 + \|g\|^2 \right) \tag{6.12}$$

(the last inequality is a consequence of the Bessel inequality).
Hence

$$N(\lambda; F_{f,g,\Omega}) \leqslant \tfrac{1}{2}\mu(\Omega)(\sup|f(x)| + \sup|g(x)|).$$

This estimate ensures that condition b) of Lemma 5.2 is satisfied for $\mu(\Omega_k)^{-1}N(\lambda; F_{f,g,\Omega_k})$. The remainder of the proof of this theorem is similar to the proof of Theorem 6.1 (here it is necessary to use Corollary 3.2 with g replaced by $\bar{g}$).

COROLLARY 6.1. *Under the hypotheses of Theorem 6.2, set for any $\xi \in \mathbf{R}^n$*

$$A_{\Omega}(\lambda, \xi) = \sum_{\lambda_l < \lambda} \left| \int \psi_l(x)e^{i\xi \cdot x}\, dx \right|^2. \tag{6.13}$$

Then

$$\mu(\Omega_k)^{-1} A_{\Omega_k}(\lambda, \xi) \underset{w}{\to} A(\lambda, \xi),$$

where $A(\lambda, \xi)$ is a c.d.f. not depending on the choice of the sequence of Ω_k and boundary conditions, and

$$\tilde{A}(t, \xi) = 2^n \int_{\mathbf{R}^n} M_z\{\Gamma(t, z + w, z - w)\} e^{2i\xi \cdot w} \, dw. \tag{6.14}$$

PROOF. Set $f(x) = g(x) = e^{-i\xi \cdot x}$ and use Theorem 6.2.

2. We proceed to consider a problem in which there arise complex-valued Radon measures in the plane, concentrated in the angular domain $[a, \infty) \times [a, \infty)$. Such a measure is given by a two-dimensional c.d.f. $F(\lambda, \mu)$ equal to the measure of the domain $(-\infty, \lambda) \times (-\infty, \mu)$. We denote by $N(\lambda, \mu; F)$ the total variation of the measure defined by the c.d.f. on the domain $\{(\lambda', \mu'): \lambda' < \lambda, \mu' < \mu\}$. The integral of a function $f(\lambda, \mu)$ with respect to this measure we denote by $\int f(\lambda, \mu) dF(\lambda, \mu)$. The Laplace transform $\tilde{F}(t, s)$ is defined by

$$\tilde{F}(t, s) = \int_{\mathbf{R}^2} e^{-t\lambda - s\mu} \, dF(\lambda, \mu), \qquad t > 0, s > 0.$$

Let $F_k(\lambda, \mu)$ $(k = 1, 2, \ldots)$ be a sequence of two-dimensional c.d.f. We say that F_k *converges weakly* (to the c.d.f. F) as $k \to +\infty$, and write $F_k(\lambda, \mu) \underset{w}{\to} F(\lambda, \mu)$, if

$$\int \varphi(\lambda, \mu) \, dF_k(\lambda, \mu) \underset{k \to +\infty}{\to} \int \varphi(\lambda, \mu) \, dF(\lambda, \mu)$$

for any continuous function φ with compact support in $\mathbf{R}^2$.

We formulate an analogue of Lemma 6.1.

LEMMA 6.2. *Suppose that there is given a sequence of c.d.f. $F_k(\lambda, \mu)$, $k = 1, 2, \ldots$, and that the following conditions are satisfied:*

a) *There exists an a such that $N(\lambda, \mu; F_k) = 0$ for $\lambda < a$ or $\mu < a$ and all $k = 1, 2, \ldots$.*

b) $|\tilde{N}(t, s; F_k)| \leqslant C(t, s), t > 0, s > 0,$
where $C(t, s)$ does not depend on k.

c) $\lim_{k \to +\infty} \tilde{F}_k(t, s)$ *exists for all $t, s > 0$.*

Then $F_k(\lambda, \mu) \underset{w}{\to} F(\lambda, \mu)$, where $F(\lambda, \mu)$ is a c.d.f., and

$$\tilde{F}(t, s) = \lim_{k \to +\infty} \tilde{F}_k(t, s), \qquad t > 0, s > 0.$$

The proof is similar to the proof of Lemma 5.2. In place of the Helly selection theorem it is necessary to use on a finite rectangle the well-known assertion regarding the weak sequential compactness of the unit ball in the space dual to a separable Banach space (see, for example, [28], Chapter VIII, §1, Theorem 2) and the Riesz theorem describing the space dual to the space $C(\Pi)$, where Π is a compact set in $\mathbf{R}^2$, as the space of finite Radon measures on Π.

THEOREM 6.3. *Suppose there are given two operators A' and A'', a sequence of domains Ω_k, and two families of boundary conditions on the boundaries of Ω_k (one for A' and the other for A''), while the hypotheses of Theorem 6.2 are satisfied for both A' and A''. Denote by ψ_l', ψ_r'', λ_l' and λ_r'' the respective eigenfunctions and eigenvalues of the operators A' and A'' with their respective boundary conditions. Finally, let L' and L'' be two differential operators with coefficients in $\mathrm{CAP}(\mathbf{R}^n)$. Form the two-dimensional c.d.f. $F_\Omega(\lambda, \mu)$ by the formula*

$$F_\Omega(\lambda, \mu) = \sum_{\substack{\lambda_l' < \lambda \\ \lambda_r'' < \mu}} (L'\psi_l', \psi_r'')(L''\psi_r'', \psi_l'). \tag{6.15}$$

Then

$$\mu(\Omega_k)^{-1} F_{\Omega_k}(\lambda, \mu) \underset{w}{\to} F(\lambda, \mu),$$

where $F(\lambda, \mu)$ is a c.d.f. not depending on the choice of the sequence of Ω_k and boundary conditions, and

$$\tilde{F}(t, s) = 2^n \int_{\mathbf{R}^n} M_z \left\{ \left[L'\left(z + w, \frac{1}{2}(D_z + D_w) \right) \Gamma'(t, z + w, z - w) \right] \right.$$
$$\left. \times \left[L''\left(z - w, \frac{1}{2}(D_z - D_w) \right) \Gamma''(s, z - w, z + w) \right] \right\} dw. \tag{6.16}$$

PROOF. It is necessary to verify the applicability of Lemma 6.2 to the sequence $\mu(\Omega_k)^{-1} F_{\Omega_k}(\lambda, \mu)$. Condition a) here follows from Lemma 4.2. We verify condition b):

$$N(\lambda, \mu; F_\Omega) = \sum_{\substack{\lambda_l' < \lambda \\ \lambda_r'' < \mu}} |(L'\psi_l', \psi_r'')| \, |(L''\psi_r'', \psi_l')|$$

$$\leqslant \frac{1}{2} \sum_{\substack{\lambda_l' < \lambda \\ \lambda_r'' < \mu}} |(L'\psi_l', \psi_r'')|^2 + \frac{1}{2} \sum_{\substack{\lambda_l' < \lambda \\ \lambda_r'' < \mu}} |(L''\psi_r'', \psi_l')|^2$$

$$\leqslant \frac{1}{2} \sum_{\lambda_l' < \lambda} \|L'\psi_l'\|^2 + \frac{1}{2} \sum_{\lambda_r'' < \mu} \|L''\psi_r''\|^2.$$

The last two sums are estimated as in the proof of Theorem 6.1. As a result, we obtain

$$N(\lambda, \mu; F_\Omega) \leqslant C\left[(1 + |\lambda|)^p N_\Omega'(\lambda) + (1 + |\mu|)^p N_\Omega''(\mu) \right], \tag{6.17}$$

where N_Ω' and N_Ω'' are the distribution functions of the eigenvalues of the operators A_Ω' and A_Ω'' with their respective boundary conditions (see §5).

The required estimate of $\tilde{N}(t, s; F_\Omega)$ now follows from the corresponding estimates for $\tilde{N}_\Omega'$ and $\tilde{N}_\Omega''$ obtained in the proof of Theorem 5.1.

We now compute $\tilde{F}_\Omega(t, s)$ in terms of the Green functions $\Gamma'_\Omega(t, x, y)$ and $\Gamma''_\Omega(t, x, y)$ of the operators A'_Ω and A''_Ω:

$$\tilde{F}_\Omega(t, s) = \sum_{l,r} e^{-\lambda'_l t - \lambda''_r s} \int_\Omega \left[L'(x, D_x)\psi'_l(x) \right] \overline{\psi''_r(x)}\, dx$$

$$\times \int_\Omega \left[L''(y, D_y)\psi''_r(y) \right] \overline{\psi'_l(y)}\, dy$$

$$= \int_\Omega \left[\sum_l e^{-\lambda'_l t} L'(x, D_x)\psi'_l(x)\overline{\psi'_l(y)} \right]\left[\sum_r e^{-\lambda''_r s} L''(y, D_y)\psi''_r(y)\overline{\psi''_r(x)} \right] dx\, dy$$

$$= \int_\Omega \left[L'(x, D_x)\Gamma'_\Omega(t, x, y) \right]\left[L''(y, D_y)\Gamma''_\Omega(s, y, x) \right] dx\, dy.$$

Therefore, part c) of Lemma 6.2 and hence all assertions of Theorem 6.3 follow from Theorem 3.3.

COROLLARY 6.2. *Let A and Ω_k be as in Theorem 6.2. Form the c.d.f.*

$$S^{ij}_\Omega(\lambda, \mu) = \sum_{\substack{\lambda_l < \lambda \\ \mu_r < \mu}} \left(\frac{\hat{\partial}\psi_l}{\partial x_i}, \psi_r \right)\left(\frac{\partial\psi_r}{\partial x_j}, \psi_l \right), \tag{6.18}$$

where $\mu_r = \lambda_r$. Then

$$\mu(\Omega_k)^{-1} S^{ij}_{\Omega_k}(\lambda, \mu) \underset{\mathrm{w}}{\to} S^{ij}(\lambda, \mu),$$

where the c.d.f. $S^{ij}(\lambda, \mu)$ does not depend on the choice of the sequence of Ω_k and boundary conditions, and

$$\tilde{S}^{ij}(t, s) = 2^{n-2} \int_{\mathbf{R}^n} M_z\Bigg\{ \left(\frac{\partial}{\partial z_i} + \frac{\partial}{\partial w_i} \right)\Gamma(t, z + w, z - w)$$

$$\times \left(\frac{\partial}{\partial z_j} - \frac{\partial}{\partial w_j} \right)\Gamma(s, z - w, z + w) \Bigg\}\, dw. \tag{6.19}$$

PROOF. In Theorem 6.3 it is necessary to set $A' = A'' = A$, $L' = \partial/\partial x_i$ and $L'' = \partial/\partial x_j$.

§7. The density of states and the relative trace

In [10] (for more details see [11]) a distribution function $N(\lambda)$ was introduced which is related to the operator A in the same way as in §5 of this paper. This function has been defined in terms of the relative trace on a certain II_∞-factor already used in [13] and [14] to construct the theory of the index. We now recall the basic definitions and prove that the function $N(\lambda)$ introduced in [10] coincides with the function $N(\lambda)$ contained in Theorem 5.1 and defining the density of states.

We denote by $B^2(\mathbf{R}^n)$ the nonseparable Hilbert space of almost periodic Besicovitch functions obtained as the completion of $\mathrm{CAP}(\mathbf{R}^n)$ in the Hilbert norm

332   M. A. ŠUBIN

induced by the inner product $(f, g)_B = M\{f \cdot \bar{g}\}$. We consider the Hilbert space $\mathcal{H} = B^2(\mathbf{R}^n) \,\hat{\otimes}\, L^2(\mathbf{R}^n)$ and the von Neumann algebra $\mathfrak{A}$ in it generated by the two families of operators $\{e_\xi \otimes e_\xi, \xi \in \mathbf{R}^n\}$ and $\{I \otimes T_\xi, \xi \in \mathbf{R}^n\}$, where e_ξ is the operator of multiplication by $e^{i\xi \cdot x}$ in $B^2(\mathbf{R}^n)$ or in $L^2(\mathbf{R}^n)$ and T_ξ is the operator of translation by the vector ξ in $L^2(\mathbf{R}^n)$, i.e., $T_\xi u(x) = u(x - \xi)$. The space $\mathcal{H}$ can be interpreted as the space of functions of two n-dimensional variables z and x, where $z \in \mathbf{R}^n_B$ and $x \in \mathbf{R}^n$ ($\mathbf{R}^n_B$ denotes the Bohr compactification of the group $\mathbf{R}^n$ —the maximal ideal space of the ring $\mathrm{CAP}(\mathbf{R}^n)$; it is important that $\mathbf{R}^n_B$ is a compact topological group, and $L^2(\mathbf{R}^n_B) = B^2(\mathbf{R}^n)$). The algebra $\mathfrak{A}$ is a II_∞-factor [14]; we denote by $\mathrm{Sp}_{\mathfrak{A}}$ the relative trace on the factor $\mathfrak{A}$.

Suppose there is given an operator A of the form (1.1) satisfying the conditions of Theorem 5.1. We form the operator

$$A^{\#} = \sum_{|\alpha| \leqslant m} a_\alpha(z + x)D_x^\alpha, \tag{7.1}$$

which we shall consider as an unbounded operator in $\mathcal{H}$ with domain $B^2(\mathbf{R}^n) \,\hat{\otimes}\, H^m(\mathbf{R}^n)$. It is then easy to verify that the operator $A^{\#}$ is selfadjoint, and $A^{\#}$ is associated to $\mathfrak{A}$, i.e., $A^{\#}$ commutes with any unitary operator in the commutant $\mathfrak{A}'$ of the algebra $\mathfrak{A}$ (we abbreviate this in the form $A^{\#}\eta\mathfrak{A}$). Therefore, if $\tilde{E}_\lambda$ are the spectral projectors of the operator $A^{\#}$, then $\tilde{E}_\lambda \in \mathfrak{A}$, and the relative trace $\mathrm{Sp}_{\mathfrak{A}} \tilde{E}_\lambda$ is defined.

THEOREM 7.1.

$$N(\lambda) = \mathrm{Sp}_{\mathfrak{A}} \tilde{E}_\lambda, \tag{7.2}$$

when $N(\lambda)$ is the distribution function of the density of states introduced in §5.

PROOF. We temporarily set $N_1(\lambda) = \mathrm{Sp}_{\mathfrak{A}} \tilde{E}_\lambda$, and we shall show that the functions $N(\lambda)$ and $N_1(\lambda)$ have the same Laplace transforms $\tilde{N}(t)$ and $\tilde{N}_1(t)$. Obviously,

$$\tilde{N}_1(t) = \int_{-\infty}^{+\infty} e^{-\lambda t}\, dN_1(\lambda) = \int_{-\infty}^{+\infty} e^{-\lambda t}\, d\big(\mathrm{Sp}_{\mathfrak{A}} \tilde{E}_\lambda\big)$$
$$= \mathrm{Sp}_{\mathfrak{A}}\bigg[\int_{-\infty}^{+\infty} \infty^{-\lambda t}\, d\tilde{E}_\lambda\bigg] = \mathrm{Sp}_{\mathfrak{A}}\, e^{-t A^{\#}} \tag{7.3}$$

(the possibility of taking the trace outside the integral sign is easily deduced from its normality).

We shall study the structure of the operator $e^{-t A^{\#}}$. It is easy to see that for an operator A_z of the form

$$A_z = \sum_{|\alpha| \leqslant m} a_\alpha(x + z)D_x^\alpha, \tag{7.4}$$

considered as an operator on $\mathbf{R}^n$ with parameter $z \in \mathbf{R}^n$, the fundamental solution of the Cauchy problem (1.3), (1.4) has the form

$$\Gamma_z(t, x, y) = \Gamma(t, x + z, y + z), \tag{7.5}$$

where $\Gamma(t, x, y)$ is the fundamental solution of the same problem for the operator A. Theorem 1.1 shows that the function (7.5) can also be defined for $z \in \mathbf{R}^n_B$ by continuity; and, again by continuity, it is clear that for any $z \in \mathbf{R}^n_B$ the function $\Gamma(t, x, y)$ is the fundamental solution of problem (1.3), (1.4) for A_z.

This implies that the operator $e^{-tA^\#}$ is given by

$$e^{-tA^\#} u(z, x) = \int_{\mathbf{R}^n} \Gamma(t, x + z, y + z) u(z, y)\, dy. \tag{7.6}$$

Noting that $\mathrm{Sp}_{\mathfrak{A}}(U^{-1}TU) = \mathrm{Sp}_{\mathfrak{A}} T$ for any $T \in \mathfrak{A}$, $T \geqslant 0$, and for any unitary operator $U \in \mathfrak{A}$ (see, for example, [29], Chapter VII), we see that the trace of the operator $e^{-tA^\#}$ coincides with the trace of the operator $L_\xi = (I \otimes T_{-\xi}) \times e^{-tA^\#}(I \otimes T_\xi)$ written in the form

$$L_\xi u(z, x) = \int_{\mathbf{R}^n} \Gamma(t, x + z + \xi, y + z + \xi) u(z, y)\, dy. \tag{7.7}$$

Proceeding formally for the time being we can write

$$\mathrm{Sp}_{\mathfrak{A}} e^{-tA^\#} = \mathrm{Sp}_{\mathfrak{A}} L_\xi = M_\xi \{ \mathrm{Sp}_{\mathfrak{A}} L_\xi \} = \mathrm{Sp}_{\mathfrak{A}} M_\xi \{ L_\xi \}. \tag{7.8}$$

But $M_\xi \{ L_\xi \}$ is an operator $\tilde{L}$ of the form

$$\tilde{L} u(z, x) = \int_{\mathbf{R}^n} M_\xi \{ \Gamma(t, x + \xi, y + \xi) \} u(z, y)\, dy. \tag{7.9}$$

We set

$$f(t, x) = M_z \{ \Gamma(t, x + z, z) \}. \tag{7.10}$$

From (7.9) it is then clear that $\tilde{L}$ can be written in the form

$$\tilde{L} u(z, x) = \int_{\mathbf{R}^n} f(t, x - y) u(z, y)\, dy. \tag{7.11}$$

It follows from Theorem 1.1 that $f(t, x) \in S(\mathbf{R}^n_x)$ (for each fixed $t > 0$). Therefore, $\tilde{L}$ can be written in the form

$$\tilde{L} = I \otimes \tilde{M}_{\tilde{f}}, \tag{7.12}$$

where $\tilde{f} = Ff$ is the Fourier transform of the function $f(t, x)$ with respect to x, and $\tilde{M}_{\tilde{f}} = F^{-1} M_{\tilde{f}} F$, where $M_{\tilde{f}}$ is the operator of multiplication by the function $\tilde{f}$. Now, for operators of the form (7.12) the relative trace is given by [14]

$$\mathrm{Sp}_{\mathfrak{A}} \tilde{L} = (2\pi)^{-n} \int_{\mathbf{R}^n} \tilde{f}(t, \xi)\, d\xi. \tag{7.13}$$

Using this formula, it is easy to justify the calculations made above ((7.8) and (7.9)). Namely, we denote by $S_1(\mathfrak{A})$ the ideal of all operators with absolutely convergent trace (i.e., operators $T \in \mathfrak{A}$ such that $\mathrm{Sp}_{\mathfrak{A}} \sqrt{T^*T} < +\infty$). It is a Banach space with respect to the norm $|T\|_1 = \mathrm{Sp}_{\mathfrak{A}} \sqrt{T^*T}$, and $\mathrm{Sp}_{\mathfrak{A}}$ is a continuous linear functional on $S_1(\mathfrak{A})$. The justification for (7.8) now follows from the fact that, as is evident from (7.13) and the estimates of Theorem 1.1, the mean value $M_\xi \{ L_\xi \}$ exists as a limit in the norm of $S_1(\mathfrak{A})$ and hence commutes with taking the trace $\mathrm{Sp}_{\mathfrak{A}}$. The computations (7.6)–(7.13) show that

$$\mathrm{Sp}\; e^{-tA^{\#}} = (2\pi)^{-n}\int_{\mathbf{R}^n} \tilde{f}(t,\xi)\, d\xi = f(t,0) = M_x\{\Gamma(t,x,x)\}. \qquad (7.14)$$

A comparison of (7.14) and (5.10) shows that $\tilde{N}(t) = \tilde{N}_1(t)$, and hence $N(\lambda) = N_1(\lambda)$. Theorem 7.1 is proved.

REMARK. The formula for $\mathrm{Sp}_{\mathfrak{A}}\, e^{-tA^{\#}}$ can also be derived from [11], Proposition 8.3.

COROLLARY 7.1. *Let the operator A satisfy the hypotheses of Theorem 5.1. Consider it as a selfadjoint operator in $L^2(\mathbf{R}^n)$ (with domain $H^m(\mathbf{R}^n)$). Then its spectrum $\sigma(A)$ coincides with the set of points of increase of the distribution function $N(\lambda)$ (points λ such that $N(\lambda + \varepsilon) - N(\lambda - \varepsilon) > 0$ for any $\varepsilon > 0$).*

PROOF. In view of the easily verified relation $\sigma(A) = \sigma(A^{\#})$, the assertion follows from Theorem 7.1 and the exactness of the trace $\mathrm{Sp}_{\mathfrak{A}}$.

§8. Explicit formulas

In this section explicit formulas are given for the distribution function $N(\lambda)$ (and the density of states) in the following two cases: the case of constant coefficients and the one-dimensional Schrödinger operator with a periodic potential.

1. *The case of constant coefficients.*

THEOREM 8.1. *Let $A = A(D)$ be an elliptic differential operator with constant coefficients such that $A(\xi) = \overline{A}(\xi)$ (the condition of selfadjointness) and $a_m(\xi)$ for $\xi \neq 0$. Denote by $N(\lambda)$ its distribution function of the density of states from Theorem 5.1, and set*

$$V(\lambda) = (2\pi)^{-n}\mu\{\xi: A(\xi) < \lambda\} = (2\pi)^{-n}\int_{A(\xi)<\lambda} d\xi. \qquad (8.1)$$

Then for all $\lambda \in \mathbf{R}$

$$N(\lambda) = V(\lambda). \qquad (8.2)$$

PROOF. It is clear that

$$\tilde{V}(t) = \int_{-\infty}^{+\infty} e^{-\lambda t}\, dV(\lambda) = (2\pi)^{-n}\int_{\mathbf{R}^n} e^{-tA(\xi)}\, d\xi.$$

At the same time, in this case

$$\Gamma(t,x,y) = (2\pi)^{-n}\int_{\mathbf{R}^n} e^{i(x-y)\cdot\xi - tA(\xi)}\, d\xi.$$

Therefore

$$\tilde{N}(t) = M_x\{\Gamma(t,x,x)\} = \Gamma(t,0,0) = \tilde{V}(t),$$

whence (8.2) follows.

REMARK. This result is also easily obtained from Theorem 7.1 by using (7.13). From (8.2) it is easy to deduce for large λ the existence also of the density itself with respect to Lebesgue measure, $\rho(\lambda) = dN(\lambda)/d\lambda$, which is an infinitely differentiable function of λ. Namely,

$$\rho(\lambda) = (2\pi)^{-n} \int_{A(\xi)=\lambda} \frac{dS_\xi}{|\operatorname{grad} A(\xi)|},$$

where dS_ξ is the element of surface area of the surface $A(\xi) = \lambda$ which has no singularities for large λ.

2. *The one-dimensional Schrödinger operator with a periodic potential.* We consider the operator

$$A = -d^2/dx^2 + q(x), \tag{8.3}$$

where $q(x) \in C^\infty(\mathbf{R}^1)$ and $q(x)$ is periodic with period $a > 0$ (it is possible to avoid the assumption of infinite smoothness, but we shall not concern ourselves with this here). We denote by $D(\lambda)$ the trace of the monodromy matrix for the equation $A\psi = \lambda\psi$ (the monodromy matrix is the matrix of the operator of translation in x by the length of the period a in the two-dimensional space of solutions of this equation).

THEOREM 8.2. *Let A be an operator of the form* (8.3) *with a function $q(x) \in C^\infty(\mathbf{R}^1)$ having period $a > 0$. Denote by $N(\lambda)$ the distribution function of Theorem 5.1 for the operator A. Then $N(\lambda)$ is absolutely continuous, and the formula*

$$\rho(\lambda) = \frac{|D'(\lambda)|}{\pi a\sqrt{4 - D^2(\lambda)}}\,\theta\!\left(4 - D^2(\lambda)\right), \tag{8.4}$$

holds for the density $\rho(\lambda) = dN(\lambda)/d\lambda$, where θ is the Heaviside function ($\theta(t) = 1$ for $t > 0$ and $\theta(t) = 0$ for $t < 0$).

REMARK. The well-known character of the dependence of $D(\lambda)$ on λ (see, for example, [30], Chapter 2 and also the beginning of the proof of Theorem 8.2) makes it possible to obtain from (8.4) a description of the behavior of $\rho(\lambda)$ (see the figure).

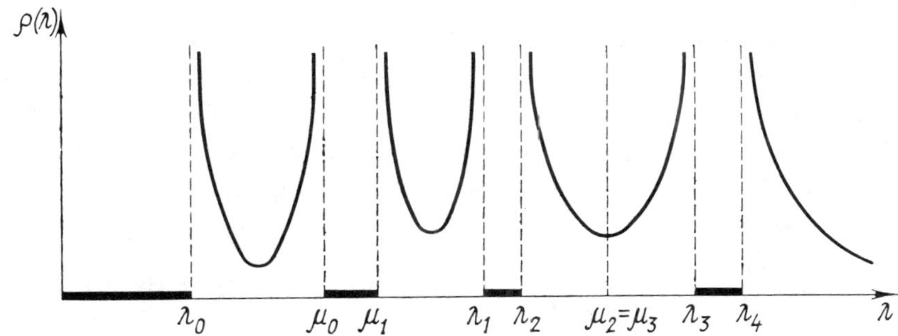

Here λ_j are the periodic and μ_j the antiperiodic eigenvalues, all the singularities at the points λ_j and μ_j are of the type $1/\sqrt{x}$ as $x \to +0$, $\rho(\lambda) = 0$ on the intervals $(-\infty, \lambda_0)$ (μ_0, μ_1), (λ_1, λ_2), (μ_2, μ_3), (λ_3, λ_4), etc., while if some such interval is empty, for example, $\mu_2 = \mu_3$ in the figure, then the singularity vanishes, and $\rho(\lambda)$ assumes at this point a finite, nonzero value.

PROOF OF THEOREM 8.2. We shall use the following facts ([30], Chapter 2), where in the formulation the notation λ_j and μ_j of the preceding remark is used:

a) $\lambda_0 < \mu_0 \leqslant \mu_1 < \lambda_1 \leqslant \lambda_2 < \mu_2 \leqslant \mu_3 < \lambda_3 \leqslant \lambda_4 < \cdots$.

b) The set $\{\lambda : |D(\lambda)| > 2\}$ coincides with the union of the intervals $(-\infty, \lambda_0)$, (μ_0, μ_1), $(\lambda_1 \lambda_2)$, (μ_2, μ_3), etc.

c) If $|D(\lambda)| < 2$, then $D'(\lambda) \neq 0$, and $D(\lambda)$ decreases from 2 to -2 on the intervals (λ_0, μ_0), $(\lambda_2, \mu_2), \cdots$ and increases from -2 to 2 on the intervals (μ_1, λ_1), $(\mu_3, \lambda_3), \ldots$.

d) If $\lambda_{2k-1} \neq \lambda_{2k}$, then $D'(\lambda_{2k-1}) > 0$ and $D'(\lambda_{2k}) < 0$, while if $\lambda_{2k-1} = \lambda_{2k}$, then $D'(\lambda_{2k}) = 0$ and $D''(\lambda_{2k}) < 0$; an analogous assertion holds for the numbers μ_j.

e) If k is an integer, then a solution of the equation $A\psi = \lambda\psi$ having period ka exists if and only if $D(\lambda) = 2\cos(2\pi l/k)$, where l is an integer.

f) If $k \geqslant 3$, k is an integer, and the equation $A\psi = \lambda\psi$ has a ka-periodic solution not having periods a and $2a$, then all solutions of this equation have ka.

Using Theorem 5.1, we compute $N(\lambda)$ by means of periodic solutions with period ka, where k is a natural number.

Let $I = [\lambda_1, \lambda_2)$ be a finite interval of the λ-axis. We denote by $P_k(I)$ the number of values $\lambda \in I$ such that the equation $A\psi = \lambda\psi$ has a solution with period ka. Since periodic solutions with periods a and $2a$ may exist only at points of a discrete subset of the λ-axis, it follows from f) that if $N(I) = N(\lambda_2) - N(\lambda_1)$ and λ_1, λ_2 are points of continuity of $N(\lambda)$, then

$$N(I) = \lim_{k \to +\infty} (2p_k(I)/ka). \tag{8.5}$$

From this and e) it follows in an obvious manner that if $|D(\lambda)| > 2$, then $\rho(\lambda) = 0$.

We now fix a point λ such that $|D(\lambda)| < 2$. We denote by $I_{\Delta\lambda}$ one of the intervals $[\lambda - \Delta\lambda/2, \lambda + \Delta\lambda/2)$, $[\lambda, \lambda + \Delta\lambda)$ or $[\lambda - \Delta\lambda, \lambda)$, where $\Delta\lambda > 0$. We shall assume that $\Delta\lambda$ is so small that $|D(\mu)| < 2$ on the closure of $I_{\Delta\lambda}$. This and c) imply that $|D'(\mu)| \geqslant \varepsilon > 0$ for $\mu \in I_{\Delta\lambda}$.

We now use assertion e), which shows that $P_k(I_{\Delta\lambda})$ is equal to the number d of points $\kappa \in I_{\Delta\lambda}$ such that $D(\kappa) = 2\cos(2\pi l/k)$, where l is an integer. Let $\kappa_1 < \cdots < \kappa_d$ be all such points, and let $l_1, \ldots, l_d$ be the corresponding values of l. We note that $\{l_1, \ldots, l_d\}$ is a segment of successive integers taken in increasing or decreasing order. We estimate the distance between neighboring points κ_r and $\kappa_{r+1}, r = 1, \ldots, d - 1$:

$$|D(\kappa_{r+1}) - D(\kappa_r)| = |2(\cos(2\pi l_{r+1}/k) - \cos(2\pi l_r/k))|$$

$$= (4\pi/k)|\sin(2\pi l_r/k)|(1 + O(1/k)). \tag{8.6}$$

Now,

$$|\sin(2\pi l_r/k)| = \sqrt{1 - \cos^2(2\pi l_r/k)} = \sqrt{1 - D^2(\kappa_r)/4}$$

$$= (1 + O(\Delta\lambda))\sqrt{1 - D^2(\lambda)/4}\,. \tag{8.7}$$

It follows from (8.6) and (8.7) that

$$|D(\kappa_{r+1}) - D(\kappa_r)| = (2\pi/k)\sqrt{4 - D^2(\lambda)}\,(1 + O(1/k))(1 + O(\Delta\lambda)). \tag{8.8}$$

On the other hand,

$$|D(\kappa_{r+1}) - D(\kappa_r)| = |D'(\lambda)|(1 + O(\Delta\lambda))|\kappa_{r+1} - \kappa_r|. \tag{8.9}$$

From (8.8) and (8.9) we obtain

$$|\kappa_{r+1} - \kappa dr| = \frac{2\pi}{k} \frac{\sqrt{4 - D^2(\lambda)}}{|D'(\lambda)|} \left(1 + O\!\left(\frac{1}{k}\right)\right)(1 + O(\Delta\lambda)).$$

From this it obviously follows that

$$p_k(I_{\delta\lambda}) = d = \frac{\Delta\lambda \cdot k\,|D'(\lambda)|}{2\pi\sqrt{4 - D^2(\lambda)}} \left(1 + O\!\left(\frac{1}{k}\right)\right)(1 + O(\Delta\lambda)).$$

Using (8.5), we find that if $\Delta\lambda$ is such that $\lambda + \Delta\lambda/2$ and $\lambda - \Delta\lambda/2$ are points of continuity of $N(\lambda)$, then for $I_{\Delta\lambda} = [\lambda - \Delta\lambda/2, \lambda + \Delta\lambda/2)$

$$N(I_{\Delta\lambda}) = \frac{|D'(\lambda)| \cdot \Delta\lambda}{\pi a\sqrt{4 - D^2(\lambda)}}(1 + O(\Delta\lambda)), \tag{8.10}$$

and from this it is evident that λ is a point of continuity of $N(\lambda)$, and hence (8.10) is true for all types of $I_{\Delta\lambda}$. Therefore,

$$\rho(\lambda) = \lim_{\Delta\lambda \to 0} \frac{N(I_{\Delta\lambda})}{\Delta\lambda} = \frac{|D'(\lambda)|}{\pi a\sqrt{4 - D^2(\lambda)}}\,.$$

We thus find that the derivative $\rho(\lambda) = dN(\lambda)/d\lambda$ exists everywhere with the possible exception of the points λ_j and μ_j, and it is given by (8.4). Noting the integrability of $\rho(\lambda)$ in a neighborhood of the points λ_j and μ_j, we see that it remains only to check the continuity of $N(\lambda)$ at these points (from what has been proved so far it follows that it can have there only discontinuities of first kind). This continuity is obtained by an analogous argument but near a point λ at which $|D(\lambda)| = 2$.

We suppose, for example, that $|D(\lambda)| = 2$ and $D'(\lambda) \neq 0$; suppose, to be specific, that $\lambda = \lambda_{2m}$, where $m \geq 0$ and either $m = 0$ or $\lambda_{2m} \neq \lambda_{2m-1}$. Thus, $D(\lambda) = 2$ and $D'(\lambda) < 0$. For $I_{\Delta\lambda}$ we use a symmetric interval $[\lambda - \Delta\lambda/2, \lambda + \Delta\lambda/2)$, assuming that $\Delta\lambda$ is so small that $\lambda - \Delta\lambda/2 > \lambda_{2m-1}$. Introducing the same notation as above, we may assume that $l_r = r - 1$, $r = 1,\ldots,d$; $\kappa_1 = \lambda$.

Obviously,

$$|D(\kappa_d) - D(\kappa_1)| = 2\left(1 - \cos\frac{2\pi(d-1)}{k}\right) = 4\sin^2\frac{\pi(d-1)}{k}. \qquad (8.11)$$

Since $\kappa_d < \lambda + \Delta\lambda/2 < \kappa_{d+1}$, it is clear that $\kappa_d - \kappa_1 = \frac{1}{2}\Delta\lambda(1 + O(1/k))$ for fixed $\Delta\lambda$. But since

$$|D(\kappa_d) - D(\kappa_1)| = |\kappa_d - \kappa_1|\,|D'(\lambda)|(1 + O(\Delta\lambda)). \qquad (8.12)$$

it follows from (8.11) that

$$4\sin^2\frac{\pi(d-1)}{k} = \frac{1}{2}\Delta\lambda \cdot |D'(\lambda)|(1 + O(\Delta\lambda))\left(1 + O\left(\frac{1}{k}\right)\right)$$

and

$$\frac{d-1}{k} = \frac{1}{\pi}\sqrt{\arcsin\left[\frac{1}{8}\Delta\lambda \cdot |D'(\lambda)|(1 + O(\Delta\lambda))\left(1 + O\left(\frac{1}{k}\right)\right)\right]}.$$

Passing to the limit as $k \to +\infty$, we obtain

$$N(I_{\Delta\lambda}) = \frac{\sqrt{\Delta\lambda}}{\pi a\sqrt{2}}\sqrt{|D'(\lambda)|}\,(1 + O(\Delta\lambda)),$$

whence the continuity of $N(\lambda)$ at the point λ follows.

Finally, to consider a point λ at which $|D(\lambda)| = 2$, $D'(\lambda) = 0$, and $D''(\lambda) \neq 0$, it is necessary to carry out an analogous argument in which in place of (8.12) we must use the fact that

$$|D(\kappa_d) - D(\lambda)| = \frac{1}{8}|D''(\lambda)| \cdot |\Delta\lambda|^2(1 + O(\Delta\lambda)).$$

Received 10/SEPT/75

BIBLIOGRAPHY

1. J. M. Ziman, *Principles of the theory of solids*, 2nd ed., Cambridge Univ. Press, 1972.

2. R. E. Peierls, *Quantum theory of solids*, Clarendon Press, Oxford, 1955.

3. N. F. Mott, *Electrons in disordered structures*, Advances in Phys. **16** (1967), 49–144.

4. N. F. Mott and E. A. Davis, *Electronic processes in non-crystalline materials*, Clarendon Press, Oxford, 1971.

5. M. V. Romerio, *Almost periodic functions and the theory of disordered systems*, J. Mathematical Phys. **12** (1971), 552–562.

6. L. A. Pastur, *The spectra of random selfadjoint operators*, Uspehi Mat. Nauk **28** (1973), no. 1 (169), 3–64; English transl. in Russian Math. Surveys **28** (1973).

7. M. S. Agranovič and M. I. Višik, *Elliptic problems with a parameter and parabolic problems of general type*, Uspehi Mat. Nauk **19** (1964), no. 3 (117), 53–161; English transl. in Russian Math. Surveys **19** (1964).

8. Avner Friedman, *Partial differential equations of parabolic type*, Prentice-Hall, Englewood Cliffs, N. J., 1964.

9. S. D. Èĭdel'man and S. D. Ivasišen, *Investigation of the Green matrix for a homogeneous parabolic boundary value problem*, Trudy Moskov. Mat. Obšč. **23** (1970), 179–234; English transl. in Trans. Moscow Math. Soc. **23** (1970).

10. M. A. Šubin, *Elliptic almost periodic operators and von Neumann algebras*, Funkcional. Anal. i Priložen. **9** (1975), no. 1, 89–90; English transl. in Functional Anal. Appl. **9** (1975).

11. ______, *Pseudodifferential almost periodic operators and von Neumann algebras*, Trudy Moskov. Mat. Obšč. **35** (1976), 103–163; English transl. in Trans. Moscow Math. Soc. **1979**, no. 1 (35).

12. ______, *Weyl's theorem for the Schrödinger operator with an almost periodic potential*, Vestnik Moskov. Univ. Ser. I Mat. Meh. **1976**, no. 2, 84–88; English transl. in Moscow Univ. Math. Bull. **31** (1976).

13. L. A. Coburn et al., *C*-algebras of operators on a half space. II, Index theory*, Inst. Hautes Études Sci. Publ. Math. No. 40 (1971), 69–79.

14. L. A. Coburn, R. D. Moyer and I. M. Singer, *C*-algebras of almost periodic pseudo-differential operators*, Acta Math. **130** (1973), 279–307.

15. B. M. Levitan, *Almost-periodic functions*, GITTL, Moscow, 1953. (Russian)

16. L. A. Bagirov and M. A. Šubin, *The stabilization of the solution of the Cauchy problem for parabolic equations with coefficients which are almost periodic in the space variables*, Differencial'nye Uravnenija **11** (1975), 2205–2209; English transl. in Differential Equations **11** (1975).

17. T. Ja. Zagorskiĭ, *Mixed problems for systems of parabolic partial differential equations*, Izdat. L'vovsk. Gos. Univ., L'vov, 1961. (Russian)

18. V. A. Solonnikov, *On boundary value problems for linear parabolic systems of differential equations of general form*, Trudy Mat. Inst. Steklov. **83** (1965); English transl., Proc. Steklov Inst. Math. **83** (1965).

19. S. D. Eĭdel'man, *Parabolic systems*, "Nauka", Moscow, 1964; English transl., Noordhoff, Groningen, and North-Holland, Amsterdam, 1969.

20. David Ruelle, *Statistical mechanics; Rigorous results*, Benjamin, New York, 1969.

21. M. A. Šubin, *Differential and pseudodifferential operators in spaces of almost periodic functions*, Mat. Sb. **95 (137)** (1974), 560–587; English transl. in Math. USSR Sb. **24** (1974).

22. L. N. Slobodeckiĭ, *Estimates in L_2 of the solutions of linear elliptic and parabolic systems. I*, Vestnik Leningrad. Univ. **1960**, no. 7 (Ser. Mat. Meh. Astr. vyp. 2), 28–47. (Russian)

23. Yoshiki Higuchi, *A priori estimates and existence theorem on elliptic boundary value problems for unbounded domains*, Osaka J. Math. **5** (1968), 103–135.

24. Jacques-Louis Lions and E. Magenes, *Problèmes aux limites non homogènes et applications*. Vol. 1, Dunod, Paris, 1968; English transl., Springer-Verlag, 1972.

25. Lars Hörmander, *Linear partial differential operators*, Academic Press, New York, and Springer-Verlag, Berlin, 1963.

26. G. E. Šilov, *Mathematical analysis: Special course*, 2nd ed., Fizmatgiz, Moscow, 1961; English transl., Pergamon Press, New York, 1965.

27. Ju. M. Berezanskiĭ, *Expansion in eigenfunctions of selfadjoint operators*, "Naukova Dumka", Kiev, 1965; English transl., Amer. Math. Soc., Providence, R. I., 1968.

28. L. V. Kantorovič and G. P. Akilov, *Functional analysis in normed spaces*, Fizmatgiz, Moscow, 1959; English transl., Macmillan, 1964.

29. M. A. Naĭmark, *Normed rings*, 2nd rev. ed., "Nauka", Moscow, 1968; English transl., *Normed algebras*, Wolters-Noordhoff, Groningen, 1972.

30. M. S. P. Eastham, *The spectral theory of periodic differential equations*, Scottish Academic Press, Edinburgh, 1973.

Translated by J. R. SCHULENBERGER